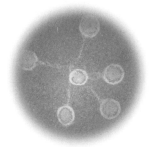

# STEM CELLS
From Bench to Bedside

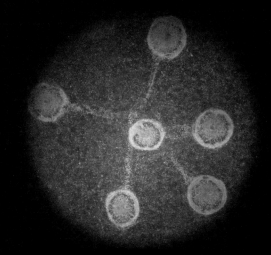

# STEM CELLS
## From Bench to Bedside

Editors

### Ariff Bongso & Eng Hin Lee

National University of Singapore

Forewords by

Sydney Brenner & Philip Yeo

**World Scientific**

NEW JERSEY • LONDON • SINGAPORE • BEIJING • SHANGHAI • HONG KONG • TAIPEI • CHENNAI

*Published by*

World Scientific Publishing Co. Pte. Ltd.

5 Toh Tuck Link, Singapore 596224

*USA office:* 27 Warren Street, Suite 401-402, Hackensack, NJ 07601

*UK office:* 57 Shelton Street, Covent Garden, London WC2H 9HE

**Library of Congress Cataloging-in-Publication Data**
Bongso, Ariff.
    Stem cells : from bench to bedside / Ariff Bongso & Eng Hin Lee.
      p. cm.
    Includes index.
    ISBN 981-256-126-9
    1. Stem cells. I. Lee, Eng Hin, 1947–    II. Title.

   QH588.S83B66 2005
   616'.02774--dc22

                      2005043553

**British Library Cataloguing-in-Publication Data**
A catalogue record for this book is available from the British Library.

For photocopying of material in this volume, please pay a copying fee through the Copyright
Clearance Center, Inc., 222 Rosewood Drive, Danvers, MA 01923, USA. In this case permission to
photocopy is not required from the publisher.

Typeset by Stallion Press
Email: enquiries@stallionpress.com

Printed by FuIsland Offset Printing (S) Pte Ltd, Singapore

# Contents

# Contributors

**Malcolm R. Alison**
Centre for Diabetes and Metabolic
Medicine
Institute for Cell and Molecular
Science
4 Newark Street, Whitechapel
London E1 2AT
UK

**Michal Amit**
Stem Cell Center, Bruce Rappaport
Faculty of Medicine
Technion — Israel Institute of
Technology
POB 9649, Haifa 31096
Israel

**Eunice Amofah**
Department of Medicine
10th floor QEQM Building
Imperial College
St Mary's Campus, W2 1NY
London
UK

**Peter W. Andrews**
Centre for Stem Cell Biology and
the Department of Biomedical
Science
University of Sheffield
Western Bank
Sheffield S10 2TN
UK

**Leonard Pek-Kiang Ang**
Singapore National Eye Centre and
Department of Ophthalmology
National University of Singapore
5 Lower Kent Ridge Road
Singapore 119074

**Ahmad R. Bahrani**
Institute of Biotechnology and
Tissue Engineering
Department of Biology
Ferdowsi University of Mashhad
Mashhad 91775
Iran

**Jennifer Batten**
Stem Cell Biology and Regenerative
Medicine
Robarts Research Institute
PO Box 5015, 100 Perth Drive
London, Ontario
Canada N6A 5K8

**Nissim Benvenisty**
Department of Genetics
Institute of Life Science
The Hebrew University of
Jerusalem
Givat Ram, Jerusalem 91904
Israel

**Barak Blum**
Department of Genetics
Institute of Life Science
The Hebrew University of
Jerusalem
Givat Ram, Jerusalem 91904
Israel

**Ariff Bongso**
Department of Obstetrics &
Gynaecology
National University of Singapore
National University Hospital
5 Lower Kent Ridge Road
Singapore 119074

**Ali H. Brivanlou**
Laboratory of Molecular Vetebrate
Embryology
The Rockefeller University
New York, NY 10021-6399
USA
&
Laboratory of Molecular Vertebrate
Embryology
The Rockefeller University
1230 York Avenue
New York, NY 10021
USA

**Justine Burley**
Graduate School for the Integrative
Sciences and Engineering
MD11, #01-10
8 Medical Drive
Singapore 117597
&
National University of Singapore

**Melissa Carpenter**
Stem Cell Biology and Regenerative
Medicine
Robarts Research Institute
PO Box 5015, 100 Perth Drive
London, Ontario
Canada N6A 5K8
&
CyThera, Inc.
3550 General Atomics Ct
San Diego, CA 92121

**Jose B. Cibelli**
Michigan State University
East Lansing, MI 48824
USA

**Chris N. Denning**
Institute of Genetics
University of Nottingham
Queens Medical Centre
Nottingham NG7 2UH
UK

**Mirella Dottori**
Laboratory of Embryonic Stem Cell
Biology
Monash Institute for Medical
Research, Building 75
The STRIP, Wellington Road
Clayton, VIC. 3800
Australia

**F. Fändrich**
Department of General Surgery and
Thoracic Surgery
University of Kiel, 24105 Kiel
Germany

**Chui-Yee Fong**
Department of Obstetrics &
Gynaecology
National University of Singapore
Kent Ridge
Singapore 119074

**Stuart J. Forbes**
Department of Medicine
10th floor QEQM Building
Imperial College
St Mary's Campus, W2 1NY, London
UK

**Sharon Gerecht-Nir**
Stem Cell Center, Bruce Rappaport
Faculty of Medicine
Technion — Israel Institute of
Technology
POB 9649, Haifa 31096
Israel

**James Cho Hong Goh**
Department of Orthopaedic
Surgery
National University of Singapore
NUS Tissue Engineering Program
DSO (Kent Ridge) Building
27 Medical Drive, #03-01
Singapore 117510

**Lisa Hoffman**
Stem Cell Biology and Regenerative
Medicine
Robarts Research Institute
PO Box 5015, 100 Perth Drive
London, Ontario
Canada N6A 5K8

**J.H.P. Hui**
Department of Orthopaedic
Surgery
National University of Singapore
Block MD 5, Level 3
12 Medical Drive
Singapore 117598

**Dietmar W. Hutmacher**
Division of Bioengineering
Faculty of Engineering
Department of Orthopaedic Surgery
Faculty of Medicine
Faculty NUS Graduate School for
Integrative Sciences and Engineering
National University of Singapore
Engineering Drive 1
Singapore 119260

**Woo Suk Hwang**
Department of Theriogenology and
Biotechnology
College of Veterinary Medicine
Seoul National University
San 56-1 Shillim-Dong
Kwanak-Gu, Seoul 151-742
Korea

**Ang Hwan Hyun**
Department of Development
Biotechnology
College of Veterinary Medicine
Chungbuk National University
48 Gaesin-dong, Heungdeok-gu
Cheongju, Chungbuk 361-763
Korea (R.O.K)

**Joseph Itskovitz-Eldor**
Department of Obstetrics and
Gynecology
Rambam Medical Center & Stem
Cell Center
Bruce Rappaport Faculty of Medicine
Technion — Israel Institute of
Technology, POB 9602, Haifa 31096
Israel

**Sung Keun Kang**
Department of Theriogenology and
Biotechnology
College of Veterinary Medicine
Seoul National University
San 56-1 Shillim-Dong
Kwanak-Gu, Seoul 151-742
Korea

**Daniel Kraft, MD**
Hematology/Oncology/BMT
Stanford Univ. School of Medicine
Beckman Center, B-265
Stanford, CA 94305
USA

**Byeong Chun Lee**
Department of Theriogenology and
Biotechnology
College of Veterinary Medicine
Seoul National University
San 56-1 Shillim-Dong
Kwanak-Gu, Seoul 151-742
Korea

**Eng Hin Lee**
Division of Graduate Medical Studies
Faculty of Medicine
National University of Singapore
Block MD 5, Level 3
12 Medical Drive
Singapore 117598

**Duncan Liew**
Centre for Stem Cell Biolog and the
Department of Biomedical Science
University of Sheffield
Western Bank, Sheffield S10 2TN
UK

**Bing Lim**
Genome Institute of Singapore
60 Biopolis Street, #02-01 Genome
Singapore 138672

**J.T.K. Lim**
Robert Jones and Agnes Hunt
Orthopaedic and District Hospital
Oswestry, Shropshire SY10 7AG
UK

**Maryam M. Matin**
Institute of Biotechnology and
Tissue Engineering
Department of Biology
Ferdowsi University of Mashhad
Mashhad 91775
Iran

**Shin Yong Moon**
Department of Theriogenology and
Biotechnology
College of Veterinary Medicine
Seoul National University
San 56-1 Shillim-Dong
Kwanak-Gu, Seoul 151-742
Korea

**Christine L. Mummery**
Hubrecht Laboratory and
Interuniversity Cardiology
Institute of the Netherlands
Heart Lung Institute Utrecht
Uppsalalaan 8
3584 CT Utrecht
The Netherlands

**Sulaiman A. Nanji**
Department of Surgery
University of Alberta
Edmonton, Alberta
Canada

**Christopher Navara**
Department of OB/Gyn/Reproductive
Sciences
University of Pittsburgh School of
Medicine and the Pittsburgh
Development Center of the
Magee-Womens Research Institute
204 Craft Avenue
Pittsburgh, PA 15213
USA

**Scott A. Noggle**
Department of Genetics
University of Georgia
Athens, 30605
USA
&
Institute of Molecular Medicine
and Genetics
Medical College of Georgia
Augusta, 30912
USA

**Hong Wei, Ouyang**
NUS Tissue Engineering Program
The Office of Life Sciences
NUS Tissue Engineering Program
DSO (Kent Ridge) Building
27 Medical Drive
#03-01
Singapore 117510

**Robert Passier**
Hubrecht Laboratory and
Interuniversity Cardiology
Institute of the Netherlands
Uppsalalaan 8
3584 CT Utrecht
The Netherlands

**Martin F. Pera**
Monash Institute of Reproduction
and Development
246 Clayton Road
Clayton, VIC 3168
Australia

**Mahendra Rao**
National Institute of Ageing
StemCell, LNS, GRC
333 Cassell Drive
Baltimore
MD 21224
USA

**Mark Richards**
Department of Obstetrics &
Gynaecology
National University of Singapore
Kent Ridge
Singapore 119074

**J.B. Richardson**
Department of Orthopaedic and
Traumatic Surgery
Keele University
Robert Jones and Agnes Hunt
Orthopaedic and District Hospital
Oswestry
Shropshire SY10 7AG
UK

**Enrique Roche**
Associate Prof of Nutrition
Institute of Bioengineering
University Miguel Hernandez
03550 San Juan, Alicante
Spain

**Dr. Maren Ruhnke**
Transplantation and Biotechnologie
Department of General and Thoracic
surgery
University Hospital Schleswig Holstein
Campus Kiel
Arnold-Heller Str.7
24105 Kiel, Germany

**Francesco Russo**
Department of Medicine
10th floor QEQM Building
Imperial College
St Mary's Campus
W2 1NY
London
UK

**Noboru Sato**
Laboratory of Molecular Vertebrate
Embryology
The Rockefeller University
1230 York Avenue
New York, NY 10021
USA

**Gerald Schatten**
Pittsburgh Development Center
Magee-Womens Research Institute
Obstetrics-Gynecology &
Reproductive
Sciences and Cell Biology-Physiology
University of Pittsburgh School of
Medicine
204 Craft Avenue
Pittsburgh, PA 15213
USA

**A.M. James Shapiro**
Clinical Islet Transplant Program
University of Alberta
2000 College Plaza, 8215
112th Street, Edmonton AB
Canada

**Calvin Simerly**
Department of OB/Gyn/
Reproductive Sciences
University of Pittsburgh School of
Medicine and the Pittsburgh
Development Center of the
Magee-Womens Research
Institute
204 Craft Avenue
Pittsburgh, PA 15213
USA

**Bernat Soria**
Instituto de Bioingenieria
Universidad Miguel Hernandez
Campus de San Juan
03550 San Juan, Alicante
Spain

**Suman Lal Chirammal Sugunan**
Division of Bioengineering
National University of Singapore
EA-03-12, Faculty of Engineering
9 Engineering Drive 1
Singapore 117576

**Donald T.H. Tan**
Singapore National Eye Centre and
Department of Ophthalmology
National University of Singapore
5 Lower Kent Ridge Road
Singapore 119074

**Patrick H.C. Tan**
Haematology and Stem Cell Transplant
Centre
Mt. Elizabeth Hospital
3 Mt. Elizabeth, Block B, Level 7
Singapore 228510

**Pamela Vig**
Department of Medicine
10th floor QEQM Building
Imperial College
St Mary's Campus, W2 1NY
London
UK

**Irving L. Weissman**
Beckman Center
Room B-257 Stanford, CA 94305
USA

**Su-Chun Zhang**
Stem Cell Research Program
Waisman Centre, Rm T613
University of Wisconsin
1500 Highland Avenue
Madison, WI 53705
USA

# Foreword

by
*Sydney Brenner*
*Nobel Laureate in Physiology or Medicine, 2002*
*Biomedical Research Council*
*Agency for Science, Technology and Research*
*Singapore*

Stem cell research has gained much prominence in recent years for its therapeutic potential in dealing with difficult diseases many of which are essentially incurable by normal therapies. These diseases are characterized by progressive cell loss and, in the nervous system, which has no regenerative potential, the degenerative process leads to Alzheimer and Parkinson diseases. These have become serious health problems as people in advanced societies now live longer. There is great variability in the occurrence and onset of these diseases and the underlying environmental and genetic factors are unknown. Cell loss can also be caused by autoimmune reactions which produce the destruction of the beta cells of pancreatic islets, the main cause of diabetes, another serious health problem.

In all of these cases, it is hoped that using stem cells with a potential to differentiate into the destroyed cells will allow a replacement of function and a cure or amelioration of the disease. A large part of stem cell research aims to identify and propagate the appropriate stem cells to recapitulate, so to speak, embryonic development for that particular cell lineage.

For therapeutic use we would want autologous cells or else we face the problem of transplantation barriers. There are organs which are capable of regeneration, such as the liver, as well several systems which are continually being renewed, such as the epithelial lining of the intestine, skin and, most prominently, the hemopoietic system. There is a large effort to identify and cultivate such stem cells and to see whether there are dormant stem cells and whether one stem cell could be converted into another.

Ideally, we would like methods to take a normal differentiated cell of an individual and drive it back in time to generate a stem cell with the appropriate differentiation potential. The only way of doing this today is to put the nucleus of a differentiated cell into a fertilized egg and allow it to develop a little further to isolate pluripotential stem cells bearing the same genetic properties of the donor. Since this is the first step in the cloning of organisms from somatic cells, this area has aroused much controversy. There are strong moves to ban human reproduction by cloning on socioethical grounds quite apart from the uncertainties of cloning that have been revealed by animal experiments. There are also moves to extend this prohibition to what is called therapeutic cloning, i.e. the use of cloning to generate stem cells. There is also the question of using human embryos to develop techniques of selecting and propagating human stem cells with specific differentiation potential.

The reader will find a complete and up-to-date coverage of this important field of stem cell research, ranging all the way from molecular biology to bioethics.

# Foreword

by
*Philip Yeo*
*Chairman, Agency for Science, Technology and Research*
*Co-Chairman, Biomedical Sciences Group*
*Economic Development Board*
*Singapore*

Stem cell research has great potential to transform modern clinical medicine, given the ability of stem cells to change themselves physiologically, morphing into distinct tissue types. These undifferentiated, self-renewing cells can provide doctors with the hope and means to tackle illnesses such as Parkinson and Alzheimer diseases, spinal cord injuries and insulin-dependent diabetes.

To take full advantage of their limitless potential requires highly-skilled, dedicated personnel and excellent facilities, all functioning under a forward-looking regulatory framework. Singapore possesses these key factors, as exemplified by our two universities, our hospitals, and our Biopolis, Singapore's unique biomedical sciences public and private research hub.

The success of the First International Stem Cell Conference 2003, co-hosted by our Biomedical Research Council of the Agency for Science, Technology and Research and the National University of Singapore, put the seal of approval on our island-state's reputation as a global player in this arena, something that it has been working towards since Prof. Ariff Bongso's discovery of human embryonic stem (ES) cells in 1994.

Prof. Bongso has since gone on to produce several ES cell lines and developed an animal-free system for growing them in culture, making him one of the world's acknowledged authorities on ES cells and an excellent choice for editor of this book.

Prof. Bongso's skills are complemented well by fellow editor Prof. Lee Eng Hin's expertise in adult stem cell-mediated physeal and articular cartilage repair. Prof. Lee and his team at the Faculty of Medicine, The National University of Singapore, were among the first in the world to bring this technology to clinical trials. They have since demonstrated that it can successfully restore damaged physeal and articular cartilage.

Together, the two co-editors bring a unique perspective to bear on Singapore's stem cell research scene; one that I hope the readers of this book will greatly enjoy.

# Stem Cells: Their Definition, Classification and Sources

Ariff Bongso* and Eng Hin Lee

## Introduction

Stem cell biology has attracted tremendous interest recently. It is hoped that it will play a major role in the treatment of a number of incurable diseases via transplantation therapy. Several varieties of stem cells have been isolated and identified *in vivo* and *in vitro*. Very broadly they comprise of two major classes: embryonic/fetal stem cells and adult stem cells. Some scientists wish to pursue research on embryonic/fetal stem cells because of their versatility and pluripotentiality, while others prefer to pursue research on adult stem cells because of the controversial ethical sensitivities behind embryonic/fetal stem cells. However, both embryonic/fetal and adult stem cells are equally important and research on both types must be enthusiastically pursued since the final objective is the application of this technology for the treatment of a variety of diseases that plague mankind. It is very possible that the findings from one stem cell type may complement that of the other.

The word "stem cell" has also been loosely used by some scientists without the demonstration of stem cell markers or confirmation of stemness via transcriptome profiling. It is their ability to self-renew and differentiate that certain cells are termed stem cells both *in vivo* and *in vitro*. It is very crucial that the correct definition and proof of stemness through proper and accepted characterization tests be addressed before a particular cell type is classified as a stem cell. Stem cell therapy has already reached the bedside in some hospitals through the transplantation of donor bone marrow stem cells into the circulatory system of leukemic patients and the transfer of

*Correspondence: Department of Obstetrics & Gynaecology, National University of Singapore, Kent Ridge, Singapore 119074. Tel.: 65-67724260, Fax: 65-67794753, e-mail: obgbongs@nus.edu.sg

umbilical cord stem cells into the circulatory system of leukemic children or their siblings produced from the same mother who had previously stored her umbilical cord cells. However, the more challenging and impactful use of stem cells would come from the directed differentiation or transdifferentiation of stem cells into other cell types and tissues to help cure a plethora of incurable diseases. It would be tremendously useful if embryonic, fetal, adult or umbilical cord stem cells could be coaxed to produce islets cells for the treatment of diabetes or neurons for neurodegenerative diseases, cardiomyocytes for heart disease, and so on. This chapter attempts to define, classify and describe the sources of the various types of stem cells that have been isolated to date.

## Definition of Stem Cells

Stem cells are unspecialized cells in the human body that are capable of becoming specialized cells, each with new specialized cell functions. The best example of a stem cell is the bone marrow stem cell that is unspecialized and able to specialize into blood cells, such as white blood cells and red blood cells, and these new cell types have special functions, such as being able to produce antibodies, act as scavengers to combat infection and transport gases. Thus one cell type stems from the other and hence the term "stem cell." Basically, a stem cell remains uncommitted until it receives a signal to develop into a specialized cell. Stem cells have the remarkable properties of developing into a variety of cell types in the human body. They serve as a repair system by being able to divide without limit to replenish other cells. When a stem cells divides, each new cell has the potential to either remain as a stem cell or become another cell type with new special functions, such as blood cells, brain cells, etc.

Most tissue repair events in mammals are dedifferentiation independent events brought about by the activation of pre-existing stem cells or progenitor cells. By definition, a progenitor cell lies in between a stem cell and a terminally differentiated cell. However, some vertebrates such as salamanders regenerate lost body parts through the dedifferentiation of specialized cells into precursor cells. These dedifferentiated cells then proliferate and later form new specialized cells of the regenerated organ. In fact, some invertebrates such as the Planarian flatworm and the hydra regenerate tissues very quickly and with precision.[1,2] The word "stem" actually originated from old botanical monographs from the same terminology as the stems of plants, where stem cells were demonstrated in the apical root and shoot meristems that were responsible for the regenerative competence of plants. Hence also the use of the word "stem" in "meristem."[3] Today, stem cells have been isolated from preimplantation embryos, fetuses, adults and the umbilical cord and under certain conditions, these undifferentiated stem cells can be pluripotent (ability to give rise to cells from all three germ layers, viz. ectoderm, mesoderm and endoderm) or multipotent (ability to give rise to a limited number of other specialized cell types).

## Classification and Sources of Stem Cells

Stem cells can be classified into four broad types based on their origin, viz. stem cells from embryos; stem cells from the fetus; stem cells from the umbilical cord; and stem cells from the adult. Each of these can be grouped into subtypes (Fig. 1). Some believe that adult and fetal stem cells evolved from embryonic stem cells and the few stem cells observed in adult organs are the remnants of original embryonic stem cells that gave up in the race to differentiate into developing organs or remained in cell niches in the organs which are called upon for repair during tissue injury. [4]

### Embryonic stem cells

In mammals, the fertilized oocyte, zygote, 2-cell, 4-cell, 8-cell and morula resulting from cleavage of the early embryo are examples of totipotent cells (ability to form a complete organism).[5] Proof that these are indeed totipotent cells comes from the fact that identical twins can be generated from splitting of the early embryo *in vitro* by micromanipulation in domestic animals. However, strictly speaking, the fertilized oocyte and blastomeres cannot be termed "stem cells" because the making of more of them is limited during early cleavage division. They, thus, cannot self-renew even though they have the potential to form a complete organism.

The inner cell mass (ICM) of the 5- to 6-day old human blastocyst is the source of pluripotent embryonic stem cells (hESCs). During embryonic development, the ICM develops into two distinct cell layers, the epiblast and hypoblast. The hypoblast forms the yolk sac which later becomes redundant in the human, and the epiblast differentiates into the three primordial germ layers (ectoderm, mesoderm and endoderm). Embryonic endoderm cells are rather restricted in their developmental pathways. A small population of multipotent cells, called the definitive endoderm, gives rise to all of the endoderm derived organs in the adult. The definitive endoderm is

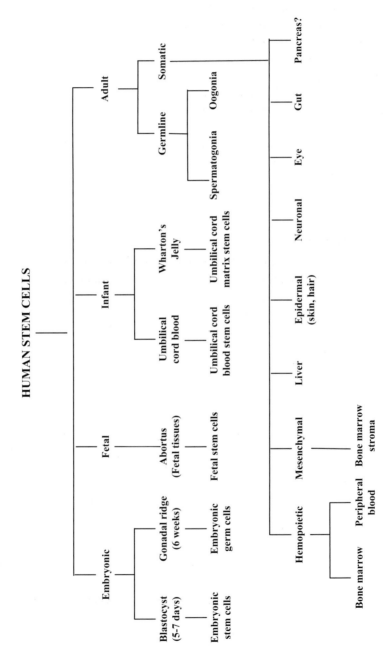

**Figure 1.** Classification of human stem cells.

separated from the pluripotent ICM during gastrulation immediately after implantation. The definitive endoderm comprises an epithelial sheet of approximately 600 cells that cover the ventral surface of the embryo. This sheet later forms the fore and hind gut. The fore gut later forms the lung, liver, stomach and pancreas, while its more posterior aspects gives rise to the intestines (mid-gut) and cloaca. The hind gut gives rise to the rectum and large intestine.[6] Knowing what drives these developmental pathways is crucial to understanding the factors and events that lead to differentiation of embryonic stem cells to desirable tissues such as the pancreas. Pluripotent embryonic stem cells can give rise to many cell types *in vitro*, including cells specific to endodermal tissues. Advances in the understanding as to how ES cells differentiate should provide answers for re-programming of stem cells from adult tissues.

## Embryonic germ cells

Primordial germ cells or diploid germ cell precursors transiently exist in the embryo before they closely associate with somatic cells of the gonads and then become committed as germ cells. Human embryonic germ cells (hEGCs) which are also stem cells, originate from the primordial germ cells of the gonadal ridge of 5- to 9-week old fetuses. hEGCs have been successfully isolated and characterized.[7] These stem cells are pluripotent and are able to produce cells of all three germ layers.

## Fetal stem cells

Fetal stem cells are primitive cell types found in the organs of fetuses. Neural crest stem cells, fetal hematopoietic stem cells and pancreatic islet progenitors have been isolated in abortuses.[8] Fetal neural stem cells found in the fetal brain were shown to differentiate into both neurons and glial cells.[9,10] Fetal blood, placenta and umbilical cord are rich sources of fetal hematopoietic stem cells.

## Umbilical cord stem cells

Umbilical cord blood contains circulating stem cells and the cellular contents of umbilical cord blood appear to be quite distinct from those of bone marrow and adult peripheral blood.[11] The characteristics of hematopoietic stem cells in umbilical cord blood have recently been clarified. The frequency of umbilical cord blood hematopoietic stem cells equals or exceeds that of bone marrow and they are known to produce large colonies *in vitro*, have different growth factor requirements, have long telomeres and can be expanded in long term culture. Cord blood shows decreased graft versus host reaction compared with bone marrow, possibly due to high interleukin-10 levels produced by the cells and/or decreased expression of the beta-2-microglobulin. Cord blood stem cells have been shown to be multipotent by being able to differentiate into neurons and liver cells.[11]

While most of the attention has been on cord blood stem cells and more specifically their storage for later use, there have also been reports that matrix cells from the umbilical cord contain potentially useful stem cells.[12] This matrix termed Wharton's jelly has been a source for isolation of mesenchymal stem cells. These cells express typical stem cell markers, such as c-kit and high telomerase activity; have been propagated for long population doubling times; and can be induced to differentiate *in vitro* into neurons.

## Adult stem cells

### Hematopoietic stem cells (bone marrow and peripheral blood)

Bone marrow possesses stem cells that are hematopoietic and mesenchymal in origin. Hematopoiesis is the production and maintenance of blood stem cells and their proliferation and differentiation into the cells of peripheral blood. The hematopoietic stem cell is derived early in embryogenesis from mesoderm and becomes deposited in very specific hematopoietic sites within the embryo.[13] These sites include the bone marrow, liver, and yolk sac. Hematopoietic stem cells can be purified using monoclonal antibodies, and recently, common lymphoid progenitor and myeloid-erythroid progenitor cells have been isolated and characterized.[13] Bone marrow stem cells may be more plastic and versatile than expected because they are multipotent and can be differentiated into many cell types both *in vitro* and *in vivo*.

### Mesenchymal stem cells (bone marrow stroma)

Mesenchymal stem cells (MSCs) are found postnatally in the non-hematopoietic bone marrow stroma. Marrow stromal tissue is made up of a heterogenous population of cells, which include reticular cells, adipocytes, osteogenic cells, smooth muscle cells, endothelial cells and macrophages.[14] In a steady state or in response to injury, turnover of stromal tissue and repair occurs through the participation of a population of stem cells found in the stromal tissue.[15] Apart from bone marrow stroma, MSCs can also be derived from periosteum, fat and skin. MSCs are multipotent cells that are capable of differentiating into cartilage, bone, muscle, tendon, ligament and fat.[16] There is some recent evidence that there is a rare cell within MSC cultures that is pluripotent and can give rise not only to mesodermal but to endodermal tissues.[17] The authors have called this a Multipotent Adult Progenitor Cell.

## Gut stem cells

The gastrointestinal epithelial lining undergoes continuous and rapid renewal throughout life. Differentiation programs thus exist in specific regions of the tract. Epithelial renewal is sustained with populations of multipotent stem cells residing in distinct anatomic sites governed by niches.[18] A major challenge is to identify these niches, the properties of these stem cells and the molecular mechanisms underlining their fate decisions in appropriate developmental pathways. These answers will provide clues as to why some patients infected with *Helicobacter pylori* are at risk in developing gastric adenocarcinoma. Many patients harbor *H. pylori* in their stomachs but only a percentage goes on to develop pathology.[19]

Epithelial cell renewal in the intestine is sustained by multipotent stem cells located in the crypts of Lieberhahn. In the small intestine, epithelial cells of enterocytic, goblet and enteroendocrine origin differentiate as they migrate from a crypt up an adjacent villus and leave the intestine once they reach the villus tip. In the colon, it is different. Epithelial cells migrate from the crypt to a flat surface cuff that surrounds its opening. The stem cell hierarchy in the gut and the fact that stem cells and their progeny are located in well defined anatomic units make the gut an ideal *in vivo* model for stem cell research.[20]

## Liver stem cells

Mammals are said to survive surgical removal of at least 75% of the liver by regeneration. The original tissue can be restored in 2–3 weeks. This is in contrast to most other organs such as the kidney or pancreas. Recent evidence strongly suggests that different cell types and mechanisms are responsible for organ reconstitution, depending on the type of liver injury. In the case of the liver, regeneration must be distinguished by transplantation (repopulation) with donor cells.[21]

## Bone and cartilage stem cells

Mesenchymal Stem Cells in bone marrow can differentiate into bone and cartilage under appropriate conditions. However, if bone or cartilage is injured, are there stem cells inherent in bone or cartilage to participate in the repair process? Bone itself has been found to have both uncommitted stem cells as well as committed osteoprogenitor cells.[22,23] In addition, when bone is fractured, there is exposed marrow and abundant bleeding with hematoma formation in the marrow space, which results in good repair potential. *In vivo*, articular cartilage has a very limited capacity for repair if injured. It is currently not clear whether there is a committed chondrocyte progenitor cell located within cartilage. In the presence of injury to cartilage, stem cells do participate in the repair process. The numbers, however, are small and the regulatory factors are limited.[24,25] It is postulated that these cells may be derived from surrounding tissues such as muscle, bone or other non-cartilaginous tissues.[26]

## Epidermal stem cells (skin and hair)

The human skin comprises the outer epidermis and underlying dermis. Hair and sebaceous glands also make up the epidermis. The most important cell type in the epidermis is the keratinocyte which is an epithelial cell that divides and is housed in the basal layer of the epidermis. Once these cells leave the basal layer they undergo terminal differentiation resulting in a highly specialized cell called a squame which eventually forms either the hair shaft or the lipid-filled sebocyte that form an outer skin layer between the harsh environment and underlying living skin cells. The epidermis houses stem cells at the base of the hair follicle and their self-renewing properties allow for the re-growth of hair and skin cells that occurs continuously. New keratinocytes are produced continuously during adult life to replace the squames shed from the outer skin layers and the hairs that are lost. Stem cells differentiate into an intermediate cell called the "transient amplifying cell" which gives rise to the more differentiated cell types inclusive of the keratinocytes and sebocytes.[27]

## Neuronal stem cells

It has been suggested that a continuous neurogenic turnover occurs in some limited areas of the central nervous system (CNS). Two neurogenic regions of the adult mammalian CNS are supposed to be involved in this process: the subventricular zone (SVZ) of the forebrain[28–30] and the dental gyrus of the hippocampus[31,32] which are considered reservoirs of new neural cells. Thus, neural stem cells (NSCs) are known to reside in these two areas and they consistently generate new neurons.[33–35] *In vivo*, endogenous NSCs seem to be able to produce almost exclusively neurons, while a single NSC *in vitro* is competent to generate neurons, astrocytes and oligodendrocytes.[36] NSCs are multipotent progenitoir cells that have self-renewal activity. Although it seems clear at present that the bona fide NSC is the subventricular zone B cells, the search for self-renewing, multipotent NSCs is in progress and conflicting information is available in the literature. There has been data to suggest that the SVZ NSC is an ependymal cell,[37] while others have demonstrated that the SVZ astrocyte is the NSC.[38] It was also demonstrated that ependymal cells were unipotent giving rise to only glial cells, whereas SVZ astrocytes were able to produce multipotent neurospheres that yielded both neurons and glia.[39] The final fate of the NSC is under tight environmental control and a stem cell niche has been postulated for the adult mammalian brain.

## Pancreatic stem cells

There has been controversy as to whether the pancreas contains true stem cells. It was reported that the endocrine cells of the rat pancreatic islets of Langerhans, including insulin-producing beta-cells, turn over every 40–50 days by processes of apoptosis and the proliferation and differentiation of new islet cells (neogensis) from progenitor epithelial cells located in the pancreatic ducts. The administration to rats of glucose or glucagon-like peptides resulted in the doubling of the islet cell mass, suggesting that islet progenitor cells may reside within the islet themselves.[40] The same authors showed that rat and human pancreatic islets contained an unrecognized population of cells that expressed the neural stem cell-specific marker nestin. These nestin-positive cells were distinct from ductal epithelium. These nestin positive cells, after isolation, had an unusually extended proliferative capacity *in vitro*, could be cloned repeatedly and appeared to be multipotential. They were able to differentiate *in vitro* into cells that expressed liver and exocrine pancreas markers. The authors proposed that these nestin-positive islet derived progenitor cells were a distinct population of cells that resided within the pancreatic islets and participated in neogenesis of islet endocrine cells.[40]

More recently, however, in an effort to pin down the source of new b cells, Dor *et al.*[41] designed transgenic mice in which insulin-producing cells were prompted to produce HPAP that is detected by blue staining. When the mice were 6–8 weeks old, the HPAP gene was turned on. Once the HPAP gene was tuned on, b cells were expected to pass on the gene to daughter cells. If the new b cells came from stem cells, then they should not be labeled by the stain. After 12 months, the percentage of blue cells was higher than that in 6-week-old mice, suggesting that the b cells replicate themselves and that the pancreas is unlikely to harbor stem cells that produce large numbers of new b cells. Later, Seaberg *et al.*[42] exposed pancreatic cells to culture media that encourage growth of neural stem cells. One out of every 5000 cells quickly multiplied into groups of cells. The authors suggested that this grouping was characteristic of stem cells. Additionally, the authors demonstrated the formation of a variety of cell types from these cell groups when the culture medium was changed to encourage the cell groups to differentiate. The cell milieu comprised neurons and pancreatic cells inclusive of b cells based on gene profiling. The b cells secreted insulin and when sugars were added to the culture medium, the b cells put out more than twice as much of insulin. The unequivocal demonstration of the existence of stem cells in the pancreas was, however, not proven.

## Eye stem cells

Stem cells have been identified in the adult mouse eye.[43] Single pigmented ciliary margin cells were shown to clonally proliferate *in vitro* to form sphere colonies of cells that can differentiate into retinal-specific cell types, including rod photoreceptors, bipolar neurons and Muller glia. The adult retinal stem cells were localized to the pigmentary ciliary margin and not to the central and peripheral retinal pigmented epithelium.

## REFERENCES

1. Wolpert L, Hicklin J, Hornbruch A. (1971) Positional information and pattern regulation in regeneration of hydra. *Symp Soc Exp Biol* **25**: 391–415.
2. Brockes JP. (1997) Amphibian limb regeneration: Rebuilding a complex structure. *Science* **276**: 81–87.
3. Kiessling AA, Anderson SC. (2003) Human embryonic stem cells. Boston: Jones and Bartlett.
4. Anderson DJ, Gage FH, Weissman IL. (2001) Can stem cells cross lineage boundaries? *Nature* **7**: 393–395.

5.  Bongso A, Richards M. (2004) History and perspective of stem cell research. In *Best Practice & Research Clinical Obstetrics & Gynaecology*, eds. N. Fisk & J. Itskovitz, London: Elsevier Ltd.

6.  Stem Cells and the Future of Regenerative Medicine. (2001) *Comm Biol Biomed Appl Stem Cell Res*. Board of Life Sciences, NRC. Washington DC: National Academy Press.

7.  Shamblott MJ, Axelman J, Wang S, *et al.* (1998) Derivation of pluripotent stem cells from cultured human primordial germ cells. *Proc Natl Acad Sci USA* **95**: 13726–13731.

8.  Beattie GM, Otonkoski T, Lopez AD, *et al.* (1997) Functional beta-cell mass after transplantation of human fetal pancreatic cells: Differentiation or proliferation? *Diabetes* **46**: 244–248.

9.  Brustle O, Choudary K, Karram K, *et al.* (1998) Chimeric brains generated by intraventricular transplantation of human brain cells into embryonic rats. *Nat Biotech* **16**: 1040–1044.

10. Villa A, Snyder EY, Vescovi A, *et al.* (2000) Establishment and properties of a growth factor dependent perpetual neural stem cell line from the human CNS. *Exp Neurol* **161**: 67–84.

11. Rogers I, Casper RF. (2004) Umbilical cord blood stem cells. In *Best Practice & Research Clinical Obstetrics & Gynaecology*, eds: N. Fisk & J. Itskovitz, London: Elsevier Ltd.

12. Mitchell KE, Weiss ML, Mitchell BM, *et al.* (2003) Matrix cells from Wharton's jelly form neurons and glia. *Stem Cells* **21**: 50–60.

13. Stem Cell and Developmental Biology Writing Group's Report. (2004) *Natl Inst Diabetes & Digestive & Kidney Dses, NIH.* 1–27.

14. Bianco P, Riminucci M. (1998) The bone marrow stroma *in vivo*: Ontogeny, structure, cellular composition and changes in disease. In *Marrow Stromal Cell Culture. Handbooks in Practical Animal Cell Biology*, ed. J.N. Beresford and M.E. Owen, p. 1025. Cambridge, UK: Cambridge University Press.

15. Owen ME. (1988) Marrow stromal stem cells. *J Cell Sci Suppl* **10**: 63–76.

16. Caplan AI. (1994) The mesengenic process. *Clin Plast Surg* **21**: 429–435.

17. Jiang Y, Jahagirder BN, Reinhardt RL, *et al.* (2002) Pluripotency of mesenchymal stem cells derived from adult bone marrow. *Nature* **418**: 41–49.

18. Wright NA. (2000) Epithelial stem cell repertoire in the gut: Clues to the origin of cell lineages, proliferative units and cancer. *Int J Exp Pathol* **81**: 117–143.

19. Burdick JS, Chung E, Tanner G, *et al.* (2000) Treatment of Menetrier's disease with a monoclonal antibody against the epidermal growth factor receptor. *New Eng J Med* **343**: 1697–1701.

20. Alison MR, Poulsom R, Forbes S, *et al.* (2002) An introduction to stem cells. *J Path* **197**: 419–423.

21. Alison MR, Vig P, Russo F, *et al.* (2004) Hepatic stem cells: From inside and outside the liver? *Cell Prolif* **37**: 1–21.

22. Gronthos S, Zannettino AC, Graves SE, *et al.* (1999) Differential cell surface expression of STRO-1 and alkaline phosphatase antigens on discrete developmental stages in primary cultures of human bone cells. *J Bone Miner Res* **14**: 47–56.

23. Nuttall ME, Patton AJ, Olivera DL, *et al.* (1998) Human trabecular bone cells are able to express both osteoblastic and adipocytic phenotype: Implications for osteopenic disorders. *J Bone Miner Res* **13**: 371–382.

24. Metsaranta M, Kujala UM, Pelliniemi L, *et al.* (1996) Evidence for insufficient chondrocytic differentiation during repair of full thickness defects of cartilage. *Matrix Biol* **15**: 39–47.

25. Nakajima H, Goto T, Horikawa O, *et al.* (1998) Characterization of cells in the repair tissue of full thickness articular cartilage defects. *Histochem Cell Biol* **109**: 331–338.

26. Shapiro F, Koide S, Glimcher MJ. (1993) Cell origin and differentiation in the repair to full thickness defects of articular cartilage. *J Bone Joint Surg Am* **75**: 532–553.

27. Blanpain C, Lowry WE, Geohegan A, *et al.* (2004) Self-renewal, multipotency, and the existence of two cell populations within an epithelial stem cell niche. *Cell* **118**: 530–532.

28. Reynolds BA, Weiss S. (1992) Generation of neurons and astrocytes from isolated cells of the adult mammalian central nervous system. *Science* **255**: 1707–1710.

29. Luskin MB. (1993) Restricted proliferation and migration of postnatally generated neurons derived from the forebrain subventricular zone. *Neuron* **11**: 173–189.

30. Lois C, Alvarez-Buylla A. (1993) Proliferating subventricular zone cells in the adult mammalian forebrain can differentiate into neurons and glia. *Proc Natl Acad Sci USA* **90**: 2074–2077.

31. Seaberg RM, Van der Kooy D. (2002) Adult rodent neurogenic regions: The ventricular subependyma contains neural stem cells, but the dentate gyrus contains restricted progenitors. *J Neurosci* **22**: 1784–1793.

32. Palmer TD, Ray J, Gage FH. (1995) FGF-2 responsive neuronal progenitors reside in proliferative and quiescent regions of the adult rodent brain. *Mol Cell Neurosci* **6**: 474–486.

33. Mckay R. (1997) Stem cells in the central nervous system. *Science* **276**: 66–71.

34. Gage FH. (2000) Mammalian neural stem cells. *Science* **287**: 1433–1438.

35. Temple S. (2001) The development of neural stem cells. *Nature* **414**: 112–117.

36. Bottai D, Fiocco R, Gelain F, *et al.* (2003) Neural stem cells in the adult nervous system. *J Hematother Stem Cell Res* **12**: 655–670.

37. Johansson CB, Momma DL, Clarke DL, *et al.* (1999) Identification of a neural stem cells in the adult mammalian central nervous system. *Cell* **96**: 25–34.

38. Doetsch F, Caille DA, Lim JM, *et al.* (1999) Subventricular zone astrocytes are neural stem cells in the adult mammalian brain. *Cell* **97**: 703–716.

39. Laywell ED, Rakic P, Kukekov VG, *et al.* (2000) Identification of a multipotency astrocytic stem cell in the immature and adult mouse brain. *Proc Natl Acad Sci USA* **97**: 13883–13888.
40. Zulewski H, Abraham EJ, Gerlach MJ, *et al.* (2001) Multipotential nestin positive stem cells isolated from adult pancreatic islets differentiate *ex vivo* into pancreatic endocrine, exocrine and hepatic phenotypes. *Diabetes* **50**: 521–533.
41. Dor Y, Brown J, Martinez OI, *et al.* (2004) Adult pancreatic b cells are formed by self-duplication rather than stem cell differentiation. *Nature* **429**: 41–46.
42. Seaberg RM, Smukler S, Kieffer TJ, *et al.* (2004) Clonal identification of multipotent precursors from adult mouse pancreas that generate neural and pancreatic lineages. *Nature Biotech* **22**: 1115–1124.
43. Tropepe V, Coles BLK, Chiasson BJ, *et al.* (2000) Retinal stem cells in the adult mammalian eye. *Science* **287**: 2032–2036.

# 2

# From Human Embryos to Clinically Compliant Embryonic Stem Cells: Blastocyst Culture, Xeno-free Derivation and Cryopreservation, Properties and Applications of Embryonic Stem Cells

Ariff Bongso*, Mark Richards and Chui-Yee Fong

## Introduction

The isolation of stem cells from human preimplantation embryos has been considered the biggest breakthrough of the 21st century. These mysterious cells aptly referred to as the "mother of all cells" hold promise in a new era of reparative medicine by providing an unlimited supply of different tissue types suitable for transplantation therapy (Fig. 1). Human embryonic stem cells (hESCs) can be isolated from: i) surplus embryos left over after *in vitro* fertilization (IVF) treatment[1–3]; ii) embryos created specifically for this purpose[4]; and iii) from embryos created by somatic cell nuclear transfer.[5] The latter two approaches are heavily charged with emotional and ethical implications while the first is less controversial. However, hESC research leading to prospective cures for a variety of incurable diseases has the potential to affect the lives of millions of people around the world, not through the prolongation of life but more through the improvement of the quality of life. Two major hurdles fundamental to hESC biology have to be overcome before purified hESC directed cells are used in clinical therapy, viz., a) hESC culture technology *in vitro* must be safe with

*Correspondence: Department of Obstetrics and Gynaecology, National University of Singapore, Kent Ridge, Singapore 119074. Tel: 65-67724260, fax: 65-67794753, e-mail: obgbongs@nus.edu.sg

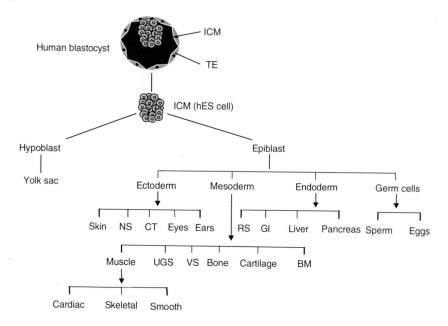

**Figure 1.** Cell lineages from the ICM of human blastocysts.

FDA approval, and b) enough hESC lines must be produced that are readily expandable to generate the required number of cells for later clinical application.

Whatever the approaches are, it is important when deriving stable hESC lines that the gametes and ensuing embryos are of high quality and free from both genetic and infectious diseases. Patients donating such material with informed consent should be screened for HIV, hepatitis B, sex-linked and autosomal genetic diseases. There is an urgent necessity for a global effort in establishing centralized hESC repositories with stringent quality control that carry panels of cell lines that are: i) xeno-free and of clinical grade, derived and grown under current good manufacturing practice (cGMP) and good tissue culture practice (GCTP) conditions; ii) diverse for gender and ethnicity; iii) HLA typed for future tissue matching; iv) fully tested for diseases and fully characterized (inclusive of molecular genetic markers); v) sustainable *in vitro* without karyotypic changes for prolonged passages; vi) grown both by the "cut and paste" (mechanical transfer) and "bulk culture" (enzymatic transfer) protocols with growth and culture conditions identified; and vii) able to differentiate with known inducers.

## Blastocyst Culture

The successful production of stable hESC lines with high success rates depends largely on the quality of embryo started with. These are usually left-over embryos from IVF clinics. Institutional Review Board (IRB) approval and informed patient consent must be sought before such frozen embryos are used for hESC derivation. Such donated embryos may have been frozen at the day-1 (pronuclear); day-2 (4 cell); day-3 (8 cell); day-4 (compacted); or day-5 (blastocyst) stages of embryonic development. Knowledge of the daily embryo scores using polarized and morphological markers and cleavage speed that will generate the best blastocyst is useful in ensuring a high success rate in hESC line derivation (Fig. 2). Once thawed, such embryos are grown to the blastocyst stage using an extended microdroplet culture protocol with two-stage sequential culture media in the presence of 5% or 6% $CO_2$ in air atmospheres.[6,7] Blastocysts with distinct inner cell masses (ICMs) and thin zona pellucidae are ideal for hESC derivation. Success rates for producing hESC lines vary from laboratory to laboratory. Taking all existing reports collectively,[2,3,8] a success rate of about 20% is expected in producing an hESC line from a single frozen day-2 to -3 embryo (Fig. 3). As more experience is gathered, these success rates are definitely expected to go up. Using the extended microdroplet culture protocol, embryo scoring, enzymatic zona removal with pronase and immunosurgery, high success rates of nearly 50% were obtained by our group (Fig. 4).

## Research Grade hESC Lines from Surplus IVF Embryos

Fertilization of a single oocyte from a natural cycle to produce a single embryo for IVF results in suboptimal pregnancy rates, and thus it has become necessary to stimulate the ovaries of such subfertile women with hormones to obtain an optimum number of oocytes that will generate at least 2–3 embryos for transfer resulting in pregnancy rates of as high as 40–50%. To obtain such high pregnancy rates, most IVF centers fertilize all aspirated oocytes, thus resulting in surplus embryos that are frozen for future use for the couple. IVF couples have several options with their surplus frozen embryos. If they have completed their families and have no personal use for such surplus embryos, they could: 1) donate the embryos to a needy childless couple; or 2) donate the embryos for research that will benefit mankind; or 3) dispose off the embryos. The embryos that are available for deriving hESC lines come from those patients who donate such

EGG & EMBRYO STAGING

| | | | |
|---|---|---|---|
| MII EGGS (denuded): | Smooth polar body (PB) | ++++ | |
| | Rough PB | +++ | |
| | Fragmented PB | ++ | |
| | Large PB | + | |
| 2PN (D1): | Nucleoli alignment | +++ | |
| | Nucleoli evenly scattered | ++ | |
| | Others | + | |
| 4–6 cell (D2): | Blastomere regularity; absence of fragments | +++ | |
| | Blastomere regularity; moderate fragments | ++ | |
| | Blastomere irregularity, abundant fragments | + | |
| 8-cell/Compacting (D3): | Compacting | +++ | |
| | 8–10 cell, no fragments | ++ | |
| | Others | + | |
| Compacting/ compacted/EC (D4): | Early cavitating | +++ | |
| | Compacted | +++ | |
| | Compacting | ++ | |
| | Others | + | |
| LC/EB/Expanding BL (D5): | Expanding blastocyst | +++ | |
| | Early blastocyst (EB) | +++ | |
| | Late cavitating (LC) | ++ | |

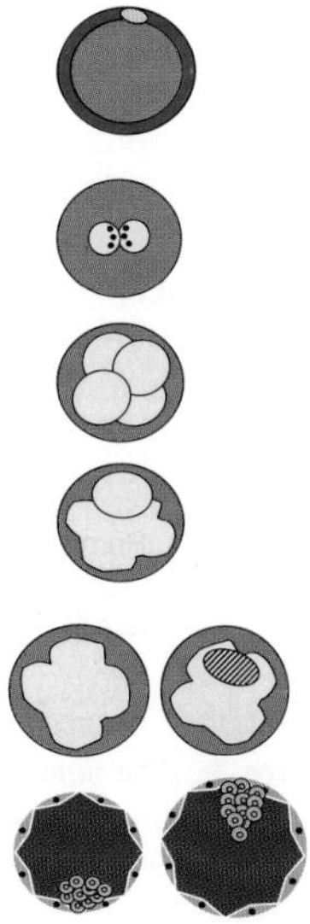

**Figure 2.**  Human egg and embryo scoring from day-0 to day-5.

embryos for useful research. Thus, derivation of hESC lines should be by informed patient consent, Institutional Review Board (IRB) approval, and by using approved hESC derivation protocols to help overcome the ethical sensitivities of hESC research.

## Derivation and propagation
### *"Cut and paste" method (mechanical transfer)*

Embryonic stem cells were first isolated from IVF human blastocysts in 1993.[1,9] After removal of the zona pellucida with pronase (Sigma, MO), 21 zona-free blastocysts from 9 IVF patients were cultured intact as a "whole embryo culture" on irradiated human adult oviductal epithelial fibroblasts in the presence of Chang's medium supplemented with 1000 IU/ml of hLIF. Nineteen of these produced healthy ICM lumps in 7 to 11 days. The lumps were mechanically separated from the peripheral trophectodermal (TE)

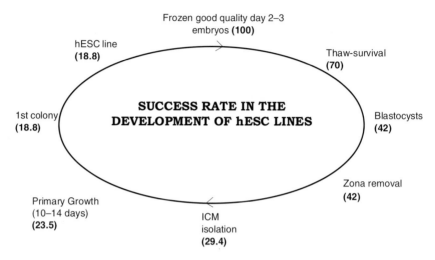

**Figure 3.** The success rates in developing hESC lines. The figures in parentheses are number of embryos. Thaw survival (70%); blastulation (60%); zona removal (100%); immunosurgery (70%); primary culture (80%); first colony (80%); hESC line (100%).

cells and underlying feeder cells with hypodermic needles, trypsinized and passed further on fresh irradiated human adult fallopian tubal feeders. hESC cell colonies were produced in the first and second passages. The hESCs of the first two passages had typical hESC morphology (high nuclear-cytoplasmic ratios and prominent nucleoli), stained positive for alkaline phosphatase and had normal karyotypes. These cells differentiated from the third passage onwards.[1]

Later, Thomson et al.[2] used immunosurgery[10] to completely separate the ICM and cultured the intact ICM on irradiated murine embryonic fibroblasts (MEFs) instead of human feeders. Parts of each hESC colony were then cut mechanically with hypodermic needles and the cell clusters grown on fresh irradiated MEFs instead of cell disassociation into single cells with trypsin. Using such a protocol, these workers were successful in producing the first hESC line. Thus, the differentiation of the hESC colonies from the third passage onwards in the reports of Bongso et al.[1,9] was perhaps due to the dissociation of the colonies into single cells very early in culture with trypsin. hESCs have been considered very "social" cells with numerous junctional complexes holding them together for their survival and hence the single cell disaggregation may have disturbed this "social" behavior resulting in differentiation. Later, Reubinoff et al.[3] used immunosurgery,

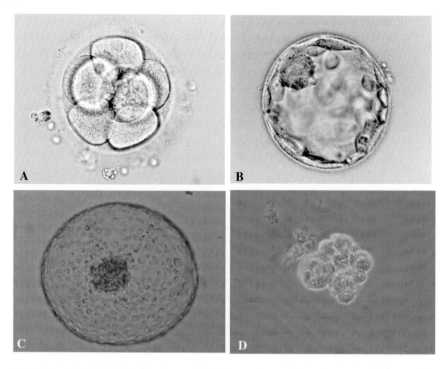

**Figure 4.**  Blastocyst culture and immunosurgery. **A.** Frozen-thawed day-3 human embryo containing 8 healthy blastomeres; **B.** Good quality blastocyst grown from (A). Note distinct ICM at 10 'o'clock position; **C.** Zona-free blastocyst after dissolution of zona pellucida with pronase; **D.** Intact ICM after immunosurgery.

mitomycin-C treated MEFs and a similar "cut and paste" mechanical transfer method to derive and propagate hESC lines that could be spontaneously differentiated into neuronal cells *in vitro*. Amit and Iskovitz[11] and Suss-Toby *et al.*[12] later confirmed that the "whole embryo culture" method worked just as well as the immunosurgery protocol to produce hESC lines. A re-visit of the work of Bongso *et al.*[1,9] to use human feeders instead of MEFs recently showed that hESC lines can be successfully derived and propagated on fetal, neonatal and adult human feeders.[13–15] hESC colonies grown on human feeders using the "cut and paste" method are more angular and rhomboid compared with the circular colonies grown on MEFs.[13,14]

## "Bulk culture" method (enzymatic transfer)

Some of the limitations of the "cut and paste" method are that it is labor-intensive and it takes a long time to scale up hESC numbers. Thus a bulk culture protocol was developed where hESC colonies are mechanically cut into clusters and transferred to MEFs in initial passages and in later passages colonies were exposed to collagenase and/or trypsin and single disassociated cells sown on mitomycin-C treated MEFs, like rice thrown on a paddy field. Several hESC colonies sprouted up and thus cell numbers could be expanded in shorter time periods. Some laboratories enzymatically treat both hESC colonies and feeders at the same time and seed the entire mixed cell suspension onto fresh MEFs, allowing the single hESCs to produce new colonies while the old feeder cells degenerate in culture.

Recently, Cowan *et al.*[8] derived 17 individual hESC lines from 97 ICMs using MEFs as feeder cells, recombinant hLIF, serum replacement and plasmanate. Each cell line was initially passaged (up to the first 5 passages) by mechanical dissociation immediately after the initial ICM outgrowth. The cell lines were then adapted to enzymatic passage with trypsin. All 17 cell lines were characterized using several molecular markers and they maintained a normal karyotype in culture. However, with prolonged passaging *in vitro* these authors observed karyotypic changes involving chromosomes 12 and 2. These chromosomal changes were associated with a proliferative advantage and a shortening in the population doubling time.

Our group recently confirmed that human feeders could also be used in the "bulk culture" protocol to successfully support hESC propagation (Richards and Bongso, unpublished data).

## Research Grade hESC Lines from Cloned Embryos

Since hESCs are derived from donor surplus IVF embryos, there has been some concern that the tissues differentiated from these hESCs may be rejected after transplantation to patients. There are several approaches, however, that could help overcome transplantation incompatibility. These include: 1) the modification of the histocompatability locus or alteration of cell surface antigens of hESCs to produce a universal cell line that would be more acceptable and rejection-free; 2) the production of large repositories of HLA-typed hESCs for tissue matching; 3) the coating of hESC derived cells with an immuno-privileged membrane during delivery; 4) the use of immunosuppressive drugs to reduce rejection; 5) the production of tailor-made customized hESCs and differentiated tissues by somatic cell nuclear transfer (SCNT) (therapeutic cloning); and 6) hESC cytoplasmic transfer into somatic cells.

Interestingly, LiLi *et al.*[16] recently demonstrated that hESCs are immuno-privileged. In some very elegant experiments these authors showed that hESCs injected into immunocompetent mice were unable to induce an immune response. Undifferentiated and differentiated hESCs failed to stimulate proliferation of alloreactive primary human T cells and inhibited third-party allogeneic dendritic cell-mediated T cell proliferation via cellular mechanisms independent of secretory factors.[16] Thus, hESCs may be behaving like donor human embryos that do not get rejected when transplanted into the uteri of surrogate mothers in IVF programs.

Recently, Hwang *et al.*[5] successfully derived hESCs from cloned human embryos using the SCNT technique. The nuclei of the patient's own cumulus cells were electrofused with her own enucleated mature oocytes to generate blastocysts, from which hESCs were derived and propagated to produce an hESC line. Of 176 oocytes that were used, 30 blastocysts were generated and from these blastocysts one hESC line was produced. The protocols for blastocyst culture, hESC derivation and propagation were the same as those described for surplus IVF embryos in this chapter. The hESC line was not xeno-free as MEFs and animal-based culture ingredients were used. Several hurdles in terms of improvement of the efficiency of the technique; the implications and consequences of the absence of gene imprinting in the absence of sperm participation; the role of the left-over mitochondrial DNA in the enucleated oocyte; the use of other cell types for nuclear re-programming because of the paucity of mature oocytes, are all issues that have to be studied and overcome before SCNT can become a routine and useful method for customizing tissues for transplantation therapy. It has been claimed that it is highly unlikely that large numbers of mature oocytes would be available for tailor-making hESC cell lines and their tissue derivatives to suit each sick patient, particularly if hundreds of oocytes are required to produce each hESC line.[17] Furthermore, it has also been suggested that epigenetic remnants of the somatic cell used as the nuclear donor may cause major functional developmental problems.[18]

## Clinical Grade hESC Lines for Research and Application

Thus far, all 78 hESC lines on the NIH registry are derived and grown on xeno-supports (MEFs) in the presence of xeno-proteins (guinea pig complement, rabbit antihuman antibodies, fetal calf serum, bovine insulin, porcine transferrin) (Table 1). hESC lines have also been propagated on cell-free matrices such as laminin and matrigel in the presence of MEF

**Table 1.** Xenosupports and Xenoproteins in hESC Lines

| | |
|---|---|
| Embryo Culture | Bovine/ovine hyaluronidase: intracytoplasmic sperm injection |
| Derivation of hESCs | Guinea pig complement |
| | Rabbit anti-human antibodies |
| | Murine embryonic fibroblasts (MEFs) |
| | DMEM supplemented with fetal calf serum (FCS) |
| | Bovine insulin-porcine transferrin-selenium (ITS) |
| | Feeder-free matrices: animal origin |
| Propagation of hESCs | Murine embryonic fibroblasts (MEFs) |
| | DMEM supplemented with fetal calf serum (FCS) |
| | Feeder cell attachments: bovine/ovine gelatine |
| | Passaging: bovine trypsin |
| | Bovine insulin-porcine transferrin-selenium (ITS) |
| | Feeder-free matrices: animal origin |

conditioned medium.[19] Unfortunately, most of these matrices are of animal origin and the hESCs may also be exposed to animal based proteins from an unconditioned medium[20] or conditioned medium.[21,22] Derivation and growth of hESCs on xenosupports in the presence of xenoproteins introduces several disadvantages with respect to exploiting the therapeutic potential of hESCs. A major drawback is the risk of transmitting pathogens such as the hantavirus and lymphocytic choriomeningitis virus from the mouse feeder cells or conditioned medium to the hESCs. A xeno-free hESC line was derived and propagated by our group using human fetal muscle as a feeder in the presence of a culture medium containing ingredients of human origin (human serum, human insulin, human transferrin) or knock out serum (KO).[13] We later evaluated a panel of in-house derived and commercial sources of human fetal and adult feeders (Table 2). Human fetal muscle (in-house derived), human fetal skin (D551, ATCC) and human adult skin (in-house derived) all supported hESCs well and ranked first, second and third, respectively.[15] Later, other reports also showed that human foreskin fibroblasts[14,23] were able to successfully support hESCs in prolonged culture. Although generally in tissue culture practice, fetal tissues perform better than adult tissues *in vitro*, the use of an adult skin biopsy from the same IVF patient donating embryos for hESC derivation has the advantages of autologous support as well as safety, since the patient would have been previously tested for HIV and hepatitis B before enrolling for IVF treatment. Also, commercially available human cell lines are usually exposed to

**Table 2.** Supportive and Non-supportive Feeders for Prolonged hESC Growth

| Feeder | Undifferentiated hESC colonies |
| --- | --- |
| Supportive | |
|    Human fetal muscle | +++ |
|    Human fetal skin (D551, ATCC) | +++ |
|    Human adult skin | ++ |
|    Human adult fallopian tubal epithelium | ++ |
|    Human foreskin (CRL-2552) | ++ |
|    Human adult muscle | + |
|    Mouse embryonic fibroblasts | ++ |
| Non-supportive | |
|    Human fetal lung (MRC-5 and WI-38) | − |
|    Human glandular endometrium | − |
|    Human stromal endometrium | − |

xenoproteins such as fetal calf serum (American Type Culture Collection, ATCC) and hence are not xeno-free. The preparation of human feeders from fetal and adult tissues is described in detail by Fong and Bongso.[24] Xeno-free derivation and propagation of hESCs without feeders would be the most ideal as it will be less labor-intensive, involving lesser risks of feeder contamination and enabling easier scaling-up of hESC numbers. However, at this point in time it is extremely difficult to grow ICMs and derive hESC cell lines without feeder cell support. Transcriptome profiling of the genetic make-up of human feeder cells will shed light on possible proteins released by these cells that will hopefully allow feeder-free growth of hESCs in the future. Since all existing NIH hESC lines are derived and propagated on MEFs, it is urgently necessary to produce panels of xeno-free clinically compliant hESC lines using cGMP and GTCP conditions.[25] The cells and tissues differentiated from such cell lines when grown under the same stringent quality control conditions will help take this science to clinical trials faster.

## Xeno-free Cryopreservation of hESCs

Safe and efficient freezing and storage methods are an important pre-requisite for hESC clinical application. These novel methods complement the production of xeno-free clinical grade hESC lines. In most stem cell laboratories, hESCs are stored in closed cryovials in the liquid phase of liquid nitrogen ($LN_2$) using the conventional programmeable slow freezing method and a mixture of 10% dimethylsulfoxide (DMSO) and 90% fetal calf serum (FCS) as cryoprotectant. The thaw-survival rates, plating efficiencies and undifferentiation rates after thaw and subsequent passage are quite low with this method. The vitrification (snap-freezing or ultra-rapid freezing) open pulled straw (OPS) protocol using ethylene glycol as cryoprotectant which is much simpler and yields better thaw-survival rates has replaced the slow freezing method.[26] Both the slow freezing and OPS vitrification protocols, however, are fraught with problems in that: 1) the slow freezing method is inefficient and not xeno-free, and hence unsafe for clinical application because of exposure of the cells to FCS; and 2) the open straw at one end in the OPS method allows the hESCs to come into direct contact with $LN_2$. Numerous reports describe the contamination of frozen blood and cells with adventitious agents, primarily viruses, in $LN_2$ storage tanks.[27] The cross-contamination of frozen cell stock between cell lines and cell types in $LN_2$ storage tanks has also been documented.[28] Also, several viruses including HIV and hepatitis B are known to survive in $LN_2$.[29,30] Thus, an efficient xeno-free protocol to freeze and store cGMP compliant clinical grade hESCs is important. Our group recently reported a safe, xeno-free cyropreservation protocol for hESCs using vitrification in closed sealed straws (heat sealed at both ends), using human serum albumin (HSA) as opposed to FCS in the cryoprotectant solution, and long term storage in the vapor phase of liquid nitrogen.[31] Post-thaw, the hESCs showed high thaw-survival rates (88%), low differentiation rates, remained pluripotent and maintained normal karyotypes throughout extended passage. The xeno-free closed straw-vitrification $LN_2$ vapor phase storage method will prove very useful for the setting up of hESC banks throughout the world.

# Nature and Properties of hESCs

## Phenotype and behavior

The various protocols used in hESC derivation and propagation are outlined in Fig. 5. hESCs *in vitro* behave like established continuous cell lines exhibiting the characteristic of immortality by indefinite growth. This feature separates them from multipotent adult somatic stem cells which show senescence in culture after a certain period of time. hESCs when frozen and

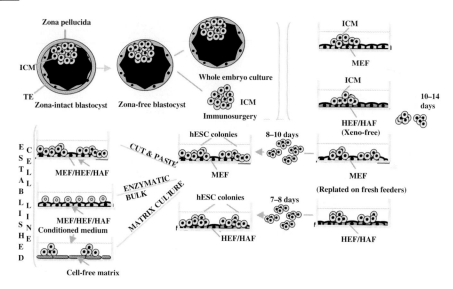

**Figure 5.** Outline of various approaches in deriving and propagating hESC lines. ICM: inner cell mass; TE: trophectoderm; MEF: murine embryonic feeder; HAF: Human adult feeder; HEF: human embryonic feeder.

then thawed continue to behave like fresh cells showing all the morphological characteristics and characterization markers for pluripotency.[26,31] hESCs supported by MEFs and grown by the "cut and paste" method have circular compact colonies that are thicker at the periphery (Fig. 6A) and take about 8 days to remain undifferentiated between passages. hESCs grown on human feeders by the "cut and paste" method are angular and rhomboid in shape, thinner (Fig. 6B) and remain undifferentiated slightly longer between passages. At high magnification, hESCs have a unique and typical morphology that distinguishes them very easily from other cells. They have scant cytoplasm, large nuclei (high nuclear-cytoplasmic ratios) and very prominent nucleoli (Figs. 6C, D). hESCs grown by the "bulk culture" method on human feeders generate several colonies per dish compared with the number of colonies obtained by the "cut and paste" method (Figs. 7A, B). The cells are tightly packed to each other and held by several junctional complexes as observed in transmission electron micrographs.[32] Transcriptome analysis using SAGE shows that the genes for claudin and connexin 43 are highly expressed in hESCs, confirming their very unique social nature.[33] If grown alone as single cells, they tend to differentiate very easily. hESCs probably possess membrane-bound receptors and secretory

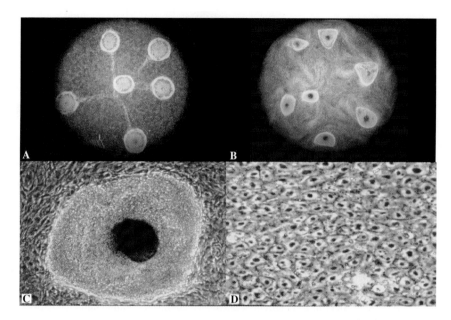

**Figure 6.** Stereo and inverted micrographs of hESC colonies and hESCs grown by the "cut and paste method." **A.** Six hESC colonies growing on MEF cells. Note circular colonies with thick edges (×100); **B.** 7 hESC colonies growing on HEF cells. Note angular rhomboid colonies with thin edges (×100); **C.** hESC colony at high magnification showing several undifferentiated small hESCs of equal size with degeneration at core of colony; **D.** High magnification of hESC colony showing individual tightly packed hESCs. Note prominent nucleolus and high nuclear-cytoplasmic ratio in each hESC.

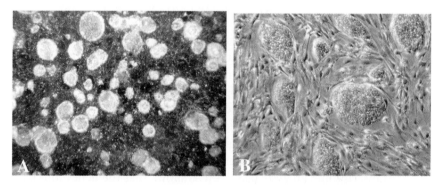

**Figure 7.** hESCs grown by the "bulk culture method." **A.** Stereo micrograph showing several hESC colonies (×100); **B.** Higher magnification phase-contrast inverted micrograph showing several large and small hESC colonies (×200).

**Table 3.** Unique Characteristics of hESCs

Prolific growth, unlimited self renewal and abundant in numbers

Cell lines available and easy to scale up with bulk culture protocol

Pluripotent, versatile with wide plasticity

Maintains normal genetic make-up in low density cultures and on feeders

Telomerase levels high and consistent

No shortening of chromosome length with serial passaging

Late apoptosis

Customization of hESC lines possible via somatic cell nuclear transfer

Ethical issues limited to some countries and institutions

Can be used for (1) transplantation therapy (2) pharmaceutical screening (3) gamete and embryo production (4) studies on human development, congenital anomalies and infant cancers

growth factors for their maintenance and self renewal. The population doubling time of hESCs usually ranges from 24 to 96 hours. Early passages have a longer population doubling time (approximately 144 to 168 hours) that gradually reduces and stabilizes with prolonged serial passage. hESCs can also be derived clonally but their clonal efficiency is low.[34] Cloning efficiency can be improved by supplementing serum-free culture media with basic fibroblast growth factor (bFGF). When hESCs are removed from their feeders and grown in suspension, they have the unique property of forming balls of cells called embryoid bodies (EBs) which give rise to a variety of cell types. hESCs differentiate spontaneously via a default pathway into their own preferred cell types such as patches of beating cardiomyocytes or neurons (axons and dendrites), while other tissues types are more difficult to derive. In some other experiments, cells arising from hESCs have shown to express genes associated with liver and pancreas function.[35] Other unique characteristics of hESCs are summarized in Table 3.

## Stability in extended culture: karyotypic changes

Certain hESC lines (H7.S6 and H14.S9, University of Wisconsin) showed karyotypic changes involving the gain of chromosomes 17q and 12 with extended culture of 22 to 60 passages (2–6 months).[36] Interestingly, these aneuploid karyotypes were very similar to those seen with human embryonal carcinoma cells that show a distinct gain of 17q and the presence of one or more isochromosomes of 12p. The authors concluded that the chromosomal changes observed in the hESCs suggest caution when designing culture conditions, in particular feeder-free conditions, because *in vitro* evolution may select for adaptive genetic changes. Pera[37] stated that it was important not to draw the conclusion that these results reflect a high intrinsic level of karyotypic unstability of hESCs grown *in vitro* because the data were seen only in two specific sublines of hESCs. Pera[37] also postulated that the results appeared to stem from particular aspects of the cell culture methods or the period for which the cells were cultured under specific conditions. Interestingly, for one subline, the cells were grown feeder-free, while for the other subline, a feeder was employed but the cells were passaged at relatively high density. It was possible that the karyotypic changes actually conferred a selective advantage of such abnormal hESCs in high density culture. Later Buzzard *et al.*[38] reported that six other NIH registered hESC lines (HES 1–6, Embryonic Stem Cell International, ESI Singapore) had been grown for 34 to 140 passages in their laboratory and they observed only one karyotype change (translocations between X, 17, 11 and 13) that occurred at an early passage in only one cell line (HES 5). They claimed they had never seen any evidence for the frequent non-cumulative aneuploidies reported by Draper *et al.*[36] Buzzard *et al.*[38] claimed that their culture methods may have had some bearing on the results. They used a mechanical "cut and paste" method for transferring peripheral colony pieces between passages unlike the other groups who used enzymatic or chemical methods of cell dissociation for passage in bulk culture. Buzzard *et al.*[38] also postulated that if hESC lines became susceptible to karyotypic change after 40 passages, one such hESC line could still generate more that $10^{35}$ cells before it became cytogenetically unstable. It is, therefore, very unlikely that such karyotypic changes will have a major impact on hESC-based therapies in the future. Similar chromosomal changes observed by Cowan *et al.*[8] in their hESC lines may also be attributed to the enzymatic bulk culture protocol they used after the mechanical disassociation of cells in early passages.

## hESC gene expression and stem cell signature

The molecular mechanisms of self renewal and differentiation of hESCs are poorly understood phenomena. It appears that hESCs and murine embryonic stem cells (mESCs) do not represent equivalent embryonic cell types. hESC lines possess heterogenous genetic backgrounds and appear to behave differently in culture. hESC lines derived by some groups do not appear to be stable and do not conform to the same kind of behavioral growth *in vitro* as other lines prepared by other groups. One good example is that not all the hESC lines on the NIH registry are equally amenable to bulk and feeder-free culture, with some lines easier to maintain than others; the population doubling times varying considerably between lines; and the degree of spontaneous differentiation *in vitro* being also diverse.[13,39] Our group suggested that a quantitative comparison of the transcriptome profiles of hESCs may allow the determination of key regulators involved in the maintenance of "stemness" as well as help identify a basis for line-specific cellular and behavioral differences.[33] Using Serial Analysis of Gene Expression (SAGE), we compared the transcriptome profiles of HES3 and HES4 (ESI) with mESCs and other human tissues. Close to 21 000 unique hESC transcripts were detected with SAGE. A large proportion of these unique transcripts appear to match to novel genes, hypothetical proteins and ESTs. hESCs and mESCs share a number of expressed gene products, although hESCs have very low levels of LIF and LIF receptor expression and almost 10-fold higher expression of POU5FI and SOX2, unlike mESCs.[33] This suggests that mESCs and hESCs differ greatly in their fundamental biology. We found that proteins involved in the translation apparatus and cytoskeletal architecture were very highly expressed in hESCs. Two proteins CLDN6 and GJA1 that form adhesion complexes between cells were also very highly expressed and examination of signaling pathways in hESCs indicated active Wnt, TGFβ and FGF signaling. A list of 21 candidate hESC marker genes responsible for stemness was also reported.[33] Eight of these genes were also identified as upregulated by microarray studies.[40] Novel candidate genes were either highly expressed in hESCs (GJA1, CLDN6, CKS1B, ERH, HMGA1) or were strongly downregulated during ES cell differentiation (POU5F1, LIN28, DNMT3B, FLJ14549/ZNF206, HESXI). These candidate genes will be very useful for studying the molecular pathways involved in the maintenance of pluripotency, self-renewal and the suppression of differentiation.[33] Intriguingly, SAGE and quantitative RT-PCR in the hESC line HES4 did not detect REX1 gene expression. Studies on mESCs indicate that the zinc finger transcription factor REX1 is important for the pluripotent phenotype. The lack of REX1 expression is surprising because HES4 forms teratomas in SCID mice and can be propagated for over 200 population doublings *in vitro*, thereby confirming it to be pluripotent and immortal. More reliable markers that distinguish truly pluripotent hESCs are necessary and one approach is to catalogue the genes that are switched on and off at various times in culture. Data from such gene expression profiles using a variety of analytical methods such as microarray, SAGE, MPSS and other profiling methods need to be compared in order to identify a clear blueprint or signature for hESCs.

## Strategies for differentiation

Differentiation is the biological process whereby an unspecialized cell acquires the properties of a specialized cell. For example, *in vivo*, a bone marrow stem cell could differentiate into a blood cell e.g. lymphocyte. Differentiation *in vitro* could either be spontaneous or controlled. Under sub-optimal conditions or in high-density cultures hESCs spontaneously differentiate into several different cell types with a preference towards neuronal and cardiomyocyte-like cells. These differentiated cell types seem to form via a default cell growth pathway. With time in culture, representatives of all three germ layers (ectoderm, mesoderm and endoderm) are produced. Some scientists have mechanically separated the desired cell type from this mixed milieu of cells, and propagated a pure line of such desired cell type e.g. neuronal cells.[41,42] Controlled differentiation can be accomplished by three approaches: 1) Biochemically treating the hESCs with specific agents (growth factors, chemicals) to generate a specific differentiated cell type e.g. retinoic acid to produce neurons; 2) growing hESCs in direct contact with companion fetal cells of the required cell type or a different cell type (coculture) with the hope that the companion cell will release certain factors that will entice the hESCs to be converted to the desired cell type. For example, cardiomyocytes have been produced *in vitro* by coculturing hESCs by direct contact with visceral endodermal cells[43]; 3) specific gene constructs are transfected into hESCs and the gene then switched on so as to direct the hESC to differentiate along a specific cell pathway, e.g. the cardiomyosin gene when transfected into mESCs and then switched on converts the mESCs into cardiomyocytes. The above three approaches could be carried out on hESCs *per se* or by allowing the hESCs to spontaneously form embryoid bodies (EBs) in culture and then exposing the EBs to the biochemical factors, companions cells or transfection.

## Bona Fide hESC Lines

Several criteria need to be satisfied before an hESC line can qualify as a bona fide hESC line. These criteria include: 1) maintenance of classical colony and individual cell morphological characteristics; 2) Oct-4 gene expression and other pluripotent marker expression; 3) teratoma formation (tissues from all 3 germ layers) in SCID mice; 4) prolonged passaging (at least 200 passages) with maintenance of normal karyotype; 5) clonality; 6) telomerase expression; and 7) alkaline phosphatase expression (Fig. 8). The fact

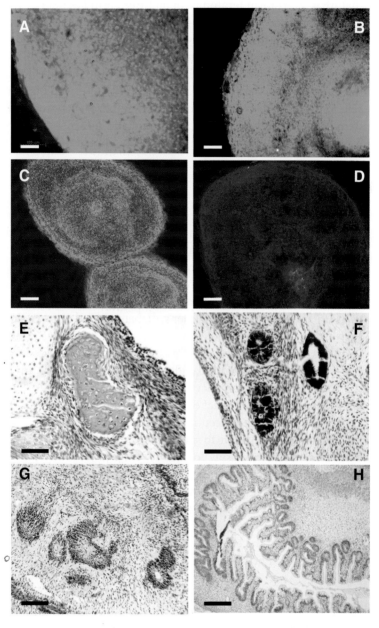

**Figure 8.** Characterization tests of hESCs derived and grown on human feeder cells. Immunohistochemistry (A–D) and SCID mice teratomas (E–H). **A.** Tra-1-60; **B.** Tra-1-81; **C.** SSEA-4; **D.** Alkaline phosphatase; **E.** Bone and cartilage; **F.** Pigmented epithelium; **G.** Neural rosettes; **H.** Columnar epithelium. Scale bars: Immunohistochemistry, bar = 200 μm; Teratomas, bar = 100 μm.

that hESCs can form teratomas of all three germ layers in SCID mice confirms that they are pluripotent. Pluripotency is the ability to produce several cell types via all three germ layers (ectoderm, mesoderm and endoderm) but not produce the entire organism. In theory, therefore, hESCs have the potential to produce all 210 cell types in the human body. Totipotency, on the other hand, is the ability to form the entire fetus and placenta and such cells in the human are limited to the early embryo. Fetal stem cells and adult stem cells such as hemopoietic, neuronal and mesenchymal stem cells are multipotent with the ability to give rise to a limited range of cell types.[25]

## Clinical Application of hESCs

hESCs have several clinical uses. Firstly, it is possible to differentiate these cells either spontaneously or in a controlled fashion into a variety of desirable replacement tissues that can be used for the treatment of several target diseases by transplantation therapy. It is a matter of finding the right inducer or signal that could trigger the hESC to form that specific desirable tissue. There is thus hope for the treatment of heart diseases, neuronal diseases such as Parkinson's and Alzheimer's, diabetes, blood cancers, skin problems such as burns and so on.

hESCs also serve as ideal *in vitro* assays for screening and testing potential drugs for the pharmaceutical industry. The directed derivatives of hESCs such as a human cardiomyocyte cell line can be used *in vitro* as an assay for a potential drug that may act on the heart. At the moment, animal cell lines are being used by pharmaceutical companies to test drugs. Similarly, hESCs themselves being sensitive to suboptimal culture conditions, could perhaps be used as a sensitive assay for quality control for water purification and microbiology laboratories.

hESCs also serve as ideal models to study early human development (through the production of germ cells, gametes and embryos *in vitro*), infant cancers and congenital anomalies. Given the fact that hESCs are pluripotent cells, the effects of potential teratogens can be studied *in vitro* using hESCs as a plating assay.

## Hurdles to Overcome in the Development of Cell-based Therapies from hESCs

There are several unanswered questions with respect to stem cell biology that need to be addressed before this science can be taken to the bedside.

Clinically compliant safe xeno-free protocols for the generation of hESCs and its directed derivatives are crucial because this technology is expected to yield a cell based therapy and as such the administered cells must be safe. All the risks of administering whole cells containing all their organelles into the portal veins or directly into malfunctioning tissues of patients need to be addressed. hESC differentiated cells should first be purified to avoid any renegade hESC being administered and thereby yielding teratomas. The administration of hESC derived "progenitor" cells (rather than terminally derived cells), hESC conditioned medium or agents from the conditioned medium, which do not induce teratomas but make use of the stem cell niche environment to trigger off normal tissue development and function, would be a short-cut to therapy and solve many of the consequential unexpected problems. It would be ideal if we could simply inject a progenitor rather than an hESC into the malfunctioning tissue or organ and let the progenitor take its cues from the surrounding environment. A stem cell niche has been recognized by some workers. This niche comprises a combination of cells and extracellular matrix components in the local tissue environment that govern stem cell behavior. The niche supports and controls stem cell activity. For example, in the hair follicles the bulge just underneath the sebaceous glands appears to be a stem cell niche and in the intestinal mucosa the pericryptal myofibroblasts that ensheath the crypts serve as niche cells.[44] It would be important to find out how differentiated hESCs once placed in a living brain, for example, will communicate with surrounding neurons. Studying the embryonic niche in organogenesis is one approach that will help identify the signals that normally instruct hESCs to choose a particular pathway of development.

The dosages and routes of administration have to be carefully worked out in animal models. This will involve studying targeted delivery by direct injection of cells into malfunctioning organs or homing of the cells into the specific malfunctioning organ after portal vein administration. The number of cells for administration and the stability of treatment in terms of the duration need to be addressed as well. Will a patient have to have repeated doses of cell therapy or would a single or few modes of treatment be adequate to bring about a cure?

It would be important to study the *in vivo* functional aspects of hESCs differentiated *in vitro*. This could only be validated in animal models and preferably the non-human primate. The development of a SCID macaque would be the most ideal to make these studies. Currently, mouse and rat models are being used which may not be the best choice.

**Table 4.** Steps in the Development of Novel Cell Therapies from hESCs

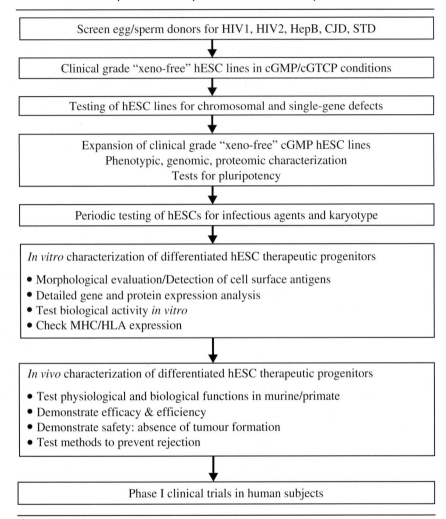

Screen egg/sperm donors for HIV1, HIV2, HepB, CJD, STD

Clinical grade "xeno-free" hESC lines in cGMP/cGTCP conditions

Testing of hESC lines for chromosomal and single-gene defects

Expansion of clinical grade "xeno-free" cGMP hESC lines
Phenotypic, genomic, proteomic characterization
Tests for pluripotency

Periodic testing of hESCs for infectious agents and karyotype

*In vitro* characterization of differentiated hESC therapeutic progenitors

- Morphological evaluation/Detection of cell surface antigens
- Detailed gene and protein expression analysis
- Test biological activity *in vitro*
- Check MHC/HLA expression

*In vivo* characterization of differentiated hESC therapeutic progenitors

- Test physiological and biological functions in murine/primate
- Demonstrate efficacy & efficiency
- Demonstrate safety: absence of tumour formation
- Test methods to prevent rejection

Phase I clinical trials in human subjects

The new approaches of generating gametes and embryos from hESCs *in vitro* are very encouraging, as this branch of science will generate alternative material in terms of oocytes for therapeutic cloning. Additionally, if it is possible to exploit the power of the hESC cytoplasm by using hESCs to re-program somatic cell nuclei, the need for mature oocytes for SCNT would not be necessary.

Finally, the more complicated the organ, such as the brain and the heart, the greater is the likelihood of the existence of a multitude of progenitor cells for different parts of the organ and hence the more difficult it will be to generate single specific hESC directed tissues for treatment. Preparing islet cells for the pancreas for the treatment of diabetes may thus be less complicated but yet challenging, given the fact that this organ is made up of a few cell types. The steps in the development of future hESC differentiated cell based therapies are outlined in Table 4.

# REFERENCES

1. Bongso A, Fong CY, Ng SC, *et al.* (1994) Isolation and culture of inner cell mass cells from human blastocysts. *Hum Reprod* **9**:2110–2117.
2. Thomson JA, Itskovitz-Eldor J, Shapiro SS, *et al.* (1998) Embryonic stem cell lines from human blastocysts. *Science* **282**:1145–1147.
3. Reubinoff BE, Pera MF, Fong CY, *et al.* (2000) Embryonic stem cell lines from human blastocysts: Somatic differentiation *in vitro*. *Nat Biotechnol* **18**: 399–404.
4. Lazendorf SE, Boyd CA, Wright DL, *et al.* (2001) Use of human gametes obtained from anonymous donors for the production of human embryonic stem cell lines. *Fertil Steril* **76**:132–137.
5. Hwang WS, Ryu YJ, Park JH, *et al.* (2004) Evidence of a pluripotent human embryonic stem cell line derived from a cloned blastocyst. *Science* **303**:1669–1674.
6. Bongso A, Fong CY, Matthew J, *et al.* (1998) Benefits to human *in vitro* fertilization of transferring embryos after *in vitro* embryonic block: Alternatives to day-2 transfers. *Asst Reprod* **9**:70–78.
7. Bongso A. (1999) *Handbook on Blastocyst Culture.* Serono Singapore: Sydney PressIndusprint (s) Pte Ltd.
8. Cowan CA, Klimanskaya I, McMahon JP, *et al.* (2004) Derivation of embryonic stem-cell lines from human blastocysts. *N Engl J Med* **350**:1353–1356.
9. Bongso A, Fong CY, Ng SC, *et al.* (1993) The growth of inner cell mass cells from human blastocysts (Abstract). *Theriogenology* **41**:161.
10. Solter D, Knowles BB. (1975) Immunosurgery of mouse blastocysts. *PNAS, USA* **72**:5099–5102.
11. Amit M, Itskovitz-Eldor J. (2002) Derivation and spontaneous differentiation of human embryonic stem cells. *J Anat* **200(Pt 3)**:225–232.
12. Suss-Toby E, Gerecht-Nir S, Amit M, *et al.* (2004) Derivation of a diploid human embryonic stem cell line from a mononuclear zygote. *Hum Reprod* **19**:670–675.

13. Richards M, Fong CY, Chan WK, *et al.* (2002) Human feeders support prolonged undifferentiated growth of human inner cell masses and embryonic stem cells. *Nat Biotechnol* **20**:933–936.

14. Hovatta O, Mikkola M, Ye Z, *et al.* (2003) A culture system using human foreskin fibroblasts as feeder cells allows production of human embryonic stem cells. *Hum Reprod* **18**:1404–1449.

15. Richards M, Tan S, Fong CY, *et al.* (2003) Comparative evaluation of various human feeders for prolonged undifferentiated growth of human embryonic stem cells. *Stem Cells* **21**:546–556.

16. LiLi M, Baroja ML, Majumdar A, *et al.* (2004) Human embryonic stem cells possess immuno-privileged properties. *Stem Cells* **22**:448–456.

17. Trounson AO. (2001) The derivation and potential use of human embryonic stem cells. *Reprod Fertil Dev* **13**:523–532.

18. Daniels R, Hall VJ, French AJ, *et al.* (2001) Comparison of gene transcription in cloned bovine embryos produced by different nuclear transfer techniques. *Mol Reprod Dev* **60**:281–288.

19. Xu C, Inokuma MS, Denham J, *et al.* (2001) Feeder-free growth of undifferentiated human embryonic stem cells. *Nature Biotech* **19**:971–974.

20. Amit M, Shariki C, Margulets V, *et al.* (2004) Feeder and serum free culture system for human embryonic stem cells. *Biol Reprod* **70**:837–845.

21. Rosler ES, Fisk GJ, Ares X, *et al.* (2004) Long term culture of human embryonic stem cells in feeder-free conditions. *Dev Dyn* **229**:259–274.

22. Carpenter MK, Rosler ES, Fisk GJ, *et al.* (2004) Properties of four human embryonic stem cell lines maintained in a feeder-free culture system. *Dev Dyn* **229**:243–258.

23. Amit M, Margulets V, Segev H, *et al.* (2003) Human feeder layers for human embryonic stem cells. *Biol Reprod* **68**:2150–2156.

24. Fong CY, Bongso A. (2004) Derivation of human feeders for prolonged support of human embryonic stem cells. In *Human Embryonic Stem Cells: Materials and Protocols.* ed. K Turksen, USA: Humana Press.

25. Bongso A, Richards MR. (2004) History and perspective of stem cell research. In *Stem Cells in Obstetrics and Gynaecology*, Best Practice Clinical Series, eds. N Fisk and J Istzkovitz, London: Elsevier.

26. Reubinoff BE, Pera MF, Vajta G, *et al.* (2001) Effective cryopreservation of human embryonic stem cells by the open pulled straw vitrification method. *Hum Reprod* **10**:2187–2194.

27. Burdon DW. (1999) Issues in contamination and temperature variation in the cryopreservation of animal cells and tissues, Revco Technologies Application Note, pp. 99–108.

28. Tomlinson M, Sakkas D. (2000) Is a review of standard procedures for cryopreservation needed? *Hum Reprod* **15**:2460–2463.

29. Tedder RS, Zuckerman MA, Goldstone, *et al.* (1995) Hepatitis B transmission from contaminated cryopreservation tank. *Lancet* **346**:137–140.

30. Clarke GN. (1999) Sperm cryopreservation: Is there a significant risk of cross-contamination? *Hum Reprod* **14**:2941–2943.

31. Richards M, Fong CY, Tan S, *et al.* (2004) An efficient and safe xeno-free cryopreservation method for the storage of human embryonic stem cells. *Stem Cells* **22**:779–789.

32. Sathananthan H, Pera MF, Trounson AO. (2002) The fine structure of human embryonic stem cells. *J Reprod Biomed Online* **4**:56–61.

33. Richards M, Tan SP, Tan JH, *et al.* (2004) The transcriptome profile of human embryonic stem cells as defined by SAGE. *Stem Cells* **22**:51–64.

34. Amit M, Carpenter MK, Inokuma MS, *et al.* (2000) Clonally derived human embryonic stem cell lines maintain pluripotency and proliferative potential for prolonged periods of culture. *Dev Biol* **227**:271–278.

35. Schuldiner M, Yanuka O, Itskovitz-Eldor J, *et al.* (2000) Effects of eight growth factors on the differentiation of cells derived from human embryonic stem cells. *Proc Natl Acad Sci* **97**:11307–11312.

36. Draper JS, Smith K, Gokhale P, *et al.* (2003) Recurrent gain of chromosomes 17q and 12 in cultured human embryonic stem cells. *Nature Biotech Online* doi:10.1038/nbt922.

37. Pera MF. (2004) Unnatural selection of cultured human ES cells? *Nature Biotech* **22**:42–43.

38. Buzzard JJ, Gough NM, Crook JM, *et al.* (2004) Karyotype of human ES cells during extended culture. *Nature Biotech* **22**:381–382.

39. Vogel G. (2002) Stem cells. Are any two cell lines the same? *Science* **295**:1820.

40. Sato N, Sanjuan IM, Heke M, *et al.* (2003) Molecular signature of human embryonic stem cells and its comparison with the mouse. *Dev Biol* **260**:404–413.

41. Reubinoff BE, Itsykson P, Turetsky T, *et al.* (2001) Neural progenitors from human embryonic stem cells. *Nat Biotechnol* **19**:1134–1140.

42. Zhang SC, Wernig M, Duncan ID, *et al.* (2001) *In vitro* differentiation of transplantable neural precursors from human embryonic stem cells. *Nat Biotechnol* **19**:1129–1133.

43. Mummery C, Ward-van Oostwaard D, Doevendans P, *et al.* (2003) Differentiation of human embryonic stem cells to cardiomyocytes: Role of coculture with visceral endoderm-like cells. *Circulation* **107**:2733–2740.

44. Allison MR, Poulsom R, Forbes S, Wright NA. (2002) An introduction to stem cells. *J Path* **197**:419–423.

## 3

# Characterization of Human Embryonic Stem Cells

Maryam M Matin, Ahmad Bahrami, Duncan Liew
and Peter W Andrews*

## Introduction

Embryonic stem cell (ESC) lines were first derived from human embryos as recently as 1998,[1] and subsequently a large number of lines have been reported by several investigators.[2] Nevertheless, the study of these cells has been built upon a long history of investigations, first with teratocarcinomas and ESC lines from the laboratory mouse, then human teratocarcinoma cell lines and finally primate ESCs (reviewed in Ref. 3). Teratocarcinomas are a subset of germ cell tumors (GCT) containing a disorganized array of somatic tissues and malignant stem cells, the embryonal carcinoma cells (ECCs), which are able to differentiate into the various cell types found in the tumor, and confer the malignant properties of the tumor.[4] These tumors may also contain tissues resembling the extraembryonic membranes of the early conceptus, typically the yolk sac and, in humans but not mice, the trophoblast. Malignant GCT occur most commonly in the testis in humans, and are the most common neoplasm of young adult men, although malignant teratocarcinomas with similar histopathology occur occasionally in the ovary or in extragonadal locations. By contrast, benign GCT occur much more commonly in the ovary where they are typically known as dermoid cysts. These benign "teratomas" may contain many cell types and sometimes well organized tissues, but they differ from the testicular teratocarcinomas and rare ovarian teratocarcinomas by lacking an ECC component.

Teratomas and teratocarcinomas in humans have provided medical curiosities for many centuries.[5] Histological studies led to the first proposals

*Correspondence: The Centre for Stem Cell Biology and the Department of Biomedical Science, University of Sheffield, Western Bank, Sheffield S10 2TN, UK. Tel.: +44 (0)114-222-4173, fax: +44 (0)114-222-2399, e-mail: p.w.andrews@sheffield.ac.uk

that these tumors in someway reflected the processes of embryonic development, and that ECCs are their malignant stem cells.[6] However, serious experimental study had to await the discovery by Stevens that male 129/J mice regularly develop testicular teratocarcinomas.[7] This provided the springboard from which further research confirmed the derivation of these tumors from germ cells,[8] established the pluripotent stem cell character of the ECCs,[9] with their ability to differentiate into all cell types found in the tumors, and demonstrated that similar tumors could be derived from early embryos placed in ectopic sites.[10,11]

ECC lines from teratocarcinomas of the laboratory mouse were established in culture, first in 1967,[12] and subsequent studies demonstrated their close relationship to the cells of the inner cell mass (ICM) of blastocyst stage embryos.[13] This demonstration was initially dependent upon a confluence of research into the development of the hematopoietic system and studies of the early embryo. By the early 1970s, surface antigens had become well established tools for defining subsets of cells in the immune system, and the concept of differentiation antigens was developed.[14] Building on these ideas, Artzt and her colleagues[15] identified an antigen, "F9" antigen, typically expressed by both ECCs from teratocarcinomas and ICM cells from early mouse embryos. However, a more explicit demonstration of the relationship of EC and ICM cells came from the finding that ECCs could participate in normal embryonic development if they were injected into a blastocyst that was subsequently transferred to a pseudopregnant mouse and allowed to develop to term.[16,17,18] Chimeric mice were born with apparently normal tissues derived from both the host blastocyst and the implanted ECCs.

The culmination of these studies eventually came in 1981, with the derivation of stable lines of Embryonic Stem cells (ESCs) directly from mouse blastocysts placed in culture.[19,20] These ESC lines appeared to be genetically normal, could be maintained in culture apparently indefinitely, where they could be induced to differentiate, and could form chimeras of all tissues, including the germ line, when placed back into a blastocyst. Indeed, if ESCs are combined with blastocysts from tetraploid embryos, then they can generate a whole mouse.[21] Such ESCs have since provided one of the most important tools for experimental mouse embryology and genetics, as the vehicle for homologous recombination and the generation of transgenic mouse strains. Nevertheless, with some notable exceptions, relatively little work over the past 20 years has been directed to the basic biology of mESCs.

## Human Embryonal Carcinoma Cells (hECCs)

The early studies of mouse EC cells (mECCs) raised the prospect that the study of their properties and differentiation *in vitro* could provide insights into early embryonic development. Not only did these cells express developmentally regulated markers characteristic of the inner cell mass and early epiblast of the mouse embryo, e.g. the F9 and later monoclonal antibody defined antigen, SSEA1, they also could be induced to differentiate in culture in ways that seemed to resemble early developmental processes. For example, the initial formation of embryoid bodies in which a layer of cells resembling primitive endoderm developed over an inner core of undifferentiated ECCs seemed to resemble the formation of primitive endoderm from the inner cell mass during the late blastocyst stage of the embryo.[22] Further, it was possible to induce differentiation of certain ECCs into particular lineages. For example, F9 ECCs differentiated to parietal endoderm if treated with retinoic acid and cyclic AMP in monolayer culture, or visceral endoderm if embryoid bodies were exposed to retinoic acid.[23,24] Other lines, e.g. P19, would differentiate into neural or muscle lineages under specific conditions.[25]

However, this work in the mouse was soon over shadowed by the derivation of mESC lines. For human development, however, human embryos and hESC lines were not available in the early 1980s; ECCs offered the only realistic prospect at that time of accessing cells that might resemble pluripotent cells of the early human embryo. At that time there were already grounds for believing that there might be significant differences between mouse and human embryos; most obviously the organization of the early germ layers of the embryo and the extraembryonic tissues was apparently significantly different. Such differences provided some justification for working with human ECC lines, but cancer biology provided other grounds. For example, the incidence of testicular germ cell tumors, already the most common solid tumor of young men, was rising substantially, having at least doubled over the preceding 50 years.[26]

The first lines of human germ cell tumors were established as retransplantable xenografts in hamster cheek pouches in the 1950s,[27] but little came out of that work. Subsequently, the first human cell lines established in culture from teratocarcinomas were the TERA1 and TERA2 lines of Fogh,[28] and SuSa by Hogan.[29] Initially, these were thought to closely resemble mECCs, and were suggested to express similar antigens, notably the F9 marker antigen in common use at that time.[30] Further lines were then derived by several investigators, most notably Bronson and his colleagues in the University of Minnesota.[31] The paradoxical problem with many of these lines was that, although unlike many mECC lines, they did not show much propensity to differentiate extensively into well defined derivatives, and did tend to produce somewhat heterogeneous cultures, in which some cells resembled mECCs, while other cells had a distinct though ill defined morphology.[32] However, some did produce xenograft tumors composed entirely of EC-like elements, suggesting that the lines were composed of hECCs. A detailed study of one such line, 2102Ep, which was cloned by picking single cells, led to the conclusion that although hECCs resembled mECCs morphologically, they had a confusingly opposite surface antigen phenotype.[33] Thus it appeared that hECCs express the surface antigen SSEA3 (which is expressed in early mouse embryos, but not by inner cell mass cells or by mECCs), but do not express the antigen SSEA1, the successor to the F9 antigen, which is expressed by mouse ICM and ECCs. In other experiments, ECC components in clinical biopsies of human teratocarcinomas were also shown to express SSEA3.[34] Further confusion was caused because, if cultured at low cell densities, many hECC lines undergo some differentiation to yield morphologically distinct cells that do express SSEA1. The nature of these SSEA1 positive derivatives was unclear, though in some cases they included trophoblastic derivatives.[35] They appeared to be stably differentiated and did not seem to revert to an EC phenotype. It became evident that an important aspect of hECC culture, and a significant difference from mECC culture, was the need to maintain cells at high densities to prevent this type of differentiation.

## Pluripotent Human ECC and ESC Lines

Although the presence of a wide range of histological cell types is the hallmark of teratocarcinomas, a significant number of germ cell tumors contain ECCs without evidence of extensive somatic differentiation. This can be rationalized in the context of tumor progression by the realization that in many cases the somatic differentiated derivatives of ECCs have a limited proliferative capacity and typically have a much reduced malignant potential. Thus, as tumors progress any genetic changes that occur to limit the capacity of ECCs to differentiate will have a selective advantage. Indeed, in several studies in which "nullipotent" mECCs (i.e. ECCs that have lost their ability to differentiate) were fused with certain differentiated somatic cells, typically from the hematopoietic system, the resulting hybrid cells exhibited an EC phenotype and, moreover these hybrid EC cells were capable of differentiation.[36,37] Apart from the dominance of the

EC phenotype in these hybrids, the results are most easily explained by the proposition that the nullipotent parental ECCs had lost the function of key genes required for pluripotency and that complementation of these mutations occurred in the hybrids with the introduction of a "normal" genome from the somatic parental cells.

As with the clinical tumors, many hECC lines in culture also appear to have lost the capacity for extensive differentiation; in fact, it may be that adaptation to *in vitro* culture places further selection pressures for the outgrowth of nullipotent lines. However, there are several exceptions, most notably TERA2, from which NTERA2 cl.D1 is the most widely studied clonal subline,[38] as well as GCT27[39] and NCCIT.[40] Of these, GCT27 is notable for its requirement for feeder cells for maintenance in an undifferentiated state. The other lines do not require feeders.

NTERA2 cells, which are perhaps the most widely used pluripotent hECCs, closely resemble the ECCs of other nullipotent human lines like 2102Ep. Morphologically, they are characterized by closely packed small cells, with little cytoplasm, pale nuclei and prominent, few nucleoli (Figs. 1A, B). However, they differentiate in response to retinoic acid, unlike

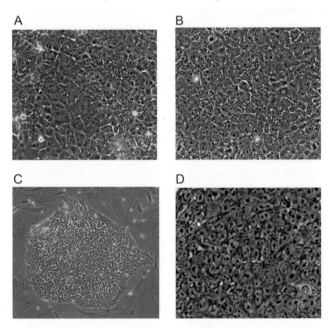

**Figure 1.** The morphology of hECCs and hESCs cells. **Panel A:** NTERA2 hECCs; **Panel B:** 2102Ep hECCs; **Panel C:** a colony of H7 hESCs; and **Panel D:** H7 cells with higher magnification. Note the similar morphology of the cells which are tightly packed and have relative high nucleus: cytoplasm ratio, and prominent nucleoli.

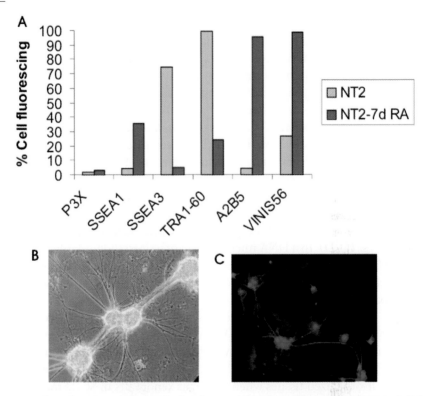

**Figure 2. Panel A:** The expression of several surface antigens by undifferentiated NTERA2 hECCs and by the same cells following differentiation induced with retinoic acid. Note the high levels of SSEA3 and TRA-1-60 antigen expression by the undifferentiated cells, the down regulation of these antigens after differentiation, and the appearance of cells expressing SSEA1, A2B5 and VINIS56 (see also Table 1). **Panels B** and **C** show clusters of neurons in retinoic acid induced cultures of NTERA2 cells, respectively by phase contrast and by immunofluorescence staining for the neural marker TUJ1 (note: these are different fields of cells).

many hECC lines, and yield a complex array of differentiated cells that include neurons (Fig. 2).[41] Other cells, most likely of a mesenchymal type, are found in retinoic acid-treated NTERA2, and may include smooth muscle, chondrocytes and others, but no definitive evidence of mesodermal derivatives has been forthcoming. NTERA2 cells also differentiate extensively in xenograft tumors or after treatment with HMBA or the BMPs in culture.[42,43] In these cases, distinct differentiation pathways appear to be activated. Apart from their value in studies of neurobiology — NTERA2 neurons have been tested in transplant models for stroke, and even in human patients in limited trials — the NTERA2 system has been used to investigate certain viruses, notably human Cytomegalovirus (HCMV)[44] and human Immunodeficiency virus (HIV),[45] neither of which can replicate in the undifferentiated human ECCs but can replicate in some of their differentiated derivatives.

However, the differentiation capacity of even the best hECCs is limited compared with mESCs, and it was not until 1995 that the first primate ESC lines were derived,[46] and 1998 when the first hESC lines were reported.[1] These hESC lines were isolated in a manner similar to the isolation of mESCs, by isolating the inner cell masses from human blastocysts by immunosurgery and plating onto feeder cells. The resulting hESC lines can be maintained in culture on mouse or human fibroblasts that have been inactivated by irradiation or mitomycin C. In such cultures, they closely resemble hECCs in morphology, tending to grow in tight colonies of small cells with little cytoplasm, large nuclei and few prominent nucleoli. (Figs. 1C, D). In a DNA microarray expression study, established hECC and hESC lines clustered separately, though they were nevertheless more closely related than to other cell and tumor types in global analyses of their gene expression profiles.[47]

The ESCs also differentiate readily if removed from the feeders or grown in suspension as aggregates. When such aggregates or "embryoid bodies" are subsequently plated and allowed to attach, a wide arrange of cells grow out (Fig. 3). They also form well differentiated teratomas with many different well-organized tissues when grown as xenografts in SCID mice.

## Surface Antigen Markers of hECCs and hESCs

Surface antigens have generally proved to be valuable markers of cell type as they can readily be detected and assessed on single cells. They can also be used readily in various immunological techniques, most notably fluorescence activated cell sorting (FACS), to isolate subsets of cells from heterogeneous mixtures, such as those typically found in cultures of ECCs and ESCs, especially when differentiation is deliberately induced. The past studies of hECCs have now provided an array of antigens (Table 1) that are also expressed and developmentally regulated in hESCs as well as ECCs.[48,49] These antigens include the globoseries glycolipid antigens SSEA3 and SSEA4, the high molecular weight proteoglycans detected such as TRA-1-60, TRA-1-81 and GCTM2, and several protein antigens such as TRA-2-54 and TRA-2-49 (liver/bone/kidney alkaline phosphatase), Thy1. All of these show marked down regulation during the differentiation of hECCs and hESCs (Figs. 2, 3).

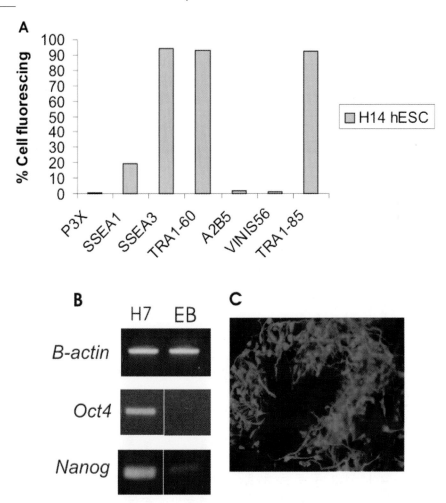

**Figure 3. Panel A**: Flow cytofluorimetric assay of surface antigen expression by undifferentiated H14 hESCs. Note, like undifferentiated NTERA2 ECCs, these cells are predominantly SSEA3 and TRA-1-60 positive and lack expression of SSEA1, A2B5 and VINIS56. Note also the expression of the antigen TRA-1-85, a pan human antigen that is not expressed by the mouse feeder cells on which the hESCs are growing; reactivity with this antibody indicates that over 90% of the culture is composed of human cells. **Panel B**: RT.PCR demonstrating the expression of Oct4 and Nanog in undifferentiated H7 hESCs and their absence following the formation of embryoid bodies (EB) by growing the cells in suspension for 12 days. This is an effective way to induce differentiation of the ESCs and many different cell types grow out when they are allowed to reattach to a substrate. **Panel C**: Neurons growing out from replated embryoid bodies of H7, stained for TUJ1.

**Table 1.** Common Surface Antigen Markers of hECCs and hESCs

| Antibody | Antigen | Glycolipid structures | Reference |
|---|---|---|---|
| *Some common markers of undifferentiated hECCs and hESCs* | | | |
| MC631 | SSEA3 | Globoseries glycolipid | $\text{Gal}\beta1\rightarrow3\text{GalNAc}\beta1\rightarrow3\text{Gal}\alpha1\rightarrow4\text{Gal}\beta1\rightarrow4\text{Glc}\beta1\rightarrow$**Cer** (62, 63) |
| MC813-70 | SSEA4 | Globoseries glycolipid | $\text{NeuNAc}\alpha\rightarrow3\text{Gal}\beta1\rightarrow3\text{GalNAc}\beta1\rightarrow3\text{Gal}\alpha1\rightarrow4\text{Gal}\beta1\rightarrow4\text{Glc}\beta1\rightarrow$**Cer** (63) |
| TRA-1-60 | TRA-1-60 | Keratan suphate proteoglycan | (64, 65) |
| TRA-1-81 | TRA-1-81 | Keratan suphate proteoglycan | (64, 65) |
| GCTM2 | GCTM2 | Keratan suphate proteoglycan | (66, 67) |
| TRA-2-49 | L-ALP | Liver/bone/kidney isozyme of alkaline phosphatase | (68) |
| TRA-2-54 | L-ALP | Liver/bone/kidney isozyme of alkaline phosphatase | (68) |
| *Some markers of differentiation* | | | |
| MC480 | SSEA1 [Le$^x$] | Lactoseries glycolipid; also associated with high mol wt glycoproteins | $\text{Gal}\beta1\rightarrow4\text{GlcNA}\beta1\rightarrow4\text{Gal}\beta1\rightarrow4\text{Glc}\beta1\rightarrow$**Cer** (69, 70, 71) |
| | | | with at position 3 ← Fucα1 |
| A2B5 | GT3 | Ganglioseries glycolipid | $\text{NeuNAc}\alpha2\rightarrow8\text{NeuNAc}\alpha2\rightarrow3\text{Gal}\beta1\rightarrow4\text{Glc}\beta1\rightarrow$**Cer** (72) |
| ME311 | 9-0-acetylGD3 | Ganglioseries glycolipid | $\text{(9-0-acetyl)NeuNAc}\alpha2\rightarrow8\text{NeuNAc}\alpha2\rightarrow3\text{Gal}\beta1\rightarrow4\text{Glc}\beta1\rightarrow$**Cer** (73) |
| VINIS56 | GD3 | Ganglioseries glycolipid | $\text{NeuNAc}\alpha2\rightarrow8\text{NeuNAc}\alpha2\rightarrow3\text{Gal}\beta1\rightarrow4\text{Glc}\beta1\rightarrow$**Cer** (42) |
| VIN2PB22 | GD2 | Ganglioseries glycolipid | GalNAcβ1 ↓4 $\text{NeuNAc}\alpha2\rightarrow8\text{NeuNAc}\alpha2\rightarrow3\text{Gal}\beta1\rightarrow4\text{Glc}\beta1\rightarrow$**Cer** (42) |

It is quite clear that generally the surface antigen phenotype of hESCs closely resembles that of hECCs and is quite different from that of mESCs.[49] Before the derivation of hESCs, it was possible that this difference between murine and human ECCs could be because hECCs were not, in fact, related to the ICM cells of the human blastocyst — perhaps they were related to another unidentified embryonic cell type. However, the similarity of hECCs and hESCs suggests that such a hypothesis is unlikely. We have since also shown that the pattern of antigen expression in the early human embryo itself is, in fact, distinct from that of the mouse embryo. In particular, the antigen phenotype of the ICM from human embryos is SSEA3(+), SSEA4(+), TRA-1-60(+), SSEA1(−), like that of hECCs and hESCs, and distinct from that of mouse ICM, EC and ES cells.[50] These observations serve to underline further the differences between mouse and human embryos, and to emphasize the importance of the direct study of human pluripotent stem cells.

## Pluripotent marker genes of hECCs and hESCs

A key issue in understanding the biology of ECCs and ESCs is the nature of the molecular mechanism that maintains the undifferentiated, pluripotent state. It certainly seems that pluripotent stem cells equivalent to those of the ICM soon disappear from the developing embryo and, apart from the specialized case of primordial germ cells, pluripotent stem cells have disappeared by the time of gastrulation. However, in culture, ESCs can apparently be maintained in a state of self-renewal indefinitely, if cultured under appropriate conditions. Such maintenance is presumably the result of removal from a source of an embryonic factor that promotes differentiation, or the continued presence in culture of some factor that promotes self-renewal, but that disappears *in vivo*. Despite various hypotheses, a definitive answer, or the identification of the key factors functioning *in vitro*, remains elusive. Clearly, the feeder cells used *in vitro* produce some factor that is essential for self renewal. For mESCs, the cytokine LIF can replace the need for feeder cells, but at least LIF is not sufficient to promote the self renewal of hESCs in the absence of feeders.[1,2] Recently, it has been shown that BMPs can also promote self renewal of mESCs in serum free conditions, but whether this is true of hESCs remains to be tested.[51]

Other factors intrinsic to the cells do seem to be commonly expressed in hESCs and mESCs, and probably to play similar functions. Thus Oct4 is strongly expressed by human and mouse ECCs and ESCs, and is down regulated upon differentiation (Fig. 3); other than in primordial germ cells, it is not known to be expressed in other tissues. Thus Oct4 provides an excellent marker for the undifferentiated ECCs and ESCs. We have now shown that indeed Oct4 is required for maintenance of the undifferentiated state of hESCs which differentiate towards trophectoderm if Oct4 expression is knocked down by RNA interference.[52]

The recently discovered gene, *Nanog* is another strongly expressed gene characteristic of undifferentiated ESCs in both mice and humans, and it is certainly required for maintenance of the undifferentiated state of mESCs.[53] It is also down regulated upon the differentiation of hESCs (Fig. 3). A variety of other genes, e.g. Sox2, Rex1, Tert, Utf1 and FGF4, have also been suggested to be expressed by hESCs and to be used as markers for the undifferentiated, pluripotent state, but they have yet to be tested widely [e.g. Refs. 50, 54] Several DNA microarray expression studies of hESCs have also been reported and analyses of these have been used to suggest various genes as specific stem cell related marker genes.[55,56,57,58] However, many of the candidates differ between the studies, which also suffer from the difficulties of interpreting results from heterogeneous cultures resulting from the typical spontaneous differentiation of the ESCs. Definition of an agreed set of markers characteristic of the undifferentiated hESCs remains to be achieved.

## Future Prospects

The study of hESCs is still in its infancy. Although the first lines were reported in 1998, development of work on these cells has been slow, in part because of the ethical issues raised, and in part because of difficulties in culture. Nevertheless, studies are now expanding rapidly with a significant number of laboratories worldwide reporting new lines. No doubt progress will accelerate, given the strong interest in using these cells to generate specific functional differentiated cell types for transplantation to patients to replace diseased or damaged tissues. But significant challenges remain.

A key issue is to establish how best to maintain the undifferentiated cells. As we have already noted, these have a tendency towards spontaneous differentiation which hampers their continued culture. More importantly, under some culture conditions, the lines are subject to karyotypic instability. Strikingly, this instability is manifest in the amplification of chromosomes 17q and 12p, chromosomes that are also characteristically amplified in ECCs in germ cell tumors.[59] At present, the factors that promote this karyoptypic change are unknown. However, we propose that it relates to the mechanisms that control the balance between self renewal, differentiation and apoptosis. Understanding the causes of this instability and the nature of the selective advantage conferred on the altered cells will help establish ways to minimize its occurrence, help elucidate the mechanisms that control self renewal, and provide insights into progression of germ cell tumors and, perhaps, also other cancers that involve stem cells.

We still know very little about the mechanisms that do control self renewal in hECCs and hESCs. Some clearly are similar to those operating in the mESCs, notably Oct4 and Nanog. But others may be different. A challenge is to characterize these mechanisms in detail. Equally important, though, will be elucidating the control of commitment to specific cell lineages. It is fairly simple to generate neurons from hESCs, but the functional characteristics of the specific types of neuron produced remains to be assessed. Cardiac muscle differentiation has also been observed in ESC culture,[60] offering hope that eventually techniques will be developed for cardiovascular clinical applications. Yet another important application could be in diabetes. However, although some tentative reports have raised hopes,[61] robust protocols for directing pancreatic differentiation remain to be developed. Clearly many significant problems remain to be overcome, not least of which will be scaling up the production of genetically stable ESCs and their functional derivatives. Nevertheless, the opportunities for developing novel and powerful treatments for hitherto untreatable diseases are certainly great.

## Acknowledgements

This work was supported in part by grants from the BBSRC, the MRC, Yorkshire Cancer Research, and the Juvenile Diabetes Research Foundation.

## REFERENCES

1. Thomson JA, Itskovitz-Eldor J, Shapiro SS, *et al.* (1998) Embryonic stem cell lines derived from human blastocysts. *Science* **282**: 1145–1147.
2. Reubinoff BE, Pera MF, Fong CY, *et al.* (2000) Embryonic stem cell lines from human blastocysts: Somatic differentiation *in vitro*. *Nature Biotechnol* **18**: 399–404.
3. Andrews PW. (2002) From teratocarcinomas to embryonic stem cells. *Phil Trans Roy Soc Lond B* **357**: 405–417.

4. Damjanov I. (1993) Pathogenesis of testicular germ cell tumors. *Eur Urol* **23**: 2–7.

5. Wheeler JE. (1983) History of teratomas. In *The Human Teratomas: Experimental and Clinical Biology*, eds. I Damjanov, BB Knowles and D Solter pp. 1–22. Clifton NJ: Humana Press.

6. Dixon FS, Moore RA. (1952) Tumors of the male sex organs. In *Atlas of Tumor Pathology*, Vol 8, pp. 31b and 32. Washington DC: Armed Forces Institute of Pathology.

7. Stevens LC, Little CC. (1954) Spontaneous testicular teratomas in an inbred strain of Mice. *Proc Natl Acad Sci USA* **40**: 1080–1087.

8. Stevens LC. (1967) Origin of testicular teratomas from primordial germ cells in mice. *J Nat Cancer Inst* **38**: 549–552.

9. Kleinsmith LJ, Pierce GB. (1964) Multipotentiality of single embryonal carcinoma cells. *Cancer Res* **24**: 1544–1552.

10. Solter D, Škreb N, Damjanov I. (1970) Extrauterine growth of mouse egg-cylinders results in malignant teratoma. *Nature* **227**: 503–504.

11. Stevens LC. (1970) The development of transplantable teratocarcinomas from intratesticular grafts of pre- and post-implantation mouse embryos. *Dev Biol* **21**: 364–382.

12. Finch BW, Ephrussi B. (1967) Retention of multiple developmental potentialities by cells of a mouse testicular teratocarcinomas during prolonged culture *in vitro* and their extinction upon hybridisation with cells of permanent lines. *Proc Natl Acad Sci USA* **57**: 615–621.

13. Jacob F. (1978) Mouse teratocarcinoma and mouse embryo. *Proc Roy Soc Lond B* **201**: 249–270.

14. Boyse EA, Old LJ. (1969) Some aspects of normal and abnormal cell surface genetics. *Ann Rev Genet* **3**: 269–290.

15. Artzt K, Dubois P, Bennett D, *et al.* (1973) Surface antigens common to mouse cleavage embryos and primitive teratocarcinoma cells in culture. *Proc Natl Acad Sci USA* **70**: 2988–2992.

16. Brinster RL. (1974) The effect of cells transferred into the mouse blastocyst on subsequent development. *J Exp Med* **140**: 1049–1056.

17. Mintz B, Illmensee K. (1975) Normal genetically mosaic mice produced from malignant teratocarcinoma cells. *Proc Natl Acad Sci USA* **72**: 3585–3589.

18. Papaioannou VE, McBurney MW, Gardner RL, Evans MJ. (1975) Fate of teratocarcinoma cells injected into early mouse embryos. *Nature* **258**: 70–73.

19. Martin GR. (1981) Isolation of a pluripotent cell line from early mouse embryos cultured in medium conditioned by teratocarcinoma stem cells. *Proc Natl Acad Sci USA* **78**: 7634–7636.

20. Evans MJ, Kaufman MH. (1981) Establishment in culture of pluripotential cells from mouse embryos. *Nature* **292**: 154–156.

21. Nagy A, Rossant J, Nagy R, *et al.* (1993) Derivation of completely cell culture-derived mice from early-passage embryonic stem cells. *Proc Natl Acad Sci USA* **90**: 8424–8428.

22. Martin GR, Wiley LM, Damjanov I. (1977) The development of cystic embryoid bodies *in vitro* from clonal teratocarcinoma stem cells. *Dev Biol* **61**: 230–244.

23. Strickland S, Smith KK, Marotti KR. (1980) Hormonal induction of differentiation in teratocarcinoma stem cells: Generation of parietal endoderm by retinoic acid and dibutyryl cAMP. *Cell* **21**: 347–355.

24. Hogan BL, Taylor A, Adamson E. (1981) Cell interactions modulate embryonal carcinoma cell differentiation into parietal or visceral endoderm. *Nature* **291**: 235–237.

25. Jones-Villeneuve EM, McBurney MW, Rogers KA, Kalnins VI. (1982) Retinoic acid induces embryonal carcinoma cells to differentiate into neurons and glial cells. *J Cell Biol* **94**: 253–262.

26. Møller H. (1993) Clues to the aetiology of testicular germ cell tumors from descriptive epidemiology. *Eur Urol* **23**: 8–15.

27. Pierce GB, Verney EL, Dixon FJ. (1957) The biology of testicular cancer I. Behaviour after transplantation. *Cancer Res* **17**: 134–138.

28. Fogh J, Trempe G. (1975) New human tumor cell lines. In *Human Tumor Cells in vitro*, ed. J Fogh, pp. 115–159. New York: Plenum Press.

29. Hogan B, Fellous M, Avner P, Jacob F. (1977) Isolation of a human teratoma cell line which expresses F9 antigen. *Nature* **270**: 515–518.

30. Holden S, Bernard O, Artzt K, *et al.* (1977) Human and mouse embryonal carcinoma cells in culture share an embryonic antigen (F9). *Nature* **270**: 518–520.

31. Wang N, Trend B, Bronson DL, Fraley EE. (1980) Nonrandom abnormalities in chromosome 1 in human testicular cancers. *Cancer Res* **40**: 796–802.

32. Andrews PW, Bronson DL, Benham F, *et al.* (1980) A comparative study of eight cell lines derived from human testicular teratocarcinoma. *Int J Cancer* **26**: 269–280.

33. Andrews PW, Goodfellow PN, Shevinsky L, *et al.* (1982) Cell surface antigens of a clonal human embryonal carcinoma cell line: Morphological and antigenic differentiation in culture. *Int J Cancer* **29**: 523–531.

34. Damjanov I, Fox N, Knowles BB, *et al.* (1982) Immunohistochemical localization of murine stage-specific embryonic antigens in human testicular germ cell tumors. *Am J Pathol* **108**: 225–230.

35. Damjanov I, Andrews PW. (1983) Ultrastructural differentiation of a clonal human embryonal carcinoma cell line *in vitro*. *Cancer Res* **43**: 2190–2198.

36. Andrews PW, Goodfellow PN. (1980) Antigen expression by somatic cell hybrids of a murine embryonal carcinoma cell with thymocytes and L cells. *Somat Cell Genet* **6**: 271–284.

37. Rousset J-P, Bucchini D, Jami J. (1983) Hybrids between F9 nullipotent teratocarcinomas and thymus cells produce multidifferentiated tumors in mice. *Dev Biol* **96**: 331–336.

38. Andrews PW, Damjanov I, Simon D, *et al.* (1984) Pluripotent embryonal carcinoma clones derived from the human teratocarcinoma cell line Tera-2: Differentiation *in vivo* and *in vitro. Lab Invest* **50**: 147–162.

39. Pera MF, Cooper S, Mills J, Parrington JM. (1989) Isolation and characterization of a multipotent clone of human embryonal carcinoma cells. *Differentiation* **42**: 10–23.

40. Teshima S, Shimosato Y, Hirohashi S, *et al.* (1988) Four new human germ cell tumor cell lines. *Lab Invest* **59**: 328–336.

41. Andrews PW. (1984) Retinoic acid induces neuronal differentiation of a cloned human embryonal carcinoma cell line *in vitro. Dev Biol* **103**: 285–293.

42. Andrews PW, Nudelman E, Hakomori S-i, Fenderson BA. (1990) Different patterns of glycolipid antigens are expressed following differentiation of TERA-2 human embryonal carcinoma cells induced by retinoic acid, hexamethylene bisacetamide (HMBA) or bromodeoxyuridine (BUdR). *Differentiation* **43**: 131–138.

43. Andrews PW, Damjanov I, Berends J, *et al.* (1994) Inhibition of proliferation and induction of differentiation of pluripotent human embryonal carcinoma cells by osteogenic protein-1 (or bone morphogenetic protein-7). *Lab Invest* **71**: 243–251.

44. Gönczöl E, Andrews PW, Plotkin SA. (1984) Cytomegalovirus replicates in differentiated but not undifferentiated human embryonal carcinoma cells. *Science* **224**: 159–161.

45. Hirka G, Prakesh K, Kawashima H, *et al.* (1991) Differentiation of human embryonal carcinoma cells induces human immunodeficiency virus permissiveness which is stimulated by human cytomegalovirus coinfection. *J Virol* **65**: 2732–2735.

46. Thomson JA, Kalishman J, Golos TG, *et al.* (1995) Isolation of a primate embryonic stem cell line. *Proc Natl Acad Sci USA* **92**: 7844–7848.

47. Sperger JM, Chen X, Draper JS, *et al.* (2003) Gene expression patterns in human embryonic stem cells and human pluripotent germ cell tumors. *Proc Natl Acad Sci USA* **100**: 13350–13355.

48. Andrews PW, Casper J, Damjanov I, *et al.* (1996) Comparative analysis of cell surface antigens expressed by cell lines derived from human germ cell tumors. *Int J Cancer* **66**: 806–816.

49. Draper JS, Pigott C, Thomson JA, Andrews PW. (2002) Surface antigens of human embryonic stem cells: Changes upon differentiation in culture. *J Anat* **200**: 249–258.

50. Henderson JK, Draper JS, Baillie HS, *et al.* (2002) Preimplantation human embryos and embryonic stem cells show comparable expression of stage-specific embryonic antigens. *Stem Cells* **20**: 329–337.

51. Ying QL, Nichols J, Chambers I, Smith A. (2003) BMP induction of Id proteins suppresses differentiation and sustains embryonic stem cell self-renewal in collaboration with STAT3. *Cell* **115**: 281–292.

52. Matin MM, Walsh JR, Gokhale PJ, *et al.* (2004) RNA interference mediated knock-down of Oct4 expression induces differentiation of human EC and ES cells. *Stem Cells* **22**: 659–668.

53. Chambers I, Colby D, Robertson M, *et al.* (2003) Functional expression cloning of Nanog, a pluripotency sustaining factor in embryonic stem cells. *Cell* **113**: 643–655.

54. Brivanlou AH, Gage FH, Jaenisch R, *et al.* (2003) Stem cells. Setting standards for human embryonic stem cells. *Science* **300**: 913–916.

55. Ginis I, Luo Y, Miura T, *et al.* (2004) Differences between human and mouse embryonic stem cells. *Dev Biol* **269**: 360–380.

56. Ivanova NB, Dimos JT, Schaniel C, *et al.* (2002) A stem cell molecular signature. *Science* **298**: 601–604.

57. Ramalho-Santos M, Yoon S, Matsuzaki Y, Mulligan RC, Melton DA. (2002) "Stemness": Transcriptional profiling of embryonic and adult stem cells. *Science* **298**: 597–600.

58. Sato N, Sanjuan IM, Heke M, *et al.* (2003) Molecular signature of human embryonic stem cells and its comparison with the mouse. *Dev Biol* **260**: 404–413.

59. Draper JS, Smith K, Gokhale PJ, *et al.* (2003) Karyotypic evolution of human Embryonic Stem (ES) cells in culture: Recurrent gain of chromosomes 17 (17q) and 12. *Nat Biotech* **22**: 53–54.

60. Mummery C, Ward-van Oostwaard D, Doevendans P, *et al.* (2003) Differentiation of human embryonic stem cells to cardiomyocytes: Role of coculture with visceral endoderm-like cells. *Circulation* **107**: 2733–2740.

61. Assady S, Maor G, Amit M, *et al.* (2001) Insulin production by human embryonic stem cells. *Diabetes* **50**: 1691–1697.

62. Shevinsky LH, Knowles BB, Damjanov I, Solter D. (1982) Monoclonal antibody to murine embryos defines a stage-specific embryonic antigen expressed on mouse embryos and human teratocarcinoma cells. *Cell* **30**: 697–705.

63. Kannagi R, Cochran NA, Ishigami F, *et al.* (1983) Stage-specific embryonic antigens SSEA3 and -4 are epitopes of a unique globo-series ganglioside isolated from human teratocarcinoma cells. *EMBO J* **2**: 2355–2361.

64. Andrews PW, Banting G, Damjanov I, *et al.* (1984) Three monoclonal antibodies defining distinct differentiation antigens associated with different high molecular weight polypeptides on the surface of human embryonal carcinoma cells. *Hybridoma* **3**: 347–361.

65. Badcock G, Pigott C, Goepel J, Andrews PW. (1999) The human embryonal carcinoma marker antigen TRA-1-60 is a sialylated keratan sulfate proteoglycan. *Cancer Res* **59**: 4715–4719.

66. Pera MF, Blasco-Lafita MJ, Cooper S, *et al.* (1988) Analysis of cell-differentiation lineage in human teratomas using new monoclonal antibodies to cytostructural antigens of embryonal carcinoma cells. *Differentiation* **39**: 139–149.

67. Cooper S, Bennett W, Andrade J, *et al.* (2002) Biochemical properties of a keratan sulphate/chondroitin sulphate proteoglycan expressed in primate pluripotent stem cells. *J Anat* **200**: 259–265.

68. Andrews PW, Meyer LJ, Bednarz KL, Harris H. (1984c) Two monoclonal antibodies recognizing determinants on human embryonal carcinoma cells react specifically with the liver isozyme of human alkaline phosphatase. *Hybridoma* **3**: 33–39.

69. Solter D, Knowles BB. (1978) Monoclonal antibody defining a stage-specific mouse embryonic antigen SSEA1. *Proc Natl Acad Sci USA* **75**: 5565–5569.

70. Gooi HC, Feizi T, Kapadia A, *et al.* (1981) Stage specific embryonic antigen involves $\alpha1 \rightarrow 3$ fucosylated type 2 blood group chains. *Nature* **292**: 156–158.

71. Kannagi R, Nudelman E, Levery SB, Hakomori S. (1982) A series of human erythrocyte glycosphingolipids reacting to the monoclonal antibody directed to a developmentally regulated antigen, SSEA1. *J Biol Chem* **257**: 14865–14874.

72. Eisenbarth GS, Walsh FS, Nirenberg M. (1979) Monoclonal antibody to a plasma membrane antigen of neurons. *Proc Natl Acad Sci USA* **76**: 4913–4917.

73. Thurin J, Herlyn M, Hindsgaul O, *et al.* (1985) Proton NMR and fast-atom bombardment mass spectrometry analysis of the melanoma-associated ganglioside 9-O-acetyl-GD3. *J Biol Chem* **260**: 14556–14563.

# 4

# Stem Cells and Their Developmental Potential

Martin F. Pera* and Mirella Dottori

## Introduction

A stem cell is a cell that can self renew to produce more stem cells, or undergo diffferentiation into particular types of specialized tissues. A critical feature of any stem cell, whether of embryonic, fetal or adult origin, is its developmental potential, defined operationally as the range of precursor or mature tissue cell types the stem cell can produce in a given experimental context. In this chapter, we will discuss the developmental potential of various stem cell populations, and compare stem cells derived from embryonic stages with stem cells from mature tissues, focusing on the bone marrow and central nervous system.

Embryogenesis is generally viewed as a series of committed events, during which the developmental capacity of the cells of the embryo becomes gradually restricted, so that once the tissues of the body have been formed, their constitutive stem cells can only give rise to a restricted range of cell types. Recent experiments have challenged this concept, but much of the evidence for this phenomenom known as tissue stem cell plasticity remains controversial.

## Definition of Terms

A **stem cell** has been defined in the introduction above. Implicit in that definition is the requirement that the experimental evidence demonstrates, through single cell cloning, that a putative stem cell can both self renew

*Correspondence: Monash Institute of Reproduction and Development, 246 Clayton Road, Clayton 3168 Victoria Australia. Tel.: 61 3 9594 7318, Fax: 61 3 9594 7311, e-mail: martin.pera@ med.monash.edu.au

and give rise to a differentiated progeny. A **progenitor cell** is a cell that is a precursor to mature differentiated cells, but is not capable of extensive self renewal. A term used to describe some progenitor cells is the **transit amplifying cell**, a committed progenitor cell capable of some division that sits in a hierarchy between stem cells and mature differentiated cells. **Embryonic stem (ES) cells** are stem cells derived from the early embryo, before the specialized tissues of the body have begun to form. **Embryonic germ (EG) cells** are stem cells derived from primordial germ cells, the precursors of sperms and eggs, before their differentiation into oogonia or prospermatogonia. The term **"adult stem cell"** is often used to denote stem cells with specific immunologically and molecularly defined phenotypes and known differentiation capabilities that reside in particular mature tissues. However, the use of the term "adult" is not entirely accurate, since stem cells with similar properties are often found in the corresponding fetal, neonatal, or pediatric tissues. We prefer the use of the term **tissue stem cell**, and would use this term to refer to stem cells from fully developed tissues after the period of organogenesis in the embryo. Tissue stem cells are sometimes referred to as **constitutive**, if they function as stem cells during routine renewal and repair, in contrast to **facultative** stem cells, which are stem cell populations that play a role in special forms of tissue repair but do not function as stem cells under normal physiological conditions.

There is some lack of clarity in the literature regarding the terms used to describe cellular developmental potential. The terms **totipotent, pluripotent, and multipotent are** commonly used to describe stem cell developmental capacity. Totipotent is a term that some use to identify a stem cell that can give rise to every extraembryonic, somatic, or germ cell known in mammalian development. The difficulty with this use of the term totipotent is that many assume that such a cell is developmentally equivalent to the zygote. Such usage has led to the confusion of embryonic stem cells with embryos in the context of ethical discussions concerning the use of human embryos in research. We prefer the use of the term **totipotent** to be confined to cells in mammals that can give rise on their own to a new organism given appropriate maternal support. **Pluripotent cells**, on the other hand, are cells that can give rise to all cells of the body including germ cells, and some of the extraembryonic tissues that function to support development in mammals, but cannot give rise on their own to a new organism. **Multipotent stem cells** are cells that can give rise to several types of mature cells. There are also examples of stem cell populations that are **bipotent, or unipotent**.

It is most important to note that a cell's developmental capacity is always operationally defined in a particular experimental context, such as in an assay designed to measure engraftment into a damaged tissue. Developmental potential, as we shall see, may be a function of the cellular environment, and therefore may vary depending upon the circumstances of the experiment.

# Developmental Potential of the Cells of the Early Embryo

## Commitment and pluripotentiality

Stem cells do not really exist in the early embryo, since embryonic development represents a state of flux in which multiplication, commitment, differentiation, and death constantly reshape and redefine the relationship between various cell populations until the body plan emerges and the major organ systems have developed.[1] Nevertheless some understanding of cell commitment in the embryo is critical to understanding of the concept of pluripotentiality, and to understanding of the developmental potential of stem cells derived from embryonic tissues.

The unfertilized oocyte may be regarded as a pluripotent cell, since artificial activation of the mammalian egg can give rise to a parthenogenetic conceptus that will develop up to the early postimplantation stages,[2] and ES cells derived from parthenogenetic embryos have been described in mice and primates. However, there are limitiations to the development of parthenotes imposed by genomic imprinting and the resulting requirement for both a maternal and paternal genome for normal development.[3] Thus, the oocyte is not truly totipotent in the sense that the zygote is.

The zygote and early blastomeres of the mouse embryo are totipotent, in the sense that we have defined the term, up to the eight-cell stage of development. Thereafter, the early stages of mammalian development are devoted largely to the formation of the extraembryonic tissues, tissues derived from the zygote that support development, including the trophoblast, yolk sac, amnion, and allantois. It is around the eight-cell stage that the formation of these extraembryonic tissues begins, with the preparation for the first commitment event in mammalian development, the formation of the trophectoderm.[4] Compaction of the embryo leads to an asymmetric division of its constituent cells to produce either inner or outer cells, leading to segregation of the future trophectoderm cells to the outer layer of the developing blastocyst.[5] The outer cells lose pluripotency and become committed to the trophectoderm fate. Stem cell lines may be established from the trophoblast in the mouse, but these cell lines participate only in the development of the placenta.[6]

The cells on the inside of the blastocyst constitute the inner cell mass, precursor of all body tissues plus extraembryonic endoderm and mesoderm. Not long after it has been formed, the inner cell mass soon loses the ability to generate a trophoblast. The next commitment event in the development is the formation of the primitive endoderm, the precursor of the yolk sac. The yolk sac can be regarded as a primitive form of placenta, functioning in the uptake processing and transport of nutrients prior to the development of the chorioallantoic placenta. The yolk sac endoderm also has important functions in the patterning of the pluripotent cells of the early conceptus as well.

Following differentiation of the extraembryonic endoderm and implantation, the inner cell mass develops into an epithelium known as the epiblast. The epiblast retains pluripotentiality, since cell lineage tracing experiments show that it will contribute to multiple tissues at this stage, and it can give rise to teratomas when transplanted into ectopic sites. A teratoma is a disorganized growth comprising multiple tissue types foreign to its anatomic site of origin.[7] This capacity to form teratomas disappears after about 7.5 days of development in the mouse. ES cells can also be derived from the mouse epiblast. However, epiblast cells, unlike inner cell mass cells, cannot colonize a host blastocyst. Whether this reflects a limitation of their developmental capacity, or a limitation of the assay in that these cells may not be able physically to integrate into the host embryo, is uncertain.

Beginning shortly after implantation, pluripotent cells of the epiblast receive signals from the surrounding extraembryonic endoderm and extraembryonic ectoderm that specify cell fate in a regionally and temporally controlled fashion.[8] These signals help to define the future anteroposterior axis of the embryo. In the pregastrula embryo, there is an anterior to posterior gradient in the intensity of Wnt, nodal, and BMP signaling that helps to determine cell fate. Before gastrulation, the fate of the particular epiblast cells correlates with their specific localization in the embryo, but the progeny of one cell can contribute to multiple germ layer derivatives, and grafting studies have shown that the fate of these cells is not yet fixed but can be altered in a different environment. However, the process of gastrulation results in the commitment of most cells in the embryo to particular fates. Thus, after gastrulation, cells become restricted in their developmental capacity. Movement through the primitive streak commits

**Figure 1.** The human blastocyst shortly after implantation, depicting the early extraembryonic lineages and the pluripotent cells of the epiblast. Amnion (blue) develops early in the human, via cavitation of the inner cell mass, and will later envelop the developing conceptus and provide it with a protective cushion. Trophectoderm (red) will shortly begin to invade the maternal tissue (purple) to help establish the definitive placenta. The primitive endoderm (green) develops below the epiblast, and some cells migrate out to line the blastocoel cavity (these latter cells, shown as green strands, are probably homologous to rodent parietal endoderm. The pluripotent epiblast (orange) is an epithelial disk that will receive signals from the surrounding tissue to drive regionally sepcific commitment of its constituent cells. Gastrulation will begin in a few days at the posterior end of the embryo, giving rise to all three embryonic germ layers. After this point, pluripotency is limited to cells of the germline.

cells to an endodermal or a mesodermal fate, depending on the time of their emergence. Cells of the distal epiblast retain pluripotentiailty for some time, but the cells in the anterior become committed to a neural fate quickly.

## Primordial germ cells

One embryonic cell population escapes some of these restrictions on development fate. Primordial germ cells, the precursors of oogonia and prospermatogonia,[9,10] are induced in the proximal epiblast by signals from the surrounding extraembryonic ectoderm and extraembryonic endoderm, in particular BMP-2, -4, and -8. Once formed, primordial germ cells proliferate as they migrate through the hindgut to the genital ridge. Prior to overt sexual differentiation of the gonad, primordial germ cells show a similar phenotype in either sex. Differentiation of the supporting cells of the gonad is accompanied by sex-specific differentiation of the primordial germ cells. Male cells undergo additional cell division before arresting in G1, whereas female cells enter the prophase of the first meiotic division. Male cells, if they fail to migrate properly and escape the environment of the developing testis, enter the first prophase of meiosis similar to oogonia.

During normal development, primordial germ cells ultimately give rise only to sperm or egg, both of which could be regarded as highly specialized cell types. But under a range of different circumstances, primordial germ cells can express pluripotentiality. Primordial germ cells are the cells of origin of spontaneously occurring teratocarcinomas that occur at low frequency in most mammals but are relatively common in certain inbred strains of mice.[7] A teratocarcinoma is a tumor similar to a teratoma, but contains malignant stem cells in addition to differentiated tissues. In those mouse strains that develop teratocarcinomas, the origin of the tumors can be traced back to primordial germ cells at around day 12.5 of development. Ectopic transplantation of the genital ridge up to day 12.5 of mouse development can also give rise to teratocarcinomas. Primordial germ cells can also give rise to EG cells when cultured *in vitro*.[11] EG cells are permanent cell lines that share many properties of ES cells, including pluripotentiality, although their genomic imprinting status reflects the erasure of the parental specific imprints that occurs during the development of the germ line. The rate of formation of teratocarcinomas from grafted genital ridges, or the rate of EG cell line establishment from primordial germ cells, drops off abruptly after differentiation of the germ cells. EG cells may also be established from embryonic or fetal gonads in man, and though the human cells are difficult to propagate, they can certainly differentiate into a variety of cell types. Recent evidence indicates that mouse male neonatal testis contains a cell population that can give rise to pluripotent cell lines. These stem cell lines, called multipotent germ cells, are not really equivalent to ES cells in developmental potential, but can form chimeras when placed in a host blastocyst.[12]

Pluripotency is not a property of primordial germ cells during normal development; EG cells and multipotent germ cells appear to require some period of cultivation *in vitro* before they express the property of pluripotentiality, and teratocarcinogenesis requires a specific genetic predisposition in mice, or exposure to an abnormal environment. Thus, some workers regard the expression of pluripotentiality in primordial germ cells as a type of developmental reprogramming. However, it is also true that primordial germ cells continue to express many molecular markers of the pluripotent state, and that they are unique amongst the cells of the postgastrulation embryo in this respect. Thus, primordial germ cells, while they may require some degree of reprogramming to express pluripotentiality, are probably much closer to the pluripotent state than are most tissue stem cells.

## Embryonic stem cells

Cells that show the expected properties of ES cells have only been developed in mice and primates.[13–15] These properties include pluripotentiality, immortality, and maintenance of a normal diploid karyotype during long term cultivation *in vitro*. In either species, ES cells are usually derived from the inner cell mass, but in the mouse, they may be developed more efficiently from the epiblast.[16] The ability of ES cells to form a wide range of tissues is best tested by chimera formation, in which small numbers of cells are introduced into a host blastocyst, which is then returned to a foster mother. The embryo is allowed to develop, and the contribution of the ES cells to various tissues is assessed through the use of genetic markers. In the mouse, the bright line test of pluripotentiality is the establishment of germ line chimerism, and the transmission of ES cell genotype through the germ cells of the chimera. Mouse ES cells are capable of forming all the body cells of a chimera, including germ cells, when they are combined with tetraploid four-cell stage embryos, formed by the fusion of diploid two-cell stage blastomeres. In such chimeras, the trophoblast and yolk sac endoderm is formed by tetraploid embryonic cells. The tetraploid embryo cells complement the developmental capacity of the ES cells, which cannot form trophoblast, and rarely form extraembryonic endoderm in the context of chimeras (though they do so *in vitro* readily).

Thus mouse ES cells on their own cannot give rise to an embryo, in part because they cannot form trophectoderm. However, the developmental limitations of ES cells go beyond their inability to form certain extraembryonic derivatives. ES cells undergoing differentiation in embryoid bodies often recapitulate temporal patterns of gene expression seen in the embryo, but the three dimensional structure of the embryoid body does not show the spatial organization of the periimplantation embryo. Thus, ES cells do not undergo patterning or axis formation, and are incapable of generating a body plan on their own. Combination of ES cells with recent progeny of a zygote is required to support participation of the ES cells in normal development. Nonhuman primate stem cells have not yet been tested for their ability to contribute to the tissues of a chimeric animal, but monkey and human ES cells can both form teratomas containing a wide range of different cell types.

## Molecular control of pluripotency

Several factors are known to be essential for the maintenance of the pluripotent state in the mammalian embryo. These include the POU domain transcription factor Oct-4, the SRY containing gene Sox-2, the homeobox gene Nanog, and the winged helix transcription factor Fox-D3.[17,18] Deletions in any of these genes will result in failure of the pluripotent lineage in the embryo to develop or be maintained, and it is impossible to derive ES cell lines from embryos that are homozygous for deletion of these genes. In addition to these functionally defined determinants of the pluripotent state, a number of studies of human and mouse embryonic stem cells have analyzed the transcriptome of these cell types, and a common set of genes found in pluripotent stem cells of both species has been identified.[19] These include DNA modifying enzymes, specific cell surface markers, transcription factors, and receptors for specific growth and differentiation factors. Comparison of gene expression in ES cells with gene expression in various embryonic stages of development revealed a number of important differences, though in global terms, the ES cells were most similar to embryos at age E6.5 or E7.5 or to blastocysts.[20]

The loss of pluripotentiality is governed not only by downregulation of expression of genes characteristic of the pluripotent stem cell, but also by the onset of expression of lineage specific transcription factors, including, for example, CDX-2 in the trophectoderm lineage and GATA-6 in the extraembryonic endoderm. Thus, commitment in the early embryo involves upregulation of some genes and downregulation of others. Recent results indicate that the expression of GATA-6 begins early (day 3.5), and suggest that commitment to extraembryonic endoderm has occurred by day 3.5 of development, well before overt differentiation.

## Multipotent Tissue Stem Cells

Stem cells have long been known to exist within the adult animal in various organs and tissues. A general feature of multipotent stem cells is that under homeostatic conditions, they remain quiescent. Upon entering the cell cycle, stem cells may divide symmetrically to give rise to two identical stem cells. Alternatively, stem cells may divide asymmetrically, resulting in one identical daughter stem cell (for self renewal) and one daughter progenitor cell, or two different daughter progenitor cells. The progenitor daughter cell then undergoes cellular proliferation and progressive differentiation, leading to an expansion of committed progenitor populations. The presence of multipotential stem cells within the adult animal may be an evolved compensatory mechanism of cell replacement in response to injury and/or an ongoing required mechanism for replacing cells in tissues that have a high cell turnover, such as the skin and blood.

Recent studies have shown that certain tissue stem cells sometimes display the capacity to differentiate into multiple cell types, including cells outside their lineage of origin, i.e. they demonstrate the property of plasticity. The mechanism of tissue stem cell plasticity is not yet well understood. Different theories advanced to account for plasticity include transdifferentiation of a stem cell to unrelated cell type, de-differentiation of a mature cell to a more primitive stem cell followed by subsequent differentiation into another cell, or perhaps for some cases, fusion between two different cells. Evidence for the presence of plastic stem cells in mature tissues extends to bone marrow, brain, liver, skin and muscle. For example, facultative tissue stem cells found within the liver, known as oval cells, are capable of differentiation into hepatocytes as well as bile duct epithelium. Multipotent precursor cells from the pancreas have also been clonally isolated, and these cells may differentiate into various pancreatic cells as well as neural and glial cells.

Tissue stem cells of the bone marrow and brain have been extensively studied and are described in greater detail below. The current controversy over the developmental potential of stem cells in these tissues[21] serves to illustrate important concepts about the biology of stem cells, their potential and their limitations, and how they may be therapeutically useful for treatment of diseases and injuries.

## Stem Cells in the Bone Marrow

The bone marrow harbors hematopoietic stem cells (HSCs) and mesenchymal or stromal stem cells (MSCs).[22,23] Hematopoietic stem cells (HSCs) give rise to the eight major hematopoietic lineages, including the two lymphoid lineages of T cells and B cells, and six myeloid lineages. HSCs are identified by purifying various populations of bone marrow cells on the basis of cell surface marker expression and then functionally assayed *in vivo*. A true HSC, intravenously injected into an irradiated recipient, is able to home to the host's bone marrow and repopulate the entire hematopoietic system long term. Markers of human and mouse HSCs include high expression of CD34 and Sca-1, respectively, and lower levels of c-kit and Thy-1 expression in both species. The frequency of occurrence of HSCs within the bone marrow is relatively low (1 in $10^4$–$10^5$ cells), although several transplant studies have demonstrated that only very few HSCs (even a single cell) are required to reconstitute the hematopoietic system. In addition to the bone marrow, low numbers of HSCs migrate in the circulation and thus they can also be found within non-hematopoietic tissues. This consideration should be taken into account in the interpretation of experiments showing the ability of various tissues to give rise to blood cells.

MSCs differentiate into osteoblasts, chondrocytes, and adipocytes, and provide supporting stromal cells for HSC proliferation and maintenance. There are no specific markers for MSCs but they are usually identified and isolated using a combination of antigenic markers, such as CD90 (Thy-1) and CD105, together with an absence of typical hematopoietic antigen. The most reliable method of identifying MSCs in culture is to test their ability to clonally expand and form colonies *in vitro* as well as differentiate under the appropriate conditions to all mesenchymal lineages.

## Plasticity of cells from bone marrow

Several studies have shown that following bone marrow transplantation, there is recruitment of bone marrow cells (BMCs) to various tissue sites and their subsequent differentiation into cells of that tissue, suggesting they are plastic.[24,25] Plasticity of BMC differentiation *in vivo* has been observed within multiple tissues, including the liver, skeletal muscle, cardiac muscle, pancreas, brain, kidney and skin. Under certain culture conditions, BMCs may be induced *in vitro* to differentiate into multiple cell types, including neural, hepatocytes, endothelial and cardiomyocytes. Plasticity of enriched populations of either HSCs or MSCs has been shown, suggesting that plasticity of BMCs may be derived from either stem cell population.

Another type of stem cell, referred to as multipotent adult progenitor cells (MAPCs), have also been identified and isolated from long term, low density cultures of bone marrow cells. These cells have shown to differentiate *in vitro* into various lineages of endoderm, mesoderm and ectoderm and furthermore, injection of MAPCs into mouse blastocysts demonstrated their contribution and differentiation into most somatic lineages. It has yet to be determined whether pluripotent MAPCs endogenously exist within the bone marrow as a subpopulation of MSCs, or whether they arise *in vitro* from specialized long term culture conditions of bone marrow. However, their potential to differentiate into such a widespread array of cell types makes them therapeutically useful candidates for cell replacement. Further investigations are needed to demonstrate which population of BMCs is relatively more efficient for differentiation into certain lineages, or it may be that engraftment of BMCs, consisting of a mixture of multiple stem cell types, is more effective.

The homing and differentiation of transplanted BMCs particularly occurs when there is injury or damage to the host tissue. In the liver, muscle, or skin, engraftment and plasticity of marrow-derived cells is greatly enhanced when the host tissues are undergoing repair. The signals required for recruitment of bone marrow stem cells to the injury site are not known, and one possible hypothesis is a response to inflammatory signals.

The mechanisms that enable BMCs to transdifferentiate into multiple cells types have not yet been identified. Furthermore, many of the earlier reports that claimed BMC plasticity have been challenged by more recent studies.[26] The fate of bone marrow cells, either unfractionated or enriched for HSCs, injected into damaged myocardium, remains controversial, with some studies showing differentiation into cardiomyocytes and others showing only formation of hematopoietic or endothelial cells from the grafted tissue. Some workers attribute cardiomyocyte formation following marrow grafts to MSCs, rather than HSCs: One possible mechanism that may explain the observed plasticity of BMCs into different cell types is fusion between donor and host cells. In several models, careful evaluation of engrafted tissue reveals fusion between graft and host cells, with apparent reprogramming of gene expression in the resulting hybrid cell. Macrophages appear to be particularly adept at fusing to other cell types.

Thus, there are many aspects one needs to consider before claiming transdifferentiation of a stem cell to an unrelated cell type, and in defining the developmental potential of stem cells or mature cells from various tissues. One needs to investigate expression of phenotypic markers, as well as morphological and functional characteristics of the differentiated cell, and to determine the absence of cell fusion between donor and host cell. There are some reports that have investigated many of these features, particularly the absence of cell fusion, and that claimed that transdifferentiation had occurred. Transdifferentiation, though often a low frequency event in response to tissue damage, would appear to be a real phenomenon, and further studies are required to determine the signals and mechanisms involved that enable such a process to occur.

## Stem Cells within the Nervous System

Multipotential stem cells also exist within the adult central nervous system.[27] There are questions over the role of these cells in normal physiology[28] and in repair of the brain and spinal cord that illustrate how the developmental potential of a cell may vary, depending upon its environment.

In the adult brain, multipotent neural stem cells (NSCs) are generated within the subgranular zone (SGZ) of the dentate gyrus in the hippocampus and the subventricular zone (SVZ) of the lateral ventricles In the SVZ.[29] The fate of these cells under routine experimental conditions is fairly restricted. NSCs (also referred to as type B cells) give rise to rapidly dividing transient amplifying cells (type C cells) and committed migratory neuroblasts (type A cells). Neuroblasts migrate in chains along the rostral migratory stream to the olfactory bulb, where they then differentiate into two types of interneurons, granule and periglomerular cells. NSCs originating from the SGZ of the hippocampus, migrate to the granule cell layer of the dentate gyrus and give rise to new granule cells.

NSCs arising from SVZ and SGZ share some common characteristics; both are slow proliferating cells; they express glial fibrillary acidic protein (GFAP); and both share morphological and electrophysiological properties of astrocytes. The origin and phenotypic nature of NSCs is still debatable, and the continuing controversy shows the difficulty in defining the developmental potential of these cells. There are a few studies that propose that NSCs are derived from radial glial cells during embryogenesis and remain in the adult SVZ as a specialized type of astrocyte with stem cell properties. During embryonic development, radial glial cells are found within the SVZ and these cells give rise to both glial and neurons; however, they are not present in the adult mammalian brain. Other studies suggested that NSCs are derived from ependymal cells which line the lateral ventricle, but these studies have been disputed by many other reports. The glial properties of NSCs[30] has led to the proposal that NSCs are a subtype of astrocytes residing within the SVZ and SGZ regions. NSCs do not express the astrocyte marker, S100, thereby suggesting that NSCs are not the same as the commonly known astrocytes. In addition, cells with properties similar to NSCs have also been identified from other regions of the CNS, such as the olfactory bulb and spinal cord, thereby suggesting multipotential NSCs may be more widespread than originally thought.

Do these cells represent a population of astrocytes that in certain specific niches *in vivo* can undergo differentiation into neuroblasts? Although they display some phenotypic characteristics of glial cells, the majority of endogenous NSCs give rise to neuroblasts *in vivo*. However, their stem cell characteristics of self renewal and multipotentiality can be demonstrated *in vitro*. A number of studies have shown that round aggregates called neurospheres can be generated from NSCs grown in the presence of EGF and FGF *in vitro*, and that cells within these neurospheres have the capacity of self renewal as well as generating neurons, astrocytes and oligodendrocytes.

Recruitment and increased proliferation of adult NSCs have been observed in the adult brain in response to acute or chronic injury, again suggesting that the environment of these cells can influence their developmental fates. Induced injuries, such as global forebrain ischemia and middle cerebral artery occulsion, induce increased neurogenesis in the SVZ and SGZ.[31] As a result of injury, the endogenous migration of NSCs to their normal destination is transiently altered by their migration to the lesion site. Rodent models of stroke injury show an increase of cell proliferation within the SVZ regions and their migration into the damaged striatum. Recruitment of NSCs to the damaged area may also be promoted by infusion of growth factors. However, in most cases of brain injury, the contribution of endogenous NSCs to tissue repair is not significant. Thus, although the presence of endogenous multipotential neural stem cells within the nervous system is promising for repair of brain damage, therapies still need to be devised that will promote endogenous NSC recruitment, migration, differentiation and their survival in the site of injury.[32,33]

## Plasticity of NSCs

There is some experimental evidence for plasticity of adult NSCs. Irradiated mice inoculated with adult NSCs had hematopoietic cells derived from the donor NSCs.[34] Differentiation of NSCs to muscle cells has been observed in low density cultures of NSCs, and in co-cultures of NSCs with muscle cells. Remarkably, incorporation of mouse adult NSCs into chick embryos or mouse blastocysts demonstrated their differentiation to cells of all three germ layers.[35] However, another study demonstrated cell fusion in co-cultures of NSCs with ES cells, and in a more recent report, NSC injection into blastocytes resulted in their differentiation to glial cells. Thus, similar to what is observed with bone marrow stem cells, further investigations are required to determine whether NSCs are capable of pluripotency, and if so, to what cell types and in what environment. It is important to remember that neural crest cells show mesoderm as well as neural differentiation as a part of their normal developmental repetoire,[36] and it is important to bear this in mind in the interpretation of studies of NSC transdifferentiation.

## Reprogramming of the Adult Cell Epigenome

The successful cloning of mammals by somatic cell nuclear transfer into an oocyte shows that an adult cell nucleus, under the right circumstances, can be reprogrammed to support the entire program of embryonic development, and thus regain totipotency.[37] The cloning process in mammals is highly inefficient and subject to stochastic error, resulting in developmental failures, but some degree of successful reprogramming of the adult nucleus is often observed.[38] Thus, cloning rarely restores totipotency, but frequently endows the donor nucleus with pluripotency.[39,40] Fusion of adult cells with pluripotent cells can also result in reprogramming and expression of some aspects of the pluripotent phenotype.[41] These observations show that in mammals, as in lower vertebrates, restriction of developmental capacity, and even terminal differentiation, is a solely epigenetic phenomenon and therefore potentially reversible. An improved understanding of the mechanisms of reprogramming may lead to an improved understanding of the basis and potential practical application of tissue stem cell plasticity.

## REFERENCES

1. Nagy A, Gertenstein M, Vintersten K, Behringer R. (2003) *Manipulating the Mouse Embryo*, New York: Cold Spring Harbor Press.
2. Rougier N, Werb Z. (2001) Minireview: Parthenogenesis in mammals. *Mol Reprod Dev* **59**: 468–474.
3. Allen ND, Barton SC, Hilton K, Norris ML, Surani MA. (1994) A functional analysis of imprinting in parthenogenetic embryonic stem cells. *Development* **120**: 1473–1482.
4. Rossant J, Chazaud C, Yamanaka Y. (2003) Lineage allocation and asymmetries in the early mouse embryo. *Philos Trans R Soc Lond B Biol Sci* **358**: 1341–1348; discussion 1349.
5. Johnson MH, McConnell JM. (2004) Lineage allocation and cell polarity during mouse embryogenesis. *Semin Cell Dev Biol* **15**: 583–597.
6. Rossant J. (2001) Stem cells in the mammalian blastocyst. *Harvey Lect* **97**: 17–40.
7. Stevens LC. (1980) Teratocarcinogenesis and spontaneous parthenogenesis in mice. *Results Probl Cell Differ* **11**: 265–274.
8. Rossant J, Tam PP. (2004) Emerging asymmetry and embryonic patterning in early mouse development. *Dev Cell* **7**: 155–164.
9. Saitou M, *et al.* (2003) Specification of germ cell fate in mice. *Philos Trans R Soc Lond B Biol Sci* **358**: 1363–1370.

10. McLaren A. (2003) Primordial germ cells in the mouse. *Dev Biol* **262**: 1–15.

11. Donovan PJ, de Miguel MP. (2003) Turning germ cells into stem cells. *Curr Opin Genet Dev* **13**: 463–471.

12. Kanatsu-Shinohara M, *et al.* (2004) Generation of pluripotent stem cells from neonatal mouse testis. *Cell* **119**: 1001–1012.

13. Smith AG. (2001) Embryo-derived stem cells: Of mice and men. *Annu Rev Cell Dev Biol* **17**: 435–462.

14. Pera MF, Reubinoff B, Trounson A. (2000) Human embryonic stem cells. *J Cell Sci* **113 (Pt 1)**: 5–10.

15. Pera MF, Trounson AO. (2004) Human embryonic stem cells: Prospects for development. *Development* **131**: 5515–5525.

16. Gardner RL, Brook FA. (1997) Reflections on the biology of embryonic stem (ES) cells. *Int J Dev Biol* **41**: 235–243.

17. Chambers I, Smith A. (2004) Self-renewal of teratocarcinoma and embryonic stem cells. *Oncogene* **23**: 7150–7160.

18. Cavaleri F, Scholer HR. (2003) Nanog: A new recruit to the embryonic stem cell orchestra. *Cell* **113**: 551–552.

19. Robson P. (2004) The maturing of the human embryonic stem cell transcriptome profile. *Trends Biotechnol* **22**: 609–612.

20. Sharov AA, *et al.* (2003) Transcriptome analysis of mouse stem cells and early embryos. *PLoS Biol* **1**: E74.

21. Raff M. (2003) Adult stem cell plasticity: Fact or artifact? *Annu Rev Cell Dev Biol* **19**: 1–22.

22. Szilvassy SJ. (2003) The biology of hematopoietic stem cells. *Arch Med Res* **34**: 446–460.

23. Short B, Brouard N, Occhiodoro-Scott T, *et al.* (2003) Mesenchymal stem cells. *Arch Med Res* **34**: 565–571.

24. Herzog EL, Chai L, Krause DS. (2003) Plasticity of marrow-derived stem cells. *Blood* **102**: 3483–3493.

25. Grove JE, Bruscia E, Krause DS. (2004) Plasticity of bone marrow-derived stem cells. *Stem Cells* **22**: 487–500.

26. Wagers AJ, Weissman IL. (2004) Plasticity of adult stem cells. *Cell* **116**: 639–648.

27. Johansson CB, *et al.* (1999) Identification of a neural stem cell in the adult mammalian central nervous system. *Cell* **96**: 25–34.

28. Kempermann G, Wiskott L, Gage FH. (2004) Functional significance of adult neurogenesis. *Curr Opin Neurobiol* **14**: 186–191.

29. Alvarez-Buylla A, Lim DA. (2004) For the long run: Maintaining germinal niches in the adult brain. *Neuron* **41**: 683–686.

30. Doetsch F. (2003) The glial identity of neural stem cells. *Nat Neurosci* **6**: 1127–1134.

31. Kokaia Z, Lindvall O. (2003) Neurogenesis after ischaemic brain insults. *Curr Opin Neurobiol* **13**: 127–132.

32. Lindvall O, Kokaia Z, Martinez-Serrano A. (2004) Stem cell therapy for human neurodegenerative disorders-how to make it work. *Nat Med* **10(Suppl)**: S42–50.

33. Lie DC, Song H, Colamarino SA, *et al.* (2004) Neurogenesis in the adult brain: New strategies for central nervous system diseases. *Annu Rev Pharmacol Toxicol* **44**: 399–421.

34. Bjornson CR, Rietze RL, Reynolds BA, *et al.* (1999) Turning brain into blood: A hematopoietic fate adopted by adult neural stem cells *in vivo. Science* **283**: 534–537.

35. Clarke DL, *et al.* (2000) Generalized potential of adult neural stem cells. *Science* **288**: 1660–1663.

36. Le Douarin NM, Creuzet S, Couly G, Dupin E. (2004) Neural crest cell plasticity and its limits. *Development* **131**: 4637–4650.

37. Pomerantz J, Blau HM. (2004) Nuclear reprogramming: A key to stem cell function in regenerative medicine. *Nat Cell Biol* **6**: 810–816.

38. Latham KE. (2004) Cloning: Questions answered and unsolved. *Differentiation* **72**: 11–22.

39. Hochedlinger K, *et al.* (2004) Nuclear transplantation, embryonic stem cells and the potential for cell therapy. *Hematol J* **5(Suppl 3)**: S114–117.

40. Allegrucci C, Denning C, Priddle H, Young L. (2004) Stem-cell consequences of embryo epigenetic defects. *Lancet* **364**: 206–208.

41. Jouneau A, Renard JP. (2003) Reprogramming in nuclear transfer. *Curr Opin Genet Dev* **13**: 486–491.

# 5

# Transcriptome Profiling of Embryonic Stem Cells

Mahendra Rao* and Bing Lim

## Introduction

The past few years have seen remarkable progress in our understanding of embryonic stem cell (ES cell) biology. The wealth of genomic data, and the multiplicity of cell lines available have enabled researchers to identify critical conserved pathways regulating self-renewal and identify markers that tightly correlate with the ES cell state. Comparison across species has suggested additional pathways likely to be important in long-term self-renewal of ESCs. In this chapter we have discussed the relative merits of various large scale analytical techniques, including microarray, EST enumeration, SAGE and MPSS. We suggest that while much has been learned, already additional information remains to be gleaned by metaanalysis of existing data. Finally, newer technologies which may complement the existing strategies are briefly discussed.

## The Transcriptome and Embryonic Stem (ES) Cells

Transcriptional regulation plays a central role in defining the state of a cell. Activating and inhibiting signals establish a dynamic cascade of coordinated gene expression changes in response to extrinsic signals and intrinsic programming. Integration of these instructions occurs in the nucleus through combinations of signal-activated and tissue-restricted transcription factors (TFs) binding to and controlling related enhancers or *cis*-regulatory modules (CRMs) of co-expressed genes. Additional regulation is provided by

*Correspondence: National Institute of Ageing, StemCell, LNS, GRC, 333 Cassell Drive Baltimore, MD 21224. e-mail: raomah@grc.nia.nih.gov

previously unappreciated epigenetic mechanisms, such as histone modulation and CpG island methylation, as well as potentially by noncoding microRNAs. The set of individual components controlling a particular biological process and the interactions between them define a regulatory network, and the sum of interacting regulatory networks (transcriptome) define the state of the cell.

## ES and other Pluripotent Stem Cells

A brief summary of blastocyst and inner cell mass (ICM) maturation is shown in Fig. 1. The primary trophoblast lineage first segregates from the lineage of the embryo proper; this is followed by segregation of the hypoblast (extraembryonic endoderm) and mesoderm, and then by the amniotic endoderm.[1,2] The epiblast or the embryo proper forms the embryonic ectoderm, followed by the development of the primitive streak that leads to differentiation of the mesoderm and endoderm, which then differentiate into specific tissues and organs.

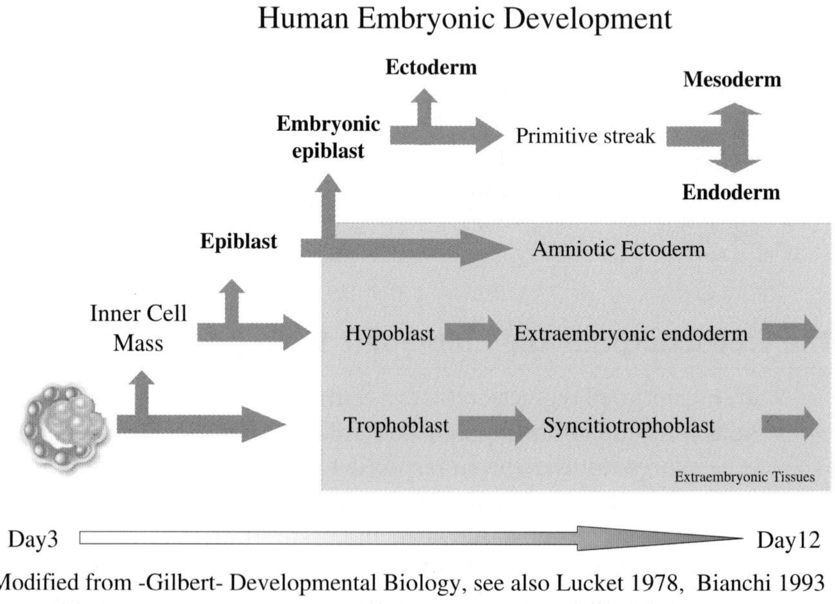

### Human Embryonic Development

Modified from -Gilbert- Developmental Biology, see also Lucket 1978, Bianchi 1993

**Figure 1.** Early differentiation in the embryo. The sequential stages of embryonic differentiation are summarized. Note that early development is dynamic with significant alterations in cell differentiation ability occurring during the preimplantation stage. (The figure has been adapted from Gilbert-Developmental Biology.)

ESCs are derived from the ICM of blastocysts prior to implantation and these cells retain many of the characteristics of ICM cells although, unlike the ICM that is a transient structure that rapidly differentiates, ES cells can be maintained relatively indefinitely in culture.[3] Mulitple ESC lines have been derived, and while some differences have been described, all ESC lines appear similar in their ability to form compact colonies, exhibit prolonged self-renewal and contribute to all germ layers.[4] ES cells recapitulate the development program of ICM cells and their differentiation is regulated by many of the same factors that regulate germ layer formation and cell type specification.[5–7]

It is important to note, however, that ICM-derived ESCs are not the only pluripotent population present in the early embryo. Pluripotent cell populations that differ from ESCs in morphology, marker expression and growth requirements have been isolated from early blastocysts, embryoid bodies and germ cells.[8–11] These cells appear similar in their expression of Oct 4 and their ability to contribute to multiple somatic tissues and germinal derivatives, but likely differ in their imprinting status,[12] their ability to differentiate into extraembryonic tissue, and their relative frequency of contribution to chimeras after blastocyst injection. In general, it has been difficult to distinguish between these cells and ICM derived ESCs.

Thus, while technologies for large scale analysis have existed, it has been difficult to apply them to these early stages of development, until recently. The first steps in such an analysis have come from examining mouse embryos and mouse ES cells (mESCs)[13–16] although recent technical advances have made multiple strategies possible (Table 1), allowing these large scale techniques to be applied to human cells. The technical advances include the ability to obtain pure populations of cells by growing cells in feeder free conditions (and) separating feeder cells from ESCs; the ability to extract RNA from single or small number of cells; enhancements in library construction; and the improvement of the quality of genomic information that allows short reads to be unambiguously mapped as well as the ability to map the information to a well annotated genomic database. These and other technical advances have allowed large scale genomic analysis to be performed by a variety of techniques (Table 1) and several reports have been released over the last year.[17–24] In subsequent sections, we discuss the various genomic methods used and their relative advantages and disadvantages, and then summarize the current understanding of human ES cells (hESCs) based on these data.

**Table 1.** Genomics Technology for ES Analysis

| Technology | Comments |
|---|---|
| Microarray | Most widely used; limited by sensitivity of probe; Limited novel gene discovery |
| EST scan | Expensive but data invaluable; small complexity And limited depth |
| SAGE | Double Sage offers increased sensitivity, precision of mapping; novel transcript discovery |
| MPSS | Highly sensitive, most depth; high discovery rate of novel genes; good for comparison; limited by annotation |
| Illumina bead technology | May be the cheapest available; adaptable to small number of cells and low RNA quantity. |
| Other Technologies | |
| Meta-analysis/ bioinformatics assessment | Softwares for merging and cross comparison across multiple large scale genomic studies |
| | In-silico prediction of binding sites and recognition sequences |
| MicroRNA expression assessment | Library sequencing; arrayed probes in chips developed for global profile of miRNAs |
| Mitochondrial sequencing | PCR sequencing; Development of Chip |
| SNP analysis | Illumina; Sequenome |
| Methylation studies | Various limited techniques for global survey of methylated versus non-methylated genes I |
| Specialized arrays | Micro Fluidics ABI system; Focused arrays |
| ChIP-to Chip | Promoter Chips and selected whole Chromosome Chip available; Total Genome Chip being develop (NimbleGen, Affymetrix) |
| Proteomics | Antibody chips (Ciphergen); SELDI; MS/MS |

# Methods of Global Analysis

## Microarray analysis

In microarray technology, probes for a large collection of genes based on cDNA clones or oligonucleotides are spotted on membranes, glass slides or chips to allow for simultaneous assessment of the expression of thousands of genes. At least five different groups have reported gene expression analysis of mESCs and hESCs using microarrays and a basic outline of the methodology is summarized in Fig. 2. The method allows very efficient gathering of data that show not only conserved and divergent pathways between species[23] but also differences between different hESC lines[18] (Tables 2 and 3).

However, there are a number of technical limitations and problems that are associated with most of the commercially available microarray chips that limit their use for studying ES or any other cells of interest. The first is the limitation of the gene coverage, since novel unknown genes will not be included in the chips. The second problem with large-scale arrays is the selection and quality of cDNA fragments. The third is the complexity of the data collection and analysis. Indeed, the cost and need for special equipment for complex data analyses are two barriers that prohibit most research laboratories from using microarray technology as a routine research tool.

## Microarray analysis

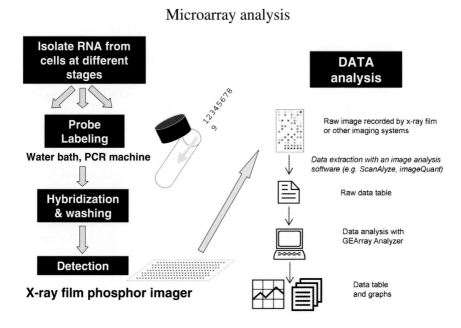

**Figure 2.** Immunosurgery of a human blastocyst for the derivation of human ES cell line. **A**. Donated human embryo produced by *in vitro* fertilization at the blastocyst stage. **B**. Human blastocyst after zona pellucida removal by Tyrode's solution, during exposure to rabbit anti-human whole antiserum. **C**. Embryo after exposure to guinea pig complement. **D**. Intact inner cell mass immediately after immunosurgery on mitotically inactivated mouse embryonic fibroblast feeder layer. Bar = 50 μm. (From Amit & Itskovitz-Eldor, *J Anat* 2002.)

**Table 2.** Signaling Pathways Conserved between Mouse and Human

| | |
|---|---|
| Stat3 signaling | Evolutionarily conserved gene, STAT binding sites present in multiple ES cell specific genes |
| Nanog | Shown to be critical in mouse and likely in human, rat genomic sequence available, not yet cloned |
| Oct-Sox | Conserved genes and co-binding sites conserved; Oct 3/4 present in fish, not in chick. |
| BMPR1a | Co-operates with LIF to sustain self-renewal, function conserved in human and mouse |
| TGFß signaling | Cripto, nodal, lefty appear to show similar patterns of expression and function; appears to be recruited more actively in hESCs |
| Igf2/H19 | Highly expressed in both mESCs and hESCs |
| MicroRNAs | Dicer required for blastocyst development; expressed in human ES cells; heterochronic gene expression (e.g. Lin-28) is conserved |
| Methylation/X Inactivation | Polycomb genes; EZH2/eed complex, Xist, TSIX, DNMT3ß and DNMT3-like show conserved patterns of expression |
| Cell cycle | While differences exist, both hESCs and mESCs show distinct Rb regulation when compared to other cell types |
| Others | DNA repair machinery, telomerase biology, some aspects of cell death, and several novel gene pathways (Dppa 2, 4 etc.), several ES specific microRNAs are likely conserved |

Cross reactivity can be minimized by careful probe selection and direct testing, and control hybridization patterns can be generated with positive and negative controls incorporated for the entire set if necessary. Several technologies have evolved which are likely to make microarrays more robust and even more reliable. These include synthesis of probes *in situ*; the use of non-overlapping longer oligonucleotide probes (40–70 mers); custom design high density arrays (e.g. NimbleGen Systems); designing of better flow and hybridization chambers (e.g. BioMicro Systems); and development of better software and data mining tools (Pathway analysis, Genesifter, etc.).

Perhaps one of the greatest problems is that microarray probes do not discriminate or even account for alternate splice variants. Indeed, many

**Table 3.** Divergent Signaling Pathways between mESCs and hESCs

| | |
|---|---|
| LIFR/gp130 | LIFR gene regulatory domains not well conserved; LIFR absent in hESCs |
| Eras and EHox | Conserved in mouse and rat, no functional ortholog in humans |
| Fox-D3 | Required in mice but expression variable in hESC lines |
| Rex-1 | Variable expression in hESC lines |
| FGF signaling | FGF2 appears high in hESCs while FGF4 is high in mESCs critical for blastocyst development in mice. Between mouse and human, differences in use of FGF pathways; several FGFRs expressed in hESCs |
| Lefty A | No orthologue in mice or rats identified |
| SSEA antigens | Differential expression seen between human and mouse |
| Tert and aging | Several components of this pathway are differentially expressed |
| Fbox15 | Does not appear to be expressed in hESCs |
| Cell cycle and cell death | Ubiquitination seems important in regulating human rb levels while mdm2 appears more important in rodent. Likewise pattern of caspases and other cell death genes expressed are quite distinct |
| Expression of trophoblast markers | MESCs do not differentiate into trophoblast while human cells appear to do so and express early trophoblast markers in maintenance conditions |
| Claudin 6 | Shows a reverse pattern of expression; high in hESCs and low in EB; low in mESCs |
| Decorin | Differentially expressed in mESCs and hESCs |
| NROB1 | Steroid (NROB1/DAX) axis appears divergent |
| Others | MPSS analysis show very low concordance in gene expression suggesting multiple additional differences exist |

genes are known to encode splice variants with different biological activities and therefore may be subject to alternate modes of transcriptional regulation. Consequently, microarray data on different platforms do not cross-validate well because probes to the same gene may be detecting different splice forms.

## EST scan

An alternative strategy that has been used successfully to assess gene expression is shotgun sequencing of cDNA clones from full length cDNA libraries. This method takes advantage of the large scale resources assembled for the genome sequencing project and the reduction in sequencing costs. In addition, the improved quality of the genome data allows short reads to be reliably mapped to the genome, providing unambiguous identification of a gene product. The number of sequences read is limited only by the resources available, and for unsubtracted libraries, the number of sequenced occurrences of a transcript is a reasonable estimate for the frequency of expression level of the corresponding gene. In general 10 000 to 1000 000 sequences are read from a prepared library. Libraries can be prepared from single cells or small amounts of RNA using linear amplification protocols.

Brandenberger and colleagues,[19] in collaboration with Celera, utilized such an approach to examine gene expression in undifferentiated ESCs as well as in differentiated populations. They obtained 148 453 expressed sequence tags (ESTs) from undifferentiated hESCs and three differentiated derivative subpopulations. Over 32 000 different transcripts expressed in hESCs were identified, of which more than 16 000 had no sequence homology to any transcript clusters in the Unigene public database. Queries of this EST database revealed 532 ESTs that were significantly up-regulated and 140 significantly down-regulated in undifferentiated hESCs. Among the differentially regulated genes identified were several representatives of signaling pathways and transcriptional regulators that likely play key roles in the control of hESC growth and differentiation. These genes were expressed at low levels and were not picked up by most microarray analyses. This study represented the first large-scale effort to capture the transcriptome of hESCs and highlighted the importance of such investigations. Details of the findings can be obtained from the manuscript but one strategy that these investigators utilized is becoming more widely used. Rather than relying on changes in gene expression in single genes, the expression data was mapped onto regulatory pathways, and patterns of coordinated gene expression were used to predict if a particular pathway was important or not. This strategy proved very useful in identifying key regulatory pathways as well as suggesting that pathways thought to be biologically active are probably inactive or tightly regulated (e.g. LIF and Wnt pathways). Comparison of this study with other in depth analyses revealed relatively good concordance.[21,24] Independent verification by semiquantitative RT-PCR or microarray has shown a high degree of reproducibility. This reproducibility provides the confidence for cross-platform comparisons, and we anticipate such meta-analyses to further enhance the utility of this and other large scale datasets. The high cost of this effort has prevented most laboratories from repeating this effort, but the data obtained are readily accessible from the NIH/NCI websites and can be searched using NCI tools such as digital differential display.

## SAGE and double SAGE

Serial analysis of gene expression (SAGE) is another directed large scale sequencing-based approach to gather quantitative and qualitative information about the transcriptome expressed by a cell.[25,26] The technique consists of isolating unique 3′-end sequence tags from individual transcripts, followed by a concatemerization of tags serially into long DNA molecules and preparing a library of the DNAs. Rapid sequencing of concatemer clones yield information about individual tags, allowing quantification and identification of a specific transcript. Typically about 150 000–200 000 tags are sequenced, from which the frequency and identity of specific transcripts can be computed (Fig. 3). Thus the method allows not only a comprehensive global snapshot of the transcriptome profile to be captured, but also provides a means of identifying transcripts of new genes.[25] Several studies have utilized SAGE to derive deep unbiased profiling of mESCs.[27] More recently, Richards et al.,[24] in a more extensive SAGE analysis, examined the transcriptome profiles for two hESC lines, and compared them with mESCs and other human tissues. Differences in the presence or absence of specific transcripts such as the LIF receptor, were extrapolated to reflect fundamental biological differences between species of what appear to be the same cell type (Table 3). This study elegantly showed that an in-depth analysis of a rare population of cells provides useful insights and, more importantly, by directly comparing two populations of ESCs showed that allelic differences could be distinguished. The data sets have been published and are available for downloading and will serve as a unique resource in the field.

3′-tags generated by SAGE and massively parallel signature sequencing (MPSS) often arise from sites several hundred bp upstream of the 3′-ends. When mapped to the genome, these "internal" tags are often ambiguous in defining transcription units and cannot identify precisely the beginning and end of the putative transcripts. To retain the efficiency of the short-tag strategy while increasing the specificity and information content of short transcript tags, Wei *et al.*[28] developed separate 5'-LongSAGE (LS) and 3'-LS protocols based on the original SAGE methodology (Fig. 3). By obtaining the first or last 18 bp of each transcript, the transcriptional initiation sites (TISs) and polyadenylation sites (PAS) can be precisely captured and mapped to the genome. Based on this experience, and to overcome its inability to connect the ends of the same transcript, Wei *et al.*[28] went on to develop a method that replaced the independent construction of two 5'- and 3'-SAGE libraries intrinsic to 5'-LS and 3'-LS, with the generation of a single paired SAGE library derived from full length cDNAs that had

## 5'- and 3'-Long Sage genome annotation

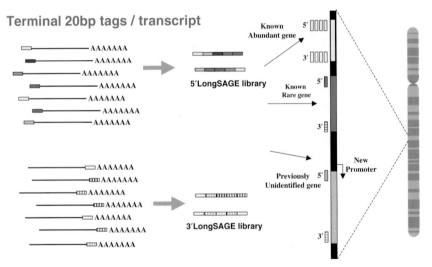

**Figure 3.**    Double SAGE. A schematic representation of the process of 5'- and 3'-Long SAGE (LS) reproduced from Wei *et al.*, 2004. An overview of the 5'-LS and 3'-LS methods for mapping transcription initiation start sites (TISs) and polyadenylation sites (PASs). **A.** The first and last 20-bp nucleotides of full-length transcripts were extracted as 5'-LS and 3'-LS tags, respectively. **B.** The 5'-LS and 3'-LS tags were concatenated and cloned as separate 5'-LS and 3'-LS libraries for sequencing analysis. **C.** The 5'- and 3'-tags were concurrently mapped to the assembled genome sequences to define the TIS and PAS of transcripts and determine expression levels. Conventional SAGE is simply 3'-SAGE with restriction enzyme used during tagging of cDNA that cut only about 16 bp from poly A tail versus Long SAGE in which an enzyme is used that cut 18 bp from poly A tail. (white arrow) Bar = 100 μm.

been tagged at both ends. This simultaneous generation of paired, linked tags representing both ends of a transcript not only immediately increased the specificity of recognizing transcripts, but also facilitated the identification of all splice variants, a very precise mapping of the first and last exons and the genomic location of genes (Fig. 3). The technique also presents a very powerful tool for discovery of novel unusual transcripts. Applied to the analysis of ESCs and other stem cells, such a method should bring the transcriptome profiling of ESCs to a deeper level of complexity and completeness not achieved before.

## Massively parallel signature sequencing (MPSS)

MPSS is another sequence-based approach to a large scale transcriptome analysis.[29] Recently, Lynx Therapeutics Inc., the pioneer of the MPSS technology, improved the methodology to obtain 20 base pair sequences from each mRNA molecule while reducing the cost and amount of required RNA (Fig. 4) These improvements have made it possible for the individual laboratory to develop comprehensive databases for particular cell types. MPSS has been used to assess multiple cell types.[30–33] The data obtained by such methods is exemplified by the work of Rao and colleagues.[20,34] Massively parallel signature sequencing (MPSS) of approximately three million signature tags (signatures) identified close to eleven thousand unique transcripts, of which approximately 25% were uncharacterized or novel genes (Fig. 5B). Since on average most cells express about 10–12 000 transcripts, such an in-depth analysis is close to the theoretical level of detection of every transcript, subject, however, to the intrinsic limitations of such technologies.

MPSS confirmed the expression of previously identified ESC markers and furthermore, multiple genes not known to be expressed by ESCs could be identified. Rao and colleagues utilized the comprehensive nature of the analysis to map expressed genes to chromosomes but could not demonstrate the presence of major regulatory loci or hotspots. Their data suggested that this may be due a difference between undifferentiated and differentiated cells because in the latter, regions of high gene expression have been shown. Comparison of MPSS data with EST scan data using identical samples showed good concordance (Fig. 5A), and identified a large number of genes that changed rapidly as ESCs underwent transition from a pluripotent to a differentiated state. These included known and unknown ESC specific genes, as well as a large number of known genes that were altered as cells

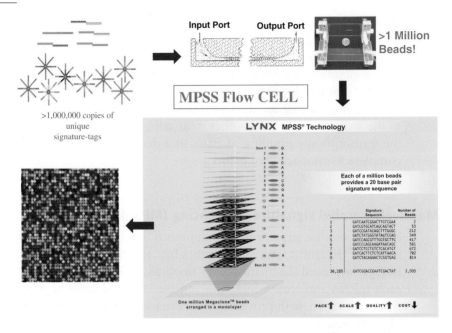

**Figure 4.** MPSS technology. MPSS, like SAGE, maps short sequences to the genome and measures the frequency of the transcript to assess the relative levels of expression. MPSS technology is based on parallel sequencing by loading beads into a flow cell such that sequences can be uniquely assigned to a particular bead and thus millions of sequences can be processed in parallel. The figure shows the flow cell (dime inserted for relative size) and the high power insert shows the loaded beads. Two Lynx MPSS flow cells are shown above with a US dime for scale. Each flow cell holds more than one million beads. The schematic shows how beads, enzymes, oligonucleotides and other reagents are delivered to the flow cell. The five micron beads flow readily into the seven micron space within the flow cell, but they back up against a dam on the right side which does not impede the flow of other reagents. Bar = 50 µm.

differentiate. MPSS analysis identified markers unique to hESCs, human embryoid bodies (hEBs), and signaling pathways that regulated differentiation. The data generated can be used to monitor the state of hESC isolated by different laboratories using independent methods and maintained under differing culture conditions. As with other methods, all of the data are available for download, including comparisons between microarray, SAGE and MPSS data sets.

Full MPSS analysis, while capable of providing a full in-depth analysis, is still a heavy undertaking for most laboratories. Lynx offers an alternative for sampling lower number of tags (350 000, and 500 000 range). The data, while still more extensive than that obtained by SAGE, are not as

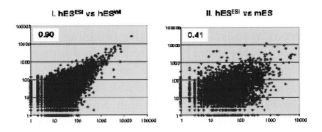

| | mES | | hES^ESI | | hES^Wi | |
|---|---|---|---|---|---|---|
| Abundance tpm | # of unique Signature | % | # of unique Signature | % | # of unique Signature | % |
| >10,000 | 13 | 0.09% | 10 | 0.05% | 9 | 0.04% |
| >5,000 | 57 | 0.41% | 27 | 0.13% | 30 | 0.13% |
| >1,000 | 168 | 1.22% | 163 | 0.81% | 160 | 0.68% |
| >500 | 307 | 2.22% | 361 | 1.80% | 303 | 1.29% |
| >100 | 1444 | 10.45% | 1947 | 9.72% | 1540 | 6.55% |
| >50 | 2759 | 19.96% | 3963 | 19.79% | 3347 | 14.24% |
| >10 | 8807 | 63.71% | 12038 | 60.11% | 13941 | 59.32% |
| Total signature count | 13824 | 100.00% | 20027 | 100.00% | 23500 | 100.00% |
| #of distinct unigene cluster | 6712 | | 9093 | | 9953 | |

**Figure 5.** MPSS scatter plot and data showing frequency of distribution of transcripts. **A**. Scatter plots for murine and human homologous genes comparing: i) 2 different hECS lines, H-9(hES^Wi) with HES-1(hES^ESI); ii) hES^ESI with E14 murine ES line (mES); iii) mES with murine EBs; iv) hES^Wi with hEB^Wi. All scatter plots were drawn after the removal of ribosomal proteins (with mitochondrial genes filtered out in the original lists). The corresponding correlation coefficients are shown in the panels. **B**. Table shows the distribution of genes with expression levels from >10,000/tpm to >10tpm in murine E14 ES cells (mES), day 4 murine embryoid bodies (mEB), human HES-2 (hES^ESI), pooled human ES cell lines H1, H7 H9 (hES^Wi) and human day 12 embryoid bodies derived from hES^Wi. Note that only about 20% of expressed genes in ES or EBs are expressed at a frequency of >50 tpm. Bar = 50 μm.

quantitative as the full MPSS in estimating relative levels of expression. Low level transcripts may also be missed. Zheng *et al.*[35] have compared lower resolution MPSS with higher resolution MPSS and confirmed that such comparisons are viable and that lower resolution MPSS, nevertheless, provides more than adequate data for the average laboratory. The data from analysis of the BGO2 line are available for download from the Lynx website.

## Illumina bead technology

More recently, a novel technology that combines the advantages of microarray and SAGE/MPSS has been described. In this case, hybridization is performed using sequences that have been attached to beads. This makes the process much cheaper and more reliable and allows hybridization stringencies to be set such that specificity is high. No results using hESCs are available for direct comparison, but readers are advised to refer to www.illumina.com for additional information on whether this technology will suit their purpose.

# Relative Strengths and Weaknesses of the Various Methods

The limitations of EST, SAGE and MPSS are the cost, sample processing time and limited genomic information available for some species. Perhaps the most important conclusion from examining the results on ESC analysis using a variety of methods is that no single methodology is optimal for all potential uses. Each method provides some important benefits and has its limitations (Table 1). Microarray is by far the most common method and provides a large database of existing information that can be used for comparison. Many of the problems associated with signal to noise ratio have been solved and array formats and density and data mining software continue to improve, while strategies to amplify, label and detect hybridization probes continue to evolve. However, microarray remains a comparative technique and hybridization, sensitivity, cross reactivity of similar/related transcripts and the variability inherent in competitive hybridization remain.

SAGE, double SAGE and EST scans continue to be good competing technologies which do not rely on competitive hybridization and offer the potential of providing an unbiased readout of known and unknown genes present in a give cell type. SAGE libraries exist and can be readily accessed. Tools to search these databases exist and new data sets are being added. More recent techniques allow longer reads and different restriction enzymes allow one to bypass digestion problems and better genome annotation reduce the number of unmapped tags. The cost of the analysis and the time taken, however, remain major issues that need to be resolved. Most SAGE analyses are limited to sequencing 50–100 000 transcripts and it often takes six months or longer to obtain and analyze all the sequences. Since a typical mammalian cell contains more than 300 000 mRNA molecules with many, including critical regulatory molecules, being expressed at only a few copies per cell, sequencing 100 000 transcripts limits sensitivity at the low end.

Massively parallel signature sequencing (MPSS™), a robust, high-throughput alternative,[29,33,36] reduces the cost of 20 base pair signature of the 3′-most DpnII sufficiently such that over a million transcripts per sample can be assessed. As a result, the method is highly sensitive with a high dynamic range, permitting the identification of high abundance transcripts as well as transcripts at single copy per cell. The time period required is also shortened to 1–2 months. However, disadvantages remain. The technique is proprietary, cannot be performed in one's own laboratory and, as with SAGE and EST scan, its utility is limited to genomes for which annotation is accurate and complete. MPSS and SAGE cannot distinguish between alternative splicing events and possible incomplete digestion during the sample preparation process. Furthermore, since signature lengths are relatively short, there are inherent ambiguities in mapping the tag to the correct genome site.[37] Sequencing errors can vitiate results and in the case of MPSS, transcripts that contain palindromic sequences (in particular double palindromes) are often undetected because of self-hybridization of single DNA strands on the bead (approximately 3% of all virtual signatures in human MGC database have double palindromes). In the GIS Analysis ditag-SAGE method,[28] the tagging of each end of a full length transcript increases many fold the efficiency and precision of mapping a transcript to the genome. Furthermore, spliced variants can be identified to an extent superior to that provided by the other competing short-tag techniques. Most intriguing, perhaps, is the possibility of capturing unusual novel transcripts arising from fusion of transcripts (trans-splicing ) and polysistronic transcription through the same locus (with intergenic splicing) (Wei *et al.*, personal communication).

## Other Large Scale Analytical Methods

A wealth of information on the properties of ESCs is available and many additional insights can be gleaned from the currently available information by comparing datasets, and examining individual genes and pathways by standard methods such as over-expression, down-regulation by knockouts or siRNA, generation of transgenics or creation of promoter reporter cell lines to perform expression screens. Such methods are in use and results will no doubt provide important information. In addition, it has become possible to begin studying aspects of cell biology that were not possible before and a few of these are discussed briefly below.

Epigenetic alterations to the DNA or to the histone packaging proteins are independent of gene sequences and over the past few years the importance of heritable epigenetic remodeling has been highlighted in regulating stem cell proliferation, cell fate determination and carcinogenesis.[38−42] Its role in regulating X inactivation and in imprinting[43,44] and in appropriate development after somatic nuclear transfer[45−48] is beginning to be better understood. It has been difficult, however, to study these events on a global scale. The ability to grow large number of cells and to differentiate them along specific pathways, coupled with the ability to perform such studies in a high throughput fashion, suggests that this will change in the near future. Global methylation studies can be performed using PCR and sequencing and methylation-specific microarray techniques have been described.

MicroRNA's are small noncoding RNA genes found in most eukaryotic genomes and are involved in the posttranscriptional regulation of gene expression. MicroRNAs appear to be processed by Dicer and double stranded RNAs appear to regulate gene expression via transcriptional, translational or protein degradation regulation.[49,50] Recent reports have described library-sequencing global strategies for identifying microRNAs and over 250 such untranslated RNAs have been identified.[51−53] These include computational analysis using sophisticated algorthims to recognize potential microRNA binding sites in targeted 3′-UTRs of mRNAs. MicroRNA chips are being tested.[54,55] Lynx therapeutics has developed sequencing protocols, analogous to MPSS, to enable quantitative data on microRNAs made by a particular cell to be obtained.

Structural and functional abnormalities in mitochondria lead to a variety of diseases and mitochondrial mutations are common in human cancers, in aging, as well as when cells are maintained in culture for prolonged periods. Mitochondrial DNA is also relevant to cloning because most animals inherit most or all of their mitochondria from the host oocyte. Assessing mitochondrial stability and accumulation of mutations is of particular importance for presidential ESC lines. Techniques to examine such mutation have been developed.[56] The recent development of a PCR-based approach for sequencing vertebrate mitochondrial genomes has attracted much attention as being more rapid and economical than traditional methods using cloned mtDNA and primer walking. A mitochondrial Custom Reseq microarray as an array-based sequencing platform for rapid and high-throughput analysis of mitochondrial DNA has been developed. The

MitoChip contains oligonucleotide probes synthesized using standard photolithography and solid-phase synthesis, and is able to sequence >29 kb of double-stranded DNA in a single assay. It is useful to note that many mutations arise in the D-loop and a simple PCR amplification and sequencing process would capture a large amount of information. No published data on baseline mitochondrial sequence and its change after culture of ESCs is currently available. Several laboratories have initiated such experiments, however, and we expect data on the long term viability of ESCs, in relation to mitochondrial stability, to be available soon.

Apart from the major chromosomal anomalies found in developmentally arrested embryos and fetuses, less detrimental rearrangements and/or mutations are likely to go unnoticed in most ESC karyotypes, as routine assessment includes only chromosomal banding analysis. More sensitive assays have been developed, including chromosome painting and comparative genome hybridization spectral genomics.

HLA typing and ST-R analysis have clearly indicated that the different ESC lines differ from each other. Further multiple studies have shown that variations in the expression of individual genes due to allelic variability can significantly alter the response of cells in culture or after transplantation.[57] This has led to the suggestion that allelic variability should be analyzed in a comprehensive fashion. Such an analysis has been difficult until recently. Illumina and Affymtrix offer competing technologies that may begin to allow better resolution of this question. Transcriptional mapping strategies have evolved to examine gene regulation as well. Some focused on in-silico computational strategy to retrieve putative genes with such binding sites. Another indirect approach has been to constitutively overexpress specific transcriptional factors and to compare transcripome profiles by microarray between cell states as a way to capture putative genes regulated by the factors. A direct and physiological approach is to perform chromatin immunoprecipation (ChIP) of factors cross-linked *in vivo* to DNA targets, followed by identification of the specific DNA binding sites.[58–60] Promoter chips from focused selection to varying degree of complete arrays are now being made. A hybridization of labeled ChIP DNA fragments to such chips may be useful for obtaining the first impression of genes targeted by the factor of interest. More useful are chips with a dense coverage of the genome for selected entire chromosomes to the total human genome, are now being made that will allow unbiased probing to locate all binding sites bound by transcriptional factors.[61] Multiple studies have shown that such ChIP-CHIP approach is feasible.

An analogous approach to identifying regulatory elements that are activated shortly after an initial stimulus is to perform labeling of early processed RNA and using it rather than total RNA to examine only genes induced after a specific stimulus. Such hybridizations, while requiring larger amounts of material, are feasible with cell lines and with ESCs and can provide a global overview of the network of changes to a specific stimulus. More importantly, they provide an element of temporal control allowing one to better place individual genes in a transcriptional network.

## Proteomic analysis

Most of our discussion has been on assessing the transcriptome. However, post-transcriptional and post-translational modifications play a crucial part in modifying genomic instructions and increasing the complexity of information that can be processed by a cell. The very complexity of the proteome has made it difficult to study, on a large scale, expression profile of cells. However, multiple breakthroughs have begun to make global protein analysis of cell populations possible. Pertinent to studies of stem cells, these include advances in sensitivity of mass spectrometry, development of variations in 2D gels and labeling techniques to identify key proteins that are altered under different conditions, and identification methods for isolating and sequencing of small quantities of proteins.[62] Proteomic analysis of ESCs has somewhat surprisingly not been reported as yet despite the fact that cell lines have been available for several years. The complex media in which these cells grow and the requirement for co-culture with feeders have made it difficult to obtain pure samples. However, improving methods for feeder-free and even defined conditions for growing hESC lines will allow such studies to be performed. Recent report on factors secreted by feeders[63] that may be important for ESC self-renewal suggests that proteomic approaches should be an integral part of the transcriptome and expression analysis of ESCs. The feasibility of this is supported by early reports of analysis of the complexity of conditioned medium from feeder cells supporting ESC growth,[64] and the potential identification of biomarkers expressed by hESCs.[65]

## Comparison Across Datasets

While a detailed analysis of results is beyond the scope of this chapter, readers are encouraged to examine the publicly available datasets and draw their own conclusions. We believe that several consensus conclusions can be drawn by comparing the datasets already available. We find that ESCs appear similar to other cells in synthesizing about 10 000–12 000 genes of an estimated 35 000 or 40 000 genes annotated in the Ref Seq database (Fig. 4).[20,34] Total RNA per million cells tends to be higher in ESCs compared with other cells (average of 5–10 µg/10 million cells), but is similar to levels seen in metabolically active cells. The distribution of transcript frequency suggests that most genes are transcribed at relatively low levels of less than 50 transcripts per cell (Fig. 4). Mitochondrial, ribosomal and housekeeping genes tend to be more abundant, while transcription factors, growth factors and other cell type specific molecules are expressed at much lower levels. This pattern of gene expression is similar to that seen in most other cell types. Comparison of data with other cell types suggests that on average the large majority of genes are shared or are in common, while approximately 20% are different (by greater than ten-fold) in any two samples. The average concordance rate between two ESC populations is 0.9 when two different lines grown under different conditions are compared,[34] and is much closer when identical samples are compared in two independent sequencing runs (0.99) (Fig. 4). In contrast, when identical samples are compared between two methods, SAGE and MPSS or EST and MPSS, then the concordance rates are much lower (0.7), suggesting that those that are common between two methods are likely to be important. However, the lack of high concordance when identical samples are analyzed by different methodologies suggests caution in assuming that the failure to detect expression by any one method negates the finding. The number of novel genes identified in most experiments are generally higher in ESCs than in other populations, as is the total number of genes, suggesting that ESCs maintain an open or globally de-repressed transcriptome and that this population is not yet well characterized, and will be a rich source of novel transcripts.

Examining ESC enriched (though not necessarily specific) genes suggests that several major pathways are active in ESCs. These include the LIF/gp130 (in rodents only); TGFß signaling pathway; the FGF signaling pathway; the cell cycle regulatory pathways, including myc and DNA repair; and anti-apoptotic pathways. Intriguingly, genes regulating timing of differentiation, anti-sense RNA, siRNA, specific methylases and chromatin remodeling enzymes appear to be present at high levels, suggesting that epigenetic remodeling is an important aspect of ESC biology. Examination of cell type specific genes suggests that several are expressed at low levels and components of many inactive pathways are present but their activity is suppressed by the expression of inhibitors, suggesting that repression plays an important role in early ESC self-renewal or differentiation. Mapping of all expressed genes does not show a chromosomal bias as has been suggested in other stem cell populations and suggests that no major hot spots of ESC expression exist. On the other hand, several cold spots (regions of low expression) can be identified, suggesting that as differentiation proceeds specific regulators are activated.[20]

## ES specific genes and genes of metabolic pathways

A good illustration of the usefulness of transcriptome profiling is the finding in ESCs of the expression of a set of genes (a few hundred based on estimates from combined data) that appear to be relatively unique to this population of cells. A small core set, often referred to as pluripotency-related or ESC related genes, that include Oct4, Nanog, Sox-2, REX-1, UTF, TERT as well as novel genes are clearly distinct from genes expressed in other "adult non-ES" stem cell populations. Only a small number of these genes have been studied and shown to be necessary for maintaining the pluripotency of ESCs. Oct 4 is an example of one of the most studied stem cell genes. In knockout zygotes, there is no development of ICM and development does not progress beyond the blastocyst.[66] Nanog was a gene uncovered by array profiling before its function was known. Selected for further investigation,[67] it was found to be a crucial pluripotency gene that works with Oct-4. This was a good illustration of the power of profiling to narrow down potential important genes to investigate. In an independent approach, Nanog was identified by expression cloning.[68] Similarly, Eras was identified by differential screening as a murine ES specific gene,[69] and further investigation showed that the gene contribute to the "tumor" like properties of ESCs (as discussed later, E-Ras is not expressed in hESCs). Genes such as Rex 1[70] and Fbox 15[71] have been studied, but their real function in stem cells remain elusive. This is true for the function of most of these "ES genes."

Other potential important genes uncovered by transcriptome profiling are transcriptional factors found to be either up or down regulated during transition of ESCs into differentiated states.[19] Presumably, these genes play relevant roles in controlling genes recruited by ESCs and a focused effort in studying these genes should yield useful information.

Examination of ESC-enriched genes, though not necessarily ES-specific, has also been very useful in revealing several major pathways that are active in ESCs. These include the LIF/gp130 (in murine cells only); TGFß signaling pathway; the cell cycle regulatory pathways, including myc and DNA repair; and anti-apoptotic pathways. A recent inter-species comparative analysis showed that, compared to mESCs, hESCs[34] express a much wider variation of FGF receptors. Such information provide a rational basis to search for other FGFs that may be useful in optimizing the culturing and propagation of hESCs for tissue engineering.

Intriguingly, genes regulating the timing of differentiation (e.g. Hox genes) and chromatin modification (e.g. Dnmt3L versus Dnmt3a) appear to have very specific patterns of expression. Together with experiments which showed that ESCs fused with other somatic cells produce heterokaryons with ESC phenotype,[72−74] there is good evidence that epigenetic remodeling may be an important aspect of ESC capability.[75]

Examination of cell type specific genes suggests that several are expressed at low levels and components of many inactive pathways are present but their activity is suppressed by the expression of inhibitors, suggesting that repression plays an important role in early ESC self-renewal or differentiation.

Future investigations aimed at identifying the network of genes regulated by these ES-genes, together with a search for the master regulatory genetic factors that orchestrate these genes, should yield very useful information that would lead to target genes and pathways that can be exploited to control precisely ESC growth and development.

## Species variation

An important finding that has become clearer from the availability of multiple data sets from both mESCs and hESCs is the species to species variability. While many key pathways are conserved, many differences have been highlighted as well. For example, Oct3/4 homologues may not exist in chicken embryos,[76] while LIF signaling which is critical for ESC self-renewal in rodents does not appear to be critical or even required for hESCs.[21,34] No paralogs of E-Hox have been identified in humans and Eras appears to be a pseudogene in humans.[18] The low overall concordance rate between human and rodent ES cells (in one comparison, it was around 40%) relative to that seen in human-to-human cell comparisons (90% between human ES cell samples) provides additional support for this hypothesis.[34] A shortlist of differences is summarized in Table 3. As can be seen, this include genes that have become redundant during evolution, factors that have been recruited to different functions and gene expression patterns that have been flipped. However, the number of differences reported from limited comparisons was quite surprising. It suggests to us that additional mechanisms driving change, perhaps evolutionary pressure for speciation may underlie the larger than expected difference observed.

Another important finding that has become obvious from such a data set examination is the lack of any common stemness genes between ESCs and other somatic stem cells. Comparing data sets[77–79] with expression in ESCs does not identify a common subset of genes, suggesting that ESCs are quite different from other somatic stem cells.

Overall, the initial efforts at profiling ESCs from rodents and human have yielded useful insights, allowing one to identify key regulatory pathways and novel genes that are likely to play a role in regulating ESC self-renewal. The data sets developed are an important resource and can be mined with readily available tools. Efforts are underway to provide all of this data in a readily accessible format that would allow even the uninitiated to be able to examine the pattern of expression of their favorite gene.

## Conclusions

Overall, large scale genomic analysis has provided unique insight into the biology of this population of pluripotent cells. The wealth of genomic data, and the multiplicity of cell lines available, have enabled researchers to identify critical conserved pathways regulating self-renewal and identify markers that tightly correlate with the ESC state. Comparison across species has suggested additional pathways likely to be important in long-term self-renewal of ESCs, and meta-analysis of existing data sets has provided additional unique insight. Newer technologies have provided sophisticated tools to probe into aspects of cell biology that were difficult to study previously. Combining these technologies and pooling information generated provides a synergistic increase in the value of the information available and provides a platform for future breakthroughs.

## Acknowledgements

This work was supported by NIH, NIA, and A-Star Singapore. We thank all members of our laboratories for constant stimulating discussions. MSR acknowledges the contributions of Dr. S. Rao that made undertaking this project possible.

# REFERENCES

1. Bianchi DW, *et al.* (1993) Origin of extraembryonic mesoderm in experimental animals: Relevance to chorionic mosaicism in humans. *Am J Med Genet* **46**(5):542–550.
2. Luckett WP. (1978) Origin and differentiation of the yolk sac and extraembryonic mesoderm in presomite human and rhesus monkey embryos. *Am J Anat* **152**(1):59–97.
3. Buehr M, Smith A. (2003) Genesis of embryonic stem cells. *Philos Trans R Soc Lond B Biol Sci* **358**(1436):1397–402; discussion 1402.
4. Carpenter MK, Rosler E, Rao MS. (2003) Characterization and differentiation of human embryonic stem cells. *Cloning Stem Cells* **5**(1):79–88.
5. Burdon T, Smith A, Savatier P. (2002) Signalling, cell cycle and pluripotency in embryonic stem cells. *Trends Cell Biol* **12**(9):432–438.
6. Loebel DA, *et al.* (2003) Lineage choice and differentiation in mouse embryos and embryonic stem cells. *Dev Biol* **264**(1):1–14.
7. Tiedemann H, *et al.* (2001) Pluripotent cells (stem cells) and their determination and differentiation in early vertebrate embryogenesis. *Develop Growth Differ* **43**:469–502.
8. Papaioannou VE, Waters BK, Rossant J. (1984) Interactions between diploid embryonal carcinoma cells and early embryonic cells. *Cell Differ* **15**(2–4): 175–179.
9. Rossant J, Papaioannou VE. (1984) The relationship between embryonic, embryonal carcinoma and embryo-derived stem cells. *Cell Differ* **15**(2-4):155–161.
10. Thomson JA, Odorico JS. (2000) Human embryonic stem cell and embryonic germ cell lines. *Trends Biotechnol* **18**(2):53–57.
11. Shamblott MJ, *et al.* (2001) Human embryonic germ cell derivatives express a broad range of developmentally distinct markers and proliferate extensively *in vitro*. *Proc Natl Acad Sci USA* **98**(1):113–118.
12. Onyango P. (2002) Genomics and cancer. *Curr Opin Oncol* **14**(1):79–85.
13. Carter MG, *et al.* (2003) *In situ*-synthesized novel microarray optimized for mouse stem cell and early developmental expression profiling. *Genome Res* **13**(5):1011–1021.
14. Ko MS. (2004) Embryogenomics of pre-implantation mammalian development: Current status. *Reprod Fertil Dev* **16**(2):79–85.
15. Tanaka TS, *et al.* (2002) Gene expression profiling of embryo-derived stem cells reveals candidate genes associated with pluripotency and lineage specificity. *Genome Res* **12**(12):1921–1928.
16. Tanaka TS, *et al.* (2000) Genome-wide expression profiling of mid-gestation placenta and embryo using a 15,000 mouse developmental cDNA microarray. *Proc Natl Acad Sci USA* **97**(16):9127–9132.

17. Abeyta MJ, *et al.* (2004) Unique gene expression signatures of independently-derived human embryonic stem cell lines. *Hum Mol Genet* **13**(6):601–608.

18. Bhattacharya B, *et al.* (2004) Gene expression in human embryonic stem cell lines: Unique molecular signature. *Blood* **103**(8):2956–2964.

19. Brandenberger R, *et al.* (2004) Transcriptome characterization elucidates signaling networks that control human ES cell growth and differentiation. *Nat Biotechnol* **22**(6):707–716.

20. Brandenberger R, *et al.* (2004) MPSS profiling of human embryonic stem cells. *BMC Dev Biol* **4**(1):10.

21. Ginis I, *et al.* (2004) Differences between human and mouse embryonic stem cells. *Dev Biol* **269**(2):360–380.

22. Sperger JM, *et al.* (2003) Gene expression patterns in human embryonic stem cells and human pluripotent germ cell tumors. *Proc Natl Acad Sci USA* **100**(23):13350–13355.

23. Sato N, *et al.* (2003) Molecular signature of human embryonic stem cells and its comparison with the mouse. *Dev Biol* **260**(2):404–413.

24. Richards M, *et al.* (2004) The transcriptome profile of human embryonic stem cells as defined by SAGE. *Stem Cells* **22**(1):51–64.

25. Velculescu VE, Vogelstein B, Kinzler KW. (2000) Analysing uncharted transcriptomes with SAGE. *Trends Genet* **16**(10):423–425.

26. Velculescu VE, *et al.* (1995) Serial analysis of gene expression. *Science* **270**(5235):484–487.

27. Anisimov SV, *et al.* (2002) SAGE identification of gene transcripts with profiles unique to pluripotent mouse R1 embryonic stem cells. *Genomics* **79**(2): 169–176.

28. Wei CL, *et al.* (2004) 5′-Long serial analysis of gene expression (LongSAGE) and 3′ LongSAGE for transcriptome characterization and genome annotation. *Proc Natl Acad Sci USA* **101**(32):11701–11706.

29. Brenner S, *et al.* (2000) Gene expression analysis by massively parallel signature sequencing (MPSS) on microbead arrays. *Nat Biotechnol* **18**(6):630–634.

30. Jongeneel CV, *et al.* (2003) Comprehensive sampling of gene expression in human cell lines with massively parallel signature sequencing. *Proc Natl Acad Sci USA* **100**(8):4702–4705.

31. Meyers BC, *et al.* (2004) The use of MPSS for whole-genome transcriptional analysis in Arabidopsis. *Genome Res* **14**(8):1641–1653.

32. Meyers BC, *et al.* (2004) Analysis of the transcriptional complexity of Arabidopsis thaliana by massively parallel signature sequencing. *Nat Biotechnol* **22**(8):1006–1011.

33. Reinartz J, *et al.* (2002) Massively parallel signature sequencing (MPSS) as a tool for in-depth quantitative gene expression profiling in all organisms. *Brief Funct Genomic Proteomic* **1**(1):95–104.

34. Wei C, *et al.* (2004) Transcriptome profiling of murine and human ES cells identifies divergent paths required to maintain the stem cell state. *Stem Cells.*

35. Zeng X, *et al.* (2004) Properties of pluripotent human embryonic stem cells BG01 and BG02. *Stem Cells* **22**(3):292–312.

36. Brenner S, *et al.* (2000) *In vitro* cloning of complex mixtures of DNA on microbeads: Physical separation of differentially expressed cDNAs. *Proc Natl Acad Sci USA* **97**(4):1665–1670.

37. Saha S, *et al.* (2002) Using the transcriptome to annotate the genome. *Nat Biotechnol* **20**(5):508–512.

38. Ohgane J, *et al.* (2004) The Sall3 locus is an epigenetic hotspot of aberrant DNA methylation associated with placentomegaly of cloned mice. *Genes Cells* **9**(3):253–260.

39. Beaujean N, *et al.* (2004) Effect of limited DNA methylation reprogramming in the normal sheep embryo on somatic cell nuclear transfer. *Biol Reprod.*

40. Vignon X, Zhou Q, Renard JP. (2002) Chromatin as a regulative architecture of the early developmental functions of mammalian embryos after fertilization or nuclear transfer. *Cloning Stem Cells* **4**(4):363–377.

41. Meehan RR. (2003) DNA methylation in animal development. *Semin Cell Dev Biol* **14**(1):53–65.

42. Huntriss J, *et al.* (2004) Expression of mRNAs for DNA methyltransferases and methyl-CpG-binding proteins in the human female germ line, preimplantation embryos, and embryonic stem cells. *Mol Reprod Dev* **67**(3):323–336.

43. Monk M. (2002) Mammalian embryonic development—insights from studies on the X chromosome. *Cytogenet Genome Res* **99**(1–4):200–209.

44. Hemberger M. (2002) The role of the X chromosome in mammalian extra embryonic development. *Cytogenet Genome Res* **99**(1–4):210–217.

45. Wutz A, Jaenisch R. (2000) A shift from reversible to irreversible X inactivation is triggered during ES cell differentiation. *Mol Cell* **5**(4):695–705.

46. Tucker M, *et al.* (1996) Preliminary experience with human oocyte cryopreservation using 1,2-propanediol and sucrose. *Hum Reprod* **11**(7):1513–1515.

47. Rideout WM. 3rd, Eggan K, Jaenisch R. (2001) Nuclear cloning and epigenetic reprogramming of the genome. *Science* **293**(5532):1093–1098.

48. Bortvin A, *et al.* (2003) Incomplete reactivation of Oct4-related genes in mouse embryos cloned from somatic nuclei. *Development* **130**(8):1673–1680.

49. Bartel DP. (2004) MicroRNAs: Genomics, biogenesis, mechanism, and function. *Cell* **116**(2):281–297.

50. Szymanski M, Barciszewski J. (2003) Regulation by RNA. *Int Rev Cytol* **231**:197–258.

51. Rajewsky N, Socci ND. (2004) Computational identification of microRNA targets. *Dev Biol* **267**(2):529–535.

52. Lewis BP, *et al.* (2003) Prediction of mammalian microRNA targets. *Cell* **115**(7):787–798.

53. Houbaviy HB, Murray MF, Sharp PA. (2003) Embryonic stem cell-specific microRNAs. *Dev Cell* **5**(2):351–358.
54. Krichevsky AM, *et al.* (2003) A microRNA array reveals extensive regulation of microRNAs during brain development. *Rna* **9**(10):1274–1281.
55. Calin GA, *et al.* (2004) MicroRNA profiling reveals distinct signatures in B cell chronic lymphocytic leukemias. *Proc Natl Acad Sci USA* **101**(32):11755–11760.
56. Du W, *et al.* (2003) Functionalized self-assembled monolayer on gold for detection of human mitochondrial tRNA gene mutations. *Anal Biochem* **322**(1):14–25.
57. Ginis I, Rao MS. (2003) Toward cell replacement therapy: Promises and caveats. *Exp Neurol* **184**(1):61–77.
58. Ren B, Dynlacht BD. (2004) Use of chromatin immunoprecipitation assays in genome-wide location analysis of mammalian transcription factors. *Methods Enzymol* **376**:304–315.
59. Weinmann AS. (2004) Novel ChIP-based strategies to uncover transcription factor target genes in the immune system. *Nat Rev Immunol* **4**(5):381–386.
60. Buck MJ, Lieb JD. (2004) ChIP-chip: Considerations for the design, analysis, and application of genome-wide chromatin immunoprecipitation experiments. *Genomics* **83**(3):349–360.
61. Cawley S, *et al.* (2004) Unbiased mapping of transcription factor binding sites along human chromosomes 21 and 22 points to widespread regulation of noncoding RNAs. *Cell* **116**(4):499–509.
62. Unwin RD, *et al.* (2003) The potential for proteomic definition of stem cell populations. *Exp Hematol* **31**(12):1147–1159.
63. Humphrey RK, *et al.* (2004) Maintenance of pluripotency in human embryonic stem cells is STAT3 independent. *Stem Cells* **22**(4):522–530.
64. Lim JW, Bodnar A. (2002) Proteome analysis of conditioned medium from mouse embryonic fibroblast feeder layers which support the growth of human embryonic stem cells. *Proteomics* **2**(9):1187–1203.
65. Hayman MW, Przyborski SA. (2004) Proteomic identification of biomarkers expressed by human pluripotent stem cells. *Biochem Biophys Res Commun* **316**(3):918–923.
66. Nichols J, *et al.* (1998) Formation of pluripotent stem cells in the mammalian embryo depends on the POU transcription factor Oct4. *Cell* **95**(3):379–391.
67. Mitsui K, *et al.* (2003) The homeoprotein Nanog is required for maintenance of pluripotency in mouse epiblast and ES cells. *Cell* **113**(5):631–642.
68. Chambers I, *et al.* (2003) Functional expression cloning of nanog, a pluripotency sustaining factor in embryonic stem cells. *Cell* **113**(5):643–655.
69. Takahashi K, Mitsui K, Yamanaka S. (2003) Role of ERas in promoting tumour-like properties in mouse embryonic stem cells. *Nature* **423**(6939):541–545.

70. Ben-Shushan E, *et al.* (1998) Rex-1, a gene encoding a transcription factor expressed in the early embryo, is regulated via Oct-3/4 and Oct-6 binding to an octamer site and a novel protein, Rox-1, binding to an adjacent site. *Mol Cell Biol* **18**(4):1866–1878.
71. Tokuzawa Y, *et al.* (2003) Fbx15 is a novel target of Oct3/4 but is dispensable for embryonic stem cell self-renewal and mouse development. *Mol Cell Biol* **23**(8):2699–2708.
72. Ying QL, *et al.* (2002) Changing potency by spontaneous fusion. *Nature* **416**(6880):545–548.
73. Tada M, *et al.* (2001) Nuclear reprogramming of somatic cells by *in vitro* hybridization with ES cells. *Curr Biol* **11**(19):1553–1558.
74. Pells S, *et al.* (2002) Multipotentiality of neuronal cells after spontaneous fusion with embryonic stem cells and nuclear reprogramming *in vitro*. *Cloning Stem Cells* **4**(4):331–338.
75. Surani A, Smith A. (2003) Differentiation and gene regulation programming, reprogramming and regeneration. *Curr Opin Genet Dev* **13**(5):445–447.
76. Soodeen-Karamath S, Gibbins AM. (2001) Apparent absence of Oct 3/4 from the chicken genome. *Mol Reprod Dev* **58**(2):137–148.
77. Fortunel NO, *et al.* (2003) Comment on 'Stemness': Transcriptional profiling of embryonic and adult stem cells and a stem cell molecular signature. *Science* **302**(5644):393; author reply, 393.
78. Ivanova NB, *et al.* (2002) A stem cell molecular signature. *Science* **298**(5593):601–604.
79. Ramalho-Santos M, *et al.* (2002) "Stemness": Transcriptional profiling of embryonic and adult stem cells. *Science* **298**(5593):597–600.

## 6

# Culture, Subcloning, Spontaneous and Controlled Differentiation of Human Embryonic Stem Cells

Michal Amit, Sharon Gerecht-Nir and Joseph Itskovitz-Eldor*

## Introduction

Embryonic stem cells (ESCs) constitute a unique type of cell derived from the inner cell mass (ICM) of the mammalian blastocyst. For a short time window, ICM cells possess the capacity to differentiate into every cell type of the adult body. This ability, known as "pluripotency," distinguishes embryonic stem cells from adult stem cells that can only differentiate into cell types of specific lineage. As mammalian embryonic development continues, the post-gastrulation embryo consists of specialized precursors of all three germ layers and the pluripotency of each individual cell in the embryo is lost. The length of this time window, during which the embryo consists of pluripotent cells, is unknown (Fig. 1). Nonetheless, if the ESCs are isolated from the embryo and cultured *in vitro* under suitable conditions, they can not only proliferate indefinitely, but also maintain their pluripotency throughout the culture period. The pluripotency of ESCs has been proven using different approaches: 1) undifferentiated mouse ESCs consistently contribute to chimeras and particularly to the germ cell line;[1] 2) some murine ESC (mESC) lines can form entire viable fetuses[2]; 3) subcutaneous injection of mESCs into severe combined immunodeficient (SCID) mice induces the formation of teratomas which may include cells of all three germ layers[3]; and 4) *in vitro* aggregations of ESCs result in the formation of embryoid bodies (EBs) with regional differentiation into embryonically distinct cell types.[4] As will be

*Correspondence: Department of Obstetrics and Gynecology, Rambam Medical Center, P.O.B. 9602, Haifa 31096, Israel. Tel.: +972-4-854-2536, Fax: +972-4-854-2503, e-mail: Itskovitz@rambam.health.gov.il

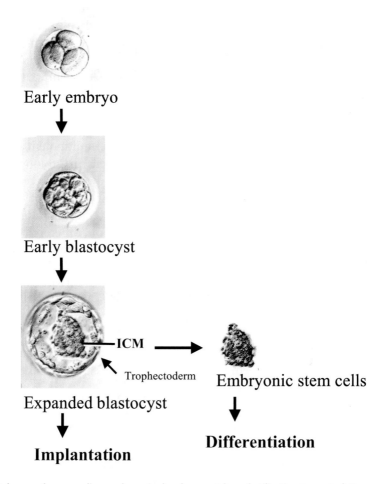

**Figure 1.** Scheme of mammalian embryonic development from fertilization to gastrulation.

discussed in detail later in this chapter, ESCs, mainly from mouse origin, have been shown to differentiate *in vitro* into hematopoietic stem cell-like cells[5,6]; neural precursors[7,8]; cardiomyocytes[9]; endothelial cells; and insulin-secreting cells.[10,11] This exceptional differentiation potential underline the cells' superior ability to serve as a model for research on early human development, lineage commitment, differentiation processes and cell-based therapy.

Compared to adult stem cells and somatic cells, ESCs exhibit many other unique characteristics in addition to their pluripotency. They maintain normal diploid karyotypes even after prolonged culture; unlike somatic cells, they remain in the S phase of the cell cycle for the majority of their lifespan and do not show X chromosome inactivation; and they express unique markers, such as Oct-4, a transcription factor known to be involved in the process of ESC self-maintenance, and *Nanog*, whose role in self-maintenance is still unclear.[12,13] ESC features are summarized in Table 1.

The first multipotent cell line was derived in the 1970s from a murine teratocarcinoma.[14] Under suitable conditions, multipotent cell lines can be maintained *in vitro* as undifferentiated cells in prolonged culture.[15] Embryonal carcinoma (EC) cells have been shown to differentiate *in vitro* into a variety of cell types, including muscle and nerve cells, to form EBs while cultured in suspension,[16] and to form teratocarcinomas following their injection into SCID mice.[17] The derivation, differentiation and culture methods developed for EC cells laid the groundwork needed for the isolation of ESCs.

The first ESC lines were derived from mouse blastocyst in 1981.[18,19] Since then, ESC lines and ESC-like lines have been derived from other rodents[20–22]; from domestic animal species[23–25]; and from three non-human primates.[26–28] However, only murine ESC lines have been reported to demonstrate the complete set of ES cell characteristics as summarized in Table 1, rendering them the most potent research model amongst existing ESC lines.

**Table 1.** Main Features of Embryonic Stem Cells

Derived from the ICM of the blastocyst

Capable of prolonged undifferentiated proliferation in culture

Exhibit and maintain normal diploid karyotype

Pluripotent

Able to integrate into all fetal tissues during embryonic development following injection into the blastocyst, including germ layer chimerism

Clonogenic, i.e. each single ES cell possesses all other features

Express high levels of Oct-4 — a transcription factor known to be involved in ES cell self-maintenance

Can be induced to differentiate after continuous culture in an undifferentiated state

Remain in the S phase of the cell cycle for the majority of their lifespan

Do not show X chromosome inactivation

Due to the contribution of mouse ESCs to the research on development, great efforts have been invested in the derivation of human ESCs (hESCs). The first step towards this goal was achieved by Bongso and colleagues who reported that ICM cells isolated from human blastocysts can be cultured with inactivated human embryonic fibroblasts (hEFs) for two passages while expressing alkaline phosphate activity, maintaining normal karyotype and demonstrating ESC-like morphology.[29] This initial study paved the way for the first isolation of hESC lines by Thomson and colleagues in 1998.[30] Accumulating knowledge shows that hESCs meet most of the criteria for ESCs listed in Table 1. hESCs were derived from embryos at the blastocyst stage[30,31] and were shown to be capable of continuous culture as undifferentiated cells while expressing high levels of Oct-4[30–32]; their morphology resembled that of mESCs, and their pluripotency has been demonstrated both *in vitro* by the formation of EBs[33] and *in vivo* in teratomas.[30,31] They have also been shown to be clonogenic, with resultant single-cell clones demonstrating all ESC features.[32] For obvious ethical reasons, the ability of hESCs to integrate into fetal tissues during embryonic development is limited. hESCs demonstrated normal diploid karyotypes even after prolonged culture.[32] Cases of karyotypic instability are scarce,[32,34,35] suggesting that they represent random changes which often occur in cell culture.

The specific stage of the cell cycle in which hESCs spend most of their time has not been reported. Recently it has been reported that hESCs, like mouse ESCs, do not exhibit X inactivation. Also, while maintained at the undifferentiated stage, both X chromosomes are active and, upon differentiation, one chromosome undergoes inactivation.[36]

To date, there are many established and well-characterized hESC lines in several laboratories worldwide.[37–39] The availability of hESC lines provides a unique new tool with widespread potential for both research purposes and clinical application.

## Derivation of hESC Lines

For the derivation of hESC lines, embryos from various sources, including surplus,[30,31] or low-quality[40,41] embryos from *in vitro* fertilization (IVF) programs and embryos produced for research purposes[42] were used. A recent report put forth the application of the human nuclear transfer technique to rabbit oocytes as an additional source for deriving hES cell lines for therapeutic application.[43] As in the production of embryos for research purposes, this method raises ethical questions with respect to therapeutic cloning that should be carefully considered. An additional potential source for the derivation of hESC lines is embryos that harbor genetic diseases from pre-implantation genetic diagnosis (PGD) programs. Models established by the use of hESC lines carrying specific genetic defects may be highly effective in developing gene therapy strategies or drugs designed to treat these diseases. Two lines, one harboring the Van Waardenburg disease (deletion at the PAX3 gene), and the other harboring Myotonic Dystropyhy, a repeated expansion mutation in the 3′-untranslated region (3′-UTR) of the myotonic dystrophy protein kinase gene ($(CTG)_n$ repeats), were established (Amit and Itskovitz-Eldor, unpublished data). These lines may also be utilized for research aimed at gaining a better understanding of the mechanisms underlying these diseases.

The increasing number of available hESC lines indicates that the derivation of these lines is a reproducible procedure, where success rates vary from less than 10%[43] to over 50%.[44]

hESC lines were derived using the methods developed in the 1970s for EC cell lines and in the 1980s for mouse ESC line derivation. Three principal methods can be used to isolate ICM cells from the blastocyst: 1) immunosurgical isolation; 2) mechanical isolation; 3) using a whole intact embryo.

## Immunosurgical isolation

Solter and Knowles designed immunosurgery as a method for deriving EC cell lines,[45] and later for the derivation of ESC lines from mouse blastocysts.[18,19] The aim of immunosurgery is to selectively isolate ICM cells from the blastocyst. This goal is achieved by the following steps:

1. The zona pellucida (ZP) is dissolved by either tyrode solution or pronase. This stage is optional as the ZP may be left intact and will not interfere with the remaining steps.
2. The exposed embryo is incubated for approximately 20 min in anti-human whole serum antibodies which attach to any human cell. Penetration of the antibodies into the blastocyst is prevented due to cell–cell connections within the outer layer of the trophoblast, thus leaving the ICM cells unharmed.
3. The embryo is incubated for up to 20 min with guinea pig complement-containing medium which lyses all antibody-marked cells. Since the ZP allows the penetration of both antibodies and guinea pig complement, it may be removed alternatively after the lysis by complement proteins.

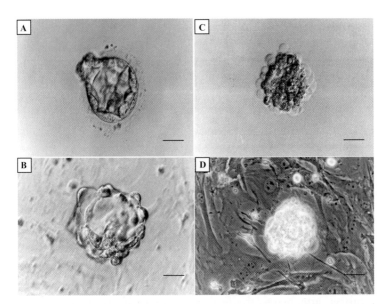

**Figure 2.**  Immunosurgery of a human blastocyst for the derivation of human ES cell line. (**A**) Donated human embryo produced by *in vitro* fertilization at the blastocyst stage; (**B**) Human blastocyst after zona pellucida removal by Tyrode's solution, during exposure to rabbit anti-human whole antiserum; (**C**) Embryo after exposure to guinea pig complement; (**D**) Intact inner cell mass immediately after immunosurgery on mitotically inactivated mouse embryonic fibroblast feeder layer. Bar = 50 μm. (From Amit and Itskovitz-Eldor, *J Anat* 2002.)

4. The intact ICM is further rinsed and cultured with mitotically inactivated MEFs. These steps are illustrated in Fig. 2.

## Mechanical isolation

Isolation of ICM cells can be achieved by the selective and mechanical removal of the trophoectoderm layer under a stereoscope. After the ZP is removed, the trophoblast layer is gently separated using 27G needles or pulled Pasture pipettes. As in the immunosurgery method, the last stage consists of proliferating the intact ICM cells with MEFs.

## Plating intact embryos in whole

Contrary to mechanical isolation and immunosurgery, this method is based on the removal of the ICM cells after one passage of culture together with the trophoblast layer.[41] Lines can be derived simply by plating a zona-free embryo in whole with mitotically inactivated MEFs. The exposed

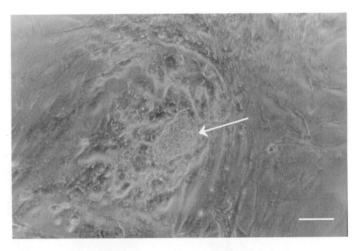

**Figure 3.** Whole embryo culture on MEFs showing ICM cell outgrowth. [White arrow: Bar = 100 μm.]

embryo attaches to the feeder layer, which in return permits the continuous growth of the ICM with the surrounding trophoblasts as a monolayer. When the ICM reaches sufficient size and is noticable, it is removed mechanically and further propagated (Fig. 3). This method bears the risk of ICM differentiation.

## Maintenance of hESCs

Following their isolation from human embryos, ESCs are further proliferated and maintained at the undifferentiated stage using methods similar to those applied earlier for mouse ESCs. Traditional culture conditions include MEFs as a feeder layer and medium supplemented with fetal bovine or calf serum (FBS).[30,31] The feeder layer plays a dual role of supporting ESC proliferation and preventing their spontaneous differentiation. When cultured in these conditions, ESCs form flat colonies with typical spaces between the cells, high nucleus-to-cytoplasm ratio, and with the presence of at least two nucleoli (Fig. 4A).

The future use of hESCs for therapeutic applications will require well-defined and animal-free conditions for their derivation and culture in order to prevent any exposure of the cells to animal photogenes. This translates into three requirements: 1) hESCs must be cultured with medium supplemented with serum replacement; 2) the culture should be prepared with human feeders and human serum; and 3) the proliferation of hESCs must be administered without feeders. A few steps toward meeting these

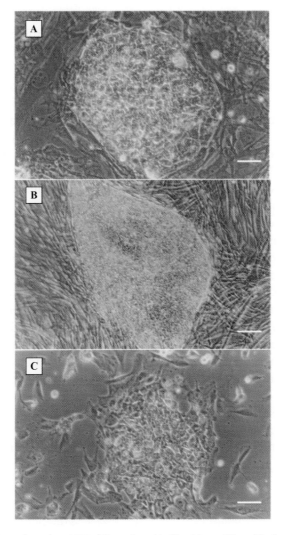

**Figure 4.**  hESC colonies cultured on MEFs (**A**), on foreskin fibroblasts (**B**) and in feeder-free conditions (**C**). [Bar = 50 μm.]

requirements have already been achieved. hESCs can be cultured with serum replacement and basic fibroblast growth factor (bFGF) as supplement, thereby providing a defined medium.[32] Several human feeder layers were found to be successful in supporting hESC maintenance and proliferation: human fetal-derived fibroblasts or Fallopian tube epithelial feeder layers[46]; foreskin fibroblasts[47,48]; and bone marrow cells.[49] An example of the morphology of hESC colony when cultured with foreskin fibroblasts is seen in Fig. 4B. Under the above culture conditions, hESCs were shown to maintain the major ESC characteristics after prolonged growth. In some cases, the conditions were found to be supportive not only of hESC maintenance but also of new hESC line derivation.[46,48]

Although these culture systems promote the goal of animal-free conditions, they cannot be regarded as well-defined. The ideal culture method would be a combination of an animal-free matrix and both serum and animal-free medium. In 2001, Xu and colleagues reported the first steps towards this solution — development of a feeder-free culture system for hESCs.[50] Their method used Matrigel, laminin or fibronectin matrix combined with 100% MEF conditioned medium, supplemented with serum replacement. Using this method, hESCs can be cultured for more than a year in continuous culture while maintaining ESC features.[51,52] The major shortcoming of this method is the use of animal based matrices and MEF-conditioned medium which may expose hESCs to mouse pathogens and is not regarded as well-defined.

Amit and colleagues eliminated the use of conditioned-medium for a feeder-free culture system using medium supplemented with serum replacement, transforming growth factor beta-1 ($TGF_{\beta 1}$) and bFGF.[53] In these culture conditions, hESCs were cultured continuously for more than 50 passages with human fibronectin, with ESC features being maintained all the while.[53] No conditioned medium was used. Examples of an undifferentiated hESC colony grown in this system are shown in Fig. 4C. The method comprises a well-defined culture system which facilitates the future use of these cells in research and provides a safer alternative for future clinical applications. However, the serum replacement used in this study contains animal products (Albumax, for example), and further improvements still need to be developed.

## Subcloning of hESCs

As mentioned earlier, hESC lines are derived from the ICM, a clump of pluripotent cells which may not represent a homogenous cell population. Therefore, there is the possibility that the pluripotency of the hESC lines reflects a collection of several distinct committed multipotential cell types. In order to eliminate this possibility, parental lines must be single-cell cloned, and the single-cell clone pluripotency has to be demonstrated. Following the failed attempt to clone hESCs in the same culture medium in which they were derived, several culture media were tested to clone the first parental hESC lines: medium supplemented with either FBS or serum replacement, either with or without human recombinant bFGF.[32] The results revealed that the highest cloning efficiency is obtained when a medium supplemented with 20% serum replacement and bFGF is used. Several parental hESC lines were cloned in our laboratory using these conditions, and the resultant clones demonstrated hESC features, including pluripotency.[32] Thus the pluripotency of human embryonic single cells has been established.

Single-cell clones may have further advantages over parental lines. Firstly, they are easier to grow and manipulate in comparison to their parental lines (Amit, unpublished data). Secondly, research models based on gene knockout or targeted recombination require homogeneous cell populations starting with a single cell harboring the desired genotype. The main disadvantage of this strategy is the relatively low cloning efficiency (1%), which, when coupled with reduced successful recombination rates,[54] results in a model which is rather difficult to obtain. This is partly overcome by the prolonged culture abilities of the resultant clone, which enable extended periods of research.

Single-cell cloning may also be required in cases of karyotype abnormalities. Any future application of hESCs for therapeutic purposes will depend on their karyotypic stability. Random karyotypic instability incidents were reported for hESCs.[32,34,35] In the case of 17q trisomy and 12 chromosome trisomy, the abnormality provides a selective advantage for the carrying cell population, resulting in a tendency to take over the culture, until 100% of the culture exhibits these trisomy.[35] If the problem is identified, single-cell cloning of clones harboring normal karyotypes may prevent the loss of cell lines. Based on the growing data on hESCs and the experience with mESCs, the periodic cloning of hESCs for the purpose of maintaining a homogenous euploid population may be needed, but will be infrequent.

## Spontaneous Differentiation of hESCs

### Teratomas

Differentiation of hESCs occurs once they are removed from culture conditions which support their pluripotency and self-renewal. *In vivo* spontaneous differentiation can be achieved by the injection of undifferentiated hESCs into immuno-compromised mice.[30] This results in the formation of a teratoma, a benign tumor containing complex structures and composed of differentiated cell types representing derivatives of the three major embryonic lineages. Figure 5 shows various tissue structures differentiated

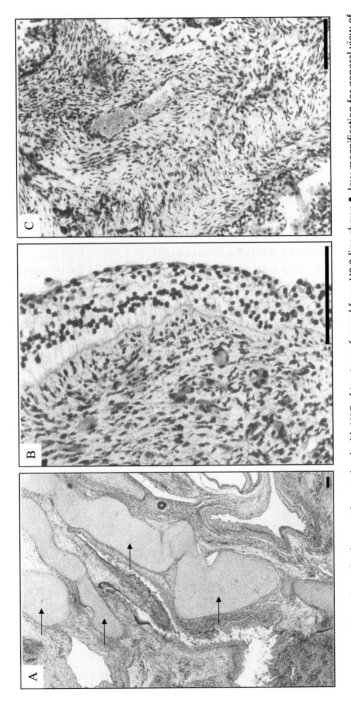

**Figure 5.** Teratoma formation. Histology sections stained with H&E of teratoma formed from H9.2 line show: **A.** low magnification for general view of cartilage formation (*arrows*) with focus on **B.** stratified epithelium and **C.** connective tissue. Bar = 100 μm.

from hESCs within a teratoma. In one study, it was demonstrated that, using a sensitive recovery method, human keratinocytes can be isolated from trypsinized cells of such a teratoma and can be grown in cell culture.[55] A more recent study suggested that a teratoma generated from hESCs may be utilized as an *in vivo* human microenviroment model for the study of tumor invasion, angiogenesis and metastasis.[56] It should be noted that the capability of undifferentiated hESCs to form teratomas is a risk factor to consider when applying hESC-derivatives to cellular therapy.

## Embryoid bodies (EBs)

### *Formation and growth*

The appeal of ESCs is their ability to differentiate *in vitro* into all three primary germ layer derivatives. After removal from their undifferentiated-supporting culture conditions, the differentiation of hESCs *in vitro* can be activated in two main ways: 1) two-dimensional (2-D) culturing on a differentiation-inducing layer (matrix or culture), or 2) formation of embryo-like aggregates in suspension, termed embryoid bodies (EBs).[33] Human EB (hEB) formation can be achieved by culturing hESC aggregates in suspension (i.e. using a non-adherent dish) or within hanging-drops, followed by their cultivation in Petri dishes. Forty eight hours after initiation of cellular aggregation, hEBs complete the agglomeration process,[57] which results in the formation of simple EBs by day 3, followed by cavitated EBs after 7–10 days, and cystic EBs after two weeks of differentiation.[33] hEBs recapitulate several aspects of early development, displaying regional-specific differentiation into derivatives of all three embryonic germ layers [Ref. 33; Fig. 6].

It was recently suggested that hEB generation and differentiation can be induced and directed by physical restriction. One study showed dynamic formation of differentiation of hEBs in rotating bioreactors.[58] The efficacy of the dynamic process compared to static cultivation in Petri dishes was analyzed with respect to the yield of hEB formation and differentiation. With the use of the Slow Turning Lateral Vessel (STLV), the hEBs formed were smaller in size and no large necrotic centers were seen, even after one month of cultivation. The appearance of representative tissues derived from the three germ layers indicated that the initial developmental events were not altered in the dynamically-formed hEBs. In general, this study defined the culture conditions in which control over aggregation of differentiating hESCs may enable scaleable cell production for clinical and industrial

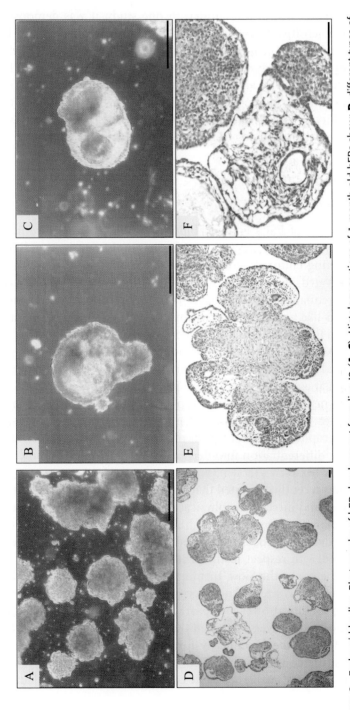

**Figure 6.** Embryoid bodies. Photographs of hEB development from line I9 (**A–C**). Histology sections of 1-month-old hEBs show: **D.** different types of cells with focus on two EBs in high magnifications (**E–F**). Bar = 100 μm.

applications. In another study, hEBs were generated directly from hESC suspensions within three-dimensional (3-D) porous alginate scaffolds.[59] The confining environments of the alginate scaffold pores enabled the efficient formation of hEBs with a relatively high degree cell proliferation and differentiation, encouraged round, small-sized hEBs, and further induced vasculogenesis in the forming hEBs to a higher degree than in other culture systems examined.[59] It was concluded that hEB differentiation can be induced and directed by physical constraints in addition to chemical cues.

## Lineage differentiation

With the use of EB formation, hESC differentiation into several lineages has been extensively studied.

*Cardiomyocytes.* Seeding hEBs on an adherent substrate may result in the appearance of beating areas. These beating areas display structural and functional properties of early-stage cardiomyocytes[60] and exhibit properties of early-stage cardiac phenotype, with gap junctions immunostained with connexin 43 and connexin 45 but not connexin 40.[60,61] High-resolution activation maps using microelectrode arrays have demonstrated the presence of a functional syncytium with stable focal activation and conduction properties along the contracting area of the human EBs.[61] Characterization of the contraction and action potential from beating EB outgrowths showed that hESCs can differentiate into multiple types of cardiomyocytes which display functional properties characteristic of embryonic human cardiac muscle.[62]

*Hematopoietic.* 15-day-old spontaneously differentiated hEBs were shown to express low levels of CD45 (1.4% ± 0.7%). Most of these cells co-express CD34 (1.2% ± 0.5%), a phenotype similar to the first definitive hematopoietic cells detected within the wall of the dorsal aorta of human embryos.[63]

*Vasculature.* During hEB differentiation, an increase in expression of several endothelial cell-specific genes and the development of extensive vasculature-resembling structures within them was shown.[64] Sorted CD31+ cells (2%) were shown to possess embryonic endothelial cell features as well as a potential for microvessel formation both *in vitro* and *in vivo*.[64] Spontaneous formation of a vessel-network of CD34+ cells within forming hEBs was further demonstrated [Ref. 65; Fig. 7]. In addition, smooth muscle cells which may be identified by specific markers, e.g. smooth muscle actin, calponin and smooth muscle myosin heavy chain

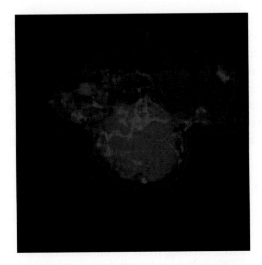

**Figure 7.** Vascular formation within EBs. Confocal analyses of 12-day-old hEBs revealed distinct areas with complex vascular structures composed of CD34+ cells (red; nuclei in blue). Total magnification ×200.

(SM-MHC), were shown to form thicker vessel structures with no observable network structure.

*Extra-embryonic tissues.* Unlike mESCs, hESCs can also form extra-embryonic tissues. The use of hESCs to derive early human *trophoblasts* is particularly valuable, because they are difficult to obtain from other sources and are significantly different from mouse trophoblasts. A recent study explored hEBs for trophoblast differentiation potential.[66] Levels of human chorionic gonadotropin (hCG), progesterone and estradiol-17beta in medium from hEBs grown in suspension culture for 1 week were found to be higher than in unconditioned culture medium or medium from undifferentiated hESCs or spontaneously differentiated hESC colonies. Another study demonstrated the formation of visceral endoderm within developing hEBs, which provides further scope to the study of early human primitive endoderm and yolk sac visceral endoderm development.[67]

## Controlled Differentiation

### Regulating differentiation via hEBs

hEB formation has been utilized as an initial step in a wide range of studies aimed at differentiating hESCs into a specific desired cell type. Alternative strategies, utilizing specific growth factor combinations and cell–cell induction systems, have enhanced differentiation of hEBs into a desired lineage. Schuldiner and colleagues were the first to examine the exposure of differentiating hESCs to growth factors.[68] Five day-old-hEBs were plated on a Petri dish and grown in the presence of eight growth factors known for their ability to direct the differentiation of mESCs basic fibroblast growth factor (bFGF); transforming growth factor $\beta_1$ (TGF$\beta_1$); activin-A, bone morphogenic protein 4 (BMP-4); hepatocyte growth factor (HGF); epidermal growth factor (EGF); $\beta$ nerve growth factor ($\beta$NGF); and retinoic acid.[68] Differentiation was examined by the expression of 24 cell-specific molecular markers that cover all embryonic germ layers and 11 different tissues. The results showed that Activin-A and TGF$\beta_1$ induced differentiation primarily into mesodermal cells; retinoic-acid, EGF, BMP-4 and bFGF induced differentiation into endodermal and mesodermal markers; and NGF and HGF allowed differentiation into three embryonic germ layers. None of the growth factors directed the differentiation exclusively to one cell type, indicating that multiple human cell types may be enriched *in vitro* by specific factors.[68]

Detailed studies have examined the use of $\beta$ nerve growth factor (bNGF) and retinoic acid (RA), showing an increase in the proportion of *neuronal cells*.[69,70] Moreover, three protocols have been established for the *in vitro* differentiation, enrichment and transplantation of neural precursors[71,72] One study showed that, upon aggregation to hEBs, differentiating hESCs formed large numbers of neural tube-like structures in the presence of FGF-2 which can be isolated by selective enzymatic digestion and further purified on the basis of differential adhesion.[71] Withdrawal of FGF-2 resulted in their differentiation into neurons, astrocytes and oligodendrocytes.[71] These progenitors further differentiated into both neurons and astrocytes after hES cell-derived neural precursors were transplanted into the neonatal mouse brain.[71] Another protocol observed a neural precursor population in hEBs grown in suspension in serum-free conditions, in the presence of 50% conditioned medium from the human hepatocarcinoma cell line HepG2.[72]

Recently it was demonstrated that, using a protocol consisting of several steps, *insulin-producing cells* can be generated from hESCs.[73] The procedure included the following phases: culturing and plating of hEBs in insulin-transferrin-selenium-fibronectin medium; supplementation of medium with N2, B27, and bFGF; reduction of glucose concentration in the medium, withdrawal of bFGF; addition of nicotinamide; and, finally, dissociation of the cells and re-growing them in suspension. This process resulted in the formation of clusters which exhibited higher insulin secretion and had longer durability than cells grown as monolayers. In addition to insulin, most cells also co-expressed glucagon or somatostatin, indicating similarity to immature pancreatic cells.[73]

Enrichment of *cardiomyocytes* was achieved when 4–6 day-old hEBs were treated with 5-aza-2'-deoxycytidine.[74] Moreover, the differentiated cultures could be dissociated and separated via Percoll gradient density centrifugation, which led to a population of 70% pure cardiomyocytes. *Hematopoietic cells* were also derived from hESCs by co-culturing them with murine bone marrow cell lines and yolk sac endothelial cell lines.[75] Treating formed hEBs with hematopoietic cytokines (SCF, Flt3L, IL-3, IL-6 and G-CSF) demonstrated a six-fold increase in the proportion of CD45+ cells co-expressing CD34.[63] The addition of these cytokines plus BMP-4 increased the frequency of CD45+CD34+ cells by six-fold. It seems that BMP-4 promotes the self-renewal of hESC-derived hemetopoietic progenitors.[63] In a continuous study, it was further shown that vascular endothelial growth factor A (VEGF-$A_{165}$) selectively promotes erythropoietic development from hESCs. Effects of VEGF-$A_{165}$ were dependent on the presence of hematopoietic cytokines and BMP-4, and could be augmented by the addition of erythropoietin (EPO). Treatment of hEBs with VEGF-$A_{165}$ increased the frequency of cells co-expressing CD34 and VEGF receptor 2 (KDR), as well as cells expressing erythroid markers. In addition to promoting erythropoietic differentiation from hESCs, the presence of VEGF-$A_{165}$ enhanced the *in vitro* self-renewal potential of primitive hematopoietic cells capable of erythroid progenitor capacity.[76]

For the induction of *trophblasts*, hEBs were explanted into Matrigel "rafts" for up to 53 days.[66] This resulted in small protrusions which appeared on the outer surface of hEBs, as well as an increase in secretion of hCG, progesterone and estradiol-17beta which remained dramatically elevated over the culture period. In comparison, hEBs maintained in suspension culture failed to demonstrate this elevation in hormone secretion.[66]

## Directed differentiation via 2-D cultures

Specific manipulation protocols of hESCs have been established in order to obtain enriched and even highly purified specific cell populations.

*Neuronal* progenitors were isolated from continuous cultures of hESCs on mouse embryonic feeder cells.[77] Patches formed in the center of differentiating hESC colonies were cut out and re-cultured in serum-free medium supplemented with known neuronal growth factors. This resulted in the generation of spheres of proliferating neuronal progenitors which were able to differentiate *in vitro* and *in vivo* into the progeny of the three neural lineages.[77] Recent studies further showed the capability to differentiate hESCs into neurons by means of co-culture with feeder cells.[78,79] One study demonstrated the derivation of midbrain dopamine neurons from hESCs by co-culture on either bone marrow or aorta-gonad-mesonephros stromal cells.[78] Using this approach, proliferating neuronal rosettes expressing neuronal progenitor markers could be obtained, and could be further dopaminergically differentiated by administration of SHH and FGF8.[78] Another study showed that culture of hESCs on astrocytes derived from embryonic striatum or on PA6 cells led to their differentiation into tyrosine hydroxylase positive neurons.[79]

Induction of *cardiac* differentiation can be achieved by 2-D differentiation of hESCs. It was shown that co-culture of hESCs on visceral endoderm-like cells results in contracting areas.[80] Characterization of these beating muscles includes sarcomeric marker proteins, chronotropic responses, ion channel expression and function, and electrophysiological features resembling human fetal ventricular cells.[81] Furthermore, cardiomyocytes generated from hESCs using this co-culture procedure were able to couple by gap junction with human fetal cardiomyocytes.[81]

A procedure for the generation of *hematopoietic cells* from hESCs was demonstrated by a co-culturing system.[75] hESCs were co-cultured with the murine bone marrow cell line S17 or the yolk sac endothelial cell line C166 with medium containing fetal bovine serum but no other exogenous cytokines. The resultant differentiated hESCs expressed known hematopoietic precursor markers. When cultured on semisolid media with hematopoietic growth factors, these hematopoietic precursor cells formed characteristic myeloid, erythroid and megakaryocyte colonies. More terminally differentiated hematopoietic cells derived from hESCs under these conditions also expressed normal surface antigens, including glycophorin A on erythroid cells, CD15 on myeloid cells, and CD41 on megakaryocytes.[75] Different studies have shown that co-culture of hESCs with OP9 bone marrow stromal cells enhanced the upregulation of CD34+ cells.[82] These CD34+ cells displayed the phenotype of primitive hematopoietic progenitors and were able to further differentiate into lymphoid as well as myeloid lineages.[82]

The use of hESCs to derive early human *trophoblasts* was demonstrated by the addition of BMP-4 to feeder-free hESC cultures.[83] DNA microarray,

RT-PCR and immunoassay analyses demonstrated that the differentiated cells expressed trophoblast markers and secreted placental hormones. When plated at low density, the BMP4-treated cells formed syncytia that expressed hCG.[83]

In order to facilitate the generation of *vascular* cells from hESC, an approach of induced mesodermal differentiation was established.[65] hESCs were differentiated via 2-D culture on type IV collagen which up-regulated endothelial progenitor markers. Re-culture under strict conditions and exposure to angiogenic growth factors resulted in a prolonged differentiation pathway into endothelial cells and vascular smooth muscle cells markers. In particular, the addition of human VEGF brought about the generation of some 20% of endothelial cells expressing von Willibrand factor, Ac-LDL and CD31. The addition of platelet-derived growth factor BB resulted in around 80% of vascular smooth muscle cells expressing smooth muscle actin, calponin and SM-MHC. The use of 3-D collagen gels and matrigel assays, for the induction and inhibition of human vascular sprouting *in vitro*, further established the vascular potential of the cells generated by the 2-D differentiation system. It was further suggested that hESCs may serve as an *in vitro* model to study angiogenic mechanisms that allow its promotion and inhibition. For this purpose, 2-D differentiated mesodermal cells were embedded into matrigel or type I collagen gels and allowed to grow. After 24–48 hours, intensive sprouting was observed, resulting in penetration of the cells into the gel and generation of tube-like structures. The formation of such structures was shown to be inhibited by adding anti-human vascular endothelial cadherin to the culture medium.[65]

## Acknowledgments

The research performed in our laboratory is partly supported by NIH grants 5R24RR018405-02 and 5R01HL073798-02.

## REFERENCES

1. Bradley A, Evans M, Kaufman MH, Robertson E. (1984) Formation of germ-line chimaeras from embryo-derived teratocarcinoma cell lines. *Nature* **309**:255–256.
2. Nagy A, Rossant J, Nagy R, *et al.* (1993) Derivation of completely cell culture-derived mice from early-passage embryonic stem cells. *Proc Natl Acad Sci USA* **90**:8424–8428.

3. Wobus AM, Holzhausen H, Jakel P, Schoneich J. (1984) Characterization of pluripotent stem cell line derived from mouse embryo. *Exp Cell Res* **152**: 212–219.

4. Doetschman TC, Eistetter H, Katz M, *et al.* (1985) The *in vitro* development of blastocyst-derived embryonic stem cell lines: Formation of visceral yolk sac, blood islands and myocardium. *J Embryol Exp Morphol* **87**:27–45.

5. Nakano T, Kodama H, Honjo T. (1994) Generation of lymphohematopoietic cells from embryonic stem cells in culture. *Science* **265**:1098–1101.

6. Kennedy M, Firpo M, Choi K, *et al.* (1997) A common precursor for primitive erythropoiesis and definitive haematopoiesis. *Nature* **386**:488–493.

7. Brüstle O, Spiro AC, Karram K, *et al.* (1997) *In vitro*-generated neural precursors participate in mammalian brain development. *Proc Natl Acad Sci USA* **94**:14809–14814.

8. Brüstle O, Jones KN, Learish RD, *et al.* (1999) Embryonic stem cell-derived glial precursors: A source of myelinating transplants. *Science* **285**:754–756.

9. Klug MG, Soonpaa MH, Koh GY, Field LJ. (1996) Genetically selected cardiomyocytes from differentiating embryonic stem cells form stable intracardiac grafts. *J Clin Invest* **98**:216–224.

10. Soria B, Roche E, Berná G, *et al.* (2000) Insulin-secreting cells derived from embryonic stem cells normalize glycemia in streptozotocin-induced diabetic mice. *Diabetes* **49**:157–162.

11. Lumelsky N, Blondel O, Laeng P, *et al.* (2001) Differentiation of embryonic stem cells to insulin-secreting structures similar to pancreatic islets. *Science* **292**:1389–1394.

12. Chambers I, Colby D, Robertson M, *et al.* (2003) Functional expression cloning of Nanog, a pluripotency sustaining factor in embryonic stem cells. *Cell* **113**(5):643–655.

13. Mitsui K, Tokuzawa Y, Itoh H, *et al.* (2003) The homeoprotein Nanog is required for maintenance of pluripotency in mouse epiblast and ES cells. *Cell* **113**(5):631–642.

14. Kahn BW, Ephrussi B. (1970) Developmental potentialities of clonal *in vitro* cultures of mouse testicular teratoma. *J Natl Cancer Inst* **44**:1015–1029.

15. Martin GR, Evans MJ. (1974) The morphology and growth of a pluripotent teratocarcinoma cell line and its derivatives in tissue culture. *Cell* **2**: 163–172.

16. Martin GR, Evans MJ. (1975) Differentiation of clonal lines of teratocarcinoma cells: Formation of embryoid bodies *in vitro*. *Proc Natl Acad Sci* **72**: 1441–1445.

17. Evans MJ. (1972) The isolation and properties of a clonal tissue culture strain of pluripotent mouse teratoma cells. *J Embryol Exp Morphol* **28**:163–176.

18. Evans MJ, Kaufman MH. (1981) Establishment in culture of pluripotential cells from mouse embryos. *Nature* **292**:154–156.

19. Martin GR. (1981) Isolation of a pluripotent cell line from early mouse embryos cultured in medium conditioned by teratocarcinoma stem cells. *Proc Natl Acad Sci USA* **78**:7634–7638.

20. Doetschman T, Williams P, Maeda N. (1988) Establishment of hamster blastocyst-derived embryonic stem (ES) cells. *Dev Biol* **127**:224–227.

21. Giles JR, Yang X, Mark W, Foote RH. (1993) Pluripotency of cultured rabbit inner cell mass cells detected by isozyme analysis and eye pigmentation of fetuses following injection into blastocysts or morulae. *Mol Reprod Dev* **36**:130–138.

22. Graves KH, Moreadith RW. (1993) Derivation and characterization of putative pluripotential embryonic stem cells from preimplantation rabbit embryos. *Mol Repro Dev* **36**:424–433.

23. Notarianni E, Galli C, Laurie S, *et al.* (1991) Derivation of pluripotent, embryonic cell lines from the pig and sheep. *J Reprod Fertil Suppl* **43**:255–260.

24. Sims M, First NL. (1994) Production of calves by transfer of nuclei from cultured inner cell mass cells. *Proc Natl Acad Sci USA* **91**:6143–6147.

25. Mitalipova M, Beyhan Z, First NL. (2001) Pluripotency of bovine embryonic stem cell line derived from precompacting embryos. *Cloning* **3**: 59–67.

26. Thomson JA, Kalishman J, Golos TG, *et al.* (1995) Isolation of a primate embryonic stem cell line. *Proc Natl Acad Sci USA* **92**:7844–7848.

27. Thomson JA, Kalishman J, Golos TG, *et al.* (1996) Pluripotent cell lines derived from common marmoset (*Callithrix jacchus*) blastocysts. *Biol Reprod* **55**: 254–259.

28. Suemori H, Tada T, Torii R, *et al.* (2001) Establishment of embryonic stem cell lines from cynomolgus monkey blastocysts produced by IVF or ICSI. *Dev Dyn* **222**:273–279.

29. Bongso A, Fong CY, Ng SC, Ratnam S. (1994) Isolation and culture of inner cell mass cells from human blastocysts. *Hum Reprod* **9**(11):2110–2117.

30. Thomson JA, Itskovitz-Eldor J, Shapiro SS, *et al.* (1998) Embryonic stem cell lines derived from human blastocysts. *Science* **282**:1145–1147 [erratum in *Science* (1998) **282**:1827].

31. Reubinoff BE, Pera MF, Fong C, *et al.* (2000) Embryonic stem cell lines from human blastocysts: Somatic differentiation *in vitro. Nat Biotechnol* **18**: 399–404.

32. Amit M, Carpenter MK, Inokuma MS, *et al.* (2000) Clonally derived human embryonic stem cell lines maintain pluripotency and proliferative potential for prolonged periods of culture. *Dev Biol* **227**:271–278.

33. Itskovitz-Eldor J, Schuldiner M, Karsenti D, *et al.* (2000) Differentiation of human embryonic stem cells into embryoid bodies comprising the three embryonic germ layers. *Mol Med* **6**:88–95.

34. Eiges R, Schuldiner M, Drukker M, *et al.* (2001) Establishment of human embryonic stem cell-transfected clones carrying a marker for undifferentiated cells. *Curr Biol* **11**:514–518.

35. Draper JS, Smith K, Gokhale P, *et al.* (2004) Recurrent gain of chromosomes 17q and 12 in cultured human embryonic stem cells. *Nat Biotechnol* **22**(1): 53–54.

36. Dhara SK, Benvenisty N. (2004) Gene trap as a tool for genome annotation and analysis of X chromosome inactivation in human embryonic stem cells. *Nucleic Acids Res* **32**(13):3995–4002.

37. Amit M, Itskovitz-Eldor J. (2004) Isolation, characterization and maintenance of primate ES cells. In: *Handbook of Stem Cells*, ed. Lanza RP, Chap. 45, pp. 419–436, Elsevier Science.

38. Cowan CA, Klimanskaya I, McMahon J, *et al.* (2004) Derivation of embryonic stem-cell lines from human blastocysts. *N Engl J Med* **25**:350(13):1353–1356. *Comments: N Engl J Med* 350(13):1275–1276; *N Engl J Med 2004* **350**(13): 1351–1352.

39. Cowan CA, Klimanskaya I, McMahon J, *et al.* (2004) Derivation of embryonic stem-cell lines from human blastocysts. *NEJM* **350**:1353–1356.

40. Mitalipova M, Calhoun J, Shin S, *et al.* (2003) Human embryonic stem cell lines derived from discarded embryos. *Stem Cells* **21**:521–526.

41. Suss-Toby E, Gerecht-Nir S, Amit M, *et al.* (2004) Derivation of a diploid human embryonic stem cell line from a mononuclear zygote. *Hum Reprod* **19**(3):670–675.

42. Lanzendorf SE, Boyd CA, Wright DL, *et al.* (2001) Use of human gametes obtained from anonymous donors for the production of human embryonic stem cell lines. *Fertil Steril* **76**:132–137.

43. Chen Y, He ZX, Liu A, *et al.* (2003) Embryonic stem cells generated by nuclear transfer of human somatic nuclei into rabbit oocytes. *Cell Res* **13**:251–263.

44. Amit M, Itskovitz-Eldor J. (2002) Derivation and spontaneous differentiation of human embryonic stem cells. *J Anat* **200**:225–232.

45. Solter D, Knowles BB. (1975) Immunosurgery of mouse blastocyst. *Proc Natl Acad Sci USA* **72**:5099–5102.

46. Richards M, Fong CY, Chan WK, *et al.* (2002) Human feeders support prolonged undifferentiated growth of human inner cell masses and embryonic stem cells. *Nat Biotechnol* **20**:933–936.

47. Amit M, Margulets V, Segev H, *et al.* (2003) Human feeder layers for human embryonic stem cells. *Biol Reprod* **68**:2150–2156.

48. Hovatta O, Mikkola M, Gertow K, *et al.* (2003) A culture system using human foreskin fibroblasts as feeder cells allows production of human embryonic stem cells. *Hum Reprod* **18**:1404–1409.

49. Cheng L, Hammond H, Ye Z, *et al.* (2003) Human adult marrow cells support prolonged expansion of human embryonic stem cells in culture. *Stem Cells* **21**:131–142.

50. Xu C, Inokuma MS, Denham J, *et al.* (2001) Feeder-free growth of undifferentiated human embryonic stem cells. *Nat Biotechnol* **19**:971–974.

51. Rosler ES, Fisk GJ, Ares X, *et al.* (2004) Long-term culture of human embryonic stem cells in feeder-free conditions. *Dev Dyn* **229**(2):259–274.

52. Carpenter MK, Rosler ES, Fisk GJ, *et al.* (2004) Properties of four human embryonic stem cell lines maintained in a feeder-free culture system. *Dev Dyn* **229**(2):243–258.

53. Amit M, Shariki C, Margulets V, Itskovitz-Eldor J. (2004) Feeder and serum free culture system for human embryonic stem cells. *Biol Reprod* **70**:837–845.

54. Zwaka TP, Thomson JA. (2003) Homologous recombination in human embryonic stem cells. *Nat Biotechnol* **21**:319–321.

55. Green H, Easley K, Iuchi S. (2003) Marker succession during the development of keratinocytes from cultured human embryonic stem cells. *Proc Natl Acad Sci USA* **100**:15625–15630.

56. Tzukerman M, Rosenberg T, Ravel Y, *et al.* (2003) An experimental platform for studying growth and invasiveness of tumor cells within teratomas derived from human embryonic stem cells. *Proc Natl Acad Sci USA* **100**:13507–13512.

57. Dang S, Gerecht-Nir S, Chen J, *et al.* (2004) Controlled scalable embryonic stem cell differentiation culture. *Stem Cells* **22**:275–282.

58. Gerecht-Nir S, Cohan S, Itskovitz-Eldor J. (2004) Bioreactor cultivation enhances the efficiency of human embryoid body (hEB) formation and differentiation. *Biotechnol Bioeng* **86**:493–502.

59. Gerecht-Nir S, Cohen S, Ziskind A, Itskovitz-Eldor J. (2004) 3-D porous alginate scaffolds provide a conducive environment for the generation of well vascularized embryoid bodies from human embryonic stem cells. *Biotechnol Bioeng* (in press).

60. Kehat I, Kenyagin-Karsenti D, Snir M, *et al.* (2001) Human embryonic stem cells can differentiate into myocytes with structural and functional properties of cardiomyocytes. *J Clin Invest* **108**:407–414.

61. Kehat I, Gepstein A, Spira A, *et al.* (2002) High-resolution electrophysiological assessment of human embryonic stem cell-derived cardiomyocytes: A novel *in vitro* model for the study of conduction. *Circ Res* **91**:659–661.

62. He JQ, Ma Y, Lee Y, *et al.* (2003) Human embryonic stem cells develop into multiple types cardiac myocytes. Action potential characterization. *Circ Res* **93**:32–39.

63. Chadwick K, Wang L, Li L, *et al.* (2003) Cytokines and BMP-4 promote hematopoietic differentiation of human embryonic stem cells. *Blood* **102**:906–915.

64. Levenberg S, Golub JS, Amit M, *et al.* (2002) Endothelial cells derived from human embryonic stem cells. *Proc Natl Acad Sci USA* **99**:4391–4396.

65. Gerecht-Nir S, Ziskind A, Cohan S, Itskovitz-Eldor J. (2003) Human embryonic stem cells as an *in vitro* model for human vascular development and the induction of vascular differentiation. *Lab Invest* **83**:1811–1820.

66. Gerami-Naini B, Dovzhenko OV, Durning M, *et al.* (2004) Trophoblast differentiation in embryoid bodies derived from human embryonic stem cells. *Endocrinology* **145**:1517–1524.

67. Conley BJ, Trounson AO, Mollard R. (2004) Human embryonic stem cells form embryoid bodies containing visceral endoderm-like derivatives. *Fetal Diagn Ther* **19**:218–223.

68. Schuldiner M, Yanuka O, Itskovitz-Eldor J, *et al.* (2000) From the cover: Effects of eight growth factors on the differentiation of cells derived from human embryonic stem cells. *Proc Natl Acad Sci USA* **97**:11307–11312.

69. Schuldiner M, Eiges R, Eden A, *et al.* (2001) Induced neuronal differentiation of human embryonic stem cells. *Brain Res* **913**:201–205.

70. Carpenter MK, Inokuma MS, Denham J, *et al.* (2001) Enrichment of neurons and neural precursors from human embryonic stem cells. *Exp Neurol* **172**:383–397.

71. Zhang SC, Wernig M, Duncan ID, *et al.* (2001) *In vitro* differentiation of transplantable neural precursors from human embryonic stem cells. *Nat Biotechnol* **19**:1129–1133.

72. Schulz TC, Palmarini GM, Noggle SA, *et al.* (2003) Directed neuronal differentiation of human embryonic stem cells. *BMC Neurosci* **4**:27.

73. Segev H, Fishman B, Ziskind A, *et al.* (2004) Differentiation of human embryonic stem cells into insulin-producing clusters. *Stem cells* **22**:265–274.

74. Xu C, Police S, Rao N, Carpenter MK. (2002) Characterization and enrichment of cardiomyocytes derived from human embryonic stem cells. *Circ Res* **91**:501–508.

75. Kaufman DS, Hanson ET, Lewis RL, *et al.* (2001) Hematopoietic colony forming cells derived from human embryonic stem cells. *Proc Natl Acad Sci USA* **98**:10716–10721.

76. Cerdan C, Rouleau A, Bhatia M. (2004) VEGF-A165 augments erythropoietic development from human embryonic stem cells. *Blood* **103**:2504–2512.

77. Reubinoff BE, Itsykson P, Turetsky T, *et al.* (2001) Neural progenitors from human embryonic stem cells. *Nat Biotechnol* **19**:1134–1140.

78. Perrier AL, Tabar V, Barberi T, *et al.* (2004) Derivation of midbrain dopamine neurons from human embryonic stem cells. *Proc Natl Acad Sci USA* **101**:12543–12548.

79. Buytaert-Hoefen KA, Alvarez E, Freed CR. (2004) Generation of tyrosine hydroxylase positive neurons from human embryonic stem cells after coculture with cellular substrates and exposure to GDNF. *Stem Cells* **22**:669–674.

80. Mummery C, Ward D, van den Brink CE, *et al.* (2002) Cardiomyocyte differentiation of mouse and human embryonic stem cells. *J Anat* **200**:233–242.

81. Mummery C, Ward-van Oostwaard D, Doevendans P, *et al.* (2003) Differentiation of human embryonic stem cells to cardiomyocytes: Role of coculture with visceral endoderm-like cells. *Circulation* **107**:2733–2740.

82. Vodyanik MA, Bork JA, Thomson JA, Slukvin II. (2004) Human embryonic stem cell-derived CD34+ cells: Efficient production in the co-culture with OP9 stromal cells and analysis of lymphohematopoietic potential. *Blood* Sep 16 (Epub ahead of print)

83. Xu RH, Chen X, Li DS, *et al.* (2002) BMP4 initiates human embryonic stem cell differentiation to trophoblast. *Nat Biotechnol* **20**:1261–1264.

# 7

# Differentiation *In Vivo* and *In Vitro* of Human Embryonic Stem Cells

Barak Blum and Nissim Benvenisty*

## Introduction

Embryonic stem cells (ESCs) are pluripotent cells generated by the *in vitro* cultivation of the inner cell mass (ICM) of a blastocyst stage embryo.[1,2] ESCs were first established from mouse embryos in 1981,[1,2] and almost two decades later, human embryonic stem cells (HESCs) were derived.[3,4] In a normally developing embryo, the pluripotency of the ICM is a transient state, and by gastrulation the ICM cells have lost their pluripotency and become committed to specific Lineages.[5] When grown *in vitro*, HESCs may remain in an undifferentiated state in the presence of a supportive feeder layer, usually mitotically inactivated mouse embryonic fibroblasts (MEF).[3,4] HESCs have the capacity for self renewal, meaning they can proliferate in an undifferentiated state indefinitely, while maintaining a normal karyotype for many passages in culture. These cells are capable of differentiating to derivates of all three embryonic germ layers either *in vivo* or *in vitro*.[3,6] Undifferentiated HESC holds some typical characteristics. They form tight colonies with defined borders (seen in Fig. 1A), in which the cells are packed together, connected by gap junctions, and have a large nucleus-to-cytoplasm ratio. They also possess some molecular characteristics, such as a high expression of telomerase, activation of the enzyme alkaline phosphatase, and the expression of typical transcription factors such as *OCT4* and *REX1*, as well as typical surface antigens such as SSEA-3, SSEA-4 and TRA-1-60.[7] Recently, it was suggested that HESCs also express a unique set of microRNAs.[8]

*Correspondence: Department of Genetics, Silberman Institute of Life Science, The Hebrew University, Jerusalem, Israel. e-mail: nissimb@mail.il.huji.ac.il

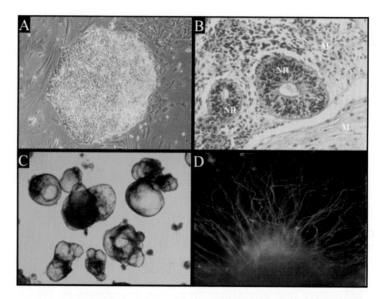

**Figure 1.** Differentiation of HESC. **A.** Phase contrast image of undifferentiated HESC colony grown on MEF feeder layer. **B.** Histology of a teratoma derived from undifferentiated HESCs. Neural rosettes (NR) can be seen within two types of mesenchymal cells (M). **C.** Cystic embryoid bodies (EBs), formed by growing HESCs in suspension for 23 days. **D.** Neuronal outgrowth of an EB grown for 20 days. Cells are immunostained with an antibody against the neurofilament heavy chain (NF-H).

Murine ESCs (MESCs) have been used as an extensive research tool in biological sciences for more than two decades. The establishment of HESC lines has opened remarkable opportunities for basic, as well as medical science. The advantages of HESC are most prominent in two fields: as a tool for modeling early differentiation events occurring in the developing human embryo, and as a virtually unlimited source of human cells for tissue transplantation and cell therapy. Here, we will present the current knowledge concerning the potential of HESCs to differentiate *in vivo*, and discuss the progress en route for reaching this differentiation potential *in vitro*.

## *In Vivo* Differentiation Of HESC

### Spontaneous differentiation of HESCs *in vivo*

When addressing the *in vivo* differentiation potential of HESCs, it is appropriate to initially discuss the fate of these cells if they would have been left to normally differentiate (i.e. during the normal course of early mammalian development). At that time, the blastocyst — at first composed only of an outer epithelial-like trophoblast layer, and an undifferentiated inner cell mass — will go through successively complex cascades of divisions and differentiation processes. While the trophoblast differentiates only to an extra embryonic tissue, the ICM — from which embryonic stem cells are derived — will differentiate into the extra embryonic endoderm, the amniotic ectoderm, and the embryonic epiblast. The latter will be separated by the process of gastrulation to the three embryonic germ layers — ectoderm, mesoderm and endoderm — which will eventually differentiate to give rise to the full repertoire of embryonic tissues (as well as the extra embryonic mesoderm).[5,9] The cells of the ICM, thus, differentiate into all types of animal tissues, aligning to form the complete body plan of an adult.

More than twenty years ago, in 1984, Bradley *et al.*[10] took the ultimate test for verifying the differentiation capabilities of murine embryonic stem cells: they injected the cells into mouse blastocysts, and produced chimera progenies. Indeed, the injected embryonic stem cells contributed to each and every tissue of the offspring, including germ cells, demonstrating a definitive proof of embryonic stem cells differentiation capacity.[10,11] In humans, however, such an assay is not feasible, due to obvious ethical reasons. The most widely accepted alternative assay for the differentiation repertoire of HESCs is, thus, ectopically injecting them into a laboratory animal, such as a mouse, to form a teratoma.

If HESCs are injected intramuscularly,[3] into the testis,[4] subcutaneously[12] or under the kidney capsule[13] of an adult immunocompromised mouse, they proliferate and differentiate to form a benign, embryo-like tumor, called a teratoma. Named after the Greek word "teraton" — a monster — the teratoma is a gross, haphazardly organized neoplasm, which includes tissue representation of all three germ layers.[14] Unlike other tumors, teratomas are not composed of a homogeneous population of neoplastic cells, but rather of a histologically defined variety of mature or pre-mature tissues,[15] which resemble a disorganized embryo.

When allowed to differentiate *in vivo* as a HESC-induced teratoma model in the mouse,[3] the cells differentiate in a manner that may resemble the series of events occurring in the developing embryo, albeit without complete pattern formation. In that sense, one can find that along with embryonic tissues of less defined nature, such as primitive neuroectoderm or embryonic mesenchyme (Fig. 1B), the tumor also displays a representation of more mature tissues. The repertoire of tissues identified within HESC-induced teratomas includes ectoderm derived neural epithelium forming neural rosettes, mesoderm derived cartilage, bone or muscle, endoderm derived gut epithelium, and even tissues whose development require inductive interaction between tissues of different origin, such as liver cells (see Table 1). In addition, many of the normal features of simple tissue architecture are imitated, and complex structures like fetal glomeruli,[3] ganglions,[4] teeth,[16] hair follicles,[13] and even brain structures such as choroid plexus[17] can be found. Yet in HESC-induced teratomas, no complete organogenesis is formed.[18]

Teratomas can be experimentally induced by other sources of pluripotent cells, such as embryonic carcinoma (see below) or embryonic germ cells, and even by grafting whole early embryos or embryonic discs to extra uterine sites. Teratomas can also form spontaneously in many animals, including humans, as congenital or adult neoplasms.[14,54,55] The presence of pluripotent cells, however, is a prerequisite in teratoma formation. This was assessed by Damjanov *et al.*, who showed that in mice, only embryos of 7-days old or less (prior to neural folding) are capable of forming teratomas.[55]

When formed spontaneously, teratomas are classified within the category of germ cell tumors (GCT), that arise mainly in the testis and ovary, but often also in extragonadal places. The most commonly accepted theory for the origin of GCTs suggests that congenital GCTs originate from totipotent primordial germ cells that were dislodged from the stream of migration toward the gonads. Adult GCTs are usually suggested to arise from parthenogenic germ cells.[14,56,57]

As a definition, teratomas contain tissues of embryonic origin only (i.e. derivatives of ectoderm, mesoderm and endoderm), and not of extra embryonic origin. The differentiation repertoire of spontaneously-occurring teratomas is extremely vast: among the cell and tissue types found are neurons, glia, skin, hair, teeth, muscle, bone, cartilage, adipose tissue, liver cells, salivary glands, etc.[14,58] In contrast to HESC-induced teratomas, spontaneous teratomas have been reported, although rarely, to contain structures of early organogenesis, such as malformed limbs, bowels, vertebrates.[14,58] and even a completely developed eye.[59] A teratoma is deemed "mature" if it contains terminally differentiated tissues; "immature" if it also contains foci of immature tissues that resemble those of the embryo; and "teratocarcinoma" if it also contains foci of undifferentiated cells, termed "embryonic carcinoma cells." These cells closely resemble the cells of the ICM, are pluripotent, express OCT-4, and can invade and

**Table 1.** Differentiation of HESC *In Vivo* and *In Vitro*. Shown is the repertoire of cells derived *in vivo* and *in vitro* from HESC in a spontaneous or directed fashion. (The classification of tissues between the germ layers is according to Ref. 9.)

| | In vivo | | In vitro | | | |
| | Spontaneous | | Directed | | | |
| | HESC-induced teratoma | EB, monolayer | By growth factors | By non-protein factors | By feeder layers | By other means |
|---|---|---|---|---|---|---|
| **Ectoderm** | | | | | | |
| Neurons | (3, 4, 13, 16, 17, 19–25) | (4, 6, 16, 26) | NGF (26, 27), EGF (28, 29), bFGF (28–30), PDGF (28), IGF-1 (28), NT-3 (28), BDNF (28), noggin (31), TGFα (32), FGF-8 (33), SHH (33) | RA (26–28) | MS5 (stromal cells) feeder layer (33), MedII (human hepatocarcinoma) condition media (34) | Transplantation into chick embryo (35, 36) |
| Glia | | | EGF (28, 29), bFGF(29, 30), PDGF (28), IGF-1 (28), NT-3 (28), BDNF (28) | RA (28) | | |
| Skin | (13, 25) | (26) | EGF (26) | | | |
| Retinal epithelia | (23) | | | | | |
| Hair | (13) | | | | | |
| Adrenal | | | | RA (26) | | |
| Dental component | (16, 25) | | | | | |

**Table 1.** (*Continued*)

| | In vivo | | In vitro | | | |
| | | Spontaneous | | Directed | | |
| | HESC-induced teratoma | EB, monolayer | By growth factors | By non-protein factors | By feeder layers | By other means |
|---|---|---|---|---|---|---|
| ***Mesoderm*** | | | | | | |
| ***Dorsal*** | | | | | | |
| Muscle | (3, 4, 13, 19, 22–24) | (26) | | | | |
| Bone | (3, 13, 19, 22, 23) | | | Dexamethasone, ascorbic acid, β-glycerophosphate (37) | | |
| Cartilage | (3, 4, 13, 16, 17, 19, 21–25) | (26) | | | | |
| Adipose tissue | (20) | | | | | |
| ***Intermediate*** | | | | | | |
| Kidney | (3, 13, 19, 25) | (26) | NGF (26) | | | |
| ***Lateral*** | | | | | | |
| Cardiomyocytes | (38) | (4, 6, 16, 26, 39) | TGFβ (26) | 5-aza-2'-deoxycytidine (40) | END-2 (visceral endoderm) feeder layer (41) | |
| Endothelial cells | (20) | (42) | VEGF, PDGF (43) | | | |
| Blood cells | | (6, 26) | Cytokines (44–46), Cytokines + BMP4 (45), Cytokines + BMP4 + VEGF-A165 (47) | | C166 (yolk sac endothelia) feeder layer (44), S17 (bone marrow) feeder layer (44) | |

**Table 1.** (*Continued*)

| | In vivo | In vitro | | | | |
| | Spontaneous | | Directed | | | |
| | HESC-induced teratoma | EB, monolayer | By growth factors | By non-protein factors | By feeder layers | By other means |
|---|---|---|---|---|---|---|
| **Endoderm** | | | | | | |
| Liver | (22, 38) | (16, 26) | NGF (26), HGF (26) aFGF (38) | Sodium butyrate (48) | Primary hepatocytes condition media (38) | |
| Pancreas | (22) | (26, 49) | NGF (26) | Various hormones, growth factors and matrix components (50) | | |
| Gut epithelia | (3, 13, 19, 21, 25) | | | | | |
| Salivary gland | (22) | | | | | |
| Germ cells | | (51) | | | | |
| **Extra Embryonic** | | | | | | |
| Ex. Embryonic endoderm | | (31) | BMP2 (31) | | | |
| Trophoblast | | (3) | BMP4 (52) | | | Growth in 3D "Matrigel rafts" (53) |

Abbreviations: NGF: nerve growth factor; EGF: epidermal growth factor; aFGF: acidic fibroblast growth factor; bFGF: basic fibroblast growth factor; FGF-8: fibroblast growth factor 8; PDGF: platelet derived growth factor; HGF: hepatocyte growth factor; IGF-1: insulin like growth factor 1; NT-3: neurotrophin 3; BDNF: brain derived growth factor; TGFα: transforming growth factor α; TGFβ: transforming growth factor β; RA: retinoic acid; BMP2: bone morphogenic protein 2; BMP4: bone morphogenic protein 4; SHH: sonic hedgehog; VEGF: vascular endothelial growth factor.

metastasize.[14,58,60,61] It was suggested that these cells are the source — or the stem cells — of all GCTs, including those of extra embryonic tissues.[14] In the mouse, teratocarcinomas are hard to obtain in an induced experimental teratoma model.[55]

Ironically, the immense opportunity residing within HESCs — their capacity to differentiate *in vivo* into all cell types — also entails their significant drawback: the risk of teratoma formation. Although not transformed in the regular fashion, the tumorogenic potential of HESC is a major issue when considered for clinical use. This is further emphasized by the fact that single mouse embryonic stem cells have been reported to be sufficient for efficiently producing teratomas in a mouse model.[62]

## Directed differentiation of HESCs *in vivo*

Nevertheless, evidence also exists for direct differentiation of HESC *in vivo*. Thus, Goldstein *et al.*[35] have transplanted undifferentiated HESC colonies instead or within the somites of chick embryos. Recapitulating inductive signals from the developing embryo, the cells differentiated to neurons, and the specific position in which they were placed determined their final appearance. Interestingly, the cells did not develop an encapsulating crust isolating them from their surroundings, as is usually seen in teratomas, but rather integrated and contributed to the host neural tissue. Additionally, the injection of undifferentiated HESCs to a chick embryo with an open neural tube defect was shown to correct the lesion. In this case, however, HESCs did not integrate into the neural tube.[36] Due to the short-timed nature of these experiments, teratoma formation could not be overruled.

Murine ESCs (MESCs) were sometimes reported to directly differentiate into neurons and integrate in mouse models (reviewed in Ref. 63). However, in one such report, an attempt to differentiate MESCs to dopaminergic neurons by injecting them into the striatum of rats resulted in an approximate 25% (5 out of 19) death rate due to teratoma formation.[64] It is possible, thus, that introduction of ESCs into the developing tissues (i.e. the embryo) will cause the stem cells to acquire the developmental-inducing signals from the surroundings and integrate with the embryonic tissues, while introducing them into adult tissues will not result in full integration, but rather in a teratoma. One might then predict that the safest way to avoid teratoma in the adult would be to completely or partially differentiate the cells *in vitro* prior to transplantation.

## *In Vitro* **Differentiation of HESC**

### Spontaneous differentiation of HESCs *in vitro*

As stated earlier, keeping HESCs in an undifferentiated state is an active process. Even when grown in standard conditions (i.e. in the presence of MEFs), HESC colonies tend to differentiate in their peripheries.[3,4] When totally deprived of MEFs or grown to full confluence, intensive differentiation is observed. In these cases, HESCs differentiate as a monolayer composed mostly of neurons and extra embryonic cells.[3,4] The variety of cell types spontaneously differentiating from HESCs can be enhanced by aggregating them into spheroid clusters, called embryoid bodies (EBs), because of their resemblance to early embryos.[6] These represent differentiation characteristics of all three embryonic germ layers.

Methods for culturing EBs from HESCs were thoroughly reviewed elsewhere.[18] In brief: HESC are grown in suspension in non-adhesive culture dishes (i.e. bacterial petri plates) without bFGF. Within two days, the cells aggregate to form tightly-packed clusters, termed "simple EBs." Soon after, the center of the EB starts to cavitate, and 14–20 days after their initial aggregation the EBs are filled with fluid, forming balloon-like "cystic EBs."[6] Such cystic EBs are seen in Fig. 1C. Another method involves growing the cells in media drops hanging upside down on a petri dish cap, enabling control over EB size.[18] Recently, a method by which EBs are grown in a rotating bioreactor was also developed.[65]

When first produced, human EBs were cultured for 20 days, after which regional clusters of specialized differentiated cells were observed, with tissue specific genes expressed within them. Among these were neurons expressing the neurofilament 68Kd protein, hematopoietic cells expressing $\zeta$-globin, endodermal cells expressing $\alpha$-fetoprotein, and even synchronically pulsing cardiomyocytes expressing $\alpha$-cardiac actin.[6] Later studies further extended the repertoire of tissues spontaneously differentiating within human EBs, and observations were reported of many cell types, such as hepatic cells,[38,39] endothelial cells,[42] germ cells,[51] etc. (See Table 1.)

Since the differentiation of EBs may imitate early events occurring during embryonic development, such as the process leading to blastocyst cavitation and the early cross-talk signals between cells conducting tissue determination, the EB system could be of major benefit as a research tool for studying early human embryogenesis.

## Directed differentiation of HESCs *in vitro*

The spontaneously differentiated EBs are not sufficient to serve as a source of differentiated human cells for clinical therapy. This is because of the stochastic nature of tissue differentiation, the non-homogenous cell population and the rather small number of specific cells that can be generated by this method, while utilization in the clinic requires mass culture of highly-purified cell populations. For that reason, methods for enrichment, selection, or direct differentiation of HESCs to specific lineages were developed.

It has been previously shown that EBs express receptors for multiple growth factors (bFGF, TGF-β1, activin-A, BMP-4, HGF, EGF, βNGF and RA).[26] When exposed to each of these factors, the enrichment of specific lineages was observed within the population of differentiating cells (see Table 1),[26] although none of the factors could obtain complete purity. Below (and listed in Table 1) is a description of the progress that was made since, in order to directly differentiate or enrich specific human cell types *in vitro*:

### *Neurons and glia*

Among the cell lineages directly differentiated from HESCs, neurons produce the largest group of reports, and the enrichment of neuronal population was achieved through various methodologies (reviewed in Ref. 63). This is perhaps because neuroectodermal precursors are the first to arise when HESCs go through spontaneous differentiation, and are among the most simple to obtain.[4,27] Thus, neurons were the first cells to be isolated from a differentiating HESC culture. As mentioned above, HESC left to spontaneously differentiate for a long period of time in a high density environment developed three-dimensional clusters, wherein neural progenitors, positively stained with antibodies against the adhesion molecule N-CAM, appear. When isolated upon their morphology and re-plated on an appropriate substrate, these cells form neurospheres, which expressed markers for mature neurons, such as β-tubulin, synaptophysin and even enzymes involved in GABA biosynthesis pathway.[4] For a list of specific markers expressed in HESC-induced neurons, please refer to Ref. 63.

The administration of the soluble growth factors bFGF and EGF to such a culture enables the propagation and expansion of the neuronal progenitors.[29] When plated on an appropriate substrate and treated with specific combinations of growth factors, it is possible to direct specific neuronal lineages. For example: when plated on laminin-coated dishes, the cells developed into mature neurons, expressing glutamate, glutamic acid decarboxylase, GABA and serotonin.[29] On the other hand, plating the cells on fibronectin coated dishes, and treating the culture with bFGF, EGF and PDGF, followed by the steroid hormone T3, leads to the differentiation of a variety of glial cells, such as astrocytes and oligodendrocytes.[29]

Similarly, the influence of growth factor administration was tested in the EB system, whereupon treating EBs with βNGF or RA resulted in the formation of complex neuronal networks when the EBs were re-plated in adhesive plates[27] (these can be seen in Fig. 1D). Administration of bFGF to re-plated EBs, followed by TGFα enriched particularly dopaminergic neurons.[32] Using a much more complex protocol, Carpenter *et al.*[28] have succeeded in enriching an almost completely pure population of functioning neuronal cells, which respond to neurotransmitters and exhibit action potentials.[28] This is done by exposing 4-day re-plated EBs to a battery of growth factors, including EGF, bFGF, PDGF, IGF-1, NT-3 and BDNF, followed by immunosorting the neuronal derivatives according to their surface antigens.

Recently, it was shown that co-culturing HESCs (un-aggregated) together with the inactivated stromal cell line MS5 leads to the formation of neural rosette-like structures.[33] Furthermore, when these neural rosettes were re-plated in the presence of FGF-8 and SHH, they differentiated specifically to dopaminergic neurons.[33] Likewise, the enrichment of neural cells within EBs may also be carried out by growing them in a serum-free media containing 50% of a medium conditioned by the human hepatocarcinoma cell line HepG2 (MedII condition media).[34]

Finally, *in vitro* differentiated neuronal precursors were shown to successfully integrate in mice brains.[29,30] Thus, neuronal precursors derived from HESCs were injected into neonatal mice brains, completed differentiation to mature neurons *in situ*, and integrated with the host tissues. Importantly, no teratoma formation was observed in any of the reports, even as late as six or eight weeks after transplantation.

### Bone

Following the treatment of re-plated EBs with an osteogenic cocktail containing the synthetic glucocorticoid hormone dexamethsone, the extracellular matrix-promoting agent ascorbic acid, and the mineralization promoter β-glycerophosphate, HESCs were reported to commit to an osteoblastic differentiation route.[37] These cells expressed features characteristic of bone-forming cells, like the deposition of hydroxyapetite minerals and transcription of osteogenic markers such as osteocalcin.

## Cardiomyocytes

Spontaneously differentiating EBs tend to display patterns of rhythmically contracting muscle areas.[6] These were further shown to display characteristics of early cardiac cells.[39,66] When thoroughly scrutinized, these cells exhibited a typical ultrastructure, containing gap junctions and Z bands, as well as a typical electrophysiological signature of several types of cardiomyocytes.[39–41,66–68]

Steps towards the enrichment of this cell population were taken, and progress reported upon the addition of TGFβ[26] or the demethylation reagent 5-aza-2'-deoxycytidine.[40] The latter method was further improved when the cardiac cells were sorted from other unwanted cell types by means of Percoll-gradient separation.[40]

Another approach for directing cardiomyocytes differentiation, came from co-culturing HESCs with the visceral endoderm-like cell line END-2.[41] Based on developmental studies, it is suggested that an interaction between mesodermal tissues and the adjacent visceral endoderm may give the inductive signals for cardiac development.[69] Indeed, when co-cultured with the visceral endoderm-like cell line, HESCs differentiated within three weeks to completely functional cardiomyocytes, as was determined by electrophysiological, morphological, and immunocytochemical methods.[41] It was recently suggested that the course of EB formation is somewhat similar to the creation of an early mammalian embryo and, after aggregation of ES to EB, the outer cells differentiate to extra embryonic endoderm.[43,70] This might explain why cardiac cells are so frequent in spontaneously differentiating EB cultures.

## Endothelial cells

The expression of endothelial specific genes, such as PECAM1/CD31 and CD34, increases during EB maturation and peaks by two weeks, indicating spontaneous differentiation toward this lineage.[42] Sorting the PECAM1/CD31 positive cells at the end of the second week by means of FACS and using fluorescent anti-PECAM antibody, results in endothelial precursors that can form a cord-like vascular structure when grown in three-dimensional Matrigel cultures. Moreover, when the cells were implanted subcutaneously into immunocompromised mice, they formed blood vessels, identified by human-specific antibodies. It has been suggested that the human blood vessels are functional and anastomized with the murine circulation network.[42]

In another report, a successful effort toward endothelial enrichment of HESC cultures was undertaken by treating the culture with various angiogenic factors, such as VEGF and PDGF.[71]

## Blood cells

In the mammalian embryo, hematopoiesis is associated with the blood islands in the ventral mesoderm surrounding the yolk sac, where the embryonic progenitors are induced to form blood cells by signals from the endothelial cells. In the adult, this process is attributed to adult stem cells which reside in the bone marrow.[5] By trying to imitate these hematopoiesis-promoting environments, differentiation of HESCs toward a fate of hematopoietic precursors was made possible. When co-cultured with either the murine bone marrow cell line S17 or the yolk sac endothelial cell line C166,[44] HESCs were shown to differentiate into a CD34 positive hematopoietic progenitor cells, also expressing specific transcription factors such as GATA-2 and TAL-1. This $CD34^+$ population could be selected by means of immunomagnetic sorting, and when cultured on semi-solid media supplemented with hematopoietic growth factors, formed colonies characteristic of myeloid, erythroid and megakaryocytic lineages. Treating EBs with a combination of hematopoietic cytokines (namely: SCF, Flt-3 ligand, IL-3, IL-6 and G-GSF) provided similar results.[45] In this experiment, however, the population developed was somewhat restricted to CD45 positive cells, indicating a leukocyte population. When treated with the same combination of cytokines, together with the addition of VEGF-$A_{165}$,[47] HESCs could be directed to a $CD34^+$ progenitor population, this time differentiating selectively to an erythropoietic fate. With a slightly different combination of cytokines (SCF, Flt-3 ligand, IL-3, G-GSF, thrombopoietin and IL-4), a $CD45^+$ population emerged, of which about 25% of the cells were MHC class II positive cells.[46] Notably, these dendritic and macrophages like cells were functional antigen-presenting cells, and were capable of directly stimulating CD4 and CD8 T-cell response in culture.[46]

## Liver cells

The differentiation of HESC into hepatocyte-like cells was suggested to be achieved by the administration of sodium butyrate to the culture.[48] This induction method provides cells with morphological features of primary hepatocytes, out of which 70–80% were claimed to express liver specific genes, such as albumin and cytokertins 8 and 18. This method, however, also results in significant cell death.

Recently, a genetic approach was established in order to select liver cells.[38] Thus, undifferentiated HESCs were genetically labeled with an eGFP coding gene under the minimal promoter of the adult-liver-specific gene albumin. Hepatocyte-derived HESCs could now be traced upon their differentiation, as they express the fluorescent protein. In a culture produced by the dissociation of a 20-day old EBs, hepatocytes composed about 6% of the cells. After sorting the cells with FACS according to their fluorescent nature, one could reach a population of almost pure hepatocytes. This homogenous culture could be further grown for several passages without losing its hepatic characteristics. By using the same genetically-labeled cells, several factors and conditions were checked,[38] whereupon the administration of media conditioned by primary hepatocytes culture, and to a lesser extent, treating with the growth factor aFGF, were shown to induce hepatic differentiation.

## Pancreatic β-cells

Cells attributed to pancreatic lineages of exocrine and endocrine nature of both mice and humans were reported to develop in spontaneously differentiating ESC cultures, and efforts were also made in order to direct their differentiation.[49,50,72] Other evidences, however, suggested that the insulin-positive immunocytochemistry is attributed to insulin uptake from the medium.[73] More research is probably needed in order to specifically direct for that illusive cell type.

## Extra-embryonic tissues

Some extra embryonic tissues, like the extra embryonic endoderm, were expected to differentiate from HESCs, since they are normally generated from cells of the ICM, ancestors of ES cells. Indeed, yolk sac visceral endoderm was shown to appear spontaneously in differentiating EB populations.[70] This process could be enhanced by the addition of BMP2 to the culture medium, and blocked by the BMP antagonist noggin.[31] Trophoblast cells, on the other hand, were not expected to arise from ESCs, based on research with murine cells. Nonetheless, a small fraction of trophoblastic cells do spontaneously differentiate in HESC cultures that are left to grow as confluent monolayers.[3] Moreover, when plated on Matrigel-coated dishes and treated with BMP4, significant amount of trophoblast differentiation is observed.[52] This was evident by the development of trophoblastic-like, multi-nucleated, syncytia, together with an elevation in specific placental gene expression. Additionally, when EBs are imbedded in three-dimensional Matrigel "rafts", and grown without conditioned media or additional factors for long periods (up to 53 days), multi-cellular projections are formed. These protrude into the substrate, and are stained positive with an antibody against hCG.[53]

## Conclusions

During normal mammalian development, gastrulation is believed to initiate both the formation of the embryo's three-dimensional axes and the segregation of the three embryonic germ layers. From the evidences gathered here, it is obvious that mature tissues of all three layers are able to differentiate without clear pattern formation. These include pure ectodermal, mesodermal and endodermal tissues, and even germ cells. It is thus possible that there is a conceptual separation between tissue differentiation and pattern formation. This might be the reason that teratomas can present the full repertoire of cell types without any detectable axis.

Since their derivation, HESCs were considered to be pluripotent,[3,4] even before enough information could be obtained directly from them, as much of the missing data was extrapolated from research done on MESCs. In the mouse, the pluripotency of ESCs was demonstrated by their ability to contribute to all embryonic tissues of a chimera offspring, after their injection into the blastocyst.[10,11] In humans, this demonstration method is not conceivable. Nonetheless, from the evidences reviewed here, it is clear that HESCs also possess similar differentiation capabilities, as was demonstrated by their differentiation to all major embryonic tissues of all three germ layers, either *in vivo* or *in vitro*, without the need for chimera formation. Evidences, however, are emerging that HESCs are somewhat different from their mouse counterparts. While MESCs do not possess the capacity to create trophoblast cells,[74] HESCs were shown to spontaneously differentiate along this route.[3,52,53] This raises questions concerning the correct placing of HESCs in the developmental cascade, and suggests that they are more primitively originated than previously considered. HESCs may therefore be regarded as "totipotent" instead of "pluripotent." While diverging from MESCs, HESCs are becoming closer to human embryonic carcinoma cells, which can form extra embryonic tissues *in vivo*. This may be alarming, since GCTs of extra embryonic nature are highly malignant.

As many of the cell types can now be regularly generated *in vitro*, the differentiation of HESCs can be used as an extensive tool for directing and enriching specific cell populations that will be utilized in the clinic. Furthermore, these cells can also be utilized for studying general and specific interaction between cell types in the course of tissue creation, enabling far-reaching understanding of early human development.

## Acknowledgment

We thank Tamar Dvash and Maya Shuldiner for their valuable assistance and Yoav Mishar for critically reading the manuscript. This research was partially supported by funds from the Israel Ministry of Science and by a grant from the Israel Science Foundation (grant No. 672/02-1).

## REFERENCES

1. Evans MJ, Kaufman MH. (1981) Establishment in culture of pluripotential cells from mouse embryos. *Nature* **292**:154–156.
2. Martin GR. (1981) Isolation of a pluripotent cell line from early mouse embryos cultured in medium conditioned by teratocarcinoma stem cells. *Proc Natl Acad Sci USA* **78**:7634–7638.
3. Thomson JA, Itskovitz-Eldor J, Shapiro SS, *et al.* (1998) Embryonic stem cell lines derived from human blastocysts. *Science* **282**:1145–1147.
4. Reubinoff BE, Pera MF, Fong CY, *et al.* (2000) Embryonic stem cell lines from human blastocysts: Somatic differentiation *in vitro*. *Nat Biotechnol* **18**:399–404.
5. Gilbert SF. (2000) *Developmental Biology*, 6th ed. Sunderland, MA: Sinauer Associates, Inc.
6. Itskovitz-Eldor J, Schuldiner M, Karsenti D, *et al.* (2000) Differentiation of human embryonic stem cells into embryoid bodies compromising the three embryonic germ layers *Mol Med* **6**:88–95.
7. Eiges R, Benvenisty N. (2002) A molecular view on pluripotent stem cells. *FEBS Lett* **529**:135–141.

8. Suh MR, Lee Y, Kim JY, *et al.* (2004) Human embryonic stem cells express a unique set of microRNAs. *Dev Biol* **270**:488–498.

9. Carlson BM. (1988) *Patten's Foundations of Embryology*, 5th ed. New York: McGraw-Hill.

10. Bradley A, Evans M, Kaufman MH, Robertson E. (1984) Formation of germ-line chimaeras from embryo-derived teratocarcinoma cell lines. *Nature* **309**:255–256.

11. Nagy A, Gocza E, Diaz EM, *et al.* (1990) Embryonic stem cells alone are able to support fetal development in the mouse. *Development* **110**:815–821.

12. Schuldiner M, Itskovitz-Eldor J, Benvenisty N. (2003) Selective ablation of human embryonic stem cells expressing a "suicide" gene. *Stem Cells* **21**:257–265.

13. Heins N, Englund MC, Sjoblom C, *et al.* (2004) Derivation, characterization, and differentiation of human embryonic stem cells. *Stem Cells* **22**:367–376.

14. Gonzalez-Crussi F. (1982) *Exrtragonadal Teratomas*, 2nd ed. Washington, DC: Armed Forces Institute of Pathology.

15. Vortmeyer AO, Devouassoux-Shisheboran M, Li G, *et al.* (1999) Microdissection-based analysis of mature ovarian teratoma. *Am J Pathol* **154**:987–991.

16. Xu C, Inokuma MS, Denham J, *et al.* (2001) Feeder-free growth of undifferentiated human embryonic stem cells. *Nat Biotechnol* **19**:971–974.

17. Reubinoff BE, Pera MF, Vajta G, Trounson AO. (2001) Effective cryopreservation of human embryonic stem cells by the open pulled straw vitrification method. *Hum Reprod* **16**:2187–2194.

18. Schuldiner M, Benvenisty N. (2003) Factors controlling human embryonic stem cell differentiation. *Methods Enzymol* **365**:446–461.

19. Amit M, Carpenter MK, Inokuma MS, *et al.* (2000) Clonally derived human embryonic stem cell lines maintain pluripotency and proliferative potential for prolonged periods of culture. *Dev Biol* **227**:271–278.

20. Tzukerman M, Rosenberg T, Ravel Y, *et al.* (2003) An experimental platform for studying growth and invasiveness of tumor cells within teratomas derived from human embryonic stem cells. *Proc Natl Acad Sci USA* **100**:13507–13512.

21. Richards M, Tan S, Fong CY, *et al.* (2003) Comparative evaluation of various human feeders for prolonged undifferentiated growth of human embryonic stem cells. *Stem Cells* **21**:546–556.

22. Park JH, Kim SJ, Oh EJ, *et al.* (2003) Establishment and maintenance of human embryonic stem cells on STO, a permanently growing cell line. *Biol Reprod* **69**:2007–2014.

23. Hwang WS, Ryu YJ, Park JH, *et al.* (2004) Evidence of a pluripotent human embryonic stem cell line derived from a cloned blastocyst. *Science* **303**:1669–1674.

24. Hovatta O, Mikkola M, Gertow K, *et al.* (2003) A culture system using human foreskin fibroblasts as feeder cells allows production of human embryonic stem cells. *Hum Reprod* **18**:1404–1409.

25. Lee JB, Lee JE, Park JH, *et al.* (2004) Establishment and maintenance of human embryonic stem cell lines on human feeder cells derived from uterine endometrium under serum-free condition. *Biol Reprod* (in press).

26. Schuldiner M, Yanuka O, Itskovitz-Eldor J, *et al.* (2000) Effects of eight growth factors on the differentiation of cells derived from human embryonic stem cells. *Proc Natl Acad Sci USA* **97**:11307–11312.

27. Schuldiner M, Eiges R, Eden A, *et al.* (2001) Induced neuronal differentiation of human embryonic stem cells. *Brain Res* **913**:201–205.

28. Carpenter MK, Inokuma MS, Denham J, *et al.* (2001) Enrichment of neurons and neural precursors from human embryonic stem cells. *Exp Neurol* **172**:383–397.

29. Reubinoff BE, Itsykson P, Turetsky T, *et al.* (2001) Neural progenitors from human embryonic stem cells. *Nat Biotechnol* **19**:1134–1140.

30. Zhang SC, Wernig M, Duncan ID, *et al.* (2001) *In vitro* differentiation of transplantable neural precursors from human embryonic stem cells. *Nat Biotechnol* **19**:1129–1133.

31. Pera MF, Andrade J, Houssami S, *et al.* (2004) Regulation of human embryonic stem cell differentiation by BMP-2 and its antagonist noggin. *J Cell Sci* **117**:1269–1280.

32. Park S, Lee KS, Lee YJ, *et al.* (2004), Generation of dopaminergic neurons *in vitro* from human embryonic stem cells treated with neurotrophic factors. *Neurosci Lett* **359**:99–103.

33. Perrier AL, Tabar V, Barberi T, *et al.* (2004) Derivation of midbrain dopamine neurons from human embryonic stem cells. *Proc Natl Acad Sci USA* **101**:12543–12548.

34. Schulz TC, Palmarini GM, Noggle SA, *et al.* (2003) Directed neuronal differentiation of human embryonic stem cells. *BMC Neurosci* **4**:27.

35. Goldstein RS, Drukker M, Reubinoff BE, Benvenisty N. (2002) Integration and differentiation of human embryonic stem cells transplanted to the chick embryo. *Dev Dyn* **225**:80–86.

36. Lee DH, Park S, Kim EY, *et al.* (2004) Enhancement of re-closure capacity by the intra-amniotic injection of human embryonic stem cells in surgically induced spinal open neural tube defects in chick embryos. *Neurosci Lett* **364**:98–100.

37. Sottile V, Thomson A, McWhir J. (2003) *In vitro* osteogenic differentiation of human ES cells. *Cloning Stem Cells* **5**:149–155.

38. Lavon N, Yanuka O, Benvenisty N. (2004) Differentiation and isolation of hepatic-like cells from human embryonic stem cells. *Differentiation* **72**:230–238.

39. Kehat I, Kenyagin-Karsenti D, Snir M, *et al.* (2001) Human embryonic stem cells can differentiate into myocytes with structural and functional properties of cardiomyocytes. *J Clin Invest* **108**:407–414.

40. Xu C, Police S, Rao N, Carpenter MK. (2002) Characterization and enrichment of cardiomyocytes derived from human embryonic stem cells. *Circ Res* **91**:501–508.

41. Mummery C, Ward D, van den Brink CE, *et al.* (2002) Cardiomyocyte differentiation of mouse and human embryonic stem cells. *J Anat* **200**:233–242.

42. Levenberg S, Golub JS, Amit M, *et al.* (2002) Endothelial cells derived from human embryonic stem cells. *Proc Natl Acad Sci USA* **99**:4391–4396.

43. Rodda SJ, Kavanagh SJ, Rathjen J, Rathjen PD. (2002) Embryonic stem cell differentiation and the analysis of mammalian development. *Int J Dev Biol* **46**:449–458.

44. Kaufman DS, Hanson ET, Lewis RL, *et al.* (2001) Hematopoietic colony-forming cells derived from human embryonic stem cells. *Proc Natl Acad Sci USA* **98**:10716–10721.

45. Chadwick K, Wang L, Li L, *et al.* (2003) Cytokines and BMP-4 promote hematopoietic differentiation of human embryonic stem cells. *Blood* **102**:906–915.

46. Zhan X, Dravid G, Ye Z, *et al.* (2004) Functional antigen-presenting leucocytes derived from human embryonic stem cells *in vitro*. *Lancet* **364**:163–171.

47. Cerdan C, Rouleau A, Bhatia M. (2004) VEGF-A165 augments erythropoietic development from human embryonic stem cells. *Blood* **103**:2504–2512.

48. Rambhatla L, Chiu CP, Kundu P, *et al.* (2003) Generation of hepatocyte-like cells from human embryonic stem cells. *Cell Transplant* **12**:1–11.

49. Assady S, Maor G, Amit M, *et al.* (2001) Insulin production by human embryonic stem cells. *Diabetes* **50**:1691–1697.

50. Segev H, Fishman B, Ziskind A, *et al.* (2004) Differentiation of human embryonic stem cells into insulin-producing clusters. *Stem Cells* **22**:265–274.

51. Clark AT, Bodnar MS, Fox M, *et al.* (2004) Spontaneous differentiation of germ cells from human embryonic stem cells *in vitro*. *Hum Mol Genet* **13**:727–739.

52. Xu RH, Chen X, Li DS, *et al.* (2002) BMP4 initiates human embryonic stem cell differentiation to trophoblast. *Nat Biotechnol* **20**:1261–1264.

53. Gerami-Naini B, Dovzhenko OV, Durning M, *et al.* (2004) Trophoblast differentiation in embryoid bodies derived from human embryonic stem cells. *Endocrinology* **145**:1517–1524.

54. Anderson GB, BonDurant RH, Goff L, *et al.* (1996) Development of bovine and porcine embryonic teratomas in athymic mice. *Anim Reprod Sci* **45**:231–240.

55. Damjanov I, Damjanov A, Solter D. (1987) Production of teratocarcinomas from embryos transplanted to extra-uterine sites. In *Teratocarcinomas and Embryonic Stem Cells: A Practical Approach*, ed. E.J. Robertson, pp. 1–18, Oxford: IRL Press.

56. Gatcombe HG, Assikis V, Kooby D, Johnstone PA. (2004) Primary retroperitoneal teratomas: A review of the literature. *J Surg Oncol* **86**:107–113.

57. Ueno T, Tanaka YO, Nagata M, *et al.* (2004) Spectrum of germ cell tumors: From head to toe. *Radiographics* **24**:387–404.

58. Scully RE, Young RH, Clement PB. (1998) *Tumors of the Ovary, Maldeveloped Gonads, Fallopian Tube, and Broad Ligment*, 3rd ed. Washington, DC: Armed Forces Institute of Pathology.

59. Sergi C, Ehemann V, Beedgen B, *et al.* (1999) Huge fetal sacrococcygeal teratoma with a completely formed eye and intratumoral DNA ploidy heterogeneity. *Pediatr Dev Pathol* **2**:50–57.

60. Stevens LC, Pierce BG. (1975) Teratomas: Definitions and terminology. In *Teratomas and Differentiation*, eds. M.I. Sherman and D. Soltereds, pp. 13–14, New York: Academic Press, INC.

61. Jones TD, Ulbright TM, Eble JN, *et al.* (2004) OCT4 staining in testicular tumors: A sensitive and specific marker for seminoma and embryonal carcinoma. *Am J Surg Pathol* **28**:935–940.

62. Lawrenz B, Schiller H, Willbold E, *et al.* (2004) Highly sensitive biosafety model for stem-cell-derived grafts. *Cytotherapy* **6**:212–222.

63. Hornstein E, Benvenisty N. (2004) The "brainy side" of human embryonic stem cells. *J Neurosci Res* **76**:169–173.

64. Bjorklund LM, Sanchez-Pernaute R, Chung S, *et al.* (2002) Embryonic stem cells develop into functional dopaminergic neurons after transplantation in a Parkinson rat model. *Proc Natl Acad Sci USA* **99**:2344–2349.

65. Gerecht-Nir S, Cohen S, Itskovitz-Eldor J. (2004) Bioreactor cultivation enhances the efficiency of human embryoid body (hEB) formation and differentiation. *Biotechnol Bioeng* **86**:493–502.

66. Kehat I, Gepstein A, Spira A, *et al.* (2002) High-resolution electrophysiological assessment of human embryonic stem cell-derived cardiomyocytes: A novel *in vitro* model for the study of conduction. *Circ Res* **91**:659–661.

67. He JQ, Ma Y, Lee Y, *et al.* (2003) Human embryonic stem cells develop into multiple types of cardiac myocytes: Action potential characterization. *Circ Res* **93**:32–39.

68. Reppel M, Boettinger C, Hescheler J. (2004) Beta-adrenergic and muscarinic modulation of human embryonic stem cell-derived cardio-myocytes. *Cell Physiol Biochem* **14**:187–196.

69. Nakamura T, Schneider MD. (2003) The way to a human's heart is through the stomach: Visceral endoderm-like cells drive human embryonic stem cells to a cardiac fate. *Circulation* **107**:2638–2639.

70. Conley BJ, Trounson AO, Mollard R. (2004) Human embryonic stem cells form embryoid bodies containing visceral endoderm-like derivatives. *Fetal Diagn Ther* **19**:218–223.

71. Gerecht-Nir S, Ziskind A, Cohen S, Itskovitz-Eldor J. (2003) Human embryonic stem cells as an *in vitro* model for human vascular development and the induction of vascular differentiation. *Lab Invest* **83**:1811–1820.

72. Lumelsky N, Blondel O, Laeng P, *et al.* (2001) Differentiation of embryonic stem cells to insulin-secreting structures similar to pancreatic islets. *Science* **292**:1389–1394.

73. Rajagopal J, Anderson WJ, Kume S, *et al.* (2003) Insulin staining of ES cell progeny from insulin uptake. *Science* **299**:363.

74. Soares MJ, Wolfe MW. (2004) Human embryonic stem cells assemble and fulfill their developmental destiny. *Endocrinology* **145**:1514–1516.

# 8

# Feeder-free Culture of Human Embryonic Stem Cells

Scott A. Noggle, Noboru Sato and Ali H. Brivanlou*

## Introduction

Since the initial derivation of mouse embryonic stem cells (mESCs) from the pluripotent inner cell mass (ICM) of blastocysts,[1,2] derivation of embryonic stem cells from other species has now been reported.[3-5] Recently, the isolation of human embryonic stem cells (hESCs) has generated the exciting possibility of both access to the basic science of human development, as well as the possibility of new hope for cell-based therapy in the clinic.[6] hESC lines provide a unique *in vitro* system in which studies of some aspects of cell fate specification in early human development are feasible. hESCs also have tremendous potential for use in cell-based therapy of degenerative diseases and injuries, as well as hold great promise for clinically applicable discoveries from hESC lines with genetic defects. Such lines could provide *in vitro* models of diseases and should be useful in screens for drug candidates directed at treating these diseases. However, the therapeutic potential of hESCs relies on their efficient derivation and maintenance in culture. This, in turn, relies on an understanding of the molecular basis of the signaling pathways that maintain ESCs in an undifferentiated, pluripotent state.

While the therapeutic potential of hESCs may be predicted by studies with mESCs, the ability to translate the available knowledge of the mechanisms involved in the maintenance of pluripotency in mESCs to hESCs is only beginning to emerge. Initial comparative molecular and cellular characterizations of hESCs seem to indicate properties both similar to and different from mESCs. For example, plating isolated inner cell mass (ICMs) on mouse embryonic fibroblast (MEF) feeder layers derives

*Correspondence: Laboratory of Molecular Vertebrate Embryology, The Rockefeller University, New York, NY 10021-6399, Tel.: (212)327-8656, Fax: (212)327-8685, e-mail: brvnlou@rockefeller.edu

mESCs. Studies of the molecular mechanisms that mESCs use to maintain the undifferentiated, pluripotent state show that signaling through the leukemia inhibitory factor (LIF)/gp130 signaling pathway is sufficient for the propagation of undifferentiated mESCs in the presence or absence of MEF feeder layers.[7–9] LIF signals through LIFR and gp130[10,11] to maintain mESC self-renewal[7–9] by activating the JAK/STAT pathway (Fig. 1).[12,13] However, while the activation of the STAT pathway seems to be sufficient for the maintenance of pluripotency in mESCs, a volume of evidence seems to suggest that it is not necessary. For example, loss of function of LIF, gp130,

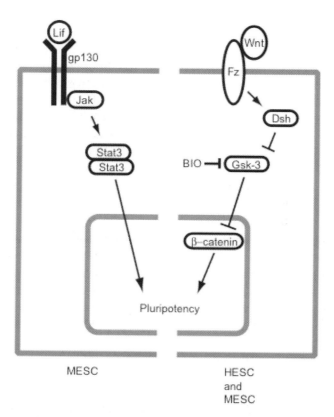

**Figure 1.** Pluripotency signaling pathways. In mESCs, the JAK/STAT pathway is activated by Il-6 family cytokines, such as LIF, resulting in maintenance of pluripotency. In hESCs (and mESCs), activation of canonical Wnt signaling is initiated by a Wnt ligand binding to the receptor Frizzled (Fz) and inhibits GSK-3 through Dishevelled (Dsh). In the absence of active Wnt signaling, GSK-3 constitutively phosphorylates β-catenin leading to degradation of β-catenin through the ubiquitin proteosome pathway. Inhibition of GSK-3 results in stabilization of β-catenin which can signal in the nucleus to maintain pluripotency.

or STAT3 does not affect the maintenance of pluripotency in mESCs. Additionally, LIF cannot sustain the undifferentiated state in hESCs; the STAT pathway has been shown not to be active in the state of pluripotency in hESCs. These arguments, taken together, strongly suggest that in mESCs, signals other than STAT3 and LIF could be at play to maintain mESCs in a pluripotent state.[14]

While the first embryonic stem cells derived from human blastocysts were also plated on MEF, it quickly became clear that the required conditions for maintaining their undifferentiated state were different from those described for mESCs. For example, basic FGF (bFGF) is required for hESCs to maintain their undifferentiated phenotype, while it is not required in mESCs. Additionally, as stated above, neither human nor non-human primate cell lines are sensitive to LIF for the maintenance of pluripotency. Despite these differences, it is clear that the MEF feeder layers provide a major, yet not fully defined, contribution to hESC maintenance. hESCs can be cultured in the absence of direct contact with the feeder cells,[15] but in the presence of feeder-conditioned medium, suggesting that soluble factors are involved. It also implies that maintenance is independent of both factors produced by feeders in response to hESC derived signals, as well as signals passed by direct contact with the feeder layer. However, their undefined properties, in addition to possible cross-contamination, make them undesirable from a clinical perspective.

While not technically prohibitive, the inclusion of xeno-derived culture components presents significant hurdles for products destined for therapy at the individual level. Animal pathogens present in the feeder layers could easily be transferred to the hESCs, which in turn might function as a carrier of these pathogens. Transmission to the individual could be efficient, as intended cell replacement therapies would bypass natural immunological barriers to these pathogens. Limiting the use of xeno-derived products in cell-based therapy will also minimize the risk of transfer of animal pathogens into the general population.

Aside from defining components of the conditioned medium that contribute to maintaining the undifferentiated state, there are regulatory issues relevant to therapeutic application, which arise from the use of feeders. The first involves the production of a consistent, predictable product. In terms of cell therapy, it will be necessary to reduce variability in quality of hESC cultures from lot to lot, cell line to cell line. While the quality of feeders is important for the maintenance of hESCs, precise indicators of feeder quality have not been found. In addition, feeder layers stimulated by serum derived factors, such as bFGF, might produce factors not found in unstimulated feeder conditioned medium. As most protocols call for bFGF in the conditioning medium, this may introduce significant variability in the quality and quantity of factors produced in response to the FGF signal. It is also possible that the strength or quality of the FGF signal might induce different factors or different amounts of active factors. For example, it has been suggested that variation between lots and feeder densities induces variable levels of TGFβ inhibitory signals that might inhibit extra-embryonic endoderm development in hESC cultures or differentially modulate their response to differentiation signals.[16] Achieving controlled application of the active components using purified compounds could reduce variability of cultured hESCs. This is critical for large-scale expansion of hESCs to meet clinical needs.

Removal of the feeder cells eliminates a major theoretical source of pathogens. Replacing the active functions, for example, with small molecule or recombinant protein pathway effectors, in hESC culture is preferable.

## Feeder-free hESC Culture System

Culture conditions for maintaining hESCs in feeder-free conditions have recently been reduced to several distinct elements. Below, we review in some detail feeder-free hESC culture and suggest future directions for possible improvement of feeder-free culture.

### hESC cell lines

We have predominantly used the following hESC lines, as they were readily available for analysis: H1 (WiCell), BG01, and BG02 (BresaGen, Inc.). The culture conditions presented below are applicable across a variety of independent hESC lines.

### Substrate requirements

The ICM of primate blastocysts is associated with a basal lamina, pointing to the possible importance of the extracellular matrix (ECM), *in vivo*, in the maintenance or function of the ICM.[17] When hESCs are grown under conditions for feeder-free growth of mESCs, such as on gelatin coated plastic dishes, they rapidly change their morphology and lose expression of pluripotency markers. hESCs grown in feeder-free conditions instead require an extracellular matrix substrate to maintain their growth in an undifferentiated state. A complex matrix, such as Matrigel (from BD Biosciences, Inc.), has routinely and successfully been used. However, purified components such as human or bovine fibronectin are also acceptable.[18] Additionally, an optimal substrate coating is essential for success, as hESCs lose their undifferentiated state when grown on inadequately prepared substrate. To prepare and optimize matrix coating with Matrigel, plates are coated with non-gelled Matrigel, usually overnight in cold conditions. Warming subsequently polymerizes the Matrigel in the humidified incubator. Excess substrate is washed from the plate with growth medium leaving a layer of extracellular matrix coating the dish. When hESCs are plated on this substrate they attach, spread, and maintain ICM morphology under non-differentiating conditions.

## Application of GSK-3 inhibitors

Using microarray analysis of differentiated and undifferentiated hESCs, as well as small molecule inhibitors, we have identified the canonical Wnt signaling pathway as instrumental in regulating the maintenance of pluripotency in both mESCs and hESCs.[19,20] In canonical Wnt signaling, Wnt ligands convey signals through Frizzled receptors at the cell surface that result in the inhibition of glycogen synthase kinase-3 (GSK-3) (Fig. 1). GSK-3 constitutively marks β-catenin in a signaling pool for degradation by ubiquitination. Inhibition of GSK-3, therefore, results in an accumulation of β-catenin that can then move to the nucleus and, in collaboration with TCFs, activate target gene expression.[21,22] Gene expression analysis showed that, among other pathways represented, all of the known members of the canonical Wnt signaling pathway are enriched in the undifferentiated hESCs. Led by this result, in addition to unbiased screens for small molecules that maintain mESCs and hESCs undifferentiated in the absence of MEF or conditioned medium (CM) derived from MEFs, we defined a functional ability of the canonical Wnt signaling pathway to maintain the pluripotent state of ESCs.

We discovered that a potent and specific pharmacological inhibitor of GSK-3, isolated from the Mollusk, is able to maintain the undifferentiated phenotype of both mouse and human ESCs by activating the canonical Wnt pathway.[20] Components of the vivid purple dye, "Tyrian purple," isolated from mollusks in various parts of the world and used by many human societies in classical antiquity and often associated with royalty and aristocracy, includes indirubins. This family of pharmacological agents has recently been shown as inhibitors of cyclin-dependent kinases (CDKs) and GSK-3.[23,24] A cell permeable synthetic derivative, 6-bromoindirubin-3′-oxime (BIO), was subsequently shown to be specific for GSK-3. We found that BIO, as opposed to the kinase-inactive derivative MeBIO, was sufficient in both mESCs and hESCs for the maintenance of the undifferentiated, pluripotent phenotype (Fig. 3). In addition to BIO, a limited number of pharmacological inhibitors of GSK-3 are also available, the best characterized of which is lithium. However, lithium is only functional at high concentrations and can have toxic side effects. BIO seems to be the most potent GSK-3 inhibitor, functioning to inhibit GSK-3 in the low nanomolar range in cell-free kinase assays. In contrast to non-substituted indirubins, BIO is over 16-fold more potent for GSK-3 than for the next most susceptible kinase, CDK5, demonstrating good selectivity for GSK-3 (Table 1). BIO is also active in cell culture, as it is able to modulate the activity of the Wnt pathway in the low micromolar range. At these concentrations related CDKs, such as CDK5, are minimally inhibited. However, it is critical to find a minimal concentration of BIO sufficient to maintain each hESC line in the undifferentiated state and that minimizes any significant effect on their viability or growth rate due to possible low-level CDK inhibition. The optimal concentration of the GSK-3 inhibitor (BIO) should be determined for each hESC line as the rate of the cell cycle correlates with the optimal dose of BIO and hESC lines can vary in their cell cycle length. Concentrations from 1 μM to 5 μM provide an appropriate starting range. After about seven days of growth, hESCs treated with BIO start showing signs of differentiation similar to hESCs grown in CM cultured for a long period. Therefore, hESCs are grown in non-CM containing BIO for five days before passaging. However, we also find that BIO-treated hESCs tend to reduce their growth rate and their ability to self-renew after three or four passages. (Adding BIO to cultures on a daily basis might improve these conditions. As the case with any drug, BIO likely undergoes a quick turnover and needs to be refurbished.) Our present goal is to determine culture

**Table 1.** Selectivity of Indirubin-derivatives

|  | GSK-3α/β | CDK1/CyclinB | CDK5/p25 |
| --- | --- | --- | --- |
| Indirubin | 1.000 | 10.000 | 10.000 |
| 6-bromoindirubin-3′-oxime (BIO) | 0.005 | 0.320 | 0.083 |
| 1-methyl-6-bromoindirubin-3′-oxime (MeBIO) | 44.000 | 55.000 | >100.000 $IC_{50}$ (μM) |

conditions that allow us to extend the BIO-mediated self-renewal in hESCs. We believe that developing this culture system is one of the most promising and necessary directions towards therapeutic applications.

Upon withdrawal from CM or BIO, hESCs show the ability to differentiate into multiple cell types *in vitro* when aggregated into embryoid bodies and *in vivo* in teratomas formed in immunodeficient mice.[20] Differentiation in these systems can be accessed morphologically by the appearance of cell types or tissues and by analysis of differentiated, tissue-specific markers.

## Passaging conditions

Two independent approaches have been reported for successful passage of hESCs. The first and most commonly used technique is to transfer cells as clumps. The second relies on cell dissociation followed by single cell passaging.[18] In both cases, the cells are traditionally cultured on MEFs or Matrigel. Time-lapse analysis of hESCs passaged in clumps illustrates consistent survival and generation of new colonies of proliferating cells. When parallel cultures with similar numbers of cells were isolated by microdissection and then broken down into a single cell suspension with Trypsin/EDTA, they survived when plated on MEFs. However, in our hands, the vast majority of these cells integrated into the MEF feeder layer and did not generate growing colonies. Rarely, however, some hESCs proliferate and generate colonies. Similar results were achieved with hESC lines BGN1, BGN2, and H1 using non-enzymatic cell dissociation buffer or EDTA alone. Therefore, some hESC lines may require cell-cell contact to remain undifferentiated. Alternative methods for passaging hESCs in cell clumps include enzymatic treatments, such as collagenase or dispase, that do not result in dissociation of the colonies into single cells. Such methods have successfully maintained hESCs, undifferentiated in culture. However, as the more technically desirable single cell passaging method is possible in at least one case,[25] further study is needed to resolve this discrepancy.

It is important to note that the maintenance of cell-cell contact in hESC colonies may be important for retaining normal karyotype as well as hESC pluripotency.[26] While aneuploidy has been detected in these cases,[27] the suggestion has been made that it is due to metabolic stress (see rebuttal in 26) instead of passaging technique. It is possible that a combination of passaging technique and metabolic stress can accelerate selection for advantageous karyotypic changes. Thus, routine karyotyping of hESC cultures is highly recommended.

# hESC Behavior in Feeder-free Culture

## Maintenance of the undifferentiated state

The morphology of hESCs maintained in MEF feeder conditioned medium and in BIO shows striking similarity to the cells of the human ICM. Cells in the ICM of high grade expanded, and the hatching human blastocysts are compacted, often appearing more tightly associated with each other than with the trophectoderm.[28] Undifferentiated hESCs, much as cells of the ICM, show a high nuclear to cytoplasmic ratio without signs of polarization. The appearance of a polarized epithelium in the ICM *in vivo* initiates the formation of the amniotic cavity and is one of the first morphological signs of differentiation in the ICM of implanting primate embryos.[17] This parallels differentiation seen in hESCs cultured on feeders and on Matrigel. When hESCs begin to differentiate, they decrease their nuclear to cytoplasmic ratio, elongate, and spread on the dish (Fig. 2). This morphological change is usually evident within days of withdrawal from CM or BIO. However, we have noted that increased density or size of the colonies can delay differentiation, indicating that hESCs may produce factors that aid in sustaining their maintenance. Different cell lines also differentiate at different rates.

While morphology is convenient for monitoring the differentiation state of any cell, including hESCs, it can be misleading if taken as conclusive evidence. Basic embryologists working with model embryos have noticed for some time that cells endowed with closely similar morphology might be expressing genetic programs that are quite distinct and, in fact, remain opposite from each other. The frog blastocyst is a good example of this conundrum, as it is morphologically identical in specific regions of the spherical embryo, yet gives rise to distinct cell fates following regional molecular clues. This led to the use of cell type-specific molecular markers as a diagnostic of cell type, giving birth to the field of molecular embryology in the latter part of the 20th century.

As these stem cells are embryonic in origin, we strongly believe that rules should be applied in the diagnosis of the differentiation state of hESCs. A number of molecular markers were originally suggested as a standard of measurement of hESC potency.[29] This list has expanded today to allow quantitative and dynamic measurements of molecular changes occurring during hESC differentiation. An example of marker expression of hESCs cultured in BIO is shown in Fig. 3. hESCs in BIO, but not a

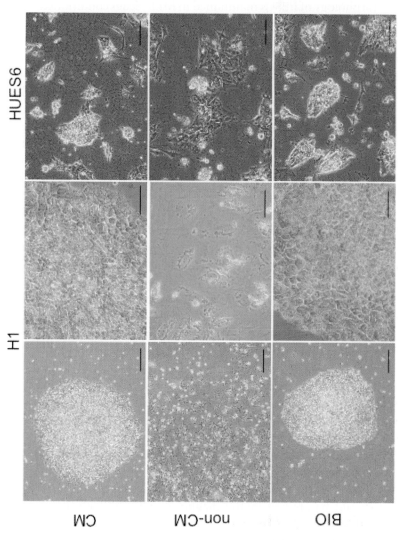

**Figure 2.** The ICM-like morphology of hESCs is maintained by BIO. H1 (*left and middle* columns) or HUES6 (25) (*right* column) cell lines were cultured in conditions indicated for 7 and 5 days, respectively. Note that both cell lines show tightly associated cells with a high nuclear to cytoplasmic ratio in MEF-conditioned medium (CM). This morphology is lost in the absence of CM (non-CM). In BIO, both cell lines maintain a morphology similar to that seen in CM. Scale bars: (*left* column) 300 μM; (*middle* and *right* columns) 100 μM.

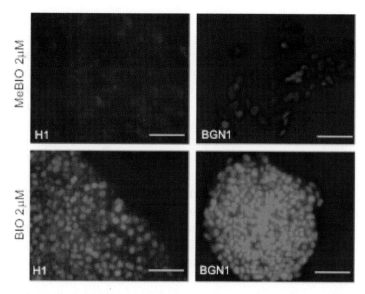

**Figure 3.** BIO maintains the undifferentiated state of hESCs. hESCs from the H1 and BGN1 cell lines were maintained in 2 μM MeBIO (*top*) or BIO (*bottom*) for 7 days and analyzed by immunofluorescence for Oct-3/4. hESCs in MeBIO have a low level of Oct-3/4, whereas those in BIO have maintained a high level of Oct-3/4 expression indicative of the undifferentiated state. Scale bars: 100 μM.

kinase-inactive derivative MeBIO, are marked by expression of Oct-3/4 by immunofluorescence.

Molecular markers of "transition state" during differentiation must be identified. As with the biochemical nature of the metabolic pathway, the discovery of the "intermediate" or "transitional" fate during hESC differentiation will be crucial to the understanding of the hierarchy of molecular decisions underlying the maintenance of the undifferentiated state, as well as specific differentiation paradigms. These types of molecular marker analysis have been instrumental in the discovery of small compounds that can direct hESCs toward different paths.

## Time-lapse microscopy of hESCs

Preliminary analysis of the dynamic behavior of hESCs in feeder-free conditions by time-lapse microscopy has also revealed unexpected single cell behavior in hESCs in BIO or CM on Matrigel. A collection of these are available on the Brivanlou Lab website at http://xenopus.rockefeller.edu.

For time-lapse photography, the culture conditions were maintained essentially as described above with consideration of temperature, evaporation, and pH of the medium by using an on-stage incubation system (Zeiss) as described below. Temperature was controlled by using a heated stage, with airflow containing 5% $CO_2$. High-resolution images were acquired at one minute intervals for 12 hours to 24 hours. Initially, phase contrast images were acquired with continuous illumination of the culture. This proved to be incompatible with BIO-treated cultures. All cells in BIO-treated hESC colonies showed signs of apoptosis within hours during continuous illumination. This underscores the fact that BIO is photosensitive, as is the dye Tyrian purple from which it is derived. Untreated cultures did not show this behavior even after several days of continuous illumination. In subsequent experiments, illumination was shuttered except during image acquisition. This was compatible with successful culture in BIO for at least 12 hours. This property of BIO is under further exploration.

One striking characteristic of hESCs revealed in the feeder-free system is the activity of filopodial projections at the borders of the colony. These projections extend and contract from cells on the edge of the colony, sometimes reaching over a cell diameter away to make contact with the ECM matrix of the Matrigel (Fig. 4). Long fibers of matrix can be seen stretching from the Matrigel substrate and attaching to the colonies as the projections pull the Matrigel matrix into the colony. This can also be seen in cultures as a clearing of the matrix around the colony. In the observed instances, this exploratory behavior happens in minutes.

Filopodial projections are a sign of cell polarity and are thought to perform a sensing or long-range signaling function. They are also indicative of the ability of a cell to migrate. During migration, cells use filopodia, lamellipodia, or a combination of the two to reach out and sense the substrate on which they are traveling. These projections regulate the activity of their cytoskeletal machinery to guide and stabilize attachment and promote movement. Close inspection by time-lapse video microscopy of hESCs cultured in MEF conditioned medium reveals progressive spreading, whereby filopodia are initiated and lamellipodia are stabilized between the filopodia. While hESCs at the border of the colonies display motile or migratory behavior, cells within the colony are less active and not polarized. In contrast, filopodia are generated in BIO treated hESC cultures but are not subsequently stabilized. BIO treated cultures, in contrast to non-treated cultures, usually display a very tight colony morphology

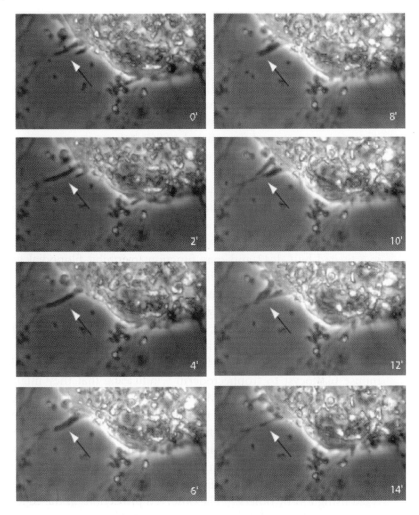

**Figure 4.** Filopodial dynamics of BGN1 hESCs in BIO. BGN1 hESCs were plated on Matrigel in the presence of BIO and imaged by time-lapse microscopy under phase contrast conditions. Images were collected at 2 min. intervals over 14 min. A filopodial projection is indicated by an arrow. During imaging, this projection extends from the edge of the hESC colony and then contracts.

with little difference seen between the edge cells of the colony and in cells in the interior of the colony (Fig. 2, BIO). BIO treated cells show less spreading on the substrate than hESCs exposed to CM. hESCs, without BIO or CM, spread and quickly differentiate (Fig. 2, non-CM). Therefore, the Wnt signaling pathway, in addition to its ability to promote pluripotency, may also delay some aspects of the acquisition of polarity in hESCs.

These observations are consistent with studies that have linked regulation of GSK-3 to the signaling pathways that control both polarity and cell migration (reviewed in Ref. 30). Our global microarray expression analysis of genes expressed during the undifferentiated state of hESCs suggests that at least some molecular components needed to regulate filopodial dynamics are present in undifferentiated hESCs.[19] Several genes are expressed in hESCs that are associated with filopodial dynamics, including Cdc42, N-WASP homologues, PAK1 and Myosin-10. Regulation of GSK-3 by Cdc42, a central regulator of cytoskeletal dynamics and polarity, is thought to modulate the activity of filopodia. In addition, PAK1 and Mysoin-10 in undifferentiated hESCs by microarray analysis are respectively enriched 5.6- and 4-fold, suggesting a function in stem cell maintenance. PAK1 is activated by Cdc42 and may function to inhibit filopodial formation by inactivating myosin light chain kinase which activates myosin filament assembly.[31] As PAK1 is enriched in undifferentiated hESCs, filopodial projections may indicate a first sign of differentiation. This is consistent with EM studies of primate blastocysts in which there is little specialization of structures within ICM cells prior to implantation.[17]

Filopodial projections have been described that connect the ICM of the mouse blastocyst to trophectoderm[32] indicating that they might have a function *in vivo*. However, EM studies have shown that these projections arise from larger projections of trophectoderm covering the juxtacoelic surface of the ICM.[33] Thus, the ICM remains a non-polarized, undifferentiated cell type. In light of these observations, polarization as evidenced by filopodia, is unexpected in hESCs. Polarization at the outer edges of hESC colonies grown on MEF feeders has been noted in hESCs analyzed using EM by others.[34] The observed projections at the edge of hESC colonies are also reminiscent of the trophectodermal projections described for rodent, bovine, equine, and human blastocysts that have been implicated in locomotion, attachment and implantation.[35] These are described as long projections originating from the late blastocyst trophectoderm. These observations may suggest that hESCs at the edge of the colony may retain some of the characteristics of trophectoderm. Indeed, hESCs in feeder-free conditions produce trophectoderm at colony borders in response to a BMP signal.[36] Thus, the projections from hESC colonies may indicate that the cells at the border may be more plastic in their differentiation state, displaying characteristics of both undifferentiated hESCs and the more differentiated trophectoderm. If true, hESCs may provide a cell culture model for studies of implantation mechanisms the failure of which are seen as a major cause of infertility.

Together, these observations suggest that inhibition of GSK-3 may, in addition to maintaining the pluripotency of hESCs, have a direct role in the polarity and motile behavior of hESCs. As the ICM is associated with a basal lamina both prior to and after implantation,[17] it is possible that integrin mediated interactions with the substrate play a role in survival as well as migration of hESCs. Mysoin-10, which is also enriched in hESCs, promotes filopodial formation and integrin relocalization in several cell types.[37,38] Mysoin-10 also interacts with integrin betas and several integrins are expressed in hESCs. This suggests an important function in ECM attachment in hESCs. Integrins are known to provide survival cues to cells through attachment to the ECM. These observations will need further study at the molecular level to understand the details of the pathways that hESCs use to regulate survival, polarity and the ability to migrate and differentiate.

While preliminary, these observations suggest a dynamic interaction of hESCs with their substrate. Using expression data as a guide, future studies should shed light on the optimal substrate conditions for hESC growth and maintenance in feeder-free conditions.

## Conclusions

Application of hESC technology in the clinic, from cell-based therapies to discoveries from lines harboring disease specific genetic defects, depends on a detailed knowledge of the basic science of hESC biology and on reliable and safe methods for their derivation, culture and expansion. Gene expression analysis by microarray technology has been productive in directing study towards unraveling the signaling pathways that maintain hESCs in an undifferentiated, pluripotent state. Further work is also needed to define markers of "transitional states" that occur during hESC differentiation. In addition to providing diagnostic markers for quality control of hESC cultures, they will be instrumental in the discovery of small compounds that can guide the molecular decisions underlying the maintenance of the undifferentiated state as well as direct hESCs toward different paths. We are optimistic that discoveries of well-defined stable compounds, such as the pharmacological GSK-3 inhibitor, BIO, will open new opportunities to direct the fate of hESCs.

For practical applications in the clinic to become a reality, the discovery of safe and dependable methods for regulating these pathways must be a priority.

# REFERENCES

1. Martin GR. (1981) Isolation of a pluripotent cell line from early mouse embryos cultured in medium conditioned by teratocarcinoma stem cells. *Proc Natl Acad Sci USA* **78**: 7634–7638.
2. Evans MJ and Kaufman MH. (1981) Establishment in culture of pluripotential cells from mouse embryos. *Nature* **292**: 154–156.
3. Thomson JA, Kalishman J, Golos TG, *et al.* (1995) Isolation of a primate embryonic stem cell line. *Proc Natl Acad Sci USA* **92**: 7844–7848.
4. Thomson JA, Kalishman J, Golos TG, *et al.* (1996) Pluripotent cell lines derived from common marmoset (Callithrix jacchus) blastocysts. *Biol Reprod* **55**: 254–259.
5. Suemori H, Tada T, Torii R, *et al.* (2001) Establishment of embryonic stem cell lines from cynomolgus monkey blastocysts produced by IVF or ICSI. *Dev Dyn* **222**: 273–279.
6. Thomson JA, Itskovitz-Eldor J, Shapiro SS, *et al.* (1998) Embryonic stem cell lines derived from human blastocysts. *Science* **282**: 1145–1147.
7. Smith AG and Hooper ML. (1987) Buffalo rat liver cells produce a diffusible activity which inhibits the differentiation of murine embryonal carcinoma and embryonic stem cells. *Dev Biol* **121**: 1–9.
8. Smith AG, Heath JK, Donaldson DD, *et al.* (1988) Inhibition of pluripotential embryonic stem cell differentiation by purified polypeptides. *Nature* **336**: 688–690.
9. Williams RL, Hilton DJ, Pease S, *et al.* (1988) Myeloid leukaemia inhibitory factor maintains the developmental potential of embryonic stem cells. *Nature* **336**: 684–687.
10. Gearing DP and Bruce AG. (1992) Oncostatin M binds the high-affinity leukemia inhibitory factor receptor. *New Biol* **4**: 61–65.
11. Gearing DP, Thut CJ, VandeBos T, *et al.* (1991) Leukemia inhibitory factor receptor is structurally related to the IL-6 signal transducer, gp130. *Embo J* **10**: 2839–2848.
12. Niwa H, Burdon T, Chambers I and Smith A. (1998) Self-renewal of pluripotent embryonic stem cells is mediated via activation of STAT3. *Genes Dev* **12**: 2048–2060.
13. Burdon T, Stracey C, Chambers I, *et al.* (1999) Suppression of SHP-2 and ERK signalling promotes self-renewal of mouse embryonic stem cells. *Dev Biol* **210**: 30–43.

14. Dani C, Chambers I, Johnstone S, *et al.* (1998) Paracrine induction of stem cell renewal by LIF-deficient cells: A new ES cell regulatory pathway. *Dev Biol* **203**: 149–162.

15. Xu C, Inokuma MS, Denham J, *et al.* (2001) Feeder-free growth of undifferentiated human embryonic stem cells. *Nat Biotechnol* **19**: 971–974.

16. Pera MF, Andrade J, Houssami S, *et al.* (2004) Regulation of human embryonic stem cell differentiation by BMP-2 and its antagonist noggin. *J Cell Sci* **117**: 1269–1280.

17. Enders AC, Schlafke S, Hendrickx AG. (1986) Differentiation of the embryonic disc, amnion, and yolk sac in the rhesus monkey. *Am J Anat* **177**: 161–185.

18. Amit M, Shariki C, Margulets V, Itskovitz-Eldor J. (2004) Feeder and serum-free culture of human embryonic stem cells. *Biol Reprod* **70**: 837–845.

19. Sato N, Sanjuan IM, Heke M, *et al.* (2003) Molecular signature of human embryonic stem cells and its comparison with the mouse. *Dev Biol* **260**: 404–413.

20. Sato N, Meijer L, Skaltsounis L, Greengard P, Brivanlou AH. (2004) Maintenance of pluripotency in human and mouse embryonic stem cells through activation of Wnt signaling by a pharmacological GSK-3-specific inhibitor. *Nat Med* **10**: 55–63.

21. van Es JH, Barker N, Clevers H. (2003) You Wnt some, you lose some: Oncogenes in the Wnt signaling pathway. *Curr Opin Genet Dev* **13**: 28–33.

22. Moon RT, Bowerman B, Boutros M, Perrimon N. (2002) The promise and perils of Wnt signaling through beta-catenin. *Science* **296**: 1644–1646.

23. Knockaert M, Greengard P, Meijer L. (2002) Pharmacological inhibitors of cyclin-dependent kinases. *Trends Pharmacol Sci* **23**: 417–425.

24. Meijer L, Skaltsounis AL, Magiatis P, *et al.* (2003) GSK-3-selective inhibitors derived from Tyrian purple indirubins. *Chem Biol* **10**: 1255–1266.

25. Cowan CA, Klimanskaya I, McMahon J, *et al.* (2004) Derivation of embryonic stem-cell lines from human blastocysts. *N Engl J Med* **350**: 1353–1356.

26. Buzzard JJ, Gough NM, Crook JM, Colman A. (2004) Karyotype of human ES cells during extended culture. *Nat Biotechnol* **22**: 381–382; author reply 382.

27. Draper JS, Smith K, Gokhale P, *et al.* (2004) Recurrent gain of chromosomes 17q and 12 in cultured human embryonic stem cells. *Nat Biotechnol* **22**: 53–54.

28. Veeck LL, Zaninovic N. (2003) *An Atlas of Human Blastocysts* (Parthenon, Boca Raton; London).

29. Brivanlou AH, Gage FH, Jaenisch R, *et al.* (2003) Stem cells. Setting standards for human embryonic stem cells. *Science* **300**: 913–916.

30. Harwood A, Braga VM. (2003) Cdc42 & GSK-3: Signals at the crossroads. *Nat Cell Biol* **5**: 275–277.

31. Sanders LC, Matsumura F, Bokoch GM, de Lanerolle P. (1999) Inhibition of myosin light chain kinase by p21-activated kinase. *Science* **283**: 2083–2085.

32. Salas-Vidal E, Lomeli H. (2004) Imaging filopodia dynamics in the mouse blastocyst. *Dev Biol* **265**: 75–89.

33. Fleming TP, Warren PD, Chisholm JC, Johnson MH. (1984) Trophectodermal processes regulate the expression of totipotency within the inner cell mass of the mouse expanding blastocyst. *J Embryol Exp Morphol* **84**: 63–90.

34. Sathananthan H, Pera M, Trounson A. (2002) The fine structure of human embryonic stem cells. *Reprod Biomed Online* **4**: 56–61.

35. Gonzales DS, Jones JM, Pinyopummintr T, *et al.* (1996) Trophectoderm projections: A potential means for locomotion, attachment and implantation of bovine, equine and human blastocysts. *Hum Reprod* **11**: 2739–2745.

36. Xu RH, Chen X, *et al.* (2002) BMP4 initiates human embryonic stem cell differentiation to trophoblast. *Nat Biotechnol* **20**: 1261–1264.

37. Zhang H, Berg JS, Li Z, *et al.* (2004) Myosin-X provides a motor-based link between integrins and the cytoskeleton. *Nat Cell Biol* **6**: 523–531.

38. Berg JS, Cheney RE. (2002) Myosin-X is an unconventional myosin that undergoes intrafilopodial motility. *Nat Cell Biol* **4**: 246–250.

# Epigenesis in Pluripotent Cells

Lisa M. Hoffman, Jennifer Batten and Melissa K. Carpenter*

## Introduction

Mendelian inheritance states that both parents contribute a set of chromosomes to their offspring, ensuring genetic equality and offering protection against deleterious recessive mutations. The importance to mammals of a proper combination of parental genes is exemplified by studies of mouse embryos that contain only maternally (parthogenetic, pg)- or paternally (androgenetic, ag)-derived chromosomes; in embryos containing only female-derived chromosomes, the extraembryonic tissues that are required to support embryonic growth develop poorly, and the embryo dies soon after implantation. In contrast, development of embryos that contain only male-derived chromosomes is retarded while extraembryonic tissues develop well.[1,2] These findings suggest that maternal and paternal genomes are not equivalent. Rather, each is endowed with different "imprints," leading to differential gene expression in the embryos. Such imprints or "epigenetic modifications" mark the two copies of each gene as being inherited either from the mother or the father. These modifications occur without alterations or deletion of any DNA sequences, but rather by heritable modifications of DNA that include changes in methylation status.

Two epigenetic phenomena result in monoallelic gene expression: genomic imprinting and X inactivation. Genomic imprinting results in the stable monoallelic expression of maternal or paternal genes. To date, numerous autosomally-imprinted genes have been identified in mice and humans, and their expression analyzed throughout early embryonic development. The importance of proper imprinted expression for normal growth and

*Correspondence: Stem Cell Biology and Regenerative Medicine, Robarts Research Institute, 100 Perth Drive, London, Ontario N6A 5K8, Canada. Voice: (519) 663-5777 x34108, Fax: (519)663-3326, e-mail: mcarpenter@robarts.ca

development is underscored by the failure of embryos containing only maternal or paternal genomes to develop normally, and by the development of human diseases and cancers due to loss of imprinting. Dosage compensation of the X chromosome in mammalian females, in comparison, is achieved through X inactivation. In somatic female cells, one of the X chromosomes is silenced in either a random or imprinted fashion, depending on cell type. These epigenetic events occur during early embryonic development, when pluripotent cells transition to lineage-restricted cells.

Genomic imprinting and X-chromosome inactivation (XCI) share a number of intriguing features that include the tendency of genes to cluster, the occurrence of differentially methylated regions (DMRs), and the presence of noncoding and antisense RNAs. As these topics have been reviewed in detail elsewhere,[3-6] we will discuss them briefly below.

## Genomic Imprinting

Over 50 imprinted genes have now been identified in mice and humans (see the Harwell imprinting web site: www.mgu.har.mrc.ac.uk/research/imprinting). Interestingly, they are often organized in clusters on specific regions of chromosomes that contain key control elements, the differentially methylated regions (DMRs). These control regions are complex with multiple functions that include gene silencing, either from the inability of methylation-sensitive transcription factors to bind to their targets and regulate gene expression, or from the binding of methylation-dependent silencer proteins. In other instances, DMRs are associated with expression of an antisense transcript whose expression subsequently leads to repression of an upstream gene. Regardless of the specific function of DMRs, the end result is a monoallelic expression of imprinted genes. For the purposes of this review, we will focus on a large imprinted cluster that has been well characterized in both mice and humans: mouse 7F/human 11p15.5. Although this gene cluster can be divided into two independent imprinting subdomains: one containing H19, Igf2/IGF2 (mouse and human, respectively), and INS2 (H19 subdomain) and the other containing CDKN1C, KCNQ1, KCNQ1OT1, and surrounding genes (KCNQ1OT1 subdomain), we will focus only on H19 and Igf2/IGF2 of the former imprinting domain (Fig. 1; for full review of this topic, see Ref. 4).

H19 encodes a nontranslated RNA that is transcribed exclusively from the maternal allele.[7-10] Igf2/IGF2, in comparison, is reciprocally imprinted to H19, being transcribed only from the paternal allele (Fig. 2;

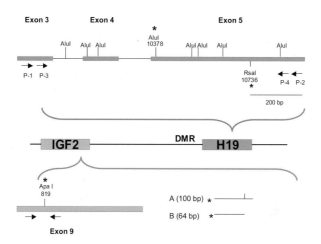

**Figure 1.** Schematic diagram of the H19/IGF2 region of chromosome 11p15.5 in humans. RsaI and AluI, and ApaI restriction site polymorphic regions are illustrated for H19 and IGF2, respectively. (Adapted from Zhang and Tycko, 1992; Ulaner *et al.*, 2003.)

Refs. 11–13). The repressed paternal H19 allele is hypermethylated over a 7 kb region[14–16]; within this region, approximately 2 kb upstream of the H19 promoter, lies a differentially methylated region (DMR)[14–19] that is believed to contain the epigenetic "mark" that distinguishes the parental alleles of H19.[18,19] Thus, this methylated region is often referred to as the imprinting control region (ICR). Deletion of this region in mice leads to biallelic expression of both H19 and Igf2.[20] Genome-wide demethylation through the use of DNA methyltransferase I (Dnmt1) mutant mice provides still further evidence that allele-specific methylation regulates monoallelic expression at this locus; prior to their death around 11 days gestation, homozygous mutant mice not only exhibit dramatically reduced levels of overall DNA methylation as might be expected, but also exhibit altered expression of various imprinted genes, including H19.[21] Further analysis of the DMR has revealed that it contains putative binding sites for the vertebrate insulator factor CTCF (CCCTC binding factor); binding of CTCF to these elements *in vitro* is inhibited by methylation of the target sequence.[22–26] Verona *et al.*[4] have postulated that on the maternal allele, where H19 is transcribed, CTCF binding to the unmethylated DMR may insulate the Igf2/IGF2 promoter from enhancers located 3′ of H19, allowing H19 exclusive access to these elements. On the paternal allele, in contrast, the methylated DMR acts to silence H19 expression; CTCF cannot bind the hypermethylated DMR, thus preventing the formation of an insulator, and allowing Igf2/IGF2

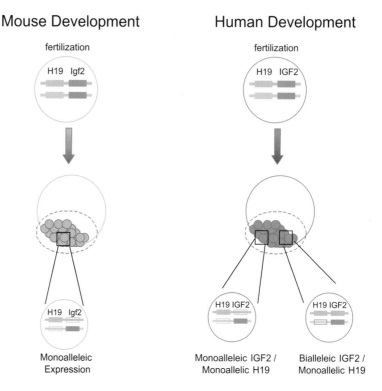

**Figure 2.** Schematic diagram of the status of imprinting of H19 and IGF2 genes in mouse and human development. At fertilization in mouse and humans, both H19 and IGF2 are expressed biallellically. H19 becomes expressed exclusively from the maternal allele at the blastocyst stage of development in mice and presumably so in humans. IGF2, in contrast, becomes expressed from the paternal allele, although in humans expression may be biallelic in certain developing embryonic tissues.

access to downstream enhancers. Cumulatively, these studies provide new insight into the molecular mechanisms that regulate genomic imprinting in mammals.

Numerous techniques have been utililized to identify imprinted genes, including targeted gene mutation, positional cloning coupled with candidate gene testing, use of parental differences in DNA methylation and expression, subtractive hybridization or differential display using cDNA from pg, ag and fertilized embryos,[27–29] genome-wide screens,[30,31] microcell-mediated cell transfer,[32] and somatic-cell hybrid systems.[33] The use of a number of these technologies, however, is not feasible in humans, and as such, examination of genomic imprinting in humans has, to date, largely been limited to analysis of *in vitro*-induced changes in DNA methylation, and examination of the effects of such changes on gene expression. Identification of single-base, or restriction site polymorphic regions within these imprinted genes provides yet another powerful means of achieving this goal (Figs. 1 and 3; Refs. 10, 34).

Considerable evidence suggests that DNA methylation is instrumental in establishing and maintaining expression patterns of epigenetically-regulated genes. It is not, however, the only determinant as changes in chromatin structure have also been shown to play a role. Briefly, euchromatin is accessible to DNA binding factors and is transcriptionally active whereas heterochromatin is generally unaccessible to transcription factors and is thus transcriptionally silent. The latter form of chromatin is characterized by three biochemical marks: histone hypoacetylation, H3-K9 methylation and, as already discussed in some detail, DNA methylation. These epigenetic "marks" appear to be interdependent, suggesting that DNA methylation and chromatin-mediated alterations of DNA act concertedly to maintain a silenced chromatin state (for review see Refs. 35, 36). Indeed, as is becoming

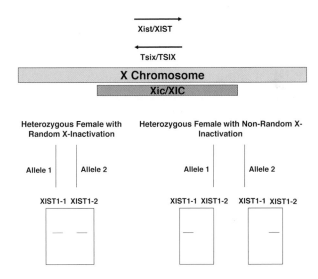

**Figure 3.** Schematic diagram of Xist/XIST on the mouse and human X chromosome. Xist/XIST and the antisense transcript, Tsix/TSIX are located within a single cis-acting region within the chromosome referred to as the X-inactivation center (Xic/XIC). The bottom left panel illustrates that X-inactivation is random since the female cDNA is amplified with both allele-specific primers. The bottom right panel, in contrast, illustrates that X-inactivation is imprinted, as cDNA is amplified by only one of the allele-specific primers. (Adapted from Plath *et al.*, 2002; Rupert *et al.*, 1995.)

increasingly obvious, regulation of autosomal imprinting is an extremely complex process; although numerous studies demonstrate that regulation occurs at many levels, much of the molecular network underlying such differential gene expression remains to be elucidated.

## Loss of Imprinting

A loss of imprinting (LOI) is often associated with a number of human genetic diseases and cancers,[37,38] one of the best-known examples being the Beckwith-Wiedemann syndrome (BWS). BWS has been mapped to chromosome 11p15 in humans, and is transmitted through the maternal chromosome. Like many diseases affected by aberrant autosomal imprinting, BWS has also been found to be associated with uniparental disomy or trisomy.[39,40] It is characterized by general overgrowth in the affected individual. Mutations in an imprinted gene, p57KIP2, accounts for about 10% of BWS patients,[41–44] whereas LOI of an antisense transcript, LIT1, accounts for 2/3 of BWS patients. LOI of LIT1 is achieved through the hypomethylation of a DMR in the promoter region of LIT1, thus converting the epigenotype from the maternal chromosome to that of the paternal chromosome and silencing the maternal expression of genes at this locus.[45–47] Another common molecular event occurring in BWS is a LOI of IGF2.[48,49] LOI at this locus, not surprisingly, may also be accompanied by the silencing of the H19 gene on the maternal allele.[50,51]

Evidence for a role for genomic imprinting in a variety of hereditary and sporadic cancers, including Wilms' tumor, come not only from the association of Wilms' tumor with BWS, but also from the parental origin-dependent LOI in Wilms' tumor. Indeed, LOI at chromosome 11p15 in sporadic Wilms' tumor involves the loss of the maternal chromosome, leading to overexpression of paternally-expressed genes and loss of expression of maternally-expressed genes.[52,53] LOI of IGF2, for example, has been observed in Wilms' tumor, as well as in a number of other forms of cancer.[54–57] These studies further attest to the importance of proper regulation of genomic imprinting.

## Imprinting in Mouse Development

Genomic imprinting occurs very early in mouse development; at the blastocyst stage, both maternal and paternal alleles of H19 are hypomethylated, and as such, expression of H19 is biallellic. Around the time of implantation, however, the paternal allele becomes fully methylated and silenced in the embryo and extraembryonic tissues, particularly the trophoblast. The maternal H19 allele, in contrast, is expressed exclusively in tissues of both pre- and post-implantation embryos (Fig. 2; Ref. 58).

Using an allele-specific fluorescence *in situ* hybridization method, Ohno *et al.*[59] have demonstrated that Igf2 transcripts are expressed biallelically in 2-cell mouse embryos until approximately the morula stage. At the blastocyst stage, however, the maternal allele is silenced and monoallelic Igf2 expression observed (Fig. 2). Interestingly, this imprinting event is affected in embryos maintained in culture, suggesting that aberrant epigenetic events and imprinting may indeed lead to developmental anomalies in *in vitro* fertilized (IVF) embryos.

## Imprinting in Mouse Pluripotent Cells

Pluripotent stem cells have been used as a model to study early developmental events in mammalian embryos. These cell lines have been used to study the mechanisms of genomic imprinting in the murine system. These pluripotent cells include embryonic stem (ES) cells derived from the inner cell mass (ICM) of preimplantation embryos, embryonic germ (EG) cells derived from primordial germ cells migrating along the gonadal ridge in embryonic development, and embryonic carcinoma cells derived from teratocarcinomas. Because these pluripotent cell lines are derived from different tissues sources, they offer a variety of developmental states in which to study these early developmental events.

The status of imprinted genes in undifferentiated and differentiated mouse ES cells (mESCs) has been assessed. In the blastocyst, the majority of the genome is demethylated; therefore, the ES cells derived from this material show low levels of methylation. Numerous studies indicate that autosomal imprinting is, at least in part, maintained in the derivation of ES cell lines. For example, expression of various imprinted genes in parthogenetic (pg) and androgenetic (ag) ES cells, and in differentiated embryoid bodies (EBs) derived by suspension of these cells in culture, is appropriate.[60] Further evidence that appropriate genomic imprinting is often maintained in ES cells comes from the demonstration that chimeras made with ag ES cells are similar in phenotype to chimeras made with ag morulae or ICMs.[61—64] In some instances, however, methylation and allele-specific expression of imprinted genes differs from that observed in ICM cells. For example, paternal-specific hypermethylation of the H19 promoter is progressively lost during ES cell passage.[65] Maternal expression of the Igf2 allele, in contrast, remains inappropriately active.[66,67] Other imprinted genes have also been demonstrated to exhibit aberrant methylation and transcript expression in ES cells.[68] Cloned embryos derived by combining early-passage ES cells with tetraploid ova similarly exhibit a dysregulation in the expression of imprinted genes,[68] as do cloned mice derived from the transplantation of ES cell nuclei into oocytes.[69]

Dysregulation of monoallelic gene expression is also often seen during the derivation of embryonic germ (EG) cell lines from migratory primordial germ cells (PGCs). Several studies have demonstrated that undifferentiated EG cell lines show biallelic expression of imprinted genes, which does not change to preferential expression of one allele, and therefore does not show a complete imprint. Furthermore, the expression patterns can be reversed from normal imprinting (for instance H19 and Igf2).[70] This inappropriate expression pattern couples with abnormal methylation patterns in both the undifferentiated EG cell lines as well as the PGCs from which they are derived.[71]

## Imprinting in Human Development

Limited information is currently available regarding autosomal imprinting in early human development. Although imprinting of H19 and IGF2 appears to be quite discrete in mouse development, it is considerably less so in humans. For example, H19 is not expressed in oocytes through to the 8-cell stage of development; rather onset of gene expression occurs at the blastocyst stage.[72] To date, there is very little information regarding the onset of H19 imprinting in humans, although there have been clear demonstrations that it does become expressed exclusively from the maternal allele.[10] Given that the expression analysis of human H19 correlates with that in mouse, it seems likely that, as in the mouse,[18,66,67] monallelic H19 expression occurs with the onset of gene transcription at the blastocyst stage of development (Fig. 2).

IGF2 transcripts, in comparison, are detectable in oocytes and in all cleavage stage embryos up to the blastocyst stage of development; and while the maternal IGF2 allele is expressed exclusively prior to the 8-cell stage, the paternal allele subsequently becomes expressed monoallellically from this stage onward.[73,74] Biallelic expression, however, has been reported in fetal neural tissues,[75] and in the developing liver (Fig. 2; Refs. 76–78). These tissue-specific differences in IGF2 expression are likely due to alterations in promoter methylation, and emphasize, once again, the complexity in the regulation of imprinting in humans.

## Imprinting in Human Pluripotent Cells

Onyango *et al.*[70] demonstrated that H19 is appropriately imprinted in human EG cell lines. Although H19 shows monoallelic expression in the differentiated cultures, other imprinted genes such as IGF2 showed a partial relaxation of imprinting, resulting in biallelic expression. To date, analysis of imprinted gene expression in human ES cells (hESCs) has unfortunately been limited. A recent study by Hwang *et al.*[79] however, has demonstrated that both maternally (H19 and UBE3A) and paternally-expressed (SNRPN and ARH1) imprinted genes are detectable in cells of an ES cell line derived from a cloned human blastocyst via somatic cell nuclear transfer. While RT-PCR conducted in this study demonstrates that various imprinted genes are indeed expressed in the cell line, it does not address the question of whether the expression pattern is appropriate, i.e. monoallelic or biallelic. As use of hESCs becomes more accessible, and more information on the culture conditions required to maintain them optimally becomes available, further analysis of imprinted gene expression in humans will ensue.

## X-inactivation

Dosage compensation of the X chromosome in mammalian females is achieved through X-chromosome inactivation (XCI), an epigenetic event that is regulated by a single *cis*-acting X-inactivation center (Xic/XIC in mouse and humans, respectively), and through the action of at least two genes that reside in this region, the X-inactive-specific transcript (Xist/XIST) and the antisense transcript (Tsix/TSIX) (Fig. 3). Inactivation is developmentally regulated, with initiation of inactivation occurring at the onset of cellular differentiation.[80] In mice, X-inactivation is imprinted in the extraembryonic trophectoderm and primitive endoderm lineages during preimplantation.[81] This imprint is labile, however, allowing cells of the epiblast to undergo random X-inactivation during postimplantation. In humans, by comparison, the choice of the chromosome to inactivate is random. Once an X chromosome is inactivated, the same X is silenced in all descendent cells; thus, females are mosaic for their X-inactivation pattern (Fig. 4; for review see Refs. 5, 6).

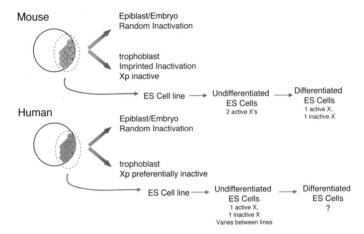

**Figure 4.** Schematic diagram of X-inactivation during cellular differentiation in mouse and human development. In trophoblast and primitive endoderm tissues, the paternal X-chromosome is inactivated. Cells of the epiblast and embryo proper, in contrast, undergo random X-inactivation. Whereas mESCs similarly exhibit random inactivation during differentiation, hESCs exhibit skewed X-inactivation in both the undifferentiated and differentiated states.

The first step in random XCI requires that a cell determine its X chromosome constitution; random XCI will occur only in those cells with two or more X chromosomes (for review see Refs. 5, 6). This process of "counting" is believed to be regulated by the Xic, and results in one X chromosome being selected to remain active (Xa), while the other is destined to be inactivated (Xi).[82–86] It has been postulated that the selection of one X chromosome as Xa is due to the presence in each cell of a blocking factor(s) with the ability to bind and repress a single Xic.[6,82] X-inactivation will then proceed from unblocked Xics. When two X chromosomes bearing identical Xic regions are present in a female cell, Xa choice occurs randomly because of the equal probability that the blocking factor will interact with either of the counting elements. Further analysis of the Xic region, however, has revealed the existence of additional *cis*-elements, termed choice elements, that can influence the counting element's affinity for blocking factor interaction.[87] Xist is a likely candidate for such a choice element since it has been shown to reduce the affinity of a *cis*-linked counting element for blocking factor.[88,89] Tsix transcription, on the other hand, promotes Xa choice, most likely by destabilizing Xist RNA and lowering the abundance of functional Xist complexes acting on the counting element in *cis*.[90–93]

Imprinted XCI differs from the random form in that parental origin dictates which chromosome is inactivated. As epigenetic marks are acquired in the germline, the X chromosome must possess marks that establish parental identity, bypassing the choice mechanism involved in random X-inactivation.

## Aberrant X-inactivation

In mice, the parental origin of a supernumerary X chromosome is of the utmost importance in assessing the viability of the embryo; XmXmY embryos have only been observed in rare circumstances,[94] whereas XmXpY embryos are fairly common.[95] Analysis of embryos disomic for the maternal X chromosome (XmXmXp; XmXmY) further demonstrates that a supernumerary Xm is embryonic lethal,[96,97] largely due to failure of extraembryonic tissues to develop. Embryonic tissues, in contrast, develop relatively well until approximately E7 when a lack of mesoderm becomes apparent.[98] Cumulatively, these findings indicate that aberrant X-chromosome inactivation is responsible for embryonic lethality and/or extreme developmental anomalies.

Aberrant gene silencing, either of autosomally-imprinted genes or of X-linked genes, has also been demonstrated to occur in aging mammalian individuals.[99] Interestingly, skewed X-chromosome inactivation is also found to occur increasingly with age,[100] as well as in recurrent loss of pregnancy,[101–103] trisomic pregnancy,[104] and in the progression of cancers.[105–107]

## X-inactivation in Mouse Development

X-chromosome inactivation is one of the earliest events in mouse development, occurring during preimplantation development. Deletion analysis indicates that both the Xist and Tsix genes are involved in regulating X-chromosome inactivation in mice; specifically, studies show that Xist expression is required for inactivation of the paternal X chromosome,[88] whereas Tsix expression is required either to maintain the active state or to prevent inactivation of the maternal X chromosome.[93,108] In support of these findings, in female blastocysts, Xist is expressed exclusively from the paternal allele, whereas Tsix is transcribed solely from the maternal X.[93,108] Female blastocysts inheriting a maternal targeted deletion of the 5′ CpG island and major promoter of Tsix fail to express the Tsix gene and express Xist biallelically, indicating that loss of Tsix transcription derepresses the normally silent maternal Xist allele in *cis.*[108] In males, this deletion results in ectopic activation of the single maternal Xist allele. As such, it has been suggested that Tsix may act as a maternal protective factor that negatively regulates Xist expression. Subsequent studies, however, indicate that a Tsix-independent mechanism silences Xist expression on the maternal X chromosome (for review see Ref. 4). As such, further studies into the mechanisms regulating XCI in mice are required.

DNA methylation also plays a key role in X-chromosome inactivation as evidenced by the heavy methylation of CpG islands associated with genes on the inactive X chromosome, and by the unmethylated status of genes on the active X chromosome. Treatment of interspecific somatic cell hybrids with demethylating 5-azacytidine confirms this, revealing a reactivation of silenced genes on the inactive X chromosome (reviewed in Ref. 109). Further evidence that methylation regulates XCI is provided by studies utilizing targeted disruptions of DNA methyltransferases.[110–113]

## X-inactivation in Pluripotent Mouse Cells

Pluripotent cell lines have also been used to study X-inactivation in early embryonic development. Evaluation of the mechanisms involved in this process, however, have to date been largely limited to studies in mESCs.[89,111,114,115] As in mouse preimplantation embryos prior to X-inactivation, Xist is expressed from both X chromosomes in female ESCs whereas it is expressed only weakly from the single X chromosome in male ESCs. At the onset of cellular differentiation, Xist expression is upregulated on one of the two chromosomes in the female (Xi), while being silenced on the remaining X chromosome (Xa) (Fig. 4; Refs. 116, 117). Expression of Xist is sufficient for XCI as demonstrated by the ability of an inducible Xist cDNA to induce gene silencing in ESCs[118]; the ability of Xist to inactivate an X chromosome, however, is dependent on the time at which it is expressed during differentiation. For example, in undifferentiated ESCs, Xist transcripts induce XCI, but silencing is reversible and dependent on continued Xist expression. At 3 days of differentiation, however, XCI becomes irreversible. Interestingly, if X inactivation is not initiated within 2 days of differentiation, ESCs are refractive to expression of Xist transcripts (for review see Ref. 113).

X-inactivation has also been assessed in a number of female mouse teratocarcinoma cell lines; the findings of these studies vary greatly between lines. For example, McBurney and Adamson[120] have demonstrated that undifferentiated teratocarcinoma lines exhibit only one active X chromosome. Martin *et al.*[119] in comparison, demonstrate that another undifferentiated teratocarcinoma cell line exhibits two active X chromosomes, and that upon differentiation, one of these X chromosomes is inactivated. The difference in these findings may be the result of the derivation of cell lines from different tissue sources. Indeed, McBurney and Adamson[120] generated teratocarcinoma cells from a tumor from a 6.5 d embryo, while Martin *et al.*[119] derived cells from spontaneous tumors. Furthermore, both labs used different culture conditions, which may have resulted in the selection of particular cell types.

X-inactivation in murine ECs has been evaluated by measuring the activities of X-linked enzymes in XX and XO cell lines. In one study, the XX line showed 2-fold higher enzyme activity in the undifferentiated EC cells; however, upon differentiation, the ratio of X-linked activity dropped.[119] In contrast, other investigators reported that enzymatic activities were similar in XX and XO cell lines, suggesting that only one of the X chromosomes is active in the female lines.[120] Because EC cell lines are derived from teratocarcinomas, there is considerable variability in starting material. These studies suggest that it is possible to isolate ECC lines which represent different stages of development.

In the female germline, the inactive X chromosome is reactivated at the onset of meiosis. The reactivated X chromosome remains in this state in oocytes throughout ovulation and fertilization, and only becomes silenced once again during preimplantation. In the male germline, in comparison, the single X chromosome is transiently inactivated during meiosis, perhaps to prevent the initiation of recombination events that might occur as a result of the presence of unpaired sites on the single X chromosome.[72,121]

## X-inactivation in Human Development

To date, the developmental regulation of XCI in humans is unclear. Studies have demonstrated, however, that XIST is detectable in oocytes, and in both male and female preimplantation embryos up until the blastocyst stage of development.[122,123] In contrast to the mouse, XIST expression is not limited to the paternal allele; findings that exclusive XCI on the paternal allele does not occur in human extraembryonic tissues are consistent with this (Fig. 4; Refs. 124, 125). Also in contrast to mouse, supernumerary X chromosomes (both Xm and Xp) are tolerated in humans, suggesting that XCI does not occur in an imprinted fashion in humans.[126,127]

## X-inactivation in Human Pluripotent Cells

As with genomic imprinting, evaluation of the mechanisms involved in the process of X-inactivation to date has largely been limited to studies in mESCs. The recent availability, however, of hESCs now provides a unique tool to assess this early developmental process in that they are derived from human blastocysts, have an apparently unlimited proliferative capacity, and can differentiate into ectoderm, mesoderm and endoderm.

In a study by Dhara and Benvenisty,[128] it was demonstrated that in undifferentiated female H9 hESCs, both X chromosomes are active. Upon differentiation, however, one of the X chromosomes is silenced; in differentiated embryonic cells, XCI is random whereas in extraembryonic cells, the silencing is non-random. We have further demonstrated that individual hESC lines exhibit distinct patterns of X-inactivation (Fig. 5 and Ref. 137). Specifically, as in human male somatic cells, expression of Xist transcripts is virtually undetectable in undifferentiated male hESC cultures. XIST transcripts are similarly expressed at extremely low or undetectable levels in the female H7 hESC line, suggesting that X-inactivation is not occurring or is occurring by an alternate, as of yet to be identified, mechanism. In contrast, undifferentiated H9 and HES25 cultures exhibit high levels of XIST RNA, indicating that XCI is occurring. hESCs are considered immortal and can be maintained over one year in continuous culture (approximately 50 passages); importantly, X-inactivation appears consistent over long term

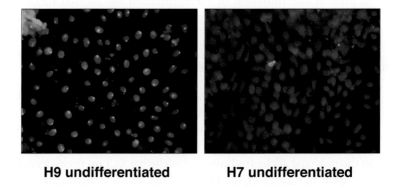

**H9 undifferentiated**          **H7 undifferentiated**

**Figure 5.** FISH analysis of XIST in hESCs. XIST transcripts are detected in undifferentiated H9 cell cultures, but not in undifferentiated H7 cell cultures, indicating that X-inactivation is not occurring in the latter.

culture, exhibiting patterns of inactivation that are consistent with those observed in low passage cultures (~passage 30). Cumulatively, these findings further reinforce the genetic stability of hESCs over long term culture.

EC cells are the pluripotent stem cells of germ cell-derived tumors which, in some cases, have the ability to recapitulate early embryonic events. While there have been reports of XCI upon differentiation in human EC cells derived from testicular tumors,[129,130] there is little to no convincing data on this process in lines derived from female tissues. Furthermore, as EC cells are aneuploid, in contrast to euploid ES cells, it seems questionable as to whether this may be an appropriate system for studying XCI.

## Summary

There are a number of striking similarities between the two epigenetic phenomena discussed herein, autosomal imprinting and X-chromosome inactivation. One of these similarities is the tendency towards gene clustering, suggesting that regional control elements coordinately regulate transcription of both autosomally imprinted and X-linked genes. Another particularly interesting similarity of imprinted domains and the Xic/XIC is the abundance of noncoding and/or antisense transcripts that include H19, Igf2sa, Xist/XIST and Tsix/TSIX; indeed, it is believed that approximately 27% of known human and mouse imprinted genes are noncoding, and their functions unclear. DNA methylation has also been shown to be instrumental in establishing and maintaining expression patterns of epigenetically-regulated genes, both in autosomal imprinting and in XCI, as have changes in chromatin structure. It is important to note, however, that there are distinct differences in both autosomal imprinting and XCI between human and mouse, as well as between various pluripotent cells; ES cells and EC cells, for example, exhibit many common features, including the expression of cell-surface markers and transcription factors. Perhaps the most significant difference, however, between these populations is that ES cells are euploid whereas EC cells are aneuploid. Furthermore, it is crucial to recognize that not all hESC lines are equivalent; although ES cells from both mouse and human typically show a normal karyotype, systematic evaluation of the karyotype of three hESC lines (H1, H7 and H9) that have been maintained in identical conditions over two years in culture reveals that 20–30% of the cultures exhibit some aneuploidy.[131] In direct comparison, a second study has shown that the H7 cell line exhibits trisomy 12 or 17, abnormalities that are typically associated with human EC cells.[132] We have further demonstrated that there are distinct differences in patterns of X-chromosome inactivation between various hESC lines. Cumulatively, the studies reviewed herein reveal that while certain parallels exist between various embryonic cell types in human and mouse, there are noteworthy differences that cannot be trivialized; these differences may be due in part to the manner in which the cell lines have been derived, and on the culture conditions in which the cells are maintained. Indeed, mouse embryos derived from high passage ESCs exhibit developmental anomalies often observed in mammalian offspring generated via nuclear transfer, including increased size and mass, and perinatal lethality.[133–135] These findings have led to the suggestion that mESCs lose their developmental potential as a result of epigenetic alterations.[136] As such, it will be important to fully characterize the pluripotent state in various human cell types, and to elucidate the molecular mechanisms by which epigenetic states become altered upon prolonged culture and cellular differentiation. Specifically assessing whether there is a loss of imprinting or otherwise inappropriate gene expression upon cell differentiation will have critical implications for the derivation of hESC lines, and for their use in cell replacement therapies.

# REFERENCES

1. Surani MA, Barton SC, Norris ML. (1984) Development of reconstituted mouse eggs suggests imprinting of the genome during gametogenesis. *Nature* **308**:548–550.
2. Barton SC, Surani MA, Norris ML. (1984) Role of paternal and maternal genomes in mouse development. *Nature* **311**:374–376.
3. Latham KE. (1999) Epigenetic modification and imprinting of the mammalian genome during development. *Curr Top Dev Biol* **43**:1–49.
4. Verona RI, Mann MR, Bartolomei MS. (2003) Genomic imprinting: Intricacies of epigenetic regulation in clusters. *Annu Rev Cell Dev Biol* **19**:237–259.
5. Heard E, Clerc P, Avner P. (1997) X-chromosome inactivation in mammals. *Annu Rev Genet* **31**:571–610.
6. Plath K, Mlynarczyk-Evans S, Nusinow DA, Panning B. (2002) Xist RNA and the mechanism of X chromosome inactivation. *Annu Rev Genet* **36**:233–278.
7. Bartolomei MS, Zemel S, Tilghman SM. (1991) Parental imprinting of the mouse H19 gene. *Nature* **351**:153–155.
8. Overall M, Bakker M, Spencer J, *et al.* (1997) Genomic imprinting in the rat: Linkage of Igf2 and H19 genes and opposite parental allele-specific expression during embryogenesis. *Genomics* **45**:416–420.

9. Rachmilewitz J, Goshen R, Ariel I, *et al.* (1992) Parental imprinting of the human H19 gene. *FEBS Lett* **309**:25–28.

10. Zhang Y, Tycko B. (1992) Monoallelic expression of the human H19 gene. *Nat Genet* **1**:40–44.

11. DeChiara TM, Robertson EJ, Efstratiadis A. (1991) Parental imprinting of the mouse insulin-like growth factor II gene. *Cell* **64**:849–859.

12. Giannoukakis N, Deal C, Paquette J, *et al.* (1993) Parental genomic imprinting of the human IGF2 gene. *Nat Genet* **4**:98–101.

13. Ohlsson R, Nystrom A, Pfeifer-Ohlsson S, *et al.* (1993) IGF2 is parentally imprinted during human embryogenesis and in the Beckwith-Wiedemann syndrome. *Nat Genet* **4**:94–97.

14. Bartolomei MS, Webber AL, Brunkow ME, Tilghman SM. (1993) Epigenetic mechanisms underlying the imprinting of the mouse H19 gene. *Genes Dev* **7**:1663–1673.

15. Brandeis M, Kafri T, Ariel M, *et al.* (1993) The ontogeny of allele-specific methylation associated with imprinted genes in the mouse. *Embo J* **12**:3669–3677.

16. Ferguson-Smith AC, Sasaki H, Cattanach BM, Surani MA. (1993) Parental-origin-specific epigenetic modification of the mouse H19 gene. *Nature* **362**:751–755.

17. Frevel MA, Sowerby SJ, Petersen GB, Reeve AE. (1999) Methylation sequencing analysis refines the region of H19 epimutation in Wilms tumor. *J Biol Chem* **274**:29331–29340.

18. Tremblay KD, Saam JR, Ingram RS, *et al.* (1995) A paternal-specific methylation imprint marks the alleles of the mouse H19 gene. *Nat Genet* **9**: 407–413.

19. Tremblay KD, Duran KL, Bartolomei MS. (1997) A 5′ 2-kilobase-pair region of the imprinted mouse H19 gene exhibits exclusive paternal methylation throughout development. *Mol Cell Biol* **17**:4322–4329.

20. Thorvaldsen JL, Duran KL, Bartolomei MS. (1998) Deletion of the H19 differentially methylated domain results in loss of imprinted expression of H19 and Igf2. *Genes Dev* **12**:3693–3702.

21. Li E, Beard C, Jaenisch R. (1993) Role for DNA methylation in genomic imprinting. *Nature* **366**:362–365.

22. Bell AC, Felsenfeld G. (2000) Methylation of a CTCF-dependent boundary controls imprinted expression of the Igf2 gene. *Nature* **405**:482–485.

23. Hark AT, Schoenherr CJ, Katz DJ, *et al.* (2000) CTCF mediates methylation-sensitive enhancer-blocking activity at the H19/Igf2 locus. *Nature* **405**:486–489.

24. Kanduri C, Holmgren C, Pilartz M, *et al.* (2000) The 5′ flank of mouse H19 in an unusual chromatin conformation unidirectionally blocks enhancer-promoter communication. *Curr Biol* **10**:449–457.

25. Kanduri C, Pant V, Loukinov D, *et al.* (2000) Functional association of CTCF with the insulator upstream of the H19 gene is parent of origin-specific and methylation-sensitive. *Curr Biol* **10**:853–856.

26. Szabo P, Tang SH, Rentsendorj A, *et al.* (2000) Maternal-specific footprints at putative CTCF sites in the H19 imprinting control region give evidence for insulator function. *Curr Biol* **10**:607–610.

27. Kaneko-Ishino T, Kuroiwa Y, Miyoshi N, *et al.* (1995) Peg1/Mest imprinted gene on chromosome 6 identified by cDNA subtraction hybridization. *Nat Genet* **11**:52–59.

28. Kuroiwa Y, Kaneko-Ishino T, Kagitani F, *et al.* (1996) Peg3 imprinted gene on proximal chromosome 7 encodes for a zinc finger protein. *Nat Genet* **12**:186–190.

29. Kagitani F, Kuroiwa Y, Wakana S, *et al.* (1997) Peg5/Neuronatin is an imprinted gene located on sub-distal chromosome 2 in the mouse. *Nucleic Acids Res* **25**:3428–3432.

30. Plass C, Shibata H, Kalcheva I, *et al.* (1996) Identification of Grf1 on mouse chromosome 9 as an imprinted gene by RLGS-M. *Nat Genet* **14**:106–109.

31. Hatada I, Sugama T, Mukai T. (1993) A new imprinted gene cloned by a methylation-sensitive genome scanning method. *Nucleic Acids Res* **21**:5577–5582.

32. Meguro M, Mitsuya K, Sui H, *et al.* (1997) Evidence for uniparental, paternal expression of the human GABAA receptor subunit genes, using microcell-mediated chromosome transfer. *Hum Mol Genet* **6**:2127–2133.

33. Gabriel JM, Higgins MJ, Gebuhr TC, *et al.* (1998) A model system to study genomic imprinting of human genes. *Proc Natl Acad Sci USA* **95**:14857–14862.

34. Rupert JL, Brown CJ, Willard HF. (1995) Direct detection of non-random X chromosome inactivation by use of a transcribed polymorphism in the XIST gene. *Eur J Hum Genet* **3**:333–343.

35. Grewal SI, Moazed D. (2003) Heterochromatin and epigenetic control of gene expression. *Science* **301**:798–802.

36. Elgin SC, Grewal SI. (2003) Heterochromatin: Silence is golden. *Curr Biol* **13**:R895–R898.

37. Lalande M. (1996) Parental imprinting and human disease. *Annu Rev Genet* **30**:173–195.

38. Hall JG. (1997) Genomic imprinting: Nature and clinical relevance. *Annu Rev Med* **48**:35–44.

39. Ping AJ, Reeve AE, Law DJ, *et al.* (1989) Genetic linkage of Beckwith Wiedemann syndrome to 11p15. *Am J Hum Genet* **44**:720–723.

40. Mannens M, Hoovers JM, Redeker E, *et al.* (1994) Parental imprinting of human chromosome region 11p15.3-pter involved in the Beckwith-Wiedemann syndrome and various human neoplasia. *Eur J Hum Genet* **2**:3–23.

41. Thompson JS, Reese KJ, DeBaun MR, *et al.* (1996) Reduced expression of the cyclin-dependent kinase inhibitor gene p57KIP2 in Wilms' tumor. *Cancer Res* **56**:5723–5727.

42. Hatada I, Ohashi H, Fukushima Y, *et al.* (1996) An imprinted gene p57KIP2 is mutated in Beckwith-Wiedemann syndrome. *Nat Genet* **14**:171–173.

43. O'Keefe D, Dao D, Zhao L, *et al.* (1997) Coding mutations in p57KIP2 are present in some cases of Beckwith-Wiedemann syndrome but are rare or absent in Wilms tumors. *Am J Hum Genet* **61**:295–303.

44. Lee MP, DeBaun M, Randhawa G, *et al.* (1997) Low frequency of p57KIP2 mutation in Beckwith-Wiedemann syndrome. *Am J Hum Genet* **61**:304–309.

45. Mitsuya K, Meguro M, Lee MP, *et al.* (1999) LIT1, an imprinted antisense RNA in the human KvLQT1 locus identified by screening for differentially expressed transcripts using monochromosomal hybrids. *Hum Mol Genet* **8**:1209–1217.

46. Lee MP, DeBaun MR, Mitsuya K, *et al.* (1999) Loss of imprinting of a paternally expressed transcript, with antisense orientation to KVLQT1, occurs frequently in Beckwith-Wiedemann syndrome and is independent of insulin-like growth factor II imprinting. *Proc Natl Acad Sci USA* **96**: 5203–5208.

47. Smilinich NJ, Day CD, Fitzpatrick GV, *et al.* (1999) A maternally methylated CpG island in KvLQT1 is associated with an antisense paternal transcript and loss of imprinting in Beckwith-Wiedemann syndrome. *Proc Natl Acad Sci USA* **96**:8064–8069.

48. Weksberg R, Shen DR, Fei YL, *et al.* (1993) Disruption of insulin-like growth factor 2 imprinting in Beckwith-Wiedemann syndrome. *Nat Genet* **5**:143–150.

49. Joyce JA, Lam WK, Catchpoole DJ, *et al.* (1997) Imprinting of IGF2 and H19: Lack of reciprocity in sporadic Beckwith-Wiedemann syndrome. *Hum Mol Genet* **6**:1543–1548.

50. Reik W, Brown KW, Schneid H, *et al.* (1995) Imprinting mutations in the Beckwith-Wiedemann syndrome suggested by altered imprinting pattern in the IGF2-H19 domain. *Hum Mol Genet* **4**:2379–2385.

51. Catchpoole D, Lam WW, Valler D, *et al.* (1997) Epigenetic modification and uniparental inheritance of H19 in Beckwith-Wiedemann syndrome. *J Med Genet* **34**:353–359.

52. Maher ER, Reik W. (2000) Beckwith-Wiedemann syndrome: Imprinting in clusters revisited. *J Clin Invest* **105**:247–252.

53. Schroeder WT, Chao LY, Dao DD, *et al.* (1987) Nonrandom loss of maternal chromosome 11 alleles in Wilms tumors. *Am J Hum Genet* **40**:413–420.

54. Rainier S, Johnson LA, Dobry CJ, *et al.* (1993) Relaxation of imprinted genes in human cancer. *Nature* **362**:747–749.

55. Ogawa O, Eccles MR, Szeto J, *et al.* (1993) Relaxation of insulin-like growth factor II gene imprinting implicated in Wilms' tumour. *Nature* **362**:749–751.

56. Kim HT, Choi BH, Niikawa N, *et al.* (1998) Frequent loss of imprinting of the H19 and IGF-II genes in ovarian tumors. *Am J Med Genet* **80**:391–395.

57. Ulaner GA, Vu TH, Li T, *et al.* (2003) Loss of imprinting of IGF2 and H19 in osteosarcoma is accompanied by reciprocal methylation changes of a CTCF-binding site. *Hum Mol Genet* **12**:535–549.

58. Sasaki H, Ferguson-Smith AC, Shum AS, *et al.* (1995) Temporal and spatial regulation of H19 imprinting in normal and uniparental mouse embryos. *Development* **121**:4195–4202.

59. Ohno M, Aoki N, Sasaki H. (2001) Allele-specific detection of nascent transcripts by fluorescence *in situ* hybridization reveals temporal and culture-induced changes in Igf2 imprinting during pre-implantation mouse development. *Genes Cells* **6**:249–259.

60. Mann JR, Szabo PE. (2004) Genomic imprinting in mouse embryonic stem and germ cells. In *Stem Cells Handbook*, ed. S. Sell, pp. 81–88. Totowa NJ: Humana Press.

61. Mann JR, Gadi I, Harbison ML, *et al.* (1990) Androgenetic mouse embryonic stem cells are pluripotent and cause skeletal defects in chimeras: Implications for genetic imprinting. *Cell* **62**:251–260.

62. Barton SC, Ferguson-Smith AC, Fundele R, Surani MA. (1991) Influence of paternally imprinted genes on development. *Development* **113**:679–687.

63. Mann JR, Stewart CL. (1991) Development to term of mouse androgenetic aggregation chimeras. *Development* **113**:1325–1333.

64. Keverne EB, Fundele R, Narasimha M, *et al.* (1996) Genomic imprinting and the differential roles of parental genomes in brain development. *Brain Res Dev Brain Res* **92**:91–100.

65. Szabo P, Mann JR. (1994) Expression and methylation of imprinted genes during *in vitro* differentiation of mouse parthenogenetic and androgenetic embryonic stem cell lines. *Development* **120**:1651–1660.

66. Szabo PE, Mann JR. (1995) Allele-specific expression and total expression levels of imprinted genes during early mouse development: Implications for imprinting mechanisms. *Genes Dev* **9**:3097–3108.

67. Szabo PE, Mann JR. (1995) Biallelic expression of imprinted genes in the mouse germ line: Implications for erasure, establishment, and mechanisms of genomic imprinting. *Genes Dev* **9**:1857–1868.

68. Dean W, Bowden L, Aitchison A, *et al.* (1998) Altered imprinted gene methylation and expression in completely ES cell-derived mouse fetuses: Association with aberrant phenotypes. *Development* **125**:2273–2282.

69. Humpherys D, Eggan K, Akutsu H, *et al.* (2001) Epigenetic instability in ES cells and cloned mice. *Science* **293**:95–97.

70. Onyango P, Jiang S, Uejima H, *et al.* (2002) Monoallelic expression and methylation of imprinted genes in human and mouse embryonic germ cell lineages. *Proc Natl Acad Sci USA* **99**:10599–10604.

71. Labosky PA, Barlow DP, Hogan BL. (1994) Mouse embryonic germ (EG) cell lines: Transmission through the germline and differences in the methylation imprint of insulin-like growth factor 2 receptor (Igf2r) gene compared with embryonic stem (ES) cell lines. *Development* **120**:3197–3204.

72. Monk M, Salpekar A. (2001) Expression of imprinted genes in human preimplantation development. *Mol Cell Endocrinol* **183**(Suppl 1):S35–40.

73. Lighten AD, Hardy K, Winston RM, Moore GE. (1997) Expression of mRNA for the insulin-like growth factors and their receptors in human preimplantation embryos. *Mol Reprod Dev* **47**:134–139.

74. Lighten AD, Hardy K, Winston RM, Moore GE. (1997) IGF2 is parentally imprinted in human preimplantation embryos. *Nat Genet* **15**:122–123.

75. Albrecht S, Waha A, Koch A, *et al.* (1996) Variable imprinting of H19 and IGF2 in fetal cerebellum and medulloblastoma. *J Neuropathol Exp Neurol* **55**:1270–1276.

76. Davies SM. (1994) Developmental regulation of genomic imprinting of the IGF2 gene in human liver. *Cancer Res* **54**:2560–2562.

77. Vu TH, Hoffman AR. (1994) Promoter-specific imprinting of the human insulin-like growth factor-II gene. *Nature* **371**:714–717.

78. Taniguchi T, Schofield AE, Scarlett JL, *et al.* (1995) Altered specificity of IGF2 promoter imprinting during fetal development and onset of Wilms tumour. *Oncogene* **11**:751–756.

79. Hwang WS, Ryu YJ, Park JH, *et al.* (2004) Evidence of a pluripotent human embryonic stem cell line derived from a cloned blastocyst. *Science* **303**:1669–1674.

80. Monk M, Harper MI. (1979) Sequential X chromosome inactivation coupled with cellular differentiation in early mouse embryos. *Nature* **281**:311–313.

81. Takagi N, Sasaki M. (1975) Preferential inactivation of the paternally derived X chromosome in the extraembryonic membranes of the mouse. *Nature* **256**:640–642.

82. Rastan S. (1983) Non-random X-chromosome inactivation in mouse X-autosome translocation embryos — Location of the inactivation centre. *J Embryol Exp Morphol* **78**:1–22.

83. Brown CJ, Ballabio A, Rupert JL, *et al.* (1991) A gene from the region of the human X inactivation centre is expressed exclusively from the inactive X chromosome. *Nature* **349**:38–44.

84. Leppig KA, Brown CJ, Bressler SL, *et al.* (1993) Mapping of the distal boundary of the X-inactivation center in a rearranged X chromosome from a female expressing XIST. *Hum Mol Genet* **2**:883–887.

85. Cattanach BM, Rasberry C, Evans EP, *et al.* (1991) Genetic and molecular evidence of an X-chromosome deletion spanning the tabby (Ta) and testicular feminization (Tfm) loci in the mouse. *Cytogenet Cell Genet* **56**:137–143.

86. Rastan S, Robertson EJ. (1985) X-chromosome deletions in embryo-derived (EK) cell lines associated with lack of X-chromosome inactivation. *J Embryol Exp Morphol* **90**:379–388.
87. Mlynarczyk SK, Panning B. (2000) X inactivation: Tsix and Xist as yin and yang. *Curr Biol* **10**:R899–903.
88. Marahrens Y, Panning B, Dausman J, *et al.* (1997) Xist-deficient mice are defective in dosage compensation but not spermatogenesis. *Genes Dev* **11**:156–166.
89. Penny GD, Kay GF, Sheardown SA, *et al.* (1996) Requirement for Xist in X chromosome inactivation. *Nature* **379**:131–137.
90. Debrand E, Chureau C, Arnaud D, *et al.* (1999) Functional analysis of the DXPas34 locus, a 3' regulator of Xist expression. *Mol Cell Biol* **19**:8513–8525.
91. Lee JT, Lu N. (1999) Targeted mutagenesis of Tsix leads to nonrandom X inactivation. *Cell* **99**:47–57.
92. Luikenhuis S, Wutz A, Jaenisch R. (2001) Antisense transcription through the Xist locus mediates Tsix function in embryonic stem cells. *Mol Cell Biol* **21**:8512–8520.
93. Sado T, Wang Z, Sasaki H, Li E. (2001) Regulation of imprinted X-chromosome inactivation in mice by Tsix. *Development* **128**:1275–1286.
94. Matsuda Y, Chapman VM. (1992) Analysis of sex-chromosome aneuploidy in interspecific backcross progeny between the laboratory mouse strain C57BL/6 and Mus spretus. *Cytogenet Cell Genet* **60**:74–78.
95. Russell LB, Chu EH. (1961) An XXY male in the mouse. *Proc Natl Acad Sci USA* **47**:571–575.
96. Shao C, Takagi N. (1990) An extra maternally derived X chromosome is deleterious to early mouse development. *Development* **110**:969–975.
97. Tada T, Takagi N, Adler ID. (1993) Parental imprinting on the mouse X chromosome: Effects on the early development of X0, XXY and XXX embryos. *Genet Res* **62**:139–148.
98. Takagi N. (2003) Imprinted X-chromosome inactivation: Enlightenment from embryos in vivo. *Semin Cell Dev Biol* **14**:319–329.
99. Bennett-Baker PE, Wilkowski J, Burke DT. (2003) Age-associated activation of epigenetically repressed genes in the mouse. *Genetics* **165**:2055–2062.
100. Hatakeyama C, Anderson CL, Beever CL, *et al.* (2004) The dynamics of X-inactivation skewing as women age. *Clin Genet* **66**:327–332.
101. Sangha KK, Stephenson MD, Brown CJ, Robinson WP. (1999) Extremely skewed X-chromosome inactivation is increased in women with recurrent spontaneous abortion. *Am J Hum Genet* **65**:913–917.
102. Lanasa MC, Hogge WA, Kubik C, *et al.* (1999) Highly skewed X-chromosome inactivation is associated with idiopathic recurrent spontaneous abortion. *Am J Hum Genet* **65**:252–254.

103. Uehara S, Hashiyada M, Sato K, *et al.* (2001) Preferential X-chromosome inactivation in women with idiopathic recurrent pregnancy loss. *Fertil Steril* **76**:908–914.

104. Beever CL, Stephenson MD, Penaherrera MS, *et al.* (2003) Skewed X-chromosome inactivation is associated with trisomy in women ascertained on the basis of recurrent spontaneous abortion or chromosomally abnormal pregnancies. *Am J Hum Genet* **72**:399–407.

105. Kristiansen M, Langerod A, Knudsen GP, *et al.* (2002) High frequency of skewed X inactivation in young breast cancer patients. *J Med Genet* **39**:30–33.

106. Feinberg AP. (2001) Cancer epigenetics takes center stage. *Proc Natl Acad Sci USA* **98**:392–394.

107. Indsto JO, Nassif NT, Kefford RF, Mann GJ. (2003) Frequent loss of heterozygosity targeting the inactive X chromosome in melanoma. *Clin Cancer Res* **9**:6476–6482.

108. Lee JT. (2000) Disruption of imprinted X inactivation by parent-of-origin effects at Tsix. *Cell* **103**:17–27.

109. Gartler SM, Goldman MA. (1994) Reactivation of inactive X-linked genes. *Dev Genet* **15**:504–514.

110. Beard C, Li E, Jaenisch R. (1995) Loss of methylation activates Xist in somatic but not in embryonic cells. *Genes Dev* **9**:2325–2334.

111. Panning B, Jaenisch R. (1996) DNA hypomethylation can activate Xist expression and silence X-linked genes. *Genes Dev* **10**:1991–2002.

112. Hansen RS, Stoger R, Wijmenga C, *et al.* (2000) Escape from gene silencing in ICF syndrome: Evidence for advanced replication time as a major determinant. *Hum Mol Genet* **9**:2575–2587.

113. Chow JC, Brown CJ. (2003) Forming facultative heterochromatin: Silencing of an X chromosome in mammalian females. *Cell Mol Life Sci* **60**:2586–2603.

114. Lee JT, Strauss WM, Dausman JA, Jaenisch R. (1996) A 450 kb transgene displays properties of the mammalian X-inactivation center. *Cell* **86**:83–94.

115. Clerc P, Avner P. (1998) Role of the region 3′ to Xist exon 6 in the counting process of X-chromosome inactivation. *Nat Genet* **19**:249–253.

116. Sheardown SA, Duthie SM, Johnston CM, *et al.* (1997) Stabilization of Xist RNA mediates initiation of X chromosome inactivation. *Cell* **91**:99–107.

117. Panning B, Dausman J, Jaenisch R. (1997) X chromosome inactivation is mediated by Xist RNA stabilization. *Cell* **90**:907–916.

118. Wutz A, Jaenisch R. (2000) A shift from reversible to irreversible X inactivation is triggered during ES cell differentiation. *Mol Cell* **5**:695–705.

119. Martin GR, Epstein CJ, Travis B, *et al.* (1978) X-chromosome inactivation during differentiation of female teratocarcinoma stem cells *in vitro. Nature* **271**:329–333.

120. Mcburney MW, Adamson ED. (1976) Studies on the activity of the X chromosomes in female teratocarcinoma cells in culture. *Cell* **9**:57–70.

121. Zuccotti M, Monk M. (1995) Methylation of the mouse Xist gene in sperm and eggs correlates with imprinted Xist expression and paternal X-inactivation. *Nat Genet* **9**:316–320.

122. Daniels R, Zuccotti M, Kinis T, *et al.* (1997) XIST expression in human oocytes and preimplantation embryos. *Am J Hum Genet* **61**:33–39.

123. Ray PF, Winston RM, Handyside AH. (1997) XIST expression from the maternal X chromosome in human male preimplantation embryos at the blastocyst stage. *Hum Mol Genet* **6**:1323–1327.

124. Harrison KB. (1989) X-chromosome inactivation in the human cytotrophoblast. *Cytogenet Cell Genet* **52**:37–41.

125. Ropers HH, Wolff G, Hitzeroth HW. (1978) Preferential X inactivation in human placenta membranes: Is the paternal X inactive in early embryonic development of female mammals? *Hum Genet* **43**:265–273.

126. MacDonald M, Hassold T, Harvey J, *et al.* (1994) The origin of 47, XXY and 47, XXX aneuploidy: Heterogeneous mechanisms and role of aberrant recombination. *Hum Mol Genet* **3**:1365–1371.

127. Quan F, Janas J, Toth-Fejel S, *et al.* (1997) Uniparental disomy of the entire X chromosome in a female with Duchenne muscular dystrophy. *Am J Hum Genet* **60**:160–165.

128. Dhara SK, Benvenisty N. (2004) Gene trap as a tool for genome annotation and analysis of X chromosome inactivation in human embryonic stem cells. *Nucleic Acids Res* **32**:3995–4002.

129. Looijenga LH, Gillis AJ, van Gurp RJ, *et al.* (1997) X inactivation in human testicular tumors. XIST expression and androgen receptor methylation status. *Am J Pathol* **151**:581–590.

130. Chow JC, Hall LL, Clemson CM, *et al.* (2003) Characterization of expression at the human XIST locus in somatic, embryonal carcinoma, and transgenic cell lines. *Genomics* **82**:309–322.

131. Rosler ES, Fisk GJ, Ares X, *et al.* (2004) Long-term culture of human embryonic stem cells in feeder-free conditions. *Dev Dyn* **229**:259–274.

132. Draper JS, Smith K, Gokhale P, *et al.* (2004) Recurrent gain of chromosomes 17q and 12 in cultured human embryonic stem cells. *Nat Biotechnol* **22**:53–54.

133. Nagy A, Gocza E, Diaz EM, *et al.* (1990) Embryonic stem cells alone are able to support fetal development in the mouse. *Development* **110**:815–821.

134. Wang ZQ, Kiefer F, Urbanek P, Wagner EF. (1997) Generation of completely embryonic stem cell-derived mutant mice using tetraploid blastocyst injection. *Mech Dev* **62**:137–145.

135. Walker SK, Hartwich KM, Seamark RF. (1996) The production of unusually large offspring following embryo manipulation: Concepts and challenges. *Theriogenology* **45**:111–120.

136. Nagy A, Rossant J, Nagy R, *et al.* (1993) Derivation of completely cell culture-derived mice from early-passage embryonic stem cells. *Proc Natl Acad Sci USA* **90**:8424–8428.

137. Hoffman LM, Batten JL, Haul, *et al.* X-inactivation status distinguishes distinct epigenetic states in hESC lines. In preparation.

# 10

# Stem Cells and Translational Medicine Ethics, Law, and Policy

Justine Burley*

## Introduction

Few developments in the biosciences have generated as much excitement as stem cell research. Disease and ill health are enemies man faces even in times of peace. The promise of stem cell research is that it may arm us to combat disease in novel ways: to cure instead of treat; to repair hitherto irreparably damaged tissue. Demonstrable benefits have already been gained from stem cell research, with one adult stem cell type — hematopoietic stem cells present in bone marrow — in widespread clinical use. Initial optimism over this success has been tempered by technical difficulties experienced in work with many other adult stem cell populations. Embryonic stem cell research, in relative infancy, has shown prowess in animal models for the amelioration of a variety of conditions. But, challenging biological hurdles have to be cleared before human ES cell research meets any of the goals which have been set for it.

It perhaps goes without saying that given we are concerned about moral as well as physical well being, science should not race forward blinkered by the lure of progress. Nor, however, should scientific progress be impeded without sound reason. In this chapter, I aim to consider various ethical, legal, and policy dimensions of stem cell research. Some familiar territory will necessarily be covered. This is because the chief ethical objection to stem research that has been voiced, and which has prompted restrictive legislation in many countries, centers on the moral permissibility of research involving human embryos. The general failure satisfactorily to address the issue of the moral status of early life in other contexts — fertility treatment, abortion, pre-implantation genetic diagnosis, and birth control — made it

*National University of Singapore, Graduate School for the Integrative Sciences & Engineering, MD11, #01-10, 8 Medical Drive. Singapore 117597. Tel: (65)6874 1281, fax: (65)6464 1148, email: ngsbjc@nus.edu.sg.

inevitable that any novel rationale for the use of embryos would see the same debate resurface. This points to the importance of directly engaging with the morality of embryo research, and also of forging a reasoned, enduring, and publicly justifiable policy stance on it. To these ends, evaluation of a subset of related issues is indicated. First, confusion reigns in ethical, and policy debates, over the respective attributes of adult and ES cells. This, I contend, has made it more difficult for some to appreciate why it is that many scientists insist that adult and ES cell research should be pursued in tandem. Second, there has been significant energy expended on the question of how embryos might ethically be sourced for research. Although such discussion has served to facilitate "compromise" legislation in some countries — ES cell research using supernumerary embryos is permitted whilst embryos deliberately made for research purposes is not — the arguments deployed in the policy context to defend this move can, at best, be described as lacking. This will be illuminated through consideration of differing views on the moral status of the embryo. The requirement of non-coerced, informed consent will also be addressed in this section, with the spotlight on two pervasive problems: the form that consent takes, and conceptual confusions in law concerning property rights in gametes, and embryos. The third issue, the one I judge to be the most crucial, is whether is it legitimate for citizens in a society to impose a particular conception of the sacredness of life, on the minority. Fourth, I shall survey national policies governing ES embryo research, and assess several of their implications for science, and the scientist. These include: the dubious wisdom of confining research to clinically unsafe stem cell lines, and basic research, the hypocrisy of certain legal regimens which impose legal penalties on scientists who transgress national stem cell laws, and whether it would be just for any nation where ES cell research is currently illegal to avail itself of research fruits plucked elsewhere. Finally, discussion shifts from the ethics of stem cell *research* to exploration of new *clinical* applications involving adult stem cells. The question to be addressed in this concluding section is whether the clinical use of a product may permissibly precede scientific understanding of its underlying mechanisms.

## The Relative Merits of Adult and ES Cells

How we define the term "stem cell" is more relevant to a discussion of the ethics of stem cell research than, at first glance, it might appear. Definitions serve as the basis for mutual understanding. If we are to conduct a sensible ethical discussion of such research in the service of sound policy-making, it is essential that we establish common factual premises for what we are arguing about. This would seem a rather basic, indeed obvious requirement. Yet loose, sometimes careless descriptions of stem cells have been adopted in both ethical and policy debates, and have even been enshrined in legislation.[1] In some measure this is due to ignorance. It also may suit certain interest groups who aim to influence policy to insist on factual inaccuracies. It does not help that scientists, themselves, do not always agree on which cells merit the label stem cell.[2] Nor does it help that scientists have not always taken pains to be more precise in the stem cell literature.

Typically, stem cells are defined by two functional properties: the ability to self-renew, and the ability to generate other cell types. The property of self-renewal is frequently described as being "unlimited" or "continuous." The property of potency is couched in terms of multipotency or pluripotency. Without further qualification, the two-part functional definition can mislead the non-specialist. It fails to highlight important differences between adult, and ES cells, the most crucial of which for present purposes are: the limited range of cell types that many adult stem cells can form; the inability of most adult stem cells to multiply outside their natural environment; the difficulties associated with identifying and purifying populations of adult stem cells — in this latter context, adult "stem cell" cultures are often mixed populations of different stem cell types or of more differentiated progenitor cells.[3] These pitfalls will be elaborated on below. Here it merits stress that the net result of the failure by ethicists and policy-makers to appreciate the important differences between the categories of stem cells is that: 1) discussions have been skewed towards consideration of the *source* of the cells; and, thereby 2) important clinical advantages that work with ES cells may bring about, have been ignored.

Many who oppose the use of embryos in research, and who advocate a research focus exclusively on adult stem cells, have indiscriminately latched on to the words "immortal" and "unlimited," and phrases like "capable of differentiating into a vast number of cell types," to make the case that adult stem cell types might serve all therapeutic needs, hence ES cell research is unnecessary. This position rests on several misunderstandings about the way in which these terms are aptly applied to adult stem cell populations in the research context, if they may be applied at all. Two confusions that we need immediately to dispel are: a) that adult stem cells all possess the same properties as one another; and b) that adult stem cell types possess the same properties as ES cells.

A plethora of studies has yielded the view that not all stem cells are equivalent. Only adult stem cells occur naturally in the body. ES cells, by contrast, are *in vitro* artefacts. The circumstances surrounding this unnatural origin may account for some of the behavioral differences in culture between ES and naturally occurring adult stem cells. Whereas ES cells have proved capable of significant proliferative capacity in culture, most adult stem cell types have not. Indeed, only adult neural stem cells (NSCs) have been induced to proliferate extensively *in vitro*,[4] and even here controversy persists in the literature as to whether NSCs are indeed "stem cells" or stem-cell derived neural progenitor cells — a cell type with an even more limited proliferative capacity.[5] This relative limitation of adult stem cells to self-renew is of no small matter for policy. When the target of stem cell research is a therapeutic one, the proliferative capacity of the cells is decidedly of import, since, depending on the clinical indication, large numbers of cells may be required. If many adult stem cell types continue to resist propagation outside of the body, ES cells, with their relatively superior performance *in vitro*, may be better research tools on this count alone. There are additional reasons for maintaining that ES cells may be preferable research tools.

A second confusion that reigns in ethical debate over stem cells concerns the functional property of potentiality. There is solid evidence showing that adult stem cells are nowhere near as potent as ES cells. The prevailing theory is that adult stem cells are generated after the stage of gastrulation, and this accounts for why it is that adult stem cell populations have a restricted differentiation potential — why they are "confined to barracks," so to speak. Recent reports (over 200 in one year) have suggested that these presumed tissue-restricted cells possess developmental capabilities more similar to pluripotent ES cells.[6] Claims to this effect have been challenged on the basis of flawed experimental design, mixed starting cell populations,[7] and for the reason that studies describing plasticity do not meet the standard criteria for stem cells. But, it remains a possibility that adult stem cells may be more plastic than was previously thought. In contrast, ES cells can differentiate into all of the cell types of the adult; here, the major technical challenge is in controlling and maximizing the desired conversion. With this important caveat in mind, the fact that adult stem cells either do not exist for certain tissues or are inaccessible or otherwise unavailable in sufficient numbers, we can conclude that, in principle, ES cells currently suggest themselves as candidates for the development of stem cell therapies for a wider range of clinical applications than do adult stem cell types.

Our overview of the relative merits of the classes of stem cells for clinical applications must take into account one further factor. Although the focus today is on the ability of stem cells to form the desired replacement tissue, clinical application of such tissues/cells requires consideration of the immunological consequences of the transplantation. One attraction of adult stem cells is that, in principle, they may be purified from the patient him/herself, thereby circumventing the need for immunosuppression. In comparison, those cell therapy products emanating from ES cells generated from embryos surplus to *in vitro* fertilization (IVF) needs, will be generally immunologically distinct from the prospective patients' cells, and will elicit an immune reaction. The avoidance of this will necessitate the application of immuno-modulation therapies, most commonly chronic use of immunosuppressive drugs. In many patients, such drugs induce moderate to severe side effects. It is this reality which motivates calls for permissive policies viz. the creation of embryos produced by somatic cell nuclear transfer (SCNT), which would enable tailor-made stem cells, immunocompatible with the patient.

## Why Conduct Research on Adult and ES Cells in Tandem?

Having elaborated the functional definition of stem cells, and reviewed some of their properties, it should now be clearer why it is that many scientists insist that human adult and ES cell research should be pursued in tandem. In the province of basic research, human ES lines provide a surrogate to the human embryo, allowing investigation of early developmental processes that otherwise can only be studied in other mammals. As mentioned above, for the purposes of pre-clinical or clinical research and applications, each area has its advantages and disadvantages. Proliferative capacity, broad range of differentiation, and ready availability, are major advantages offered by *ES cells.*Control over differentiation, possible autologous sourcing, a better safety profile, and universal ethical acceptability constitute the advantages of *adult stem cells.* It is unclear which approach will offer the most rapid clinical solutions. The likelihood is that neither will have a monopoly over all clinical indications. The only resounding therapeutic success using adult stem cells is bone marrow transplantation. Whilst this demonstration confirms the potential of adult stem cell use, the fact that this solitary success was first achieved over 40 years ago highlights the need for new thinking as well as alternative approaches.

Taken jointly, the reasons given above support the claim that pursuing adult and ES cell research in tandem is less speculative than it is commonsensical. This in-tandem recommendation also extends to embryonic germ (EG) cells.[8] EG cells possess similar characteristics to ES cells. For example, they can be propagated in culture over a long period (though less read-ily than ES cells — 200 population doublings over two years for EG cells[9] versus 700 for ES cells), and can, with the correct prompts, spontaneously differentiate into derivatives of all three primary germ layers — endoderm, mesoderm, and ectoderm.[10] It has been established that EG cell lines could serve as reliable sources for cell-based interventions.[11] There has been so little work carried out on EG cells in comparison to both ES and adults stem cells, that it is impossible to pronounce on their relative clinical appli-cability with any authority — only more research shall determine whether and for which indications EG cells might be suited.

# The Provenance of Stem Cells: Does Source Matter?

In spite of the fact that good scientific reasons can be supplied to support the claim that adult, ES and EG cell research should be carried out in parallel, ethical concerns over the provenance of the latter two cell types has resulted in outright bans or policy restrictions on their use, in many nations. Does the source of stem cells matter morally quite as much as people tend to think?

Embryonic germ (EG) cells are obtained by laboratory cultivation of primordial germ cells found in the fetal gonadal ridge.[12] These can be isolated from fetuses following pregnancy termination (artificially induced in cases of unwanted pregnancy or the product of miscarriage). The source of EG cells, immature reproductive organs of aborted fetuses, may repel some, but it seems right that some good come of the frustration of life. Indeed, given sensitivities about the destruction of early human life in ES cell research, studies on EG cells ought to be more ethically palatable for the obvious reason that the fetal source is moribund. Some anti-abortion activists have argued that research involving EG cells will create demand for fetal tissue, and this in turn will lead to the offering of financial incentives to women to terminate pregnancies that might otherwise be carried to term. This objection is effete: like it or not, there is an ample, ongoing supply of fetal tissue. The worry tabled, however, points to the need for strict prohibitions against paying donors for fetal tissue to avoid the merest hint of any financial inducement to abortion (see below). Although I omit dedicated discussion of EG cell research below, certain of the arguments advanced provide further defence of it.

Adult stem cells can be isolated from tissues in an adult organism, or from fetal tissue (excluding primordial germ cells). Mammals appear to contain some 20 major types of somatic stem cells, including brain, liver, bone, blood, cartilage, and arguably corneal.

In ES cell research embryos are sourced in two main ways: from fertility clinics, and by creating them specifically for research purposes. Either way the entity is sacrificed at some point prior to gastrulation. The creation of embryos in the laboratory setting can be performed either by inseminating an unfertilized egg, or by splitting a fertilized egg, or by the technique of SCNT — misleadingly dubbed therapeutic cloning. SCNT consists in the removal of the nucleus from an unfertilized egg, substituting it with the nucleus of a somatic cell, and then prompting the egg to undergo embryogenesis. As with embryo splitting, SCNT skirts the process of fertilization. The major benefit and driving force behind ES cells so generated is to provide an immunological match to the patient's tissues as any replacement tissue formed from ES cells would not elicit a hostile immune response.

Where research on gametes, embryos, or fetal tissue for research purposes, is permitted by law, uncoerced and informed consent of the donors is usually sought. Many institutions, national and pan national ethics boards have devised consent guidelines to ensure consent is faithfully solicited and given. A full review of these is beyond the scope of the present inquiry. I confine myself instead to making one point about written consent. A great deal has been published about the difficulty of eliciting truly informed consent in the clinical context. However, it strikes me that most of the standardly cited problems either do not obtain, or can be circumvented in cases of consent for use of biological materials in research. A cursory glance at a representative consent form gives quite the opposite impression. I judge there to be strong reasons supporting a move away from increasingly detailed, research-specific consent forms, to blanket consent or at least to forms which stipulate in only broad terms the area of intended research. The main reason behind this recommendation is that not all specific research or clinical uses of donated materials are likely to be envisaged in advance, especially in the case of fledgling technologies like ES cell research. Researchers have already been, and clinicians may be placed in the position of having to re-solicit consent to use the materials for purposes not originally specified (e.g. the creation of chimeras using the line, or a therapeutic application). Soliciting consent anew is not always logistically possible, and when it is, there is no guarantee that consent will be granted the second time around. Consequently, research can be slowed or significant expenditure on the production of a therapeutic product can be wasted. The abandonment of detailed forms may not be a suggestion popular with lawyers, but, the fact is, continued usage of them in *inappropriate* circumstances invites more problems than they solve.

Whether individuals might reasonably expect payment for the materials is a matter of some controversy, and is the second pervasive problem connected to consent. Financial incentives, it is argued, constitute a form of coercion or exploitation. Whatever the force of this objection is in relation to body parts like kidneys, it lacks punch in the case of gametes, which are replaceable, and non-essential to health. It is true that egg harvesting carries some risk to the woman (e.g. perforation of the ovary), and so that risk should be spelled out. However, it is not true that the magnitude of risk is such as to warrant state interference in the form of a ban on either donation or trade. And, we should note that risk cannot consistently be used as an objection to sale, in those countries where donation is permitted. The notion that someone might abort a foetus *because* they had been offered payment for fetal tissue, is rightly appalling. Abortion is defensible if good reasons for it can be supplied (when carrying the pregnancy to term would have special and grave consequences for the mother, when the life of the resulting child would be *profoundly* compromised or not worth living). The frustration of a life *for* monetary gain, under ordinary circumstances, does not constitute a morally defensible reason. With respect to the sale of embryos, the claim that individuals whose fertility treatment has resulted in embryos surplus to requirements are justified in seeking remuneration from researchers, invites the objection that this would be to commodify life, to treat life as a mere piece of property.

In fact, the law is conceptually confused over ownership titles in gametes and embryos. They are plausibly regarded in law as *species* of property. The problem that arises is that the corpus of law in most countries contains conflicting legal attitudes on *which* species of property, they are. Control ownership of gametes is recognized in relation to, for example, procreation, placement of gametes in banks prior to chemotherapy, non-procreative sexual acts, and directives stipulating the fate of a couple's supernumerary IVF embryos in the event of divorce. At the same time, in countries where control in these circumstances is recognized, the right to alienate our control over gametes/embryos through donation, is not always permitted. And, in many nations where donation is allowed, the right to alienate control of sperm, eggs, and embryos, through *sale* is disallowed. Are such apparently conflicting legal attitudes defensible?

What justifications are there for proscribing commercial trade in gametes, and embryos, especially in those nations where donation of all three is legal? It has been argued that if we own our gametes, and embryos then it follows that we also own our children. However, this objection (of the slippery slope variety, a style of argument about which we shall say more in another context, below), can be dodged by insisting that we cannot own human entities which possess interests of a morally relevant kind: infants, children, and adults indisputably have morally protectable interests. Gametes have no interests let alone such interests, and therefore I see no problem with unfettered property rights in them. As to whether embryos have interests, the ensuing discussion makes clear this a matter of deep disagreement about the sacred, the kind of disagreement over which government may only legitimately take a stand, if it has strong reasons for doing so, and it is most unclear that such reasons are available. The same discussion also makes clear that even if we accept that embryos have interests, these interests are permissibly overridden in many circumstances. In advance of that discussion, to round off our reply to the view that property rights in embryos entails property rights in fully formed persons, the point here, to repeat, is that there is no such entailment. Furthermore, recognition of some property rights in embryos does not amount to free license; consistent with the norms of property law, is that not all property rights can be exercised with impunity.[13]

The final objection we shall consider to donation or sale in gametes and embryos, is the woolly objection from reverence and respect for the human body.[14] It does not, however, seem to me, that we either do or should revere/respect all bodily materials. As is obvious, we hold rather cavalier attitudes to many "bits" of the body — hair, for example. And, alienation of control over some body materials is widely practiced (donation/sale of blood). Thus the objection from respect and reverence stands or falls not on our attitudes to the human body but on our attitudes to gametes and embryos. People typically, and permissibly, do not regard gametes as objects of reverence, hence any objection to unfettered property rights in them, on the basis that they do, is a non-objection. There are compelling reasons to think that embryos merit respect when the intention is that they shall become persons. When this is not the intention, the respect requirement does not necessarily figure as a moral demand. The question of whether embryos are the property of individuals, and can, on that basis, be donated, bought and sold, hinges on what interests they might plausibly be said to have. This is the subject to which we now turn, first by way a challenge to the claim that the source of embryos for ES cell research matters morally.

Emphasis has been placed on the merits of using embryos in research which would otherwise remain in deep freeze and ultimately perish, in preference to those deliberately created for research. On the face of it this makes sense — if the entity is to expire anyway it seems better that benefit be derived. But, people who oppose the destruction of any and all human life are not persuaded that the source makes any moral difference whatsoever. Nor should they, given that they believe that as soon as human life begins, it should be afforded the same protection of security of the person as adult individuals. They therefore regard the sacrifice of supernumerary embryos for research purposes to be worse than simply allowing the entity to degrade gradually in its frozen-down state, just as they would regard it as being morally worse to sacrifice an adult in failing health who was destined to die, to serve the greater good. This position is consistent although, as I shall shortly show, I strongly doubt that anyone is wholly wedded to it.

For others who subscribe to the view that human life assumes value incrementally, hence early life has little value or is less valuable than adult lives, the source of the embryos does not much matter either. If no early life has value, then why trouble over the source of that early life? And, people who think that early life has only marginal value, then, again, it is of little import to them what the source of early life is. These positions are also consistent.

There is one concern about source that people holding any of the views just described, might harbor, which is not, on the face of it, underpinned by any story of moral status. The argument is that had we not had IVF, and if we did not have SCNT to generate ES cells from research embryos, then it would be less likely that SCNT could be refined as a procreative method (i.e. to produce cloned children). This can be conceded. It is worth stressing, however, that it is definitely not the case that there is hard and fast causal relationship between a technology and all of its possible uses. Slippery slope arguments are ever current though easily discredited. There are many technologies which might be abused, but which regulation and/or legislation manages to keep in check. And, there are many technologies which are abused but which nonetheless we would not want to disinvent. Consider an analogy with information technology (IT). The Internet makes possible global distribution and commerce in child pornography. Yet, few lament the invention of IT. Thus the power of the objection that IVF and ES cell research using SCNT make reproductive cloning more likely, rests heavily on whether one can substantiate the view that IVF and deriving embryos by SCNT are themselves immoral. If this is the argument, the objector is

smuggling in the notion either that early life has morally protectable interests or that any non-natural manipulation of gametes is immoral because
*naturalness* in sex and procreation are fundamental to being human. This
constitutes, quite clearly, a conception of what is sacred about life. How the
state should respond when citizens hold divergent deep seated conceptions
of the sacred, is *the issue* in the controversy surrounding research involving
embryos, and merits sustained discussion. This we shall proffer following
scrutiny of different positions held on the moral status of early life.

## The Moral Status of the Embryo, and Conceptions of the Sacred

The moral status of the embryo has been the chief preoccupation of many
ethicists, religious groups, citizens, and policy-makers in the debate over
the moral permissibility of ES cell research. Arguments to the effect that
research on embryos is wrong because it violates human dignity have gained
currency but have virtually no value. Simply to utter the words as if, on their
own, they constitute an argument is reasoning at its sloppiest. And, to use
the concept of human dignity to place on the legislative agenda one's own
private moral view, debases the whole idea of human dignity. Many mission
statements issued by ethics committees contain "human dignity," yet for all
of the "whereas" clauses, and definitions found in these documents, no one
ever attempts to spell out what human dignity is, and how it might be
offended by ES cell research. We should be suspect of anyone who cannot
articulate any content for human dignity but who professes to know a
violation of it. In addition, it is insufficient to state that human dignity is
transgressed when the vital interests of a human are acted on in inhumane,
and degrading ways. It must be established what these vital interests are,
and who might possess such interests.

It is said that research involving embryos turns "nascent human life
into a natural resource to be mined and exploited, eroding the sense of
worth and dignity of the individual."[15] Human conceptuses are human
*entities* from the time of conception. This is indisputable (at least from
a biological/genetic standpoint). However, it is controversial as to what
follows from this fact. Some argue that because the entity is human, it merits
the same legal protections that might be afforded an adult under threat of
physical violation. Does the property of humanness, in and of itself, really
matter morally? Let us reflect on the following example.[16] There are two
islands, a violent typhoon will imminently strike both, and there is only one

rescue boat. On one of the islands there are 100 human blastocysts housed in a freezer, on the second island, there are 100 adult humans. Which island population should the captain be ordered to rescue? Surely the boat should be deployed to the second island. That the vast majority of people one might survey, would likely endorse this answer, suggests that even those who say that blastocysts have as much value as fully formed individuals who possess sentience, etc. may not in fact believe in what they say.

Mere humanness does not capture what troubles people about research involving embryos. Why would we prefer to save the adults? Is it because they are sentient, and therefore possess the capacity for suffering, and have awareness of their impending plight that (under no plausible understanding of early development) any human blastocyst has? One riposte is that although embryos are not sentient, they should be accorded full or weighty moral status on the grounds that they have the potential for sentience.

It is held by objectors to stem cell research that destroying embryos for the purposes of stem cell cultivation destroys a potential person. However, if the embryo is created specifically for the purpose of research, then does it make good sense to consider it a potential person? In what sense is x a potential y if the conditions required for the transition from x to y do not obtain? For blastocysts to become persons, the necessary conditions of environment and nurture must be satisfied. In the case of natural procreation, many embryos die owing to chromosomal abnormalities. We are told that a potential person becomes such at conception. But these embryos could never have become a person because they lacked the requisite genetic complement to survive. The point is that potentiality is necessarily a contingent concept. A blastocyst created by a stem cell researcher would always have been intended for research and/or clinical uses. Thus, from the outset, there was no possibility of the blastocyst becoming, e.g. Jane Doe, because there was never any intention of providing it with the correct environment (and so points to the conclusion that, contrary to popular wisdom, it is not better to use supernumerary embryos, because these were possible candidates for implantation, which embryos created for research never are). This begs the obvious question of whether creating life with the intention of destroying it, can be justified.

So much is made of intention, we ought to give it due consideration. One possible strategy is that intention need play no part at all in the defence of ES cell research involving SCNT. This is because SCNT, were it applied to humans for the purposes of reproduction, arguably creates no potential person. Why? Think of the reasons behind opposition to reproductive cloning. SCNT has a low efficiency rate, and is associated with severe abnormalities. The incidence of both has been traced to faulty gene imprinting. The imprinting problem is deemed to be so severe by some scientists, that they doubt that a human with a normal range of powers could emerge from the technique.[17] Furthermore it has been suggested, on the basis of defects in gene imprinting, that SCNT in humans for reproductive purposes would likely require the creation of huge numbers (hundreds or even thousands) of embryos before a single cloning success might be achieved. Hence the experimentation on humans that would be required to establish SCNT as a viable procreative method, speaks against deploying it for that purpose. My own point is that since there is so much uncertainty surrounding the nature of the abnormalities that would present, it is reasonable to state that it is possible no person could *ever* emerge through SCNT. Hence SCNT in ES cell research involves no creation of life with the intention of destroying a potential person. The factual premise is that no person can be created by the technique, therefore there is no potential person whose integrity, dignity or other such is violated. It is, however, ill advised for a philosophical argument to rest on contingent facts about science. Within the realm of the possible is that imprinting difficulties may be overcome. A more robust rejoinder to the notion that it is wrong to create life with the intention of destroying it, must be sought. One is available — the argument from consistency in moral reasoning.

Any potentiality argument which incorporates the view that life begins at conception, were it followed to its logical conclusions, denies the moral permissibility of women using intrauterine devices for birth control, IVF (which involves embryo wastage), and abortion. Perversely, it also implies that natural procreation is morally problematic. We know that on average three out of four human conceptuses die before one survives to the fetal stage.[18] When we make the choice to procreate in the light of knowledge about this attrition rate, are we not complicit in bringing about the death of early *lives* in order to produce one child? Critics will immediately object that there is a meaningful distinction to be drawn between an act which is non-maleficent (such as procreation) and one which is maleficent (creation of life with the intention of destroying it).[19] Critics also point out that whereas in the case of procreation there is the clear intention to benefit any resulting child, in the case of embryo research there is no intention to benefit the entity created. These distinctions have less purchase than they think, however. This is because much turns on how the respective acts are described. The normative description of the deliberate creation of research embryos being "maleficent" is inapt when we pitch the act in terms of: "Researchers unavoidably destroy embryos in pursuit of cures for millions of very ill people." If this is a linguistic fudge, we can note that in any country where embryo research is permitted but only to aid reproduction or our understanding of it, policy-makers happily indulge such fudges. The normative descriptive terms "non-maleficent" and "benefit" seem less apt with respect to procreation when it is described as "knowingly risking loss of some early lives in the pursuit of the selfish preference to have a child." It is hard to erase, at least in convincing fashion, complicity in, if not intended, death of some lives, from the procreative story. Moreover it is difficult to sustain the idea that procreation is *motivated* by the desire to benefit a new life. Certainly, the lives that perish in the process have not been benefited by being brought into existence. Furthermore the desire to have offspring is really about benefiting the parent, it is not about other-regarding benefit — the child's or that of the human species. The point is not that we do wrong when we procreate. The point is that we procreate without a second thought as to what it involves. Even if the arguments we have just advanced fail to convince, it remains the case that it cannot be so much the loss *tout court* of early lives that bothers individuals, but the non-natural uses for which these early lives might be employed.

In the end, the defence of natural procreation from the objection that people actively conspire in the death of early life when they procreate, boils down to the claim that it defensible because it is an unavoidable, natural state of affairs. Again, we find that the "natural" is being trumpeted as the "good." The view that *natural* procreation is desirable whatever happens in the course of it, fundamentally, is a conception of the sacred.[20] This leads us back to the question we asked above: How do we resolve disputes in the policy arena, over differing conceptions of what is sacred about human life?

## Resolving Disputes About the Sacred

Entrenched disagreement about when life assumes value has, what Ronald Dworkin calls, an "essentially religious" character.[21] What Dworkin means by the nomenclature "essentially religious" is not that the beliefs about whether the lives of embryos matter morally, and the ways in which they do, are religious *per se*. Rather he means that they are *akin* to beliefs about what the true religious path is. Both are instantiations of deep seated moral convictions about the sacred. What is the appropriate response by government in the face of such disagreements? Well, think of religious worship.

Would it be proper for government to mandate that citizens worship one God and not another? The answer that would be given in most parts of the modern world is no. A hallmark of modern societies is that freedom of religion must be protected, that the State must never seek to enforce faith or style of religious worship on its citizens. If people's views about the sanctity of life do not differ conceptually from religious beliefs, then it follows we should not abandon government neutrality when the status of early life is the subject. Once the dispute over embryonic stem cell research is recast in the way suggested, we need no longer argue about or the point at which life assumes value, potentiality, or whether, indeed, we are God's or the Blind Watchmaker's children. Government policy must be publicly justifiable. The right *policy* answer to the permissibility of embryo research is not that it is definitively right or wrong. Rather the right answer is that given *reasonable* disagreement over the permissibility of ES cell research, government has a duty not to promote the rightness of one understanding of the sacred over another. Reasonable disagreement can be had about the moral status of early human life in a way it cannot be had over the moral status of human persons. The failure of contemporary policy-making viz. ES cell research, is the failure to recognize that to take a stand on which conception of the sacred is the true one, rocks the principled foundations of good government. It is not mindless or unreasonable to object to ES cell research on the grounds that embryos have value (although, as I have argued, it defies commonsense to insist on equal weight between them and, say, adults). What is unreasonable is the expectation that it is legitimate to enlist government to impose one's own conception of the sanctity of life on other citizens. No one who takes seriously the demand of religious tolerance in modern society would champion the abandonment of government neutrality on free religious worship. It follows they should also endorse neutral political concern as regards that which is conceptually alike.

## Ethical Implications of National Policies Governing ES Cell Research for Science and the Scientist

At the time of writing, nine countries (comprising approximately 2.5 billion of the world's population) allow the deliberate creation of embryos for ES cell research using SCNT: Belgium, China, India, Israel, Japan, Singapore, South Korea, Sweden, and the United Kingdom.[22] Currently, in all but one country where the Shari'ah forms the legal system in part or whole, there is no clear policy on ES cell research. No one unified juridical body represents all of Islam. But, Muslim jurists do draw a distinction between the early stage of pregnancy (first 40 days) and its later stages. Thus, under most interpretations of Islamic law, the embryo is not given as weighty a status as are fetuses. It has therefore been suggested that not only may ES cell research be permitted by the Shari'ah, but also the deliberate creation of embryos for research (so long as that research is well intentioned).[23] By implication from what has been said, it is unclear whether or not EG cell research would be endorseable by Muslim adjudicators — the answer would appear to depend on the fetal stage reached at the time an abortion/miscarriage occurred.

A number of countries have adopted different versions of what has been called flexible or compromise policy, according to which ES cell research is permitted but not using embryos created for that purpose[24]: Australia, Canada, The Czech Republic, Denmark, Estonia, Finland, France, Greece, Hong Kong, Hungary, Iceland, Iran, Latvia, The Netherlands, New Zealand, Russia, Slovenia, South Africa, Spain, Switzerland, and Taiwan.

The United States of America (USA) has a rather unique legislative situation. In 2001, federally funded stem cell research was allowed for the first time in the USA, on a stipulated number of ES cell lines. (It turned out that over two thirds of these lines were non-lines.) The private sector, however, was left free to create new ES cell lines, including through SCNT. In addition, owing to the USA's unique federal political system it was left open to individual states to introduce state-specific laws, as California has just done.[25]

Our survey of extant legislation/policy indicates that there is little that is united about nations in this debate. The deep divisions between countries on the issue of whether it is permissible to create embryos using SCNT was recently given expression in the United Nations (UN). By a margin of one vote the UN General Assembly's legal committee in November 2004, accepted a proposal to delay further debate on SCNT. This decision effectively defeated a resolution tabled by Costa Rica, and backed by the US and 60 other countries, which called for an international treaty to ban all uses of SCNT in humans for being morally reproachable. It also supplanted a second motion, emanating from Belgium, backed by 30 nations, which sought an open-door policy for medical research on stem cells taken from cloned human embryos.

# Implications of Current Law for Science and the Scientist

The different legal situations that obtain, give rise to a set of ethical problems for science, and the scientist. First, how a stem line has been derived is a relevant policy consideration in any country which has or which seeks to introduce restrictive legislation. For example, in the USA, all of the human ES cell lines approved in 2001 for use by federally funded stem cell researchers, were developed using mouse embryonic fibroblast (mEF) feeders. Such xenogeneic systems run the risk of cross-transfer of animal pathogens from the animal feeder cells to the ES cells. Any later clinical application of the currently approved lines is therefore potentially dangerous, and would be subject to the U.S. Public Health Service Guideline on Infectious Disease Issues in Xenotransplantation.[26] As of 2001, its definition of "xenotransplantation" "include[s] any procedure that involves the transplantation, implantation, or infusion into a human recipient of either: (a) live cells, tissues, or organs from a non-human animal source; or (b) human body fluids, cells, tissues or organs that have had *ex vivo* contact with live nonhuman animal cells, tissues, or organs. Furthermore, xenotransplantation products have been defined to include live cells, tissues or organs used in xenotransplantation." "It would also be subject to stringent FDA guidelines".[27] Both sets of guidelines were drafted with the potential for catastrophic zoonotic pathogens (particularly latent retroviruses) emerging from animal to human organ transplants, in mind. It is important to note, however, that it is not, in fact, the case that the FDA would regard ES cell-derived therapeutic products as necessarily unsafe, nor is the FDA unwilling to consider for approval the use of human ES cell-derived products which have been exposed to mEFs. But, in the absence of any test case, it is unclear what the FDA's *ultimate* decision will be on any ES cell-derived therapy. It can, therefore, be insisted that it is not sensible to disallow federal funding for research on the non-xenogeneic lines that were created two years after the so-called Bush lines were approved.[28] Yet, USA policy continues to disallow federally-funded studies involving such newly created lines. Germany, where researchers can only work on cell lines created before January 2002, is sailing in much the same kind of boat, and so is Austria. In the European context, from the perspective of clinically targeted stem cell research (for example, participation in the European Stem Cell Bank Initiative) Germany and Austria may be left out in the cold. Basic research, of course, continues, but it appears that scientists have decided that the climate is so uncertain that few have applied for approval to import lines.

The USA policy, taken in its entirety, is at best described as morally bipolar, and it is not the only country with legislation so afflicted. Although Austria disallows the creation, by whatever method, of embryos for stem cell research,[29] the import of pre-existing stem cell lines is not prohibited.[30] In Germany, where The Embryo Protection Act 1990 explicitly prohibits creation and utilization of embryos for any purpose other than reproduction, and where The Stem Cell Act (Stammzellgesetz) prohibits the use of funds — public or private — to derive new human ES-cell lines, the importation of and work on embryonic stem cells produced from supernumerary embryos (for which there was no payment) that date before 1 January 2002, is legal upon approval by a supervisory body.

We can immediately note that if it is wrong to use embryos deliberately created for research, then it is wrong. The fact that these embryos have been created in the private sector (USA) or elsewhere (outside of Germany or Austria) or before a certain date (USA, Germany, Austria), should not make research involving them any more right. One might argue, as people have done in relation to the use of data collected by Nazi doctors who performed gruesome experiments on humans, that since wrong has already been done, it is desirable that some good come out of it.[31] Although I have some sympathy with that line of thinking, surely the time at which we "right a wrong" in this manner, matters a great deal. It would have been unthinkable (one hopes) for any scientist residing outside of Nazi Germany during the war years, to use data that was gathered in violation of every known law of acceptable treatment of human beings, for their own research whilst the Nazi experiments were still being carried out. To have done so would have been tantamount to collusion. Likewise, to pass a law denying that embryos may be created using SCNT on the grounds that this is immoral, and yet to allow use of embryos created by SCNT elsewhere, only a short while ago, is an act of collusion. In sum, the policies of the USA, Austria, and Germany suffer from, what philosophers call, the "dirty hands" problem. Even accepting that the policies in these countries are intended to placate different groups through compromise, and are illustrative of different political systems, and cultural histories, they are still an embarrassment if we are correct in thinking that consistent and sound reasons should underpin national legislation. The described curtailments of ES cell research are both hypocritical and contradictory at the level of moral principle.

The last subset of issues that I wish to consider in this section is the impact of legislation on the scientist. First, some countries have passed laws which impose criminal sanctions on scientists who violate laws governing ES cell research. Everyone has an obligation to obey the law. However, at the very least the law should be clear on what is and what is not allowed, and it should be consistent. In Germany this is far from being the case.

As noted, two laws govern research on embryos: The Embryo Protection Act 1990, and the Stem Cell Act 2002. After the passage of the latter utter confusion reigned over the reach of the law. To help clarify current law, the German Research Foundation (DFG) published an expert opinion in an extensive report in mid 2003. The findings of the report are that the regulations in the Stem Cell Act which cover existing stem cell lines, are only applicable to research conducted on German soil, and apply to all German scientists, including public servants. By contrast, the regulations on newly created lines appear to endorse criminal sanctions on German *Professors* (who, unlike their junior colleagues, are classified as public servants) if: They move abroad and derive human embryonic stem cells in another country, collaborate with colleagues abroad whose work involves the generation of human stem cell lines, sit on the advisory board of an international company, or are on peer review committee of a foreign research council which advises on funding for projects involving the derivation of new stem cell lines, and engage in e-mail correspondence about new lines. One effect of the uneven legal situation has been the departure of ES cell researchers from Germany.

Brain drain, described in the context of ES cell research by one journalist as an "exodus for embryos"[32] is not a new phenomena, and certainly is not one confined to scientists whose area of research involves the use of embryos. However, one observes that the adverse impact of restrictive national science policy on the complexion of the scientific community, and the calibre of research, are lessons that have been repeatedly taught in the past, but apparently have not been learnt.

## The Injustice of Anti-ES Cell Research Policies

We have so far been discussing countries where ES cell research is permitted in some form or another. What of the position of those nations where no research on embryos is allowed at all? Can these nations uphold the view that it is morally wrong to conduct research on embryos yet avail themselves at some future date of ES cell-derived therapies? If a country has not participated in ES cell research when it possessed the financial and scientific resources to do so, and if a country actively sought to impede others from conducting such research (for example, by pushing for a ban on certain forms of it at the UN, or pushing for a pan-national policy in the European Community prohibiting any research on embryos) why should it be allowed to benefit from such research? Putting aside the obvious reality of how the free market will work in favor of such nations, we pose the questions: Is the anti-ES cell research position a just one? Can it be feasibly maintained if and when ES cell-derived therapies are in the clinic? A possible way to secure fairness is to require all anti-ES cell research nations to be signatories to a charter stipulating that they shall never seek to benefit from any ES cell-derived technology. Can such a move be defended? Why not? One difficulty here is that the policy could not be extended *down* through generations. Think of the legal cases in the USA involving Jehovah's Witnesses in which the court has thwarted attempts by parents to deny their children blood transfusions. The judicial decisions rendered all appeal to the notion of the best interests of the child. This example is not as parochial as it might appear. It can be reasonably stated that no government currently opposing ES cell or other research on embryos which results in products for the cure or treatment of devastating diseases, could deny such products to their citizens, and yet maintain that the interests of citizens were being served. It can be reasonably predicted that governments currently holding moral objections to research on embryos, would manage to overcome them in this scenario. But on what basis? Simply arguing that proven ES-cell derived treatments would benefit their citizens places them in a rather awkward position. Surely something does not become moral simply because benefit is to be had. A more cogent defence of the projected policy switch would be the argument that the instrumentalisation of early life is *less bad* than children dying unnecessarily, therefore making ES-cell derived therapies available is morally permissible. Although speculative, the preceding discussion suggests that the anti-embryo research legislation of today may be myopic in the extreme.

## Clinical Applications of Novel Adult Stem-Cell Derived Therapy

In 2001 a study was published claiming that stem cells derived from bone marrow could be used to repair damaged heart muscle following a severe cardiac event.[33] This sparked a number of clinical studies involving intracoronary injection of autologous bone-marrow-cells.[34] The leap into clinical action has been criticized on several counts. Although, historically, drugs have been approved for general use without any knowledge of their mechanism of action, nowadays regulators prefer to see some evidence for the underlying restorative mechanisms, both *in vitro* and in animal models, prior to any move to the clinic. Neither was done in the case under consideration. Second, critics charge that the interpretation of the results has ignored a number of possible explanations for patient improvement which either have nothing to do with the injected cells, or which have something do with the cells injected but not with stem cells *per se*.[35] Third, and relatedly, claims of efficacy are normally based on the results of randomized, and where possible, double-blinded, controlled trials. Such trials have not been carried out.

All that being said, analysis of the risk/benefit ratio of the treatment, justifies these studies. Importantly, only minor adverse side-effects have been reported, and a number of clinical studies have shown patient improvement. If an approach works, and does no harm, then why not champion its use in the clinic? Here the question arises as to whether the studies should have been commenced in the first place. It could be argued that no clinical study should precede scientific insight regarding the underlying mechanisms of the treatment. But, think of aspirin. It took 70 years to understand the operating mechanisms of aspirin. Should we have waited until we possessed that understanding to administer it? The bottom line is that if the benefits gained significantly outweigh the risks, then the move to the clinic is defensible.

However, there are a number of identifiable problems with the existing studies: They are not blinded or controlled or standardized either in methodology or interpretation, and they are all small studies.[36] This reality has prompted calls for clarificatory clinical trials. Following a heart attack, cardiac function tends to improve over time, hence controlled trials, in which the control reproduces the same conditions of the test but does not use bone marrow stem cells, are wanted. It goes without saying that the failure to perform controlled trials can produce misleading outcomes. In addition, it is the case that patients are treated for cardiac crises in many different hospitals. An array of different practices will not serve to test efficacy fully. Therefore trials should be designed that use protocols which are practicable across the spectrum of participating institutions. Moreover the protocols shall need to be standardized, and training of medics to perform the techniques in uniform fashion is indicated.

Double-blinded studies which accord with the aforementioned recommendations will be highly costly. Who should pay for them? It has been suggested that there will be little appetite for this on the part of industry[37] mainly because autologous stem cells, in themselves, have no value as intellectual property. Their use in a trial will only attract commercial sponsorship if it is combined with a delivery system, or a cell preparation that is patentable. It may be that this prognostication is unduly gloomy. However, in non-stem cell therapy applications, like the use of fetal neural tissue for Parkinson's disease, comprehensive clinical trials were made possible in part thanks to government support. Thus the assistance of national or pan-national governmental institutions may therefore be required to determine the efficacy of novel treatments using autologous adult stem cells. There is an especially strong moral argument for insisting that in any country where ES cell research is banned or restricted, government has a duty to step in. These countries have very much pinned their hopes (somewhat misguidedly, we pointed out in the first section of this chapter) on adult stem cells. It would therefore be remiss were a novel, apparently promising treatment involving adult stem cells, not to receive the blessing of government in the form of generous financial support. On the same ground it can be insisted that any government which has placed its faith in the power of adult stem research, should boost resources for basic research in the area, particularly if a double-blinded controlled study involving the use of stem cells present in bone marrow for heart repair, proves incontrovertibly to be of benefit. Funding, too, is a moral issue.

## Conclusion

Translational medicine involving stem cells promises watershed advances in the cure and treatment of disease, and in the regeneration of insulted tissue. The current wisdom is that there may be no one stem cell system that will take us from bench to bedside for all indications. Efforts are therefore appropriately concentrated not only on different adult stem cell types, but also on ES as well as EG cells. Many of the moral objections to the latter two categories of stem cells lack force. But they have muddied the waters of debate so thoroughly that clarity shall only be achieved through sound policy. This reality calls attention to the need for policy-makers to *guide* opinion and not simply reinforce bad arguments in the form of legislation. Many national policies in place have adverse implications for science, and scientists, and the timely delivery of stem cell-derived therapies. They also raise questions of justice with respect to the fair distribution of the spoils of research. It is the duty of all modern governments to resist legislating in accordance with particular conceptions of the sacred, and to promote the view that, in the public sphere, citizens cannot *reasonably* impose their understanding of what is sacred on others. Additional territory charted above yielded the conclusions that consent forms, and laws governing the body as property require further review. Lastly, there is a pressing need for careful evaluation of recent reports of plasticity of adult stem cells in the clinical context. Government, especially those which restrict ES cell research, should be prepared to fund generously clarificatory randomized controlled trials of apparently promising ongoing clinical studies involving adult stem cells.

I wish to acknowledge very helpful comments made by Alan Colman, Jeremy Buzzard, and Ray Dunn on earlier drafts of this chapter. I wish also to thank Arif Bongso and Eng Hin Lee for their editorial comments/suggestions.

## NOTES

1. One glaring example can be found in the Republic of South Africa's, National Health Bill, 2004.
2. For example, are cells which are self-renewing and unipotent, adult stem cells or precursor cells? Slack JM. (2000) Stem cells in epithelial tissues. *Science* **287**: 1431–1433. Cf. Melton D. (2004) Stemness: definitions and criteria, and standards. *Handbook of Stem Cells*, Vol. 2, Adult and Fetal Stem Cells, ed. R. Lanza, p. xxiii, Elsevier.
3. Penvy L, Rao S. (2003) The stem cell menagerie. *Trends Neurosci* **26**: 351–359.
4. Smith AG. (2001) Embryo-derived stem cells: Of mice and men. *Annu Rev Cell Devel Biol* **17**: 387–403; Morrison SJ, Sha NM, Anderson DJ (1997) Regulatory mechanisms in stem cell biology. *Cell* **88**: 287–298.
5. Anderson DJ, Gage FH, Weissman IL. (2001) Can stem cells cross lineage boundaries. *Nature Med* **7**: 393–395.
6. Verfaillie CM. (2004) Adult stem cells: Tissue specific or not. *Handbook of Stem Cells*, Vol. 2, ed. R. Lanza, pp. 13–20, Elsevier.
7. Dor Y, Melton D. (2004) Pancreatic stem cells. In *Handbook of Stem Cells*, Vol. 2, ed. R. Lanza, pp. 513–520, Elsevier.
8. Shamblott MJ, Axelman J, Wang S, *et al.* (1998) Derivation of pluripotent stem cells from cultured human primordial germ cells. *Proc Natl Acad Sci USA* **95**: 13726–13731.
9. Shamblott MJ, Axelman J, Littlefield JW, *et al.* (2001) Human embryonic germ cell derivatives express a broad range of developmentally distinct markers and proliferate extensively *in vitro*. *Proc Natl Acad Sci USA* **98**: 113–118.
10. Onyango P, Jiang S, Uejima H, *et al.* (2002) Monoallelic expression and methylation of imprinted genes in human and mouse embryonic germ cell lineages. *Proc Natl Acad Sci USA* **99**: 10599–10604.
11. Gearhart J, Armstrong CM, Prepared for the President's Council on Bioethics. Human Embryonic Germ Cells, June 2001–July 2003.

12. Samblott MJ, Axelman J, Wang S, *et al.* (1998) Derivation of pluripotent stem cells from cultured human primordial germ cells. *Proc Natl Acad Sci USA* **95**: 13726–13731.
13. For example, in the UK if you own a Grade II listed heritage property, make alterations to it without Council approval.
14. Bayne T. (2003) Gamete donation and parental responsibility. *J Appl Phil* **20**(1): 77–87. Bayne's reply is mirrored in this paragraph.
15. UN Position Paper, United States Mission to the United Nations. (2003) www.un.int/usa/cloning-paper.htm
16. This example is drawn from my "Controversial developments in the biosciences: Xenotransplantation, human reproductive cloning, and embryonic stem cell research. *SGH Proceedings* (in press).
17. Imprinting does not pose a problem for ES cell research because no fetus is formed during the stage at which cells would be harvested, and because functional cells could be selected *in vitro*, and because derivation of cells involves a loss of "epigenetic memory." Rudolph Jaenisch.
18. See: Boklage CE. (1990) Survival probability of human conceptions from fertilization to term. *Int J Fert* **35**(2): 75–94; Leridon H. (1977) *Human Fertility: The Basic Components.* Chicago: University of Chicago Press.
19. Harris J, Savulescu J. (2003) The Great Debate. *Cambr Health Care Quart.*
20. Kass L. (1997) The wisdom of repugnance. *The New Republic.*
21. Dworkin R. *Life's Dominion: An Argument about Abortion, Euthanasia, and Individual Freedom.*
22. Belgium: Service Public Federal Santé Publique, Securité de la Chaine Alimentaire et Environnement, 11 MAI 2003. Loi relative à la recherche sur les embryons *in vitro*; China: Ministry of Health, "Ethical Principles on Assisted Reproductive Technologies for Human Beings and Human Sperm Bank" (October 2003); India: Indian Council of Medical Research (ICMR), Draft Guidelines for Stem Cell Research/Regulation in India, 2004; Singapore: Singapore, Parliament of Singapore, "Human Cloning and Other Prohibited Practices Bill," bill no. 34/2004, first read 20 July 2004; South Korea: "Life Ethics Law," January 29, 2004; Sweden: Government Bill 2003/04:148 and Committee on Health and Welfare Report 2004/05:SoU7; United Kingdom: 1990 Human Fertilisation and Embryology Act, 22 January 2001, the House of Lords passed a law (already approved in December 2000 by the House of Commons) that permits the cloning of human embryos to derive stem cells, thus allowing the possibility of therapeutic cloning.
23. Wecklerly M. "The Islamic View on Stem Cell Research" (and various cited therein) www-camlaw.rutgers.edu/publications/law-religion/new_devs/ RJLR_ND_56.pdf
24. Australia: Research Involving Human Embryos Act 2002; Canada: Bill C-6 An Act Respecting Assisted Human Reproduction and Related Research;

Netherlands: Bill containing rules relating to the use of gametes and embryos (Embryos Bill), Parliamentary Documents II, 2000/01, 27 423, nos. 1–2; Denmark: Act No. 460 on Medically Assisted Procreation in connection with medical treatment, diagnosis and research, June 10, 1997 (amended September 1, 2003) & Act No. 503 on a Scientific, Ethical Committee System and the Handling of Biomedical Research Projects; Estonia: Embryo Protection and Artificial Fertilisation Act (1997); Finland: Law 488/1999. New Zealand: Human Assisted Reproductive Technology Bill (HART), October 2004; Greece: Law No. 3089 on Medically Assisted Reproduction, 2002; Hong Kong: Human Reproductive Technology Ordinance (2000); Iceland: Act on Artificial Fertilization (55/1996); Republic of South Africa, National Health Bill, 2004, Taiwan: Spain: Spain. Boletín Oficial del Estado: Ley 45/2003, de 21 de noviembre, por la que se modifica la Ley 35/1988, de 22 de noviembre, sobre Técnicas de Reproducción Asistida. [En relación a las células troncales embrionarias]; Taiwan Department of Health, "Ethical Regulations for Embryonic Stem Cell Research," 2002.

25. For example, Proposition 71, approved by California voters on November 2, 2004, establishes a state constitutional right to pursue stem cell research, including through SCNT.

26. US Public Health Service Guideline on Infectious Disease Issues in Xenotransplantation, August 24, 2001/50(RR15);1-46. www.cdc.gov/mmwr/preview/mmwrhtml/rr5015a1.htm

27. Guidance for industry: Source Animal, Product, Preclinical and Clinical Issue Concerning the Use of Xenotransplantation Products in Humans. April 2003. Available at: http://www.fda.gov/cber/gdlns/clinxeno.pdf (accessed October 2004).

28. Richards M, Fong CY, Chan WK, *et al.* (2002) Human feeders support prolonged undifferentiated growth of human inner cell masses and embryonic stem cells. *Nat Biotechnol* **20**(9): 933–936.

29. Article 9 of The Law on Medically Assisted Human Reproduction (1992) states that fertilized human oocytes and cells derived therefrom may not be used for other purposes than medically assisted reproduction; and any intervention into the germ-line is strictly prohibited.

30. Neither by the Austrian Reproductive Medicine Act (FMedG) nor by the Austrian Pharmaceuticals Act (AMG).

31. Tyson P. "Results of death camp experiments: Should they be used?", NOVA Online, October 2000. http://www.pbs.org/wgbh/nova/holocaust/experiments.html
Frohnmayer K. (1995) "Unethically obtained data: Publish or reject?" STS Newsletters: Techné Issues, pp. 1–4.

32. "Exodus zu den Embryonen" *Der Spiegel*, August 23, 2004.

33. Orlic D, Kajstura J, Chimenti S, *et al.* (2001) *Nature* **410**: 701–705.

34. Cited in Chien KR. (2004) Lost in translation. *Nature* doi:10.1038/nature02500, www.nature.com/nature.

35. Balsam LB, Wagers AJ, Christensen JL, *et al.* (2004) Haematopoietic stem cells adopt mature haematopoietic fates in ischaemic myocardium. *Nature* **428**: 668.

    Murry CE, Soonpaa MH, Reinecke H, *et al.* (2004) Haematopoietic stem cells do not transdifferentiate into cardiac myocytes in myocardial infarcts. *Nature* **428**: 664.

36. Mathur JFM. Stem cells and repair of the heart www.thelancet.com Vol 364 July 10, 2004.

37. Mathur, JFM. Stem cells and repair of the heart www.thelancet.com Vol 364 July 10, 2004.

# 11

# Therapeutic Cloning: Derivation and Propagation of Human Embryonic Stem Cells by Somatic Cell Nuclear Transfer

Woo Suk Hwang*, Byeong Chun Lee, Sung Keun Kang, Shin Yong Moon and Jose B. Cibelli

## Introduction

Before Dolly the cloned sheep was announced to the world in 1997, mammalian de-differentiation was only reported in certain kinds of malignant tumors. Although described as de-differentiation, cancer cells proliferate in an unorganized fashion and rarely produce tissues that are representative of all three different germ layers. We would like to propose that the only process known to induce physiological de-differentiation is nuclear transfer-cloning. Here we describe our experiments that led to the generation of an undifferentiated-pluripotent embryonic stem cell line, SCNT-hES1, from a somatic adult human cell.

The idea of reactivating embryonic cells in somatic cells by nuclear transplantation was first put forward by Spemann, using newt oocytes in 1914.[1] This concept was later applied to more terminally differentiated cells in amphibian by Gurdon *et al.* and has culminated in the currently accepted idea that mammalian somatic cells can be turned into a whole new individual when placed in the oocyte of the same species.[2] In 1996, we designed an experiment using bovine oocytes as recipients and human somatic cells as nuclear donors to try to generate human embryonic stem cells using nuclear transfer. We observed some cell divisions and one blastocyst was produced using this cross-species approach; however, no permanent embryonic stem cell line was ever isolated.[3] Cross-species nuclear transfer, a term wined by the authors, has proven to be rather unpredictable. Bovine-bovine

*Correspondence: Department of Theriogenology and Biotechnology, College of Veterinary Medicine, Seoul National University, Seoul 151-742, Korea. Tel: +82-2-880-1280, Fax: +82-2-884-1902, e-mail: hwangws@snu.ac.kr

nuclear transfer, i.e. the use of bovine fetal fibroblasts as nuclear donors and bovine enucleated oocytes as recipients, however, has been more successful. We generated 37 blastocysts from 330 reconstructed oocytes (11%) and isolated 22 embryonic stem (ES)-like cell lines from them (59%). When we injected these ES-like cells into host non-transgenic bovine embryos, six out of seven calves were found to have at least one transgenic tissue in them. We concluded that in the bovine model, somatic cells can be de-differentiated into embryonic ones when fused with an oocyte. Subsequently in 2000, Munsie *et al.* showed similar results using mouse cumulus cells as nuclear donors.[4] Furthermore, these mouse nuclear transfer-derived ES cells were capable of *in vitro* differentiation. Wakayama *et al.* in 2001 demonstrated that de-differentiated cumulus cells could not only make chimeric animals (by coat color), but could also go to the germline and produce offspring. This same group also demonstrated that neurons derived from somatic-cell-cloned-ES cells can produce dopamine and serotonin.[5] In 2002, Rideout *et al.* showed that somatic cells isolated from a Rag (-) mouse, i.e. an animal that lacks T and B cells, can be transformed into ES cells genetically corrected for the Rag mutation and then turned into blood progenitors that when reintroduced into the mutant animal, will generate B and T cells.[6] This experiment elegantly demonstrated that somatic cell nuclear transfer (SCNT) for therapeutic purposes can also be used as a reliable tool for *ex vivo* gene therapy. Having thoroughly proved the concept of "physiological de-differentiation" of somatic cells — in contrast to "pathological de-differentiation" during cancer development — in animals, we set out to test whether SCNT, for the purpose of making embryonic stem cells, was feasible in the human.

## Generation of hESC Lines from Human Somatic Cells

In 2001, we published a paper on our first attempt to generate cloned embryos using human eggs and human somatic cells with the purpose of generating ES cells. Out of 19 reconstructed oocytes (11 with male adult fibroblasts and 9 with female cumulus cells) one 6-cell human embryo was produced but no further development was obtained (Fig. 1). Later on in 2002, we began a series of experiments with a significantly larger pool of human oocytes. These experiments were systematically designed to determine: 1) the appropriate reprogramming time; 2) best oocyte activation method; and 3) appropriate *in vitro* embryo culture conditions. We define reprogramming time, as the necessary period of time between cell fusion and oocyte activation that would allow proper embryonic development.

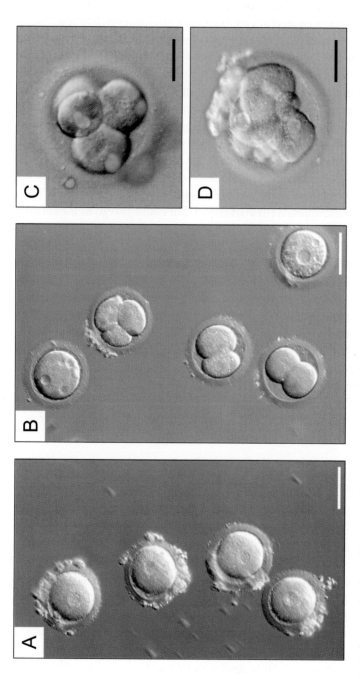

**Figure 1.** Nuclear-transfer derived human embryos reconstructed with cumulus cells. (**A**) Pronuclear stage embryos 12 h; (**B**) 36 h; (**C**) a four-cell embryo at 72 h; and (**D**) a six-cell embryo at 72 h. Scale bars=100 μm.

Our experience with domestic animals indicates that such a period plays a critical role in chromatin remodeling and it is known to determine the developmental competence *in vivo* and *in vitro* of SCNT embryos. The underlying hypothesis is that time is needed to return the gene expression pattern of the somatic cell to one that is appropriate and necessary for the development of the embryo. Our studies indicated that two hours of reprogramming time is the ideal period for development and we managed to obtain development to blastocyst amounting to 25%.

## Egg activation

Egg activation is the natural role of the sperm. During fertilization, the spermatozoa will trigger calcium release inside the egg of a particular magnitude and frequency that will lead to a cascade of events culminating in the first embryonic cell division. SCNT does not require any sperms. We must then mimic its actions using artificial mechanisms. We know from studies in mice, rabbit and pig that the closer we can resemble these events, the higher the rate of successful embryonic development will be. Different chemical, physical, and mechanical agents can induce development in oocytes of several different mammalian species[6]; however, data on human parthenogenesis is limited. Oocyte activation using calcium ionophore, ionomycin and puromycin has been shown to induce parthenogenetic development of human oocytes at different efficiencies.[7] We found that, 10 μM ionophore for 5 min followed by incubation with 2.0 mM 6-dimethyl amino purine (DMAP), has proven to be the most efficient chemical activation protocol for human SCNT embryos.

## Culture medium

Culture medium is another important factor for the development of a successful nuclear transfer protocol. It has been shown in mice that embryo culture systems that work for *in vitro* fertilized embryos not necessarily will work on SCNT embryos. We implemented a sequential culture system tailored to the different stages of embryo development. The recent development of serum-free sequential media, formulated according to the carbohydrate composition of the oviduct and adjusted for the changing physiology and metabolic requirements of the human embryo, has led to considerable improvements in the rate of pregnancies generated using assisted reproductive technologies. In this study, the human modified synthetic oviductal fluid with amino acids (hmSOFaa) was prepared by supplementing mSOFaa[8] with human serum albumin and fructose instead of bovine serum albumin and glucose, respectively. The replacement of glucose with fructose has shown to improve the developmental competence of bovine SCNT embryos.[8,9] It was apparent that culture of human SCNT-derived embryos in G1.2 for the first 48 hours, followed by hmSOFaa media, produced more blastocysts when compared with other culture regimes. The protocol described here produced cloned blastocysts at a rate of 19 to 29% (as a percentage of reconstructed oocytes) and was comparable to those from our established SCNT system in cattle (~25%)[8] and pigs (~25%).[10,11] A total of 30 SCNT-derived blastocysts were cultured, 20 inner cell masses (ICMs) were isolated by immunosurgical removal of the trophoblast, and one human ES cell line was derived (SCNT-hES1).[12] The SCNT-hES-1 cells had a high nucleus to cytoplasm ratio and prominent nucleoli (Fig. 4C). The cells express pluripotency markers such as alkaline phosphatase, SSEA-3, SSEA-4, TRA-1-60, TRA-1-81, and Oct-4, but not SSEA-1.[12] When induced to differentiate into embryoid bodies, SCNT-hES-1 can turn into derivatives of endoderm, mesoderm, and ectoderm.[12] Teratomas generated in immunodeficient mice contained tissue representative of all three germ layers. In particular we observed neuroepithelial rosettes, pigmented retinal epithelium, smooth muscle, bone, cartilage, connective tissues and glandular epithelium.[12] The identical location of polymorphisms for each STR confirms that the SCNT-hES-1 cells were originated from the cloned blastocysts reconstructed from the donor cells and not from parthenogenetic activation.[12] The biparental expression of imprinted genes in SCNT-hES-1 cells further confirmed that the cell line originated from the donor cells.[12]

# Detailed Protocol

## 1. Oocyte collection and transportation

1.1. *Informed consent.* Our donors were asked to sign a consent form the most important aspects of which are described below.

1.1.1. Oocytes and cumulus cells (including DNA) are voluntarily donated for therapeutic cloning research (SCNT, ICM excision, or ES cell growth and differentiation) and its applications (cell therapy or transplantation medicine). Egg donors would receive no benefits of any kind (financial or otherwise) for their contributions to this project.

1.1.2. Donors are aware that oocytes and cumulus cells (including DNA) would be used only for therapeutic cloning research and its applications, and that reproductive cloning would not be performed with their donated materials.

1.1.3. SCNT embryos with arrested growth during culture would be destroyed according to the "Guidelines of the American Fertility Society and Korean Society for Obstetrics and Gynecology."

1.1.4. Donors' identity would be confidential and anonymous.

1.1.5. Neither donors nor their family, relatives or associates may benefit from this research.

1.1.6. Donors could cancel their donation at any time for any reason.

1.1.7. The cell line/s will be deposited in the National Research Institute on Stem Cells and can be used to treat other patients with no financial payments to donors.

1.2. *Medical examination of oocyte donors*

Before ovarian stimulation, oocyte donors underwent medical examinations for suitability, according to the "Guidelines for Oocyte Donation" by the American Fertility Society and Korean Society for Obstetrics and Gynecology. These include chest X-ray, complete blood count, urine analysis and liver function tests. Donors were also screened for human immunodefficiency virus, hepatitis B virus, hepatitis C virus, and syphilis.

1.3. *Oocyte collection and transportation*

Human oocytes are extremely sensitive to temperature change and exposure to light[9]; special care must be taken during transportation.

1.3.1 Transport media G1.2-HSA was prepared using G1.2 version 3 (Vitrolife; cat. no. 10091) with addition of 5% human serum albumin (HSA; Vitrolife, Goteberg, Sweden; cat. no. REF 10064) equilibrated at 37°C in a humidified atmosphere of 5% $CO_2$ and 95% air under mineral oil (Sigma, St. Louis, MO; cat. M3536) 1 day before oocyte pick-up.

1.3.2 Cumulus-oocyte complexes (COCs) were placed in a 5-ml round tube and incubated with 5 ml G1.2-HSA for 40 min at 37°C in a humidified atmosphere of 5% $CO_2$ and 95% air.

1.3.3 Tubes with COCs were placed in a portable $CO_2$ incubator (MTG, Altdorf, Germany; model no. D-84032) and transferred to the SCNT laboratory within 1 hour of retrieval.

## 2. SCNT and embryo culture

During these experiments, autologous SCNT was performed, i.e. the donor's own cumulus cell was isolated from the COC and transferred back into the donor's own enucleated oocyte. The whole SCNT procedure, from oocyte pickup to injection of donor cells, was accomplished within 4 to 6 hours. Unless specified otherwise, all drops were prepared in 35-mm Petri dishes and overlaid with mineral oil (Beckton Dickinson, Heidelberg, Germany; cat. no. 1008). Manipulations were performed in warming plates at 37°C.

2.1. *Oocyte preparation and denuding*

2.1.1. Micromanipulation drops of 25-$\mu$l of G1.2-HSA were equilibrated at 37°C in 5% $CO_2$ in a humidified incubator (Heraeus, Hanau, Germany; model no. BBD6620) for 1 to 2 hours prior to the procedure.

2.1.2. Collected COCs were placed in the equilibrated drops using a capillary pipette (inner diameter 250- to 300-$\mu$m) and incubated at 37°C in 5% $CO_2$ for 30 min.

2.1.3. After incubation, COCs were washed in 25-$\mu$l drops of G1.2-HSA, followed by incubation in 25-$\mu$l drops of G1.2-HSA containing 0.1% (v/v) hyaluronidase (diluted from 0.5% hyaluronidase stock solution in G1.2-HSA; Sigma; cat. no. H3506) at 37°C in 5% $CO_2$ for 1 min.

2.1.4. Cumulus cells were removed from the oocyte by pipeting in and out of a denuding pipette (inner diameter 120- to 150-$\mu$m) under a stereo dissecting microscope (Leica, Bensheim, Germany; model no. MZ8). Denuded oocytes were placed in a 25-$\mu$l drop of G1.2-HSA at 37°C in 5% $CO_2$ for 5 min (Fig. 2A).

2.1.5. After incubation, oocytes were moved into 25-$\mu$l drops of G1.2-HSA containing 5 $\mu$g/ml bisbenzimide (prepared in $dH_2O$) for 10 min at 26°C. DNA localization was examined using an epifluorescent microscope (Fig. 2B, before enucleation). In cases where cumulus cells were not completely detached, step 2.1.4 was repeated. At times, severe mechanical denudation could cause dislocation of the oocyte chromosomes (metaphase plate) from the polar body. In such instances, the localization of the DNA was based solely on bisbenzimide staining.

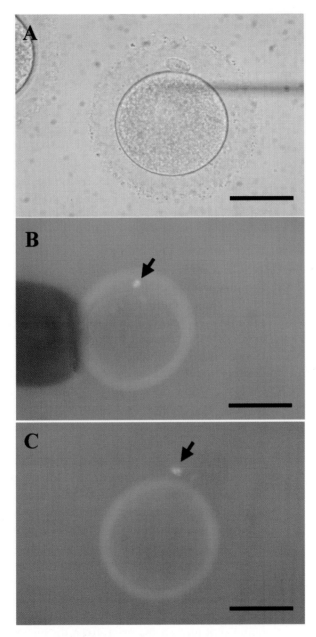

**Figure 2.** Morphology of human oocytes at metaphase II (**A**, ×200) and images (×200) of extruded DNA-MII spindle complexes (arrows) from oocyte before (**B**) and after enucleation (**C**). Scale bar = 100 μm.

2.2. *Donor cell preparation*

2.2.1. Cumulus cells were isolated as described in step 2.1.4., placed in a 1.5 ml tube (Beckton Dickinson; cat. no. 2003) containing 1 ml of 0.25% Trypsin [1 mM EDTA (Life Technologies, Karlsruhe, Germany; cat. no. 25200-056)].

2.2.2. Cells were incubated for 1 min at 26°C, centrifuged (Hanil Science, Inchon, Korea; model Micro-12) at 350 g for 3 min and washed once with 1 ml 1X Dulbecco's phosphate buffered saline (DPBS; Life Technologies; cat. no. 14190-144).

2.2.3. Cumulus cells were resuspended with 400 μl 1X Dulbecco's modified Eagle's medium (DMEM; Life Technologies; cat. no. 12800-058) containing 1% NEAA (100X; Life Technologies; cat. no. 11140-050) and P/S (100X; Life Technologies; cat. no. 15140-122). Only those cells with a modal diameter of 10- to 12 μm were used for SCNT.

2.3. *Enucleation and injection of donor cells*

2.3.1. Five denuded oocytes at a time were placed into 4 μl drops of G1.2-HSA containing 7.5-μl/ml cytochalasin B [Stock solution = 8.1 mM (Sigma; cat. no. C-6762) in demethyl sulfoxide (DMSO; Sigma; cat. no. D2650)].

2.3.2. Under the Differential Interference Contrast (DIC) microscope (Nikon, Tokyo, Japan; model no. ECLIPSE TE300) equipped with a micromanipulation system (Narishige; model no. IM-88), oocytes were secured in place using a holding pipette of 100 microns in diameter. A slit was cut in the zona pellucida (ZP) adjacent to the polar body using a fine glass needle by rubbing the perforated ZP against the holding tip.

2.3.3. The oocyte was released from the holding pipette and squeezed between the cutting pipette and the holding pipette. A fraction of the egg's cytosol, presumably containing the metaphase II chromosomes (about 10% of total cytoplasm) along with the polar body was expelled through the slit made in the ZP. All the remaining oocytes underwent the same procedure.

2.3.4. Oocytes and extruded cytoplasm were stained with bisbenzimide (see subheading 2.1.5.) and observed under an inverted microscope equipped with epifluorescence to confirm complete removal of nuclear materials from the oocytes (Fig. 2C).

2.3.5. Once successful enucleation was confirmed, eggs were washed 3 times in 25 µl drops of G1.2-HSA and placed into a new dish containing a 4 µl drop of G1.2-HSA with 100 µg/ml phythemaglutinin (PHA; Sigma; cat. no. M2643). Donor cells (concentration $1 \times 10^6$ cell/ml) were placed in a different 4 µl drop of 1% FBS (Life Technologies; cat. no. 16000-044)-PBS (v/v).

2.3.6. Using an injection needle with an inner diameter of 20 µm, cumulus cells were aspirated and deposited into the perivitelline space using the same opening in the ZP made during enucleation. One cell per egg was delivered.

2.4. *Fusion of donor cells and oocytes*

2.4.1. After transferring donor cells, reconstructed oocytes were washed three times in 25 µl drops of G1.2-HSA on the warming plate.

2.4.2. Three serial 60 µl drops (33%, 66% and 100%) of fusion buffer were prepared using G1.2-HSA.

2.4.3. Oocytes were washed in 33% and 66% G1.2-HSA/fusion buffer sequentially, and kept in 100% fusion buffer for 1 min.

2.4.4. Fusion of cumulus cell with oocyte (NT-couplet) was performed in a fusion chamber containing 2 stainless steel electrodes 3.4-mm apart containing 100% fusion buffer at room temperature.

2.4.5. Alignment of the NT-couplet was performed with a fine, mouth-controlled pipette. *Egg and donor cell membranes* were in parallel to the electrodes.

2.4.6. Two DC pulses of 1.75 kV/cm for 15 µsec/each using a BTX Electro-cell Manipulator 2001 were delivered to the NT-couplets. Minutes later, if fusion was not observed, another set of identical pulses were delivered.

2.5. *Activation of reconstructed oocytes*

2.5.1. Immediately after electrical fusion, oocytes were rinsed with serial dilutions of fusion buffer (100%, 66% and 33%), followed by a 3 rinses with G1.2-HSA.

2.5.2. To determine an optimal reprogramming time (the lapse of time between cell fusion and egg activation), four different treatments were tested: 2, 4, 6 or 20 hours of incubation in 25 µl drops of G1.2 media at 37°C in a humidified incubator before chemical activation.

2.5.3. Two different concentrations of ionomycin A23187 or two concentrations of calcium ionophore were evaluated. To prepare 10 mM ionophore (Sigma; cat. no. C7522) 1 ml of DMSO was added into a 1 mg bottle of ionophore. To prepare 10 mM ionomycin A23187 (Sigma; cat. no. I0634) 1.34 ml DMSO was added into a 1 mg bottle of ionomycin. Both stock solutions were aliquoted and stored at $-20°C$ until use (for up to one year). Working solutions of A23187 (5 or 10 $\mu M$) or ionomycin (5 or 10 $\mu M$) were made by adding 995 or 990 $\mu l$ G1.2 medium to 5 or 10 $\mu l$ of 10 mM stock solution, respectively, in a 1.5 ml tube (activation media).

2.5.4. Reconstructed eggs were placed into 1 ml of activation media [A23187 (5 or 10 $\mu M$) or ionomycin (5 or 10 $\mu M$)] for 5 min at 26°C in a dark room.

2.5.5. Eggs were subsequently rinsed at least 3 times in 25 $\mu l$ drops of G1.2-HSA. During the washing step, fusion rates were recorded under a stereomicroscope at 40 X magnification.

2.5.6. After washing with G1.2-HSA, oocytes were placed into 25 $\mu l$ drops of mSOFaa containing 2 mM 6-DMAP and incubated at 37°C in 6% $CO_2$, 5% $O_2$, 89% $N_2$ for 4 hour.

2.6. *Culture of reconstructed embryos*

2.6.1. In order to develop an optimal culture medium for the reconstructed human oocytes, hmSOFaa culture medium was prepared by supplementing mSOFaa with human serum albumin (10 mg/ml) and fructose (1.5 mM; Sigma; cat. no. F0127) instead of bovine serum albumin (8 mg/ml) and glucose (1.5 mM), respectively.

2.6.2. After the 6-DMAP treatment, the reconstructed oocytes were rinsed in 25 $\mu l$ drops of G1.2-HSA on the warming plate and were cultured in 25 $\mu l$ drops of G1.2-HSA at 37°C in 6% $CO_2$, 5% $O_2$, 89% $N_2$ for 48 hour. On the third day of culture, cleaved embryos were transferred to the second medium (hmSOFaa or G2.2) in 25 $\mu l$ drops (5 to 6 embryos per drop) and cultured at 37°C in 6% $CO_2$, 5% $O_2$, 89% $N_2$ for another 6 days (Table 1). One group of activated oocytes was cultured throughout in hmSOFaa (Table 1).

2.6.3. Embryo development was monitored under an inverted microscope (Fig. 3).

**Table 1.** Conditions for Human Somatic Cell Nuclear Transfer

| Experiment | Activation Condition | Reprogramming Time (hrs) | 1st Step Medium | 2nd Step Medium | No. of Oocytes | No. (%) of Cloned Embryos Developed to | | |
|---|---|---|---|---|---|---|---|---|
| | | | | | | 2-Cell | Compacted Morula | Blastocyst |
| 1st set | 10 μM Ionophore | 2 | G 1.2 | hmSOFaa | 16 | 16 (100) | 4 (25) | 4 (25) |
| | 10 μM Ionophore | 4 | G 1.2 | hmSOFaa | 16 | 15 (94) | 1 (6) | 0 |
| | 10 μM Ionophore | 6 | G 1.2 | hmSOFaa | 16 | 15 (94) | 1 (6) | 1 (6) |
| | 10 μM Ionophore | 20 | G 1.2 | hmSOFaa | 16 | 9 (56) | 1 (6) | 0 |
| 2nd set | 10 μM Ionophore | 2 | G 1.2 | hmSOFaa | 16 | 16 (100) | 5 (31) | 3 (19) |
| | 5 μM Ionophore | 2 | G 1.2 | hmSOFaa | 16 | 11 (69) | 0 | 0 |
| | 10 μM Ionomycin | 2 | G 1.2 | hmSOFaa | 16 | 12 (75) | 0 | 0 |
| | 5 μM Ionomycin | 2 | G 1.2 | hmSOFaa | 16 | 9 (56) | 0 | 0 |

**Table 1.** (*Continued*)

| Experiment | Activation Condition | Reprogramming Time (hrs) | 1st Step Medium | 2nd Step Medium | No. of Oocytes | No. (%) of Cloned Embryos Developed to | | |
|---|---|---|---|---|---|---|---|---|
| | | | | | | 2-Cell | Compacted Morula | Blastocyst |
| 3rd set | 10 μM Ionophore 6-DMAP | 2 | G 1.2 | hmSOFaa | 16 | 16 (100) | 4 (25) | 3 (19) |
| | 10 μM Ionophore 6-DMAP | 2 | G 1.2 | G 2.2 | 16 | 16 (100) | 0 | 0 |
| | 10 μM Ionophore 6-DMAP | 2 | Continuous hmSOFaa | | 16 | 16 (100) | 0 | 0 |
| 4th set | 10 μM Ionophore 6-DMAP | 2 | G 1.2 | hmSOFaa | 66 | 62 (93) | 24 (36) | 19 (29) |

## 3. Isolation of inner cell masses (ICM) and maintenance of a cloned ES cell line

### 3.1. *ICM isolation by immunosurgery*

3.1.1. Blastocysts were processed (Fig. 3F) on day 6 or 7 and incubated with 25 μl of human cloned ES cell culture medium containing 0.1% pronase (Sigma; cat. no. P5147) for 1 to 2 min at 26°C to remove the ZP.

3.1.2. ZP-free blastocysts were washed 2 times for 3 min in 500 μl of DMEM/F12 culture medium.

3.1.3. ZP-free blastocysts were subsequently incubated in 25 μl 100% anti-human serum antibody (Sigma; cat. no. H8765) for 20 min at 37°C in 5% $CO_2$.

3.1.4. Blastocysts were washed 2 times for 3 min in 500 μl DMEM/F12 culture medium and incubated with 100% guinea pig complement (Sigma; cat. no. S1639) in 50-μl drops for an additional 30 min at 37°C in 5% $CO_2$ in order to lyze the trophoblast cells.

3.1.5. The remains of the blastocysts were washed 2 times for 3 min in 500 μl DMEM/F12 culture medium. Inner cell masses (ICMs) were placed on a feeder layer in a 4-well culture dish (Fig. 4A).

### 3.2. *Establishment of a human cloned ES cell line (SCNT-hES-1)*

3.2.1. Feeder layers were prepared 6 to 48 hours before ICM seeding as previously described by others.[13] When possible, we have used fresh feeder cells at low passage number.

3.2.2. Isolated ICMs were cultured on feeder layers in ES cell culture medium at 37°C and 5% $CO_2$. After 6 days in culture, we observed the formation of a colony under the microscope.

3.2.3. Subculture was performed mechanically every 5 to 7 days by dissecting undifferentiated cell colonies into 200–300 cell clumps using a hooked Pasteur pipette under an inverted microscope. Dissected cell clumps were passaged to newly prepared feeder layers.

3.2.4. After continuous proliferation *in vitro*, one ES cell line (SCNT-hES-1) was derived (Fig. 4B).

3.2.5. During the early stage of SCNT-hES-1 cell culture, human cloned ES cell culture medium was supplemented with 2000 units/ml human leukemia inhibitory factor (hLIF; Chemicon, Temecula, CA).

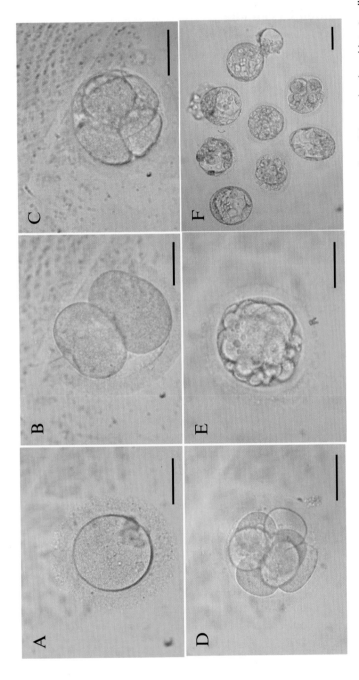

**Figure 3.** Preimplantation development of embryos after somatic cell nuclear transfer (SCNT). The fused SCNT embryo (**A**) was developed into 2-cell (**B**), 4-cell (**C**), 8-cell (**D**), morula (**E**) and blastocyst (**F**). Magnification = 200× (A to E) and 100× (F). Scale bar = 100 μm (A to E) and 50 μm (F).

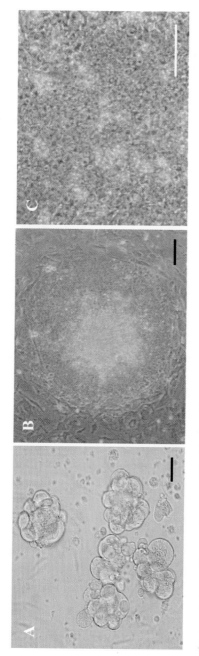

**Figure 4.** The morphology of inner cell masses (ICMs) isolated from cloned blastocysts (**A**, ×100) by immunosurgery and the phase contrast (**B**, ×100) micrographs of a colony of SCNT-hES-1 cells, and higher magnification (**C**, ×200). Scale bar = 50 μm (A) and 100 μm (B and C).

## 4. SCNT-hES-1 freezing and thawing

4.1. *Vitrification*

    4.1.1. SCNT-hES-1 cells were mechanically dissociated into 200–300 cells/clump.

    4.1.2. Five clumps at a time were moved into 25 μl VS1 solution and incubated at 37°C in 5% $CO_2$ for 1 min.

    4.1.3. Cell clumps were moved into 25 μl VS2 solution and incubated at 37°C in 5% $CO_2$ for 30 sec.

    4.1.4. Using a capillary pipette, clumps (5 clumps/grid) were carefully mounted on an EM-grid (Pelco, Redding, CA; cat. no. 3 HGC #400) under an inverted microscope and immediately submerged in liquid nitrogen. The EM-grid was placed inside the cap of a cryovial (Nunc; cat. no. 368632) and stored at −196°C.

4.2. *Thawing*

    4.2.1. The EM-grid was submerged in 25 μl TS1 solution and incubated at 37°C in 5% $CO_2$ for 1 min.

    4.2.2. Cell clumps were moved to 25 μl TS2 solution and incubated at 37°C in 5% $CO_2$ for 5 min.

    4.2.3. Cell clumps were rinsed in 25 μl ES cell culture medium and incubated at 37°C in 5% $CO_2$ for 10 min.

    4.2.4. Cells were seeded on a newly prepared feeder layer. Up to 10 clumps were seeded onto a new feeder layer and cultured at the same time. Dishes were placed at 37°C in 5% $CO_2$ and left undisturbed for 48 hours.

## 5. Expression of cell surface markers

5.1. SCNT-hES-1 cell colonies (8 to 10 colonies) were grown in a 4-well dish with 500 μl. At the time of analysis, ES culture medium was removed and the cells rinsed with 1X DPBS containing 0.9 mM $Ca^{2+}$ ($CaCl_2$) and 0.5 mM $Mg^{2+}$ ($MgCl_2$) (Life Technologies; cat. no. 14080055).

5.2. Colonies were fixed with 500 μl of 4% paraformaldehyde (Sigma; cat. no. P-6148) at 4°C for 30 min and subsequently rinsed 3 times for 3 min in 500 μl of 1X DPBS with $Ca^{2+}$ and $Mg^{2+}$.

5.3. Endogenous peroxidase activity was quenched by incubating fixed colonies with 500 μl of 0.3% $H_2O_2$ (Sigma; cat. no. H3410) for 30 min at 26°C.

5.4. Colonies were rinsed 3 times with 500 μl of 1X PBS for 5 min at 26°C.

5.5. Vectastain ABC kit (Vector, Burlingame, CA; cat. no. PK-6103) was used to stain for cell surface markers according to the manufacture's suggested protocol. The kit contains the Blocking serum (normal serum), Biotinylated antibody and VECTASTAIN ABC regent.

5.6. Primary monoclonal antibodies were diluted using blocking solution as follows: SSEA-1 (1:1000 dilution; Developmental Studies Hybridoma Bank, Iowa, IL; cat. no. MC-480); SSEA-3 (1:1000 dilution; Developmental Studies Hybridoma Bank); SSEA-4 (1:1000 dilution; Developmental Studies Hybridoma Bank; cat. no. MC-813-70); TRA-1-60 (1:1000 Dilution; Chemicon; cat. no. MAB4360); TRA-1-80 (1:1000 dilution; Chemicon; cat. no. MAB4381); and Oct-4 (1:1000 dilution; Santa Cruz Biotechnology, Sata Cruz, CA; cat. no. SC-5279).

## 6. Formation of embryoid bodies (EBs) and immunohistochemical staining

6.1. SCNT-hES-1 were dissociated using enzymatic (0.1% trypsin/1 mM EDTA) or mechanical disaggregation.

6.2. EBs were cultured for 14 days in DMEM/F12 culture medium without hLIF and bFGF.

6.3. EBs were placed in a small drop of 1% molten low melting point agarose prepared in 1X PBS and cooled to 42°C.

6.4. Agarose containing EBs was fixed in 4% paraformaldehyde in 1X PBS and embedded in paraffin.

6.5. Individual (2-μm) sections were made and placed on glass slides.

6.6. Sections were deparaffinized and hydrated through the following solutions: Xylene 2 times, 5 min; 100% EtOH 2 times, 3 min; 95% EtOH 2 times, 3 min; 85% EtOH 2 times, 3 min; 75% EtOH 2 times, 3 min; 55% EtOH 2 times, 3 min and finally 1X PBS 2 times, 3 min.

6.7. Antibodies used were as follows: alpha-1-fetoprotein (18-0003; 1:500 dilution); Cytokeratin (18-0234; 1:500 dilution); Desmin (18-0016; 1:500 dilution); Neurofilament (18-0171; 1:500 dilution); and S-100 (18-0046; 1:500 dilution) supplied by Zymed (South San Francisco, CA). HNF-2-alpha (SC-6556; 1:500 dilution); BMP-4 (SC-6896; 1:500 dilution); Myo D (SC-760; 1:500 dilution); and NCAM (SC-7326; 1:500 dilution) were purchased from Santa Cruz Biotechnology. Primary antibodies were localized with biotinylated secondary anti-rabbit, anti-mouse, or anti-goat and finally with avidin-conjugated horseradish peroxidase or alkaline phosphatase complex (see step 5.8).

6.8. 300 μl of freshly prepared NBT-BCIP working solution (Roche, Mannheim, Germany; cat. no. 1681451) was applied onto the slide and incubated for 30 min (or until desired intensity developed) at 26°C. Reaction was stopped by adding tap water.

6.9. Sections were dehydrated through the following solutions: 95% EtOH 3 min; 100% EtOH 3 times, 3 min; and Xylene 2 times, 5 min. Slides were mounted using VectorMount (Vector; cat. no. H5000) and examined under the microscope.

## 7. Formation of teratomas and H & E staining

7.1. SCID mice (CB17 strain at 6 to 8 weeks; Jackson Laboratory, Bar Harbor, ME) were anesthesized by injecting intraperitoneally 0.2 ml 2.5% Avertin (Sigma; T4 840-2) using a 1-ml syringe (Beckton Dickenson; cat. no. REF 302100).

7.2. Testes were sprayed with 70% ethanol.

7.3. Using forceps and scissors, the skin was cut and left testis exposed.

7.4. Mechanically harvested SCNT-hES-1 cell clumps (~100 of them of about ~100 cells each) with undifferentiated morphology were injected using 1-ml disposable syringe in a minimal volume of G1.2-HSA.

7.5. Mice were euthanized by cervical dislocation 10 to 12 weeks after injection and testis collected.

7.6 Testes were fixed in 10% neutral buffered formalin, embedded in paraffin, and examined histologically after hematoxylin and eosin (H & E) staining.

## 8. DNA fingerprinting assay

8.1. Genomic DNA from the donor cells, SCNT-hES-1 cells, teratomas, and lymphocytes from unrelated donors was isolated using a QIAamp DNA Mini kit (Qiagen, Chatsworth, CA; cat. no. 51304) according to manufacturer recommendations.

8.2. Genomic DNA (5 μl) was amplified with human STR markers using an STR Amp FLSTR PROFILER Kit (Perkin Elmer Corp., Wellesley, MA; cat. no. 4303326) on an automated ABI 310 Genetic Analyzer (Applied Biosystems, Foster City, CA; cat. no. 106PB07-01).

## 9. Analysis of imprinted genes

9.1. Ten to 20 SCNT-hES-1 colonies were collected in 200 μl of TRIzol (Life Technologies; cat. no. 15596-026) and total RNA isolated according to the manufacturer's suggested protocol.

9.2. RNA was resuspended in 10 µl RNase free water (diethylpoly-carbonate-treated water, Life Technologies; cat. no. 750024) in a 0.5-ml tube.

9.3. RNA solution was heated to 65°C for 10 min, and then chilled in ice for 2 min.

9.4. mRNA was converted into first-strand cDNA using a First-Strand cDNA synthesis kit (Amersham-Pharmacia Biotech. Piscataway, NJ, cat. no. 27-9261-01) in a total volume of 15 µl (use 10 µl of total RNA solution for template) according to the manufacturer's suggested protocol.

9.5. Specific primers were designed for paternally-expressed [hSNRPN (accession number: AF400432) and ARH1 (accession number: NM_004675)] and maternally-expressed [UBE3A (accession number: U84404) and H19 (accession number: M32053)] genes, and the house-keeping gene GAPDH (accession number: NM_002046).

9.6. Polymerase chain reaction was done in a total volume of 50 µl for 30 cycles using a Thermocycler (Perkin Elmer; cat. no. 9700) with the following program: denaturation at 94°C for 30 sec; annealing at 55°C for 30 sec; extension at 72°C for 30 sec.

9.7. PCR products were fractionated in 1.2% agarose gel by electrophoresis.

## Stock Solutions and Culture Medium

### Fusion buffer

4.76 g mannitol (0.28 M; Sigma; cat. no. M1902), 0.0119 g Hepes (0.5 mM pH 7.2; Sigma; cat. no. H6147), and 0.05 g fatty acid-free BSA (0.05% wt/vol; Sigma; cat. no. A6003) where dissolved to 99.9 ml dH$_2$O. Subsequently, 100 µl of a solution of 100 mM of 0.12 g MgSO$_4$ (100 mM; Sigma; cat. no. M2643) was added to the 99.9 ml fusion solution. We stored this working solution at 4°C for up to 7 days.

### Modified synthetic oviductal fluid with amino acids (mSOFaa)

Stock Solutions:

Stock T (for 50 mL): 3 g NaCl (Sigma; cat. no. S5886), 0.2669 g KCl (Sigma; cat. no. P5405), 0.081 g NaH$_2$PO$_4$ (Sigma; cat. no. S5011), 0.28 mL sodium lactate (Sigma; cat. no. L-1375), 0.0375 g kanamycin monosulfate (Sigma; cat. no. K1377).

Stock B (for 20 mL): 0.42124 g NaHCO$_3$ (Sigma; cat. no. S5761).

Stock C (for 5 mL): 0.0182 g sodium pyruvate (Sigma; cat. no. P4562).
Stock D (for 10 mL): 0.2514 g CaCl$_2$. 2H$_2$O (Sigma; cat. no. C7902).
Stock M (for 10 mL): 0.0996 g MgCl$_2$. 6H$_2$O (Sigma; cat. no. M2393).
Glucose stock (for 10 mL): 0.27024 g glucose (Sigma; cat. no. G7021).
Glutamine stock (for 10 mL): 0.14618 g L-glutamine (Sigma; cat. no. G1517).

## mSOFaa culture medium

100 mM NaCl, 7.2 mM KCl, 1.2 mM NaH$_2$PO$_4$, 3.3 mM sodium lactate, 2 mM CaCl$_2$. 2H$_2$O, 0.5 mM MgCl$_2$. 6H$_2$O, 2% (v/v) EAA (Life technology; cat. no. 11130-051), 1% (v/v) NEAA, 0.3 mM sodium pyruvate, 1.5 mM glucose, 8 mg/mL BSA (Sigma; cat. no. A6003), 1 mM, L-glutamine, 0.5% (v/v) insulin-transferrin-selenium (ITS) (Sigma; cat. no. I-3146), 0.128 mM kanamycin monosulfate. pH/Osmolarity = 7.2/275 ~285. Culture medium was stored at 4°C for up to 7 days.

## 2 mM 6-DMAP

(Sigma; cat. no. D2629) solution was prepared by dissolving 0.0031 g of 6-DMAP in 10 ml mSOFaa. 500 μl aliquots were made and stored at −20°C for up to 7 days until use.

## ES cell culture medium

DMEM/F12 (1:1, Life Technologies; cat. no. 11320-033) supplemented with 20% Knockout SR (Life Technologies; cat. no. 10828-028), 0.1 mM NEAA, 0.1 mM β-mercaptoethanol (Sigma; cat. no. M-7522), 0.5% P/S and 4 ng/ml recombinant human basic fibroblast growth factor (bFGF; Invitrogen, Carlsbad, CA; cat. no. 13256-029).

## Vitrification solutions

**VS1** 10% DMSO (v/v) and 10% ethylene glycol (v/v; Sigma; cat. no. E9129) in human cloned ES cell culture medium.

**VS2** 20% DMSO (v/v), 20% ethylene glycol (v/v) and 0.5 M sucrose (Invitrogen; cat. no. 15503-022).

## Thawing solutions

**TS1**: 0.2 M sucrose in human cloned ES cell culture medium.
**TS2**: 0.1 M sucrose in human cloned ES cell culture medium.

## Discussion

We have obtained the first human embryonic stem cell line from a living person. This discovery can potentially impact the lives of millions of people. SCNT-derived ES cells are:

1) immunocompatible with the patient;
2) immortal;
3) pluripotent; and
4) rejuvenated.

The issue of immunocompatibility is one of considerable significance. It has been proposed that human *in vitro* fertilized-ES cells could be transplanted back to the patients to cure numerous diseases; however, it is not clear to the medical community what strategy will be used to avoid the surveillance of the patient's own immune system. During the early stages of ES cell research, it was thought that these cells, due to their embryonic phenotype, will not express MHC proteins. This hypothesis was rejected when MHC class I was observed in the cells after differentiation and exposure to interferon gamma.[7] One possibility being currently tested is the use of bone marrow chimerism. This procedure entails the infusion of ES cell-derived bone marrow cells (hematopoietic stem cells) into the patient prior to transplantation in an attempt to obtain a state of blood chimerism. It is thought that a permanent source of donor antigen will purge all the alloreactive lymphocytes, building tolerance for future cell/organ transplants. Although possible in mice, studies in human have not yet been completed and there is the concern that depletion of T cells will lead to long term immunodeficiency. Another possibility being proposed is the creation of a cell bank of human ES cells that will accommodate all different tissue types for all potential patients. This proposal is flawed. Using the current American bone marrow bank as an example, there are approximately 3 million donors registered. *Considering the major histocompatibility complex these bone* marrow donors are potentially useful for 80% of the Caucasian population: If minor hitocompatibility complexes are included, the tissue match possibilities are dramatically reduced. Will SCNT-hES cells be compatible with the donor from where the somatic cell came? The answer can be found in a recent study performed in cloned pigs. Martin *et al.* recently showed that skin grafts transplanted among cloned animals of the same genotype were not rejected, whereas grafts from unrelated donors were acutely rejected.[8] It is important to point out that in this study, mitochondrial DNA among clones was different since they were originated from separate enucleated eggs. Nevertheless, this seems to be of no relevance from the tissue compatibility standpoint.

When cultured under proper conditions, primate ES cells are immortal and genetically stable. There have been a few recent reports describing chromosomal translocations or aneuploidy in human ES cells. This phenomenon is apparently related to the way cells are cultured in different laboratories. In our hands, primate ES cells are genetically stable and express the enzyme telomerase, conferring on them the capacity to replicate indefinitely. When these ES cells are induced to differentiate, telomerase is downregulated and the cells then behave as a primary cell line, eventually reaching their limit in their replicative capacity. This observation gives credence to the notion that telomerase expression is tightly regulated in ES cells. It is safe to assume that cells and tissue-derivatives from ES cells will not generate malignant tumors.

One characteristic that makes SCNT-hES even more desirable is the rejuvenation aspect. The first report trying to address the question as to whether nuclear transfer cloning will rejuvenate a cell line or not was based on telomere measurements made on Dolly the cloned sheep and her original cell line.[9] This report indicated that Dolly had shorter telomeres and as such, she would be biologically older than her chronological age. Later studies done by our group and others have clearly demonstrated that when proper reprogramming is successfully achieved, telomeres are restored to the same and sometimes longer length than control animals. We found that the average telomere length of cows cloned from a senescent line of fibroblasts was indeed longer than that of age matched controls.[10] Later Wakayama *et al.* found that serially cloned mice had their telomeres slightly longer in each generation.[11] Kubota *et al.* found that serially cloned bulls have restored telomeres as well.[12] Interestingly, Miyashita *et al.* found different levels of telomere restoration depending on the cell type used. When cows were cloned from muscle cells, the cloned animal had longer telomeres; conversely, when oviductal or mammary cells were used, shorter telomeres were observed in the clones.[13] Taken together, this data indicates that nuclear reprogramming using an oocyte has the potential to rejuvenate an individual and cells derived from it.

## Future Challenges

Considering all the advantages SCNT offers, it is not difficult to envision a great future for this technology as a therapeutic tool in humans. Still, there are stringent standards to be met before reaching the patients. These are the same standards that the US National Institute of Health has put forward for IVF-derived ES cells. The cells have to:

1) proliferate extensively and generate sufficient quantities of tissue;
2) differentiate into the desired cell type(s);
3) survive in the recipient after transplant;
4) integrate into the surrounding tissue after transplant;
5) function appropriately for the duration of the recipient's life; and
6) avoid harming the recipient in any way.

Furthermore, SCNT faces 3 other significant challenges at the moment. Besides ethics there are the technical, logistical and scientific issues. From the technical standpoint, our results indicate that we can generate three blastocysts for every 10 reconstructed eggs. These results are considered excellent for nuclear transfer efficiency. However, only one stable SCNT ES cell line was obtained out of 20 SCNT-derived blastocysts. We must improve the efficiency of transition from the ICM to an embryonic stem cell line. The logistical challenge is to be able to obtain enough donated human eggs in order to produce tailor-made ES cells for all the patients that need them. Scientifically, we now know that we can turn an already differentiated human cell into a rejuvenated-pluripotent one; however, at present, we are unable to understand how this process works. We would like to propose that the greatest legacy of our results is the demonstration that this is possible in humans and it is up to the whole scientific community to explain the mechanism that will help turn any somatic cell into ES cells without having to create embryos.

Patient customized, tailor-made hESC differentiated tissues are of great value in regenerative medicine to avoid immunological rejection of the transplanted tissues. Recently, Hwang *et al.* (2005) established 11 hESC lines by somatic cell nuclear transfer (SCNT) of skin cells from patients with disease or injury, into donated human oocytes. These lines, grown on human feeders from the same SCNT-donor or genetically unrelated individuals, were established at high efficiency, regardless of SCNT-donor sex or age. The SCNT-hESC lines were pluripotent, karyotypically normal, and matched the SCNT-patient's DNA. Major histocompatibility complex (MHC) identity of each SCNT-hESC line with the patient's showed immunological compatibility, crucial for eventual transplantation. The generation of these SCNT-hESCs allows evaluations of genetic and epigenetic stability. More work is required to develop reliable directed differentiation of xeno-free tissues from such hESCs. Before clinical application of these cells, preclinical evaluation is required to prove that transplantation of differentiated SCNT-hESCs is safe, effective and can be tolerated.

# REFERENCES

1. Spemann H. (1914) Ver. Deutsch. Zool. Ges. Leipzig (Freiburg) **24**:216–221.
2. Gurdon JB, Laskey RA, Reeves OR. (1975) The development capacity of nuclei transplanted from keratinized skin cells of adult frogs. *J Embryol Exp Morphol* **34**(1):93–112.
3. Chang KH, *et al.* (2003) Blastocyst formation, karyotype, and mitochondrial DNA of interspecies embryos derived from nuclear transfer of human cord fibroblasts into enucleated bovine oocytes. *Fertil Steril* **80**(6):1380–1387.
4. Munsie MJ, *et al.* (2000) Isolation of pluripotent embryonic stem cells from reprogrammed adult mouse somatic cell nuclei. *Curr Biol* **10**(16):989–992.
5. Wakayama T, *et al.* (2001) Differentiation of embryonic stem cell lines generated from adult somatic cells by nuclear transfer. *Science* **292**(5517):740–743.
6. Rideout WM 3rd, *et al.* (2002) Correction of a genetic defect by nuclear transplantation and combined cell and gene therapy. *Cell* **109**(1):17–27.
7. Drukker M, *et al.* (2002) Characterization of the expression of MHC proteins in human embryonic stem cells. *Proc Natl Acad Sci USA* **99**(15):9864–9869.
8. Martin MJ, *et al.* (2003) Skin graft survival in genetically identical cloned pigs. *Cloning Stem Cells* **5**(2):117–121.
9. Shiels PG, *et al.* (1999) Analysis of telomere lengths in cloned sheep. *Nature* **399**(6734):316–317.
10. Lanza RP, *et al.* (2000) Extension of cell life-span and telomere length in animals cloned from senescent somatic cells [see comments]. *Science* **288**(5466):665–669.
11. Wakayama T, *et al.* (2000) Cloning of mice to six generations. *Nature* **407**(6802):318–319.
12. Kubota C, Tian XC, Yang X. (2004) Serial bull cloning by somatic cell nuclear transfer. *Nat Biotechnol* **22**(6):693–694.
13. Miyashita N, *et al.* (2002) Remarkable differences in telomere lengths among cloned cattle derived from different cell types1. *Biol Reprod* **66**(6): 1649–1655.
14. Hwang WS, Roh SL, Lee BC, *et al.* (2005) Patient-specific embryonic stem cells derived from human SCNT blastocysts. *Science* online 19 May 2005: DOI10.1126/science.1112286.

# 12

# Hurdles to Improving the Efficiency of Therapeutic Cloning

Christopher S. Navara*, Calvin Simerly, Sang-Hwan Hyun, and Gerald Schatten

## Introduction

Therapeutic cloning, the derivation of tissue matched embryonic stem cells from blastocysts generated by somatic cell nuclear transfer using a patient as the nucleus donor, has the potential to revolutionize medicine by providing designer tissue matched replacement cells for any number of diseases ranging from diabetes ($\beta$-islet cells) to Parkinson's (dopaminergic neurons). The proof of principle experiment for therapeutic cloning has been demonstrated by the successful "curing" of an immunodeficient mouse.[1] The landmark paper by Hwang et al.[37] provides the first evidence in humans that this novel technique can be used for the derivation of embryonic stem cells but their results also reflect the pattern of this currently inefficient technique, raising questions of medical feasibility and requiring further optimization before usefulness. Perhaps no two biomedical technologies are more controversial or more hotly debated publicly than cloning and embryonic stem cells. Therapeutic cloning weds these two controversial topics and thus draws interest and criticism from both. In this chapter, we will discuss the promise and limitations in current technology for therapeutic cloning. As reproductive cloning and therapeutic cloning share many of the same experimental methodologies and limitations and because of extensive studies on reproductive cloning in agricultural, laboratory animals and non-human

*Correspondence: Department of Obstetrics, Gynecology and Reproductive Sciences, University of Pittsburgh, 204 Craft Avenue, Pittsburgh, PA 16066. Tel.: 412-641-2430, fax: 412-641-2410, e-mail: Cnavara@pdc.magee.edu
† In accordance with the International Society for Stem Cell Research's (ISSCR) Nomenclature Statement (September 2, 2004), 'Nuclear Transfer (NT)' is used instead of 'therapeutic cloning,' and 'NT Stem Cells' or NTSC describe stem cells derived from NT embryos. The terms 'Reproductive Cloning' and 'Therapeutic Cloning' are used herein as well for clarity.

primates, we include a summary of these results to enhance our discussions of therapeutic cloning.

## Reproductive cloning

The first successful studies in reproductive cloning of mammals were performed in agriculturally important species as a means of increasing production.[2,3,4] Scientists sought to minimize the inherent production variation among animals by generating identical animals, ideally using adults of known quality. Embryonic blastomeres from early cleavage stages were the first donor cells and the cytoplasts formed were cultured *in vitro* for several days before being transferred to surrogate animals resulting in births of genetically (but not phenotypically) identical animals. This was a significant advancement in technology but was limited to producing a small number of identical offspring and the genetic quality of the donor was not known at the time of cloning. This procedure also did not lend itself to prior genetic manipulation of the offspring. Wilmut *et al.* (1997) advanced the field to the current level with the seminal report on cloning of an adult sheep using somatic cells. Now identical animals could be produced from an animal of known genetic quality. The ethical, legal and societal implications currently debated regarding therapeutic cloning began at this time with the discussions of theoretical reproductive cloning in humans.[5,6,7] Additional mammalian species were cloned in rapid succession including: mice[8]; cattle[9,10]; goat[11]; pig[12,13]; cat[14]; rabbits[15]; mule[16]; horse,[17] rhesus monkeys[18] and rats.[19] A central supposition of therapeutic cloning is not only the successful derivation of embryonic stem cells (ESCs) after nuclear transfer, but also the ability to correct genetic defects prior to nuclear transfer. This is similar to the technology used for the generation of transgenic animals and this has been successful in a number of species: Cattle[20,21]; pigs[22]; goat[23] and mice.[1]

## Therapeutic cloning

The isolation of mESCs (murine embryonic stem cells) from blastocysts over 20 years ago revolutionized the use of mice for biomedical research.[24,25] These pluripotent cells were able to self-renew and maintain pluripotency indefinitely; they were able to form tissues from all three germ layers in teratomas or embryoid bodies. These cells could be genetically modified and recombined with routine embryos to form chimeras, contribute to the germ line and thus pass on the altered genotype to their offspring.

Designer mice quickly became the model of choice for studying human disease and immeasurable advances in biomedical research have resulted. Similarly, the isolation of human embryonic stem cells from excess embryos derived by *in vitro* fertilization[26] heralded a new era in biomedical research. These cells maintain pluripotency and undergo self-renewal and are able to form all three germ layers in teratomas. As it is unethical to create human chimeras, these cells have not been tested for their ability to contribute to the germ line.

The potential utility of these cells for cell therapy was recognized and the scenario shown in Fig. 1 was developed. Somatic cells are isolated from a patient (in this case a patient with a spinal cord injury) (Fig. 1.1) and are grown *in vitro*; additionally, oocytes are collected from the patient or a donor after a superstimulation protocol similar to those currently being performed in fertility clinics worldwide (Fig. 1.2). The cumulus cells are removed from the oocyte (Fig. 1.3) and the oocyte chromosomes (Fig. 1.4, highlighted) are removed by aspiration (not shown) or by squeezing the oocyte following cutting of the proteinaceous zona pellucida surrounding the oocyte using a micro needle (Figs. 1.5–1.7). It is important to note that in addition to the chromosomes, a small amount of cytoplasm and the entire meiotic spindle apparatus is removed during this procedure. This may be important in the inefficiency of somatic cell nuclear transfer as described below. Next, the donor cell is placed adjacent to the enucleated oocyte (Fig. 1.8) and the two cytoplasms fused by a brief electrical pulse (Fig. 1.9). Alternatively, an isolated nucleus can be injected directly into the cytoplasm of the enucleated oocyte as is done during mouse cloning (not shown[8]). The fused "clone" is artificially activated, simulating the function of a sperm at fertilization and the embryo undergoes cleavage *in vitro* (Figs. 1.10–1.13). The steps described up to this point are shared between reproductive and therapeutic cloning. In order to isolate the embryonic stem cells, the blastocyst is either plated directly onto feeder cells or the inner cell mass of the blastocyst is isolated by immunosurgery (Fig. 1.13). These isolated cells can then be maintained and propagated *in vitro* (Figs. 1.14–1.16) and as they are pluripotent, they can be differentiated into the necessary therapeutic cells (motor neurons in our example (Fig. 1.17) and transferred back to the patient (Fig. 1.18). The proof of principle of curing a patient's disease using transplanted differentiated embryonic stem cells has not yet been demonstrated. Here, we will discuss the progress that has been made in mouse models of therapeutic cloning and discuss the limitations described in all species but with special emphasis on non-human primates and humans.

## Therapeutic Cloning

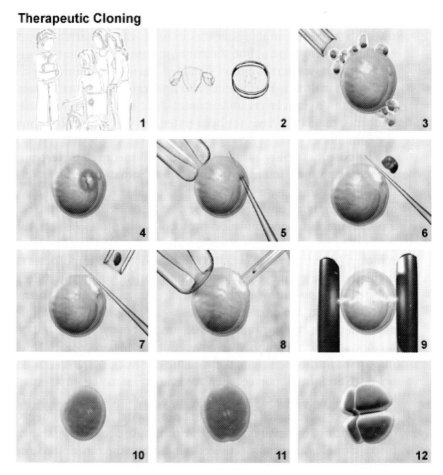

**Figure 1.** *One Hypothetical Therapeutic Cloning Procedure.* Somatic cells are isolated from a patient (Fig. 1.1). Oocytes are collected from the patient or a donor after superstimulation (Fig. 1.2). The cumulus cells are removed from the oocyte (Fig. 1.3) and the oocyte chromosomes (Fig. 1.4, highlighted) are removed by squeezing the oocyte following cutting of the proteinaceous zona pellucida surrounding the oocyte using a micro needle (Figs. 1.5–1.7). The donor cell is next placed inside the zona pellucida (Fig. 1.8) and the two cytoplasms fused by a brief electrical pulse (Fig. 1.9). The fused "clone" is artificially activated and the embryo undergoes cleavage *in vitro* (Figs. 1.10–1.13). In order to isolate ESCs, the inner cell mass of the blastocyst is isolated by immunosurgery (Fig. 1.13). These isolated cells are then maintained and propagated *in vitro* (Figs. 1.14–1.16) and as they are pluripotent, they can be differentiated into the necessary therapeutic cells (Fig. 1.17) and transferred back to the patient (Fig. 1.18).

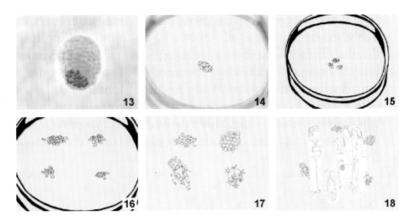

**Figure 1.** (*Continued*)

The earliest mESCs were isolated in 1981.[24,25] The first mouse was cloned in 1998[8] and this same group derived the first mESCs after somatic cell nuclear transfer.[27] Rideout and colleagues (2002) assembled these technologies along with the technology to differentiate mESCs into hematopoietic stem cells (HSCs),[28] necessary for long-term engraftment of the immune system of an immunodeficient mouse. Cells were isolated from immunodeficient Rag2-/- mice. These were cloned into mouse oocytes and ESCs (embryonic stem cells) were derived from the cloned blastocysts. Standard molecular biology techniques were used to repair the Rag2 gene by homologous recombination in the ESCs. From this point, two complementary techniques were used to test the ability of these ESCs to rescue the immunodeficient mice. The first involved using the ESCs to generate mice by tetraploid complementation chimera formation and embryo transfer.[29] This technique results in the entire mouse being derived from ESCs in one generation. These mice were used as HSC (hematopoietic stem cell) donors for the Rag2-/- mice and successfully engrafted the deficient mice. The second methodology involved *in vitro* differentiation, isolation and transplant of the repaired ESCs. The ESCs were transduced with the HoxB4 gene and cultured on OP9 stromal cells to produce HSCs for transplantation. These cells succeeded in limited engraftment but only after removal of endogenous NK cells, by antibody treatment or by using mice naturally lacking NK cells due to a deletion of the IL2 common cytokine receptor gamma chain. This work was the first successful application of therapeutic cloning but also pointed to unexpected hurdles and underscores the need for extensive, meaningful evaluation of preclinical animal models of therapeutic cloning.

The first progress towards therapeutic cloning in humans was recently published by Hwang *et al.* (2004). These authors, using the procedure detailed in Fig. 1, derived a single hESC (human embryonic stem cell) line after somatic cell nuclear transfer (SCNT), a first for any primate. As reproductive cloning of humans is unethical, legitimate attempts have only been made in non-human primates. Despite intensive efforts from several labs, only two monkeys have been derived by nuclear transfer, using embryonic blastomeres rather than somatic cells, and this has yet to be successfully replicated.[18,30]

## Inefficiencies of Cloning

It has been seven years since Dolly was born (Wilmut *et al.*, 1997). In the most efficient treatment group, Wilmut *et al.* reported that 3.4% of embryos transferred resulted in live pregnancies using somatic cells as the donor. Since that time, despite intensive effort the efficiency of reproductive cloning in all species remains at approx. 5% (for recent review see Ref. 31). If we include the number of oocytes manipulated, the success falls to 1–2%,[8,32] an important consideration when discussing the precious and limited resource of human oocytes necessary for therapeutic cloning. The isolation of ESCs in mice from embryos derived by SCNT is slightly more successful when measured as oocytes manipulated (1–6%[1,27]) but far below the average derivation rate from control blastocysts (approx. 20%) and perhaps as high as 50% under certain optimal conditions.[33] Reported efficiency of deriving ESCs from inner cell masses isolated from normally fertilized human blastocysts are 35.8%,[26] 50%,[34] 16.7%,[35] and 17.5%.[36] If the total numbers of embryos used are counted, the derivation percentage becomes 13.9%,[26] 4.9%,[36] and 6.0%.[35] It is not reasonable to compare isolation rates between groups, as the numbers are often low and not statistically significant. But even when embryos are generated specifically for ESC derivation and not cast-offs from IVF programs, stem cell derivation efficiency hovers at or below 10%.[35] The sole example of derivation of hESCs from SCNT derived blastocysts is reported as 5% (1/20),[37] but as the best development to blastocyst reported in this singular paper was 29%, the most optimistic derivation of a therapeutically useful embryonic stem cell line is 1.5%. Currently an average of 67 human oocytes would be required to derive each therapeutic line of ESCs. As multiple lines may be needed for successful therapy, this impractical number quickly becomes improbable without further refinement and improvements of the nuclear transfer technologies.

When we discuss the exquisite regulation of early embryonic development required for successful reproduction and the precise control of gene expression found in somatic cells, the question should not be why cloning does not work more often, but instead why it works at all. Successful nuclear transfer requires both genetic and epigenetic success. The donor nucleus must be reprogrammed from a somatic pattern of gene and protein expression to an embryonic pattern almost immediately (probably within one cell cycle for mouse clones). The donor nucleus or the activating conditions must replace the sperm in activating the egg and organizing the cytoplasm after cloning. In non-murine mammals, the sperm brings with it at fertilization important factors for organizing the microtubule cytoskeleton and the cytoplasm of the egg. Fertilization with sperm which poorly organize the cytoplasm after fertilization is associated with reduced developmental outcomes.[38] We will discuss our understanding of reprogramming, particularly as new technologies may improve the efficiency of cloning. We will also address what is known about epigenetic reprogramming, particularly mitotic spindle formation during cloning, and the relationship this has to aneuploidy in cloned embryos.

## Reprogramming errors

Epigenetic modification of the genome (i.e. DNA and histone methylation) ensures faithful tissue specific expression of appropriate genes. It is necessary to change the expression pattern of the nucleus from the somatic phenotype to the embryonic phenotype and to do so within one or two cell divisions. The discrepancies between gene expression of nuclear transfer derived embryos and control embryos have been well reviewed elsewhere and will not be considered in this chapter. Instead we will discuss two novel approaches to selecting and improving correct gene expression using methods appropriate for later ESC derivation.

Boiani and colleagues (2002)[39] have generated a fusion protein between the promoter for the POU transcription factor and embryonic stem cell marker Oct4 and GFP, resulting in a vital cell marker for cells expressing Oct4. Oct4 is found in cells of the developing morula but is lost from the trophectodermal cells during blastocyst formation. When SCNT was performed with transgenic cells, clones had lower expression of Oct4-GFP, including 18% of cloned blastocysts which had no expression of this pluripotency marker. They also examined correct localization of the expression. 93% of *in vivo* fertilized blastocysts have Oct4-GFP expression restricted to the inner cell mass. This compares with only 34% of cloned embryos. To test the pluripotency of those embryos expressing high levels of Oct4-GFP, indicative of correct endogenous expression of Oct4, the embryos were allowed to outgrow on feeder cells for stem cell derivation. Stem cell derivation was only observed for those outgrowths with strong Oct4-GFP expression. Utilizing this system, it is possible to select the SCNT derived embryos with the highest potential to form embryonic stem cells. This may result in increased production of therapeutically derived ESCs.

Byrne *et al.* (2003)[40] have examined the ability of amphibian germinal vesicles to reprogram isolated mammalian nuclei. When murine thymocytes are injected into the germinal vesicles of *Xenopus laevis* oocytes, the expression of the differentiated thymocyte marker thy-1 is lost prior to 5.5 days. On the same time course, the expression of murine Oct-4 is initiated at levels comparable to those observed in mESCs. This raises the possibility that mammalian nuclei could be reprogrammed prior to nuclear transfer, perhaps increasing the efficiency of embryonic stem cell derivation.

## Microtubule and mitotic spindle defects in clones

The embryo formed during nuclear transfer must successfully complete many of the same functions during the first cell cycle that an embryo formed by normal fertilization must perform. While knowledge of the cytoplasmic events of the mammalian zygotic first cell cycle is far from complete, we do know a great deal about the necessary microtubule regulation during the first cell cycle of the fertilized zygote. Except for the mouse, all mammals studied[41–46] adhere to the paternal inheritance of the dominant microtubule organizing center (MTOC). Briefly, the only microtubules present in the mature unfertilized oocyte are those found in the meiotic spindle. After insemination, microtubules are found in the cytoplasm as a small tuft adjacent to the incorporated sperm nucleus and nucleated from the sperm centrosome.[47] As this microtubule aster develops, the microtubules contact the female pronucleus and are necessary for pronuclear apposition. By the time of mitosis, the sperm centrosome duplicates and splits to form the poles for the first mitotic spindle. As cytokinesis takes place, the asters fill the cytoplasm and become the interphase microtubule array with each pole of the mitotic spindle serving as the centrosome for the daughter blastomeres. The organization of the microtubules relative to the incorporated sperm plays a critical role in development as poor organization of these micro-tubules is related to poor developmental outcome, as shown in analysis of return to estrous after embryo transfer in the bovine.[38] Few studies have been conducted to determine microtubule patterns and centrosome fate during nuclear transfer. Most of these zygotes have been examined after transfer of embryonic cells (ECNT) rather than somatic cell trans-fer (SCNT): bovine[44] (ECNT),[48] (SCNT); rabbit[49,50] (ECNT); porcine[51] (SCNT) and non-human primate[52] (ECNT and SCNT),[53] (SCNT).

In the case of bovine ECNT zygotes, an astral array of microtubules (Fig. 2A, green) is associated with the donor nucleus (Fig. 2A, blue) nucle-ated from the centrosome of the donor cell. Occasionally, multiple asters (Fig. 2B, green) form associated with the donor nucleus (Fig. 2B, blue). This latter may represent defects in microtubule formation or a donor cell with a replicated centrosome. These results serve as a reminder that nuclear trans-fer is a misnomer and that in the case of cloning utilizing cell fusion (the vast majority of non-murine cloning), all of the donor cell's organelles are trans-ferred into the recipient oocyte. In the case of the centrosome, this organelle remains active in the zygote in domestic species. Similar results have been observed during bovine somatic cell nuclear transfer (Navara *et al.*, unpub-lished results).[48] At a similar time point during rabbit ECNT, Pinto-Correia *et al.* (1995) described no microtubule formation associated with the pronucleus (pronuclei) in the cytoplasm of the nuclear transfer zygote. In this same study, Pinto-Correia *et al.* (1995) examined the mitotic spindle structure of rabbit nuclear transfer embryos generated from embryonic

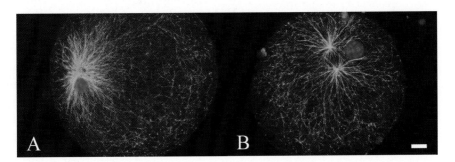

**Figure 2.** *Laser-Scanning Confocal Microscopy of two zygotes after Nuclear Transfer.* At 6.5 hours after activation, a large microtubule aster (**A**, green) is seen associated with the donor nucleus (**A**, blue). Two microtubule asters are sometimes associated with the donor nucleus. Fusion of the 32-cell stage donor blastomere with an enucleated oocyte activated with ionomycin and DMAP occurred at 1.5 hours after activation. Green = microtubules; blue = DNA; Bar = 10 μm. Reprinted with permission from Navara *et al., Dev Biol* **162**:29–40 (1994).

blastomeres. They reported the very striking result that 5/7 nuclear transplant zygotes had misaligned chromosomes on the metaphase plate and spindle errors. While only a small number of zygotes were examined, if this percentage holds true, this represents a potential for gross errors in development. Normal mitotic spindles are predominant during bovine cloning (Navara *et al.* unpublished results),[48] but spindle abnormalities are commonplace during non-human primate somatic cell nuclear transfer.[52,53] Ng *et al.* (2004) examined the spindle formed during premature chromatin condensation immediately after donor cell fusion. They found only 13.5% of these formed normal bipolar spindle structures in the meiotic cytoplasm. These authors did not examine the organization of mitotic spindles. In our experience,[52,54] primate NT embryonic and somatic constructs do not differ from control embryos when imaged using Hoffman optics. However, when microtubule patterns are examined at interphase, primate NTs display either unfocused microtubules or multiple arrays emanating from several foci. At first mitosis, all 116 ECNTs and all 30 SCNTs examined by immunocytochemistry displayed abnormal spindle morphology and poorly aligned chromosomes.[52] Despite these mitotic spindle defects, NT constructs cleave but unequal division and aneuploidy result.

Aberrant mitotic spindle assembly during nuclear transfer (Fig. 3A) suggests key mitotic proteins are absent entirely, not localized correctly or otherwise non-functional in the NT construct cytoplasm. NuMA (Nuclear-Mitotic Apparatus), a nuclear matrix protein also responsible for spindle pole assembly in somatic cells,[55,56] is observed at meiotic (Fig. 3B) and mitotic spindle (Fig. 3C) poles in primate eggs and zygotes. The somatic donor cell nucleus contains NuMA (not shown) but it is not observed in enucleated oocytes nor is it typically detected at disorganized mitotic spindles after NT (Fig. 3D). The kinesins HSET and Eg5, oppositely directed mitotic microtubule motors, play key roles in proper assembly of the mitotic spindle. HSET, the human homologue of the Kar3 kinesin-like family of minus end directed motors, is found at the minus ends of the microtubules, while Eg5, a bimC kinesin-like protein with plus end directionality, is typically found at the spindle poles.[57,58] HSET, which is observed at the spindle poles of meiotic and mitotic spindles in primates, is not detected in NT mitotic spindles (Fig. 3E). Conversely, Eg5 detects centromere pairs at meiosis and mitosis, and remains present on misaligned chromosomes on NT spindles (Fig. 3F). Collectively, these observations suggest that meiotic spindle removal either depletes the ooplasm of NuMA and HSET or results in their inaccurate positioning during mitotic spindle pole formation.

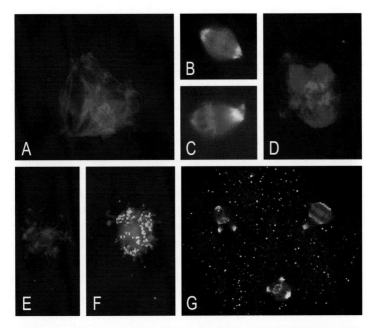

**Figure 3.** *Faulty Mitotic Spindles Produce Aneuploid Embryos After Primate Nuclear Transfer.*
**A.** Defective NT mitotic spindle with misaligned chromosomes. Centrosomal NuMA at meiosis **B.** and mitosis **C.**, but not NT-mitosis **D.** Centrosomal kinesin HSET is also missing after NT **E.**, but not centromeric Eg5 **F.** Bipolar mitotic spindles with aligned chromosomes and centrosomal NuMA after NT into fertilized eggs **G.** Blue (DNA); red (β-tubulin); green (B, C, G: NuMA; D: HSET; F: Eg5). Bar = 10 μm. Reprinted with permission from: Simerly *et al., Science* **300**:297 (2003).

To rule out the possibility that the invasive manipulations of enucleation were damaging the oocytes, the following experiment was performed. Oocytes were enucleated by needle aspiration and the autologous karyoplast carrying the meiotic spindle and maternal chromosomes was fused back into the oocyte ($\sum = 95$; 67.1% success; Simerly *et al.*, 2003). Intracytoplasmic sperm injection (ICSI) was then performed on the reconstituted oocyte to restore the normal diploid complement of DNA from both the sperm and the egg, along with the respective cytoplasmic components. These "FertClones" develop more successfully than either embryonic or somatic NTs, with one pregnancy established after 16 embryo transfers into eight surrogates. However, the pregnancy was "blighted" (an implantation attempt lacking fetal development). Perhaps not surprisingly, apoptotic rates were higher in these embryo constructs as compared with ICSI fertilized embryos.[54]

In a complementary series of experiments, the meiotic spindle was not removed and NT was performed concurrently with ICSI generating tetraploid constructs (55 oocytes; 54.4% success).[52] These constructs, when cultured until first mitosis, organized aligned chromosomes on bipolar spindles with centrosomal NuMA (Fig. 3G). The NT mitotic spindles could be distinguished from the fertilized spindle by the presence of the sperm tail visible at one pole by DIC optics.

Taken together, normal bipolar spindles found in tetraploids suggest meiotic spindle removal as a primary source of NT anomalies at first mitosis. Since "FertClones" gave apparently normal divisions, the application of the enucleation step alone could not account for observed NT mitotic defects. Proper mitotic spindles can organize around somatic chromosomes when the meiotic spindle is not removed. However, current approaches using enucleation may remove vital mitotic spindle assembly components in primates.

It is not entirely clear why primate oocytes have a stricter requirement for recycled meiotic spindle proteins than other species. Experiments with cattle and mouse oocytes suggest that the mitotic kinesins and NuMA are not exclusively concentrated on the meiotic spindle in these species and thus may be available to more properly organize a mitotic spindle after enucleation (Simerly *et al.*, 2004).[59]

Before considering the possibilities of "therapeutic cloning" as a potential source of tissue matched, ESCs the inefficiencies of NT must be overcome. These inefficiencies are observed across species and are manifested in failed or misguided reprogramming of somatic gene expression and in improper microtubule organization in the first cell cycle, resulting in aneuploidy and poor *in vitro* development. The extraordinary achievement of Hwang *et al.*[37] encourages the endless possibilities "therapeutic cloning" might bring, but the vast experience of somatic cell cloning in other species details the numerous scientific problems remaining to be addressed.

# Acknowledgements

We are indebted to many wonderful colleagues and collaborators for their contributions to the current practice of nonhuman primate cloning, including Drs. Buddy Capuano, Christa Chace, Kowit-Yu Chong, Tanja Dominko, Gabby Gosman, Laura Hewitson, Ethan Jacoby, Dave McFarland, Melanie O'Malley, Hina Qidwai, Carrie Redinger, Jody Mich-Basso, Christopher Payne, and Diana Takahashi. We also thank Dr. D. Compton (Dartmouth) for NuMA and the kinesin molecular motor antibodies. We gratefully acknowledge Serono Reproductive Biology Institute, One Technology Place, Rockland, MA for donation of gonadotropins (r-hLH, r-hCG, and/or r-hFSH) and Antide for rhesus monkey stimulation. All protocols were approved by the Magee-Women's Research Institute and the University of Pittsburgh Human Subjects Institutional Review Boards and Research Animal Review Committees. Nonhuman primate investigations are supported by grants from the National Institute of Health (NCRR, NICHD).

# REFERENCES

1. Rideout WM 3rd, Hochedlinger K, Kyba M, *et al.* (2002) Correction of a genetic defect by nuclear transplantation and combined cell and gene therapy. *Cell* **109**:17–27.
2. Robl JM, Prather R, Barnes F, *et al.* (1987) Nuclear transplantation in bovine embryos. *J Anim Sci* **64**:642–647.
3. Prather RS, Barnes FL, Sims MM, *et al.* (1987) Nuclear transplantation in the bovine embryo: Assessment of donor nuclei and recipient oocyte. *Biol Reprod* **37**:859–866.
4. Willadsen SM. (1986) Nuclear transplantation in sheep embryos. *Nature* **320**:63–65.
5. Beyleveld D, Brownsword R. (1998) Human dignity, human rights, and human genetics. *Mod Law Rev* **61**:661–680.
6. Andrews LB. (1998) Is there a right to clone? Constitutional challenges to bans on human cloning. *Harv J Law Technol* **11**:643–681.
7. Andrews LB. (1998) Human cloning: Assessing the ethical and legal questions. *Chron High Educ* **46**:B4–B5.
8. Wakayama T, Perry AC, Zuccotti M, *et al.* (1998) Full-term development of mice from enucleated oocytes injected with cumulus cell nuclei. *Nature* **394**:369–374.
9. Wells DN, Misica PM, Tervit HR. (1999) Production of cloned calves following nuclear transfer with cultured adult mural granulosa cells. *Biol Reprod* **60**:996–1005.
10. Kato Y, Tani T, Sotomaru Y, *et al.* (1998) Eight calves cloned from somatic cells of a single adult. *Science* **282**:2095–2098.
11. Baguisi A, Behboodi E, Melican DT, *et al.* (1999) Production of goats by somatic cell nuclear transfer. *Nat Biotechnol* **17**:456–461.
12. Polejaeva IA, Chen SH, Vaught TD, *et al.* (2000) Cloned pigs produced by nuclear transfer from adult somatic cells. *Nature* **407**:86–90.
13. Onishi A, Iwamoto M, Akita T, *et al.* (2000) Pig cloning by microinjection of fetal fibroblast nuclei. *Science* **289**:1188–1190.

14. Shin T, Kraemer D, Pryor J, *et al.* (2002) A cat cloned by nuclear transplantation. *Nature* **415**:859.

15. Chesne P, Adenot PG, Viglietta C, *et al.* (2002) Cloned rabbits produced by nuclear transfer from adult somatic cells. *Nat Biotechnol* **20**:366–369.

16. Woods GL, White KL, Vanderwall DK, *et al.* (2003) A mule cloned from fetal cells by nuclear transfer. *Science* **301**:1063.

17. Galli C, Lagutina I, Crotti G, *et al.* (2003) Pregnancy: A cloned horse born to its dam twin. *Nature* **424**:635.

18. Meng L, Ely JJ, Stouffer RL, Wolf DP. (1997) Rhesus monkeys produced by nuclear transfer. *Biol Reprod* **57**:454–459.

19. Zhou Q, Renard JP, Le Friec G, Brochard V, *et al.* (2003) Generation of fertile cloned rats by regulating oocyte activation. *Science* **302**:1179.

20. Arat S, Rzucidlo SJ, Gibbons J, Miyoshi K, Stice SL. (2001) Production of transgenic bovine embryos by transfer of transfected granulosa cells into enucleated oocytes. *Mol Reprod Dev* **60**:20–26.

21. Zakhartchenko V, Mueller S, Alberio R, *et al.* (2001) Nuclear transfer in cattle with non-transfected and transfected fetal or cloned transgenic fetal and postnatal fibroblasts. *Mol Reprod Dev* **60**:362–369.

22. Park KW, Lai L, Cheong HT, *et al.* (2001) Developmental potential of porcine nuclear transfer embryos derived from transgenic fetal fibroblasts infected with the gene for the green fluorescent protein: Comparison of different fusion/activation conditions. *Biol Reprod* **65**:1681–1685.

23. Reggio BC, James AN, Green HL, *et al.* (2001) Cloned transgenic offspring resulting from somatic cell nuclear transfer in the goat: Oocytes derived from both follicle-stimulating hormone-stimulated and nonstimulated abattoir-derived ovaries. *Biol Reprod* **65**:1528–1533.

24. Evans MJ, Kaufman MH. (1981) Establishment in culture of pluripotential cells from mouse embryos. *Nature* **292**:154–156.

25. Martin GR. (1981) Isolation of a pluripotent cell line from early mouse embryos cultured in medium conditioned by teratocarcinoma stem cells. *Proc Natl Acad Sci USA* **78**:7634–7638.

26. Thomson JA, Itskovitz-Eldor J, Shapiro SS, *et al.* (1998) Embryonic stem cell lines derived from human blastocysts. *Science* **282**:1145–1147.

27. Wakayama T, Tabar V, Rodriguez I, *et al.* (2001) Differentiation of embryonic stem cell lines generated from adult somatic cells by nuclear transfer. *Science* **292**:740–743.

28. Kyba M, Perlingeiro RC, Daley G. (2002) HoxB4 confers definitive lymphoid-myeloid engraftment potential on embryonic stem cell and yolk sac hematopoietic progenitors. *Cell* **109**:29–37.

29. Nagy A, Rossant J, Nagy R, Abramow-Newerly W, Roder JC. (1993) Derivation of completely cell culture-derived mice from early-passage embryonic stem cells. *Proc Natl Acad Sci USA* **90**:8424–8428.

30. Mitalipov SM, Yeoman RR, Nusser KD, Wolf DP. (2002) Rhesus monkey embryos produced by nuclear transfer from embryonic blastomeres or somatic cells. *Biol Reprod* **66**:1367–1373.

31. Mullins LJ, Wilmut I, Mullins JJ. (2003) Nuclear transfer in rodents. *J Physiol* **554**:4–12.

32. Wilmut I, Schnieke AE, McWhir J, Kind AJ, Campbell KH. (1997) Viable offspring derived from fetal and adult mammalian cells. *Nature* **385**: 810–813.

33. Brook FA, Gardner RL. (1997) The origin and efficient derivation of embryonic stem cells in the mouse. *Proc Natl Acad Sci USA* **94**:5709–5712.

34. Reubinoff BE, Pera MF, Fong CY, Trounson A, Bongso A. (2000) Embryonic stem cell lines from human blastocysts: Somatic differentiation *in vitro*. *Nat Biotechnol* **18**:399–404.

35. Lanzendorf SE, Boyd CA, Wright DL, *et al.* (2001) Use of human gametes obtained from anonymous donors for the production of human embryonic stem cell lines. *Fertil Steril* **76**:132–137.

36. Cowan CA, Klimanskaya I, McMahon J, *et al.* (2004) Derivation of embryonic stem-cell lines from human blastocysts. *N Engl J Med* **350**:1353–1356.

37. Hwang WS, Ryu YJ, Park JH, *et al.* (2004) Evidence of a pluripotent human embryonic stem cell line derived from a cloned blastocyst. *Science* **303**:1669–1674.

38. Navara CS, First NL, Schatten G. (1996) Phenotypic variations among paternal centrosomes expressed within the zygote as disparate microtubule lengths and sperm aster organization: Correlations between centrosome activity and developmental success. *Proc Natl Acad Sci USA* **93**:5384–5388.

39. Boiani M, Eckardt S, Scholer HR, McLaughlin KJ. (2002) Oct4 distribution and level in mouse clones: Consequences for pluripotency. *Genes Dev* **16**:1209–1219.

40. Byrne JA, Simonsson S, Western PS, Gurdon JB. (2003) Nuclei of adult mammalian somatic cells are directly reprogrammed to Oct-4 stem cell gene expression by amphibian oocytes. *Curr Biol* **13**:1206–1213.

41. Yllera-Fernandez MM, Crozet N, Ahmed-Ali M. (1992) Microtubule distribution during fertilization in the rabbit. *Mol Reprod Dev* **32**:271–276.

42. Wu GJ, Simerly C, Zoran SS, Funte LR, Schatten G. (1996) Microtubule and chromatin dynamics during fertilization and early development in rhesus monkeys, and regulation by intracellular calcium ions. *Biol Reprod* **55**: 260–270.

43. Simerly C, Wu GJ, Zoran S, Ord T, *et al.* (1995) The paternal inheritance of the centrosome, the cell's microtubule-organizing center, in humans, and the implications for infertility. *Nat Med* **1**:47–52.

44. Navara CS, First NL, Schatten G. (1994) Microtubule organization in the cow during fertilization, polyspermy, parthenogenesis, and nuclear transfer: The role of the sperm aster. *Dev Biol* **162**:29–40.

45. Long CR, Pinto-Correia C, Duby RT, *et al.* (1993) Chromatin and microtubule morphology during the first cell cycle in bovine zygotes. *Mol Reprod Dev* **36**:23–32.

46. Crozet N. (1990) Behavior of the sperm centriole during sheep oocyte fertilization. *Eur J Cell Biol* **53**:326–332.

47. Navara CS, Hewitson LC, Simerly CR, Sutovsky P, Schatten G. (1997) The implications of a paternally derived centrosome during human fertilization: Consequences for reproduction and the treatment of male factor infertility. *Am J Reprod Immunol* **37**:39–49.

48. Shin MR, Park SW, Shim H, Kim NH. (2002) Nuclear and microtubule reorganization in nuclear-transferred bovine embryos. *Mol Reprod Dev* **62**:74–82.

49. Collas P, Pinto-Correia C, Ponce de Leon FA, Robl JM. (1992) Effect of donor cell cycle stage on chromatin and spindle morphology in nuclear transplant rabbit embryos. *Biol Reprod* **46**:501–511.

50. Pinto-Correia C, Long CR, Chang T, Robl JM. (1995) Factors involved in nuclear reprogramming during early development in the rabbit. *Mol Reprod Dev* **40**:292–304.

51. Lai L, Tao T, Machaty Z, *et al.* (2001) Feasibility of producing porcine nuclear transfer embryos by using G2/M-stage fetal fibroblasts as donors. *Biol Reprod* **65**:1558–1564.

52. Simerly C, Dominko T, Navara C, *et al.* (2003) Molecular correlates of primate nuclear transfer failures. *Science* **300**:297.

53. Ng SC, Chen N, Yip WY, *et al.* (2004) The first cell cycle after transfer of somatic cell nuclei in a non-human primate. *Development* **131**:2475–2484.

54. Dominko T, Simerly C, Martinovich C, Schatten G. (2002) Cloning in non-human primates, In *Principles of Cloning*, eds. J.B. Cibelli, R. Lanza, K.H. Campbell, and M. West, pp. 419–431. San Diego: Academic Press.

55. Merdes A, Heald R, Samejima K, Earnshaw WC, Cleveland DW. (2000) Formation of spindle poles by dynein/dynactin-dependent transport of NuMA. *J Cell Biol* **149**:851–862.

56. Levesque AA, Howard L, Gordon MB, Compton DA. (2003) A functional relationship between NuMA and kid is involved in both spindle organization and chromosome alignment in vertebrate cells. *Mol Biol Cell* **14**:3541–3552.

57. Blangy A, Lane HA, d'Herin P, *et al.* (1995) Phosphorylation by p34cdc2 regulates spindle association of human Eg5, a kinesin-related motor essential for bipolar spindle formation *in vivo*. *Cell* **83**:1159–1169.

58. Mountain V, Compton DA. (2000) Dissecting the role of molecular motors in the mitotic spindle. *Anat Rec* **261**:14–24.

59. Simerly C, Navara C, Hyun SH, *et al.* (2004) Embryogenesis and blastocyst development after somatic cell nuclear transfer in nonhuman primates: Overcoming defects caused by meiotic spindle extraction. *Dev Biol* **276**:237–252.

# 13

# Hematopoietic Stem Cells: Basic Science to Clinical Applications

Daniel L. Kraft* and Irving L. Weissman

## Introduction

The hematopoietic stem cell (HSC) is characterized by its dual ability to both self-renew and reconstitute and differentiate into progenitors of all the mature blood cell lineages.[1] The HSC was the first stem cell to be prospectively identified, initially in murine systems and shortly thereafter in man.[2] Hematopoietic stem cell biology has since become a model for other stem cell systems, particularly as hematopoiesis demonstrates distinct, phenotypically identifiable populations, which differentiate via several possible lineage pathways to generate mature cell populations.

The HSC is perhaps the best characterized stem cell, and is currently the only stem cell population with extensive applications for the treatment of a broad array of diseases. Clinical utilization of autologous or allogeneic donor HSC is through the field of bone marrow transplantation (BMT) or mobilized peripheral blood (MPB) transplantation, with applications ranging from treatment of malignant diseases (primarily leukemias and lymphomas), to genetic diseases, immunodeficiencies, and potentially broader roles in the newly evolving field of regenerative medicine. The field of hematopoietic cell transplantation, in particular, has been driven by bench to bedside research, in which principles and innovations worked out in laboratory animal models have had direct translation to the clinic, and observations in the clinic lead to laboratory investigations.

Unlike pluripotent embryonic stem cells (discussed elsewhere in this textbook) which clearly have the ability to differentiate into cells of all three germ layers, HSC have generally been considered multipotent, generating

*Correspondence: Beckman Center, Room B-257 Stanford, CA, 94305. Tel.: (650) 723-6520, Fax: (650) 723-4034, E-mail: daniel.kraft@stanford.edu

the hematolyphoid lineage of red blood cells, white blood cells and platelets. Recently, there has evolved significant debate as to the "plasticity" of HSC, defined in terms of the potential of HSC to contribute directly through transdifferentiation to the generation of muscle, hepatic, or neuronal tissues. More recent data is inconsistent with transdifferentiation, albeit very rare fusion events between HSC and damaged tissue cells can be found, and some authors have proposed that such cell fusions play a role in the repair of damaged tissues.

In this chapter, we seek to summarize the biology of HSC as defined initially in the mouse and later in the human system, and discuss the current, evolving and potential future applications for human hematopoietic stem and progenitor cell populations.

# Biology of Hematopoietic Stem Cells

## Definition and isolation of hematopoietic stem cells

The ability of cells within the bone marrow to reconstitute the hematolyphoid system was first demonstrated by experiments in the 1950s, in which mice receiving lethal doses of radiation (experience from Japanese atomic bomb victims having demonstrated the radiation sensitivity of bone marrow) could be rescued by bone marrow transplantation, providing donor-derived hematopoiesis.[3] The concept that a single cell which could give rise to multiple hematopoietic and lymphoid outcomes was further advanced in the 1960s, when a population of clonogenic bone marrow cells was described which could give rise to both myeloid and erythroid progeny (through generation of colony forming unit-spleen (CFU-S)), and which also contained a subset of cells which could reconstitute the hematolymphoid system of lethally irradiated mice.[4,5,6]

In the two decades following these fundamental experiments, biomedical tools which included monoclonal antibody technology enabling detection of molecules expressed on bone marrow cells, and high-speed fluorescence-activated cell sorting (FACS) technology were developed. Researchers were then able to prospectively isolate and test rare, phenotypically defined marrow populations (based on the presence and absence of certain monoclonal antibody-defined surface determinants). The next step was to develop assays for clonogenic precursors of blood cell types; separate quantitative assays for myeloerythroid lineage, then, T cell lineage, and B cell lineages were developed. Through *in vitro* and *in vivo* assays of clonogenic hematopoietic precursor mouse bone marrow cell populations, the phenotypic and functional characteristics of mouse HSC, and later of committed myeloid and lymphoid progenitors, were successfully defined.[7]

A key initial finding in the search for HSC was that the population of cells with clonogenic potential for the generation of myeloerythroid, B lineage and T lineage cells was contained in a very rare population of bone marrow cells ($\sim$0.01% of all marrow cells) which had none to very low levels of the lineage markers which mark the various types mature blood cells. This lack of mature cell lineage markers defines cells which are "lineage negative" (Lin-) for the markers found on T cells, B cells, granulocytes, monocytes/macrophages, erythroid cells, natural killer cells, etc. Two key markers were found to be present on murine HSC: Stem Cell Antigen-1 (Sca-1) and Thy-1[8]; both glycophosphatidyl inositol-linked immunoglobulin superfamily molecules. HSCs were also found to express the c-kit receptor tyrosine kinase[9] (see Fig. 1).

Using these markers (lack of lineage markers (Lin-) and the expression of the c-Kit, Sca-1, and Thy-1 markers), mouse hematopoietic cells were prospectively enriched or isolated by FACS using specific phenotypic combinations, and these FACS purified populations were tested for their ability to reconstitute the hematolymphoid system of lethally irradiated mice. The mouse HSC was defined with the "KTLS" phenotype: [c-Kit$^{high}$(K), Thy1.1$^{low}$ (T), Lin$^{neg}$ (L), Sca-1$^{+}$(S)], or with slightly different markers, including [Lin$^{neg}$, Sca-1$^{+}$, Rhodamine 12$^{low}$][10] or [Lin$^{neg}$, CD34$^{-/int}$, c-Kit$^{+}$, Sca-1$^{+}$].[11] These cells also rigorously exclude hydrophobic dyes such as Hoechst 33342 via an ABC transporter, and a combination of hematopoietic markers with Hoechst exclusion can be used to isolate HSC.[12]

Subsequent investigations demonstrated that the KTLS (c-Kit$^{high}$, Thy1.1$^{low}$, Lin$^{neg}$, Sca-1$^{+}$) HSC population could be subdivided into three subpopulations by surface markers and by their degree of self-renewal[13]: Long-term HSCs (LT-HSCs) perpetually self-renew, as well as give rise to short-term HSCs, which self-renew *in vivo* following transplantation for approximately 6 weeks, and which generate multipotent progenitors (MPP), each step being marked by limited, or no self-renewal.

1) **Long-term HSC (LT-HSCs):** consist of perpetually self-renewing HSCs; these HSCs lack cell surface Flk2 expression (fetal liver kinase 2, a receptor tyrosine kinase[14]) and have very low levels of Mac-1 (a macrophage differentiation antigen) expression and are CD4$^{-}$.

**Hematopoietic Stem Cells**

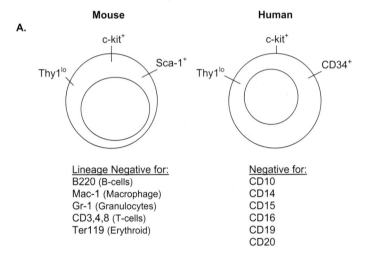

**Mouse**          **Human**

A.

Lineage Negative for:          Negative for:
B220 (B-cells)                 CD10
Mac-1 (Macrophage)             CD14
Gr-1 (Granulocytes)            CD15
CD3,4,8 (T-cells)              CD16
Ter119 (Erythroid)             CD19
                               CD20

B.                    **Stem cell subpopulations**

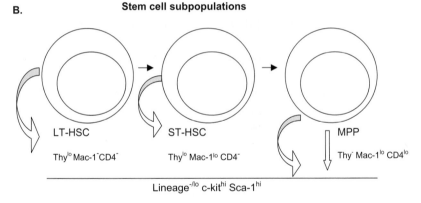

LT-HSC                ST-HSC                MPP

$Thy^{lo}$ Mac-1⁻CD4⁻   $Thy^{lo}$ Mac-1$^{lo}$ CD4⁻   Thy⁻ Mac-1$^{lo}$ CD4$^{lo}$

Lineage$^{-/lo}$ c-kit$^{hi}$ Sca-1$^{hi}$

**Figure 1.** **A.** Hematopoietic stem cells have distinctive surface phenotypes (shown on mouse and humans). **B.** Thy-1.1, Mac-1 and CD4 expression delineate a pathway of hematopoietic stem cell (HSC) differentiation. Surface expression of Mac-1 is upregulated and Thy1.1 is down regulated as HSC self-renewal capacity diminishes. LT-HSC, long-term hematopoietic stem cells; MPP, multipotent progenitors; ST, short-term; Sca-1, stem cell antigen.

2)  **Short term (ST-HSCs):** which self-renew *in vivo* following transplantation for approximately 6 weeks, express low surface levels of Mac-1 and are CD4⁻ to CD4$^{lo}$.

3)  **Multipotent progenitors (MPPs):** have short self-renewal potential, and are difficult to detect under transplant conditions. They experience

loss of Thy 1.1 expression, and have low-level expression of Mac-1 and CD4.[15]

A single Long-term hematopoietic stem cell (LT-HSC) transplanted *in vivo* with supporting host bone marrow can give long-term multilineage (red cell, white blood cell and platelet) engraftment as well as expansion of the LT-HSC, ST-HSC and MPP pool.[16] LT-HSCs give rise in a linear manner to ST-HSC, which in turn give rise to MPP (Fig. 1B). Reverse differentiation (i.e. ST-HSC giving rise to LT-HSC) has never been detected.[17]

KTLS hematopoietic stem cells are the only cells within mouse bone marrow which will both radioprotect and reconstitute all blood cell lineages.[18] Increasing the dose of purified KTLS cells transplanted into lethally irradiated hosts, shortens the time for donor cell engraftment (measured by donor derived neutrophil and platelet count), but only to a point. Once the KTLS-HSC dose required for engraftment within 10 days has been attained, further several fold increases in KTLS dose does not shorten engraftment time (see Fig. 2). This demonstrates that only stem cells are required for rapid engraftment, and that the cellular output from transplanted HSC is regulated.[19]

As few as 25 purified HSCs are capable of rescuing more than 50% of lethally irradiated animals in the syngeneic/autologous transplant setting, while 100 HSC radioprotect 95% of irradiated hosts. In lethally irradiated mice from genetically different (allogeneic) donors, higher numbers of purified HSC are required to radioprotect lethally irradiated recipients. When the donor and recipient combination has any degree of major

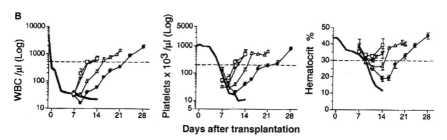

**Figure 2. Engraftment of purified hematopoietic stem-cells:** Hematopoietic recovery kinetics of white blood cells (WBC), platelets and red blood cells (hematocrit) in lethally irradiated C57BL/Ka mice transplanted with syngeneic doses of purified HSC. 100 HSC (*solid circles*), 1000 (*triangles*), 5000 (*solid triangles*) or 10 000 (*squares*). The dashed horizontal line represents recovery of blood levels to 500 WBCs/ul, 20 000 platelets/ul, and 30% hematocrit (Uchida *et al.,* 1998).

histocompatibility (MHC) mismatch, this dose requirement increases to 500–3000 donor hematopoietic stem cells.[20]

## Hematopoietic progenitor cells

Hematopoiesis takes place through the step-wise differentiation of multipotent HSC to generate a hierarchy of progenitor populations with progressively restricted developmental potential, none of which can dedifferentiate or show self-renewal capacity. This leads to the production of multiple lineages of mature effector cells (see Fig. 3). The same methodology used for the prospective isolation of mouse HSC (testing the downstream production of various purified marrow derived populations) has been utilized to identify and isolate HSC-derived lineage restricted progenitors. Downstream of the multipotent progenitor (MPP), a lineage potential decision is made, and differentiation occurs via either the common myeloid progenitor (CMP) or the common lymphocyte progenitor (CLP), each of which has been isolated to homogeneity, and each expresses a distinct gene expression profile[21] and is a clonal precursor of a limited subset of progeny.

When differentiation occurs along the lymphoid lineage, murine CLPs, which express cell surface IL-7 receptor, differentiate into T-cells, B-cells and $CD8\alpha^+$ and $CD8\alpha^-$ dendritic cells, as well as natural killer (NK) cells.[22] HSCs also differentiate via the myeloid lineage, giving rise first to the common myeloid progenitor (CMP) and then to megakaryocyte-erythroid progenitors (MEP), granulocyte-macrophage progenitors (GMP) and $CD8\alpha^+$ and $CD8\alpha^-$ dendritic cells.[23,24] GMPs are the only source of monocytes, macrophages and neutrophils, while a restricted megakaryoctye progenitor population (MkP) contributes to megakaryocyte production.[25]

## Defining and Isolating the Human HSC

### *In vitro* models

With much of the mouse hematopoietic progenitor system described, several groups focused on examining and defining similar human hematopoietic stem cell and progenitor populations. Given the obvious difficulties of examining the differentiation of HSC and committed progenitors in humans, *in vitro* and *in vivo* animal models were developed to identify and characterize human HSC and progenitors. *In vitro*, long-term culture-initiating cell (LTC-IC) assays can be used to test candidate human hematopoietic stem cells, which are cultured for 5–10 weeks on adherent,

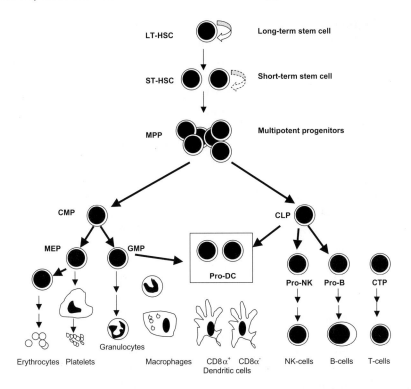

**Figure 3.   Model of hematopoietic stem cell differentiation:** Long-term hematopoietic stem cells (LT-HSCs, c-kit$^+$Thy1.1$^{lo}$Lin$^{neg}$Sca-1$^+$) are highly self-renewing cells that give rise to all mature blood cells. The process of HSC differentiation involves the loss of self-renewal with functionally irreversible maturation steps. Multipotent progenitors differentiate via either the Common Lymphocyte (CLP) or Common Myeloid Progenitor (CMP).

bone marrow-derived stromal cells in order to simulate a microenvironment similar to bone marrow.[26] Cells are then transferred to a semisolid, cytokine containing media, and cells from the primary culture which maintain their proliferative capacity and generate erythroid, myeloid and B cell colonies will define candidate human cells with self-renewal capacity or multilineage differentiation potential.[27]

## *In vivo* models

*In vivo* models for human hematopoiesis have focused on the transplantation of candidate HSC into immunodeficient hosts which do not have the ability to eliminate or immunologically challenge xenografts through an immune response, yet retain a microenvironment which allows for engraftment and multilineage differentiation of human-derived donor cells. Initial studies were done in severe combined immunodeficient (SCID) mice[28] (which have a T- and B cell defect), and beige/nude/xid (bnx) mice.[29] Residual host macrophage and NK cells activity in these mice can, however, mediate rejection of the human xenograft. NOD-SCID mice which have deficient NK, and macrophage functions improved the microenvironment for human cells readout with 10–20-fold better engraftment than in SCID mice. Engrafting cells are termed SCID repopulating cells (SRC).[30,31]

The SCID-*hu* model, in which SCID mice were surgically transplanted with human fetal bone, thymus, liver or lymph nodes,[32] has played a key role in enabling the further study and identification of human candidate stem cells. The human derived thymus, grown under the kidney capsule through engraftment of human fetal liver and thymus fragments, has enabled the *in vivo* study of human T-cell development[33] as well as the study of infectious agents such as HIV and their effect upon the human thymocytes.[34,35]

SCID and NOD-SCID mice share the disadvantages of limited life-span, high radiation sensitivity, with very rare T-cell development (bias towards B-cell development). Newer models include the NOD-SCID ß2 microglobulin knockout mice, and the RAG2/common cytokine receptor gamma chain double-mutant mice (RAG2/CG DKO).[36] Most recently, it was shown that RAG2/CG DKO pups, sublethally irradiated on their first day of life, and injected intrahepatically with CD34$^+$ cells isolated from human umbilical cord blood, leads to *de novo* development of B cells, T cells (with a normal V-beta repertoire), and dendritic cells; formation of structured primary and secondary lymphoid organs, as well as production of functional immune responses to vaccination.[37] While engraftment levels with human CD45$^+$ cells remain somewhat low and variable, this model holds promise for the further study of the human hematopoietic and adaptive immune response *in vivo*.

In addition to murine xenotransplantation, intraperitoneal transplant of low numbers of human hematopoietic stem cells into unconditioned, early gestational sheep fetuses has demonstrated that selected progenitors can engraft, with long-term (several years) of human myeloerythroid, T and B cell engraftment.[38,39]

## Isolation of human hematopoietic stem cells

The CD34 molecule, identified in 1984,[40] is recognized as a major positive marker for human hematopoietic stem and progenitor cells. CD34 is expressed on a heterogenous population of cells which make up approximately 0.5–5% of human hematopoietic cells in human fetal liver, cord blood, and bone marrow.[41] The function of CD34 on hematopoietic cells, however, is not well understood. On non-hematopoietic tissues, CD34 is expressed on endothelial cells of small vessels and is a ligand for L-selectin (CD62L).[42] In murine models, CD34 plays a role in adhesion to the stromal microenvironment,[43] although CD34 mutant mice show no significant abnormalities, and consequently it is does not appear to be essential for hematopoiesis. LT-HSC in the mouse are CD34 low; mice containing the human CD34 gene from the BAC clone have LT-HSC that are human CD34$^+$ and murine CD34$^{-/low}$.[44]

A small fraction (1–10% of CD34$^+$ cells or 0.05–0.1% of human fetal bone marrow) contains multipotent hematopoietic precursors which are lineage negative for cell markers found on mature cells (CD3, CD4, CD8, CD19, CE20, CD56, CD11b, CD14 and CD15) and contains cells which, *in vitro*, will derive lymphoid (B and NK cells), and myeloid differentiation.[45] The population of human precursor cells that expresses Thy-1 and CD34 but no known lineage markers, is enriched for clonogenic activity that establishes long-term, multi-lineage (myelomonocytic and B lymphoid) cultures on mouse marrow stromal lines, and also takes up little of the fluorescent mitochondrial dye rhodamine 123. Thy-1$^+$, CD34$^+$, Lineage$^-$ adult and fetal bone marrow cells were shown to engraft all lineages and differentiate into T-lymphocytes in SCID-*hu* mice with human thymic grafts, and were shown to engraft intact human fetal bone marrow grown in SCID mice, resulting in donor-derived myeloid and B cells.

Ninety to ninety-nine per cent of CD34$^+$ cells co-express the CD38 antigen. It is, however, the CD34$^+$, CD38$^-$ population (not the CD34$^+$CD38$^+$) which contains the SCID-*hu* and NOD-SCID[46,47] repopulating cells. Lin$^-$CD34$^-$38$^-$ cells do contain low levels of SRC and fetal-sheep repopulating ability (1 in 10$^8$ cells as compared with 1 in 10$^6$ SRC in cord blood cells), but their activity could be derived from rare, contaminating CD34$^+$38$^-$ cells.

CD133 is a recently described glycoprotein, which may be a marker of more primitive human hematopoietic precursors, as it is expressed on all CD34$^+$CD38$^-$, and on some CD38$^+$, progenitors, as well as 0.2% of Lin$^-$CD34$^-$CD38$^-$ cells.[48,49] Only the CD133$^+$ fraction of Lin$^-$CD34$^-$CD38$^-$ cells have SRC potential.[50] CD133$^+$ cells can be enriched from bone marrow and mobilized peripheral blood[51] and such strategies for collection are beginning to be used clinically in the positive selection in patients receiving matched unrelated and mismatched-related allogeneic transplants.[52] CD133$^+$ marrow-derived cells, likely distinct from HSC, have been postulated to play a role in angiogenesis, and are also being investigated for their ability to participate in the repair of injured muscle. Freshly isolated, circulating AC133$^+$ cells have been induced to undergo myogenesis when co-cultured with myogenic cells or exposed to Wnt-producing cells *in vitro* and when delivered *in vivo* through the arterial circulation or directly into the muscles of transgenic scid/mdx mice fuse into skeletal myofibers and appear to contribute to functional recovery of injected muscle,[53] although it is not clear if this is a pathologic manifestation or maturation through a skeletal muscle stem or progenitor cell.[54] Autologous marrow-derived AC133$^+$ cells have been injected into the infarcted regions of patient's myocardium during coronary bypass surgery, with evidence of improved ejection fraction and improved perfusion in injected areas,[55] although careful analysis in a similar mouse models failed to show any evidence of donor-derived cardiomyoctyes, smooth muscle, or vascular endothelium.[56]

Recently, identification of HSC on the basis of conserved stem cell function rather than cell surface phenotype has been postulated as a new and reliable way to isolate various functional stem cell populations.[57] High levels of aldehyde dehydrogenase (ALDH) have been utilized as a means to select cells by FACS sorting from lineage negative cord blood. ALDH$^{hi}$Lin$^-$ cells were found to be predominantly CD34$^+$CD38$^-$CD133$^+$, with greater than 10-fold repopulating ability compared with ALDH$^{(low)}$ cells when engrafting NOD/SCID mice.

## Hematopoietic stem cell migration

HSC, while predominantly residing in the marrow, are in a constant state of migration between the marrow and the bloodstream. In mice, approximately 100 HSC are in transient flux (with less than 5 minutes of residence time in the blood) between the marrow and the circulation at any one time, which equates to approximately 30 000 HSC circulating through the blood each day. Parabiosis experiments (in which two mice are sutured together such that their vascular system undergoes anastamosis) have demonstrated that HSC from one partner mouse will leave the marrow, circulate through the blood and home to the marrow of the paired mouse.[58,59] The presence of HSC in the blood was initially discovered when clinical researchers described increased hematopoietic activity in the blood of patients treated with cytotoxic drugs (classically cyclophosphamide).[60] Cytotoxic drugs alone, or in combination with cytokines such as granulocyte-macrophage colony-stimulating factor (GM-CSF) and G-CSF subsequently, was shown to mobilize hematopoietic cells into the bloodstream.[61] HSC were demonstrated to be present in high numbers in mobilized peripheral blood (MPB) as transplantation into lethally irradiated recipients led to rapid and sustained hematopoiesis, initially in murine models, and later as translated to the marrow transplant ward. Mobilization appears to occur through HSC proliferation and division (which occurs only within the marrow), followed by emigration of HSC into the blood spleen and liver.[62] HSCs express a number of integrins and other adhesion molecules, as well as the CCR4 chemokine receptor, play a role in cell proliferation, detachment from marrow stroma, movement through the bloodstream, margination and migration through vessels at distant sites, as well as localization to niches within the marrow. The details of this process are under active investigation.

## Human early lineage committed progenitors

As clonal lymphoid and myeloid committed progenitors have been described in murine hematopoiesis, similar populations have been sought in the human system (see Table 1). While assays for measuring human myeloid development from single cells exist, to read out T, NK and B cells derived from small numbers of candidate human progenitor populations remains difficult. Additional candidate antigens have been investigated as a means to segregate early lymphoid or myeloid progenitors. A potential human CLP population from fetal and adult marrow has the $Lin^-CD34^+CD38^+CD10^+$ phenotype as these cells generate B, NK and dendritic cells (and T-cells when transferred to SCID-hu thymus), but not myeloid progeny.[63] In subsequent studies, cord blood $CD34^+CD38^-CD7^+$ cells were shown to contain 40% of clonal B, NK, and dendritic cell precursors.[64]

Multiple reports of myeloid-colony forming cells collectively show that $CD34^+$ cells, which are $CD45RO^-$, $CD45RA^+$, $CD64^+IL3Receptor-\alpha^+$, $Flt3^+$, or CCR1+, are enriched for CFU-GM-forming cells, while $CD34^+$ cells, which are $CD45\ RO^+$, $IL-3R\alpha^-$, $Flt3^-$, or $CCR1^-$, are enriched for erythroid colony forming cells.[65,66] Three populations have been identified that are likely counterparts of the murine CMPs, GMPs and MEPs.[67] These cells are all $Lin^-CD34^+CD38^+$, and are further distinguished by expression of CD45RA, and IL-3Rα. CMP are $CD45RA^-IL-3R^{lo}$, GMPs and $CD45RA^+$

**Table 1. Mouse and Human Stem and Progenitor Populations**
Surface markers utilized to isolate these populations as prospectively identified by flow cytometry and *in vivo* read-out.

| Population | Murine | Human | Output |
|---|---|---|---|
| Hematopoietic Stem Cell (HSC) | $Thy1.1^{lo}$, $c\text{-}Kit^+$, $Sca\text{-}1^+$, Lin- (Long-term HSC) | $CD34^+$ $CD38^-$ $CD90^+$ [$Thy1^+$], Lin- | All mature hematopoietic populations |
| Common Lymphocyte Progenitoir (CLP) | $IL\text{-}7R^+$, $c\text{-}Kit^{lo}$ $Sca\text{-}1^{lo}$ $CD34^+$ $FcgR^-$, Lin- | $CD34^+$, $CD38^+$, $CD10^+$, $CD34^+$, $CD38^-$, $CD7^+$, Lin- | T and B cells, NK cells, Dendritic cells |
| Common Myeloid Progenitor (CMP) | $IL\text{-}7R^-$, $c\text{-}Kit^+$ $Sca\text{-}1^-$ $CD34^+$ $FcgR^-$, Lin- | $CD34^+$, $CD38^+$ $CD45RA^-$, $IL\text{-}3R\alpha^{+\ lo}$, Lin- | MEP and GMP |
| Granulocyte Macrophage Progenitor (GMP) | $c\text{-}Kit^+$ $Sca\text{-}1^-$, $CD34^+$ FcGRhi | $CD34^+CD38^+$, $CD45RA^+$ $IL\text{-}3R\alpha^{lo}$, Lin- | Granulocytes, Macrophages |
| Megakaryocyte erythroid Progenitor (MEP) | $c\text{-}Kit^+$ $Sca\text{-}1^-$, $CD34^-$ $FcgR^{lo}$ | $CD34^+$, $CD38^+$ $CD45RA^-$, $IL\text{-}3R\alpha^+$, Lin- | Erythrocytes and Platelets |

IL-3Rα$^{lo}$, while MEP are CD45RA$^-$ and IL-3Rα$^-$. These each have high cloning efficacies and show no *in vitro* B or NK cell read out.

## Clinical Use of Hematopoietic Stem Cells

The role of hematopoietic stem cells is central to the biological basis of bone marrow transplantation, subsequently referred to as hematopoietic cell transplantation. Hematopoietic cell transplantation (HCT) has developed from a treatment of "last resort" to an effective therapy for patients with a variety of malignant and non-malignant disorders (See Table 2). The field was pioneered by E. Donall Thomas initially in canine model systems in the 1950s and 1960s, with the first patient transplant performed in 1956 (treatment of a patient with total body irradiation followed by an infusion of marrow from an identical twin resulted in complete remission of leukemia). In 1968, the first transplantation from a related, matched donor for a non-malignancy (SCID) was successfully performed, and in 1973, the first transplant was carried using an unrelated donor. In 2002, over 45 000 transplants were performed worldwide for a wide variety of indications.[68]

It is important to note that while HCT is often termed "stem cell transplantation," clinically these are not purely "stem cell" transplants, as the cells commonly transplanted are not pure HSC but heterogeneous populations of stem, progenitor and mature cell populations contained within unfractionated donor bone marrow, mobilized peripheral blood or cord blood.

### Autologous transplantation

There are two primary forms of hematopoietic cell transplantation defined by the donor graft source: autologous and allogeneic. The principle of autologous hematopoietic transplantation allows the use of myeloablative doses of combination chemotherapy and/or radiation to optimize tumor kill, followed by transplantation with the patient's own previously collected and stored bone marrow or mobilized peripheral blood to regenerate or "rescue" the hematopoietic system.[69] This treatment modality rapidly evolved in the 1980s and has became the treatment of choice for patients with relapsed or refractory lymphoma, being increasingly used for other chemotherapy-sensitive but infrequently cured cancers, including acute leukemia, multiple myeloma, and selected solid tumors. Generally, autologous transplants have

**Table 2.** Diseases in which Autologous and/or Allogeneic Hematopoietic Cell Transplants have been used

| Malignant | Non-Malignant |
|---|---|
| Leukemia/preleukemia<br>  Chronic myeloid leukemia (CML)<br>  Chronic lymphocytic leukemia (CLL)<br>  Acute myeloid leukemia (AML)<br>  Acute lymphoblastic leukemia (ALL)<br><br>Myeloproliferative syndromes<br>  Juvenile chronic myeloid leukemia<br><br>Myelodysplastic syndromes<br><br>Therapy-related myelodysplasia/leukemia<br>  Kostmann agranulocytosis<br><br>Non-Hodgkin's and Hodgkin's lymphoma<br><br>Multiple myeloma<br><br>Solid tumors<br>  Breast cancer<br>  Neuroblastoma<br>  Ovarian cancer<br>  Renal cancer<br>  Brain tumors<br>  Testicular cancer | Severe aplastic anemia<br><br>Paroxysmal nocturnal hemoglobinuria<br><br>Hemoglobinopathies<br>  Thalassemia major<br>  Sickle cell disease<br><br>Congenital disorders of hematopoiesis<br>  Fanconl's anemia<br>  Diamond-Blackfan syndrome<br>  Familial erythrophagocytic histiocytosis<br>  Dyskeratosis congenital<br>  Shwachman-Diamond syndrome<br><br>SCID and related disorders<br>  Wiskott-Aldrich syndrome<br><br>Inborn errors of metabolism<br>  Storage Diseases<br><br>Autoimmmune disorders<br>  Systemic lupus erythematosis<br>  Rheumatoid arthritis |

lower transplant-related mortality as compared with allogeneic transplantation as engraftment and immune recovery is more rapid and there is no graft versus host disease (GVHD) related morbidity or mortality.

## Allogeneic transplantation

Allogeneic hematopoietic cell transplantation by definition involves transplantation of hematopoietic and hematolymphoid cells from a related or unrelated donor. Related donors who share the full HLA type of the recipient are termed "HLA matched" and are preferred donors due to diminished rates of graft rejection and decreased incidence and less severe graft versus host disease (GVHD) (wherein donor mature T cells respond to host transplantation antigens). Full siblings of a prospective transplant patient have a 25% chance of being a full HLA match, and if so are termed matched related donor (MRD), although single HLA mismatch donors are utilized at some transplant centers when no complete match can be identified. If no MRD donor can be identified, unrelated donors are sought, termed matched unrelated donors (MUDs), and are generally identified through marrow donor registries such as the American-based National Marrow Donor Program (NMDP). The first anecdotal reports of successful HCTs using HLA-matched unrelated donor occurred in the early 1970s; however, the polymorphism of human HLA made the feasibility of finding such a donor for an individual low. With the advent of several national and international groups such as the NMDP with organized marrow donor registries, there are now over eight million HLA-typed volunteer donors worldwide.[70] More recently, as immunosuppressive techniques have improved, haploidentical parents or sibling donors (which share only a 50% HLA match, or one of the two HLA containing chromosomes) have been utilized as donors when no MRD or MUD donors can be identified. Haploidentical and single HLA mismatched transplants may be most successful if T-cell depleted grafts containing high numbers of CD34$^+$ cells are administered.[71]

## Indications for HCT

The most common indications for allogeneic and autologous HCT are malignancy. The diseases treated in 2002 in North America are shown in Fig. 4. 69% of allogeneic HCTs were for leukemia or preleukemia; 28% for AML; 17% for ALL; 11% for CML; 9% for myelodysplastic or myeloproliferative syndromes; and 4% for other leukemias. Twenty percent are for other cancers, including non-Hodgkin's lymphoma (12%); multiple myeloma (3%); Hodgkin's disease (<1%); and other cancers (4%). The remainder are for aplastic anemia (3%; immune deficiencies (2%); and other diverse non-malignant disorders (6%)). The most common indications for autologous transplantation were multiple myeloma (34%); non-Hodgkin's lymphoma (33%); Hodgkin's disease (12%); leukemia (5%); neuroblastoma (3%); and other cancers.[72]

Allogeneic HCT most often involves high-dose radiation and chemotherapy with the aim to eliminate any residual malignancy. This therapeutic effect is often augmented by the graft versus malignancy or "graft versus

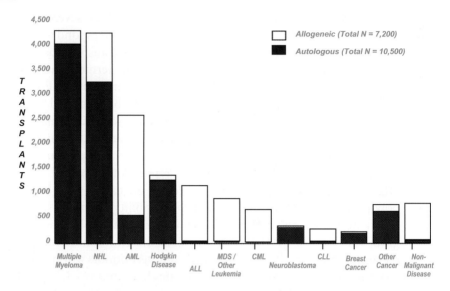

**Figure 4. Indications for hematopoietic cell transplantation** in North America, 2002. Abbreviations: Non-Hodgkins lymphoma (NHL), acute myelogenous leukemia (AML), acute lymphoblastic leukemia (ALL), myelodysplastic disease (MDS), chronic myelogenous leukemia (CML), chronic lymphocytic leukemia (CLL). Courtesy of the Statistical Center the IBMTR and ABMTR.

tumor" effect (GVT) mediated by donor T-cells.[73] Standard allogeneic bone marrow grafts contains $\sim 10^7$ CD3$^+$ cells per kilogram of recipient weight. GCSF mobilized peripheral blood contains 10-fold greater numbers of T cells per kilogram (approximately $10^8$ CD3$^+$ cells/kg). T cells within the graft can mediate a graft versus tumor effect (targeting, for example, minor histocompatability antigens expressed on residual donor hematolyphoid tissues and any residual malignant cells), but often also mediate graft versus host disease (GVHD).[74] GVHD causes significant morbidity and mortality through the action of mature, donor derived T cells which recognize host antigens as foreign and mount an immune attack, primarily manifested clinically on the patient's skin, gut, and liver.[75] The degree of HLA disparity between donor and host is a primary predictor for the occurrence of acute GVHD (within 100 days of transplant), and chronic GVHD (greater than 100 days post transplant).[76] All allogeneic recipients are therefore placed on immunosuppression during conditioning, and usually for at least 6 months post transplantation. Common immunosuppressive drugs include cyclosporine, methotrexate, anti-thymocyte globulin (ATG) and prednisone. More recently utilized suppressors of donor T-cell activity

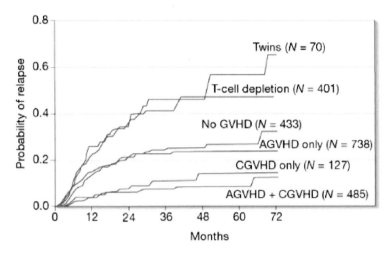

**Figure 5. Relapse rates following allogeneic and syngeneic marrow transplantation.** Allogeneic T cells in marrow grafts mediate both GVHD and graft versus leukemia effect. Relapse rates are least in patients who develop both acute and chronic graft-versus-host disease (AGVHD + CGVHD); higher in those who develop no clinically evident GVHD; and higher still if T cells are depleted from the marrow graft or in recipients of identical twin transplants.

include mycophenolate mofetil and tacrolimus. Treatment of GVHD flares often requires increasing immunosuppression, leading to increased risk of morbidity and mortality from bacterial and fungal infections.[77] Depletion of T cells from hematopoietic grafts significantly reduces GVHD; however, this depletion also results in significantly increased incidences of graft failure,[78] as well as loss of the graft-versus-tumor activity. The GVHD and Graft vs Tumor effect is a "double edged sword," in that patients with the lowest relapse rates are those who develop both acute and chronic GVHD, and disease relapse rates are higher in those who develop no clinically evident GVHD, and higher still if T cells are depleted from the marrow graft or in recipients of identical twin transplants[79] (see Fig. 5).

## Nonmyeloablative allogeneic transplantation

The demonstration of the beneficial effects of immune mediated graft versus tumor effect has led to the concept of using relatively less toxic preparative regimens for use in allogeneic recipients with relatively indolent disease or perhaps most importantly in patients with genetic or autoimmune conditions. This approach, variably termed "non-myelablative," "low intensity," "mini-transplantation" or "mixed chimeric" transplantation now contributes up to 30% of allogeneic transplants performed, and was first pioneered in experimental canine[80] and murine systems and rapidly translated to the clinic in the mid-1990s.[81] A non-myeloablative approach enables most recipients to undergo transplantation from a matched related or unrelated donor as an outpatient, as recipient marrow and blood count suppression is generally no greater than with standard dose chemotherapy, and thus this approach has broadened the indications of allotransplantation to older, and more medically frail patients.[82] Following engraftment, the recipient will have a degree of donor chimerism, in that their hematolymphoid system will be made up from both donor and host elements. Conversion to complete donor engraftment can be facilitated, most commonly through unrelated donor leukocyte infusions (DLI), which tends to convert the recipient to 100% donor chimerism via GVH mechanisms as well as mediate a GVT effect.[83] The graft versus tumor effect in non-myeloablative transplantation can take months to take effect, which generally limits this approach to slower growing malignancies. For patients with genetic diseases, such as sickle cell or beta thalassemia, where only partial donor chimerism can effect a cure of the underlying disease, a non-myeloablative approach may be ideal, in that this approach avoids much of the significant morbidity and associated mortality with a fully myeloablative conditioning preparative regimen. GVHD, however, remains a significant problem in patients who have undergone non-myelablative preparative regimens, with serious side effects approaching 25% of patients in one large series.[84] T-cell depletion of the donor graft decreases GVHD and is most appropriate for transplants treating genetic disease which do not require a graft versus tumor effect to contribute to a cure.

## Sources of HSCs

There are three primary sources of hematopoietic stem cells for clinical use: bone marrow, mobilized peripheral blood, and more recently, umbilical cord blood.

**Bone marrow:** Marrow is the traditional source of HSC for both allogeneic and autologous transplantation. Bone marrow is harvested by large bore needle aspiration from the donor's posterior ileac crests in the operating room, the donor being under general anesthesia. The harvest process generally requires 100–200 separate small volume marrow aspirates and proceeds until 1–2 liters of marrow containing a cell dose of more than $2 \times 10^8$ nucleated cells per kilogram of recipient body weight has been obtained. The marrow is filtered to remove any bone spicules, and can then given to the recipient via an intravenous infusion.

**Mobilized peripheral blood:** HSC normally circulate in the peripheral blood at low levels. HSC can be "mobilized" from the marrow following the administration of colony-stimulating factors and/or chemotherapy. Several colony stimulating factors are effective mobilizing agents, including G-CSF, GM-CSF, IL-3 and thrombopoietin. Most commonly, mobilization involves twice-daily subcutaneous administration of G-CSF at 10ug/kg/day, followed by apheresis on the 5th day.[85,86] CD34+ cells in the apheresis product are counted to ensure that the autologous or allogeneic recipient receives a minimal dose predicted to safely mediate engraftment. Apheresis sessions are repeated on subsequent days as needed until targeted numbers of CD34+ cells are obtained (typical goal of $5 \times 10^6$ CD34+ cells/kg with the minimum acceptable number generally being $2 \times 10^6$/kg for allogeneic recipients). Mobilized peripheral blood harvests are now used in the majority (>80%) of autologous transplants and in recent years for increasing numbers of allogeneic transplants. Newer mobilization agents are undergoing clinical trials, such as CXCR4 chemokine receptor blockade, which, by preventing the stimulation of HSC by stromal derived factor-1 (SDF-1) on marrow stromal cells, results in the immediate release of HSC into the bloodstream.[87]

As mobilized peripiheral blood (MPB) as a donor source has become increasingly utilized in allogeneic transplants, the rate of acute and chronic GVHD has been examined, particularly due to the 10-fold excess of donor T-cells in an unmanipulated MPB donor graft.[88] Retrospective trials indicate increased rates of chronic GVHD from MPB, and prospective clinical trials are currently underway to better compare outcomes between MPB and BM as allogeneic donor sources.

Trials comparing GCSF-stimulated bone marrow with GCSF-mobilized peripheral blood have shown that each source resulted in similar engraftment times, but patients receiving MPB have developed more steroid-resistant GVHD.[89] Additional preliminary studies in haploidentical transplants from family members suggested that GCSF stimulated allogeneic bone marrow may confer early engraftment similar to mobilized peripheral blood but possibly without the increased risk of GVHD.[90]

**Umbilical cord blood:** Cord blood, like mobilized peripheral blood contains a high frequency of HSC and hematopoietic progenitor cells[91] and has become an appealing alternative source of HSC for patients undergoing transplant for a wide variety of indications.[92] The first transplant of human umbilical cord blood (UCB) was successfully carried out in 1988, for a 5-year-old boy with Fanconi's anemia.[93] The success of this procedure led to the rapid development of UCB banks which collect and cryopreserve cord blood units for potential future use in HSC transplantation. Several private enterprises have been developed to offer this banking service for healthy newborns, as a sort of "insurance policy" for the donor or potentially other siblings into the future. Limitations of cord blood transplantation primarily relate to cell dose (CD34$^+$ cells/kg) in a UCB unit and available to be infused into the recipient, a serious factor especially in larger recipients. Relatively low total cell dosage translates to prolonged periods while awaiting engraftment, with concomitant increases in morbidity and mortality. Recently, approaches to the problem have included transplantation of two unrelated UCB units to the same recipient. While the speed of engraftment is generally slower than bone marrow and peripheral blood transplants, this is counterbalanced by the rapid availability of UCB if urgent transplant is required, and the lower incidence of severe chronic graft versus host, even with one to two HLA mismatches.[94]

## Enrichment and graft engineering technologies

With the further characterization of human HSC, technologies for graft engineering have been developed.[95] CD34 has primarily been used as a means to enrich for HSC, which can be positively selected by immunomagnetic beads (in some cases followed by high speed FACS sorting of highly enriched CD34$^+$, CD90$^+$ HSC), and in so doing reduce any contaminating tumor cells within autologous grafts.[96] In allogeneic grafts, HSC enrichment decreases the numbers of T cells in the transplant product (and subsequently rates of GVHD). Negative selection and purging of malignant cells contained in autografts has also been pursued in attempts to decrease the incidence of post-transplant relapse. Figure 6 summarizes current and proposed approaches to engineering the HSC containing transplant graft.[92] This includes for the autologous graft: gene transduction for the treatment of genetic or infectious disease; purging the graft for the removal of residual malignant cells; and the isolation, expansion and addition of primed anti-tumor cell populations. For allogeneic HCT grafts, proposed graft engineering includes expansion of HSC when there is an inadequate cell dose, gene therapy to introduce particular transgenes, removal of donor T-cells to ameliorate GVHD, and addition of facilitating cells to enhance engraftment or induce tolerance.

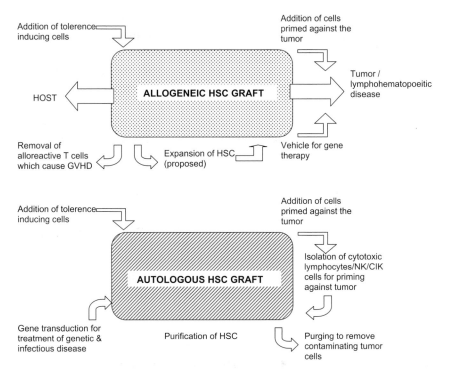

**Figure 6. Engineering the HSC containing graft:** (*top*) Examples of proposed modifications to the allogeneic graft to effect increased graft versus tumor effect, and decreased GVHD (*Bottom*). Examples of autologous HSC graft engineering include gene transduction, purging of contaminating malignancy, and expansion or addition of autologous cytotoxic populations within the graft.

## Rationale and clinical experience with purified HSCs

As previously emphasized, many studies of unpurified bone marrow and MPB cells refer to themselves as "Hematopoietic Stem Cell transplants," when in fact, the transplant graft itself contains a minority fraction of HSC, the majority consisting of a large fraction of hematopoietic progenitors and mature cells (i.e. T-cells). Of note in the setting of autografts for malignancies such as breast cancer and lymphomas, a majority fraction of hematopoietic transplant grafts contain detectable numbers of malignant cells.

Post-transplant relapse from remaining minimal residual disease remains a significant challenge and is the largest contributor to post-transplant mortality. Relapse following HCT may be secondary to persistence of residual disease, the infusion of tumor cells within the autograft, or a combination of the two. The benefits incurred from the graft versus leukemia effect in allogeneic transplantation is not present in autotransplantation. The increasing utilization of autologous peripheral blood and bone marrow transplantation for the treatment of malignancies raises the concern that re-infused tumor cells may be responsible for many post-transplant relapses.

In breast cancer patients receiving autologous grafts, the proportion of grafts with evidence of breast cancer cells is as high as 86% in some studies as detected by immunomagnetic tumor cell enrichment.[97] The clinical significance of tumor contamination in autografts[98] has also recently been shown to play a role in relapse following autologous transplant multiple myeloma.[99] The presence of gene-marked, infused tumor cells at sites of disease relapse has been demonstrated in neuroblastoma, CML and AML.[100,101,102] Such findings lead to the concept of utilizing highly purified HSC, free from contaminating tumor cells, to prevent the infusion of tumor cells, and thereby improve the incidence of disease-free survival post autotransplantation.

Three pilot clinical trials have been conducted to date utilizing highly purified autologous HSC in the setting of malignancy (breast cancer, multiple myeloma, and lymphoma), following the development of high-speed flow cytometer technology to enable the rapid isolation of transplantable human HSC from peripheral blood or bone marrow.[103] In a phase I/II study of 22 women with advanced, metastatic, stage-IV breast cancer patients performed at Stanford,[104] autologous mobilized peripheral blood was collected by apheresis and highly purified CD34$^+$ Thy-1$^+$ HSC were obtained through high-speed FACS cell sorting. Tumor cell depletion below the limits of detection of a sensitive immunofluorescence-based assay was accomplished in all patients who had detectable tumor cells in their apheresis products before processing. Compared with historic controls who received traditional, non-purged peripheral blood transplants, these patients have done extremely well, with freedom from disease progression (FFP) of 40% at 32 months post transplant. This compares favorably with contemporary transplants of analogous stage-IV breast cancer patients with unpurified MPB grafts who had only a 6% FFP at 32 months.[105]

In a related trial, 26 patients with relapsed indolent non-Hodgkins lymphoma (NHL) received purified HSC.[106] The 3-year event-free survival was 55%, and overall survival was 78%. Delays in T-cell reconstitution may have resulted in a higher rate of post-transplant infectious complications.

In each of these studies, the minimum dose of purified CD34$^+$Thy$^+$ HSC to achieve patient engraftment with an absolute neutrophil count $>500$ was

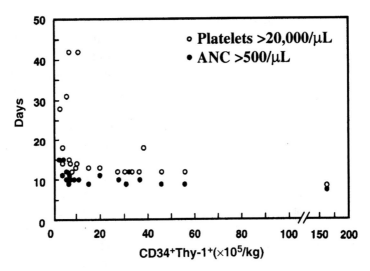

**Figure 7. Purified HSC engraftment:** Time to platelet and neutrophil engraftment in patients with metastatic breast cancer that underwent myeloablative transplantation and rescue with purified, autologous HSC. Each dot represents for one patient days to neutrophil engraftment (*solid circles*), and platelet engraftment (*open circles*), based on the number of CD34⁺Thy-1⁺ cells infused. (Vose *et al.*, 2001.)

$2$–$4 \times 10^5$ HSC/kg (10 fold less than the $2$–$5 \times 10^6$ CD34$^+$ MPB cells used in standard autologous transplants); the dose to achieve a rapid neutrophil engraftment (10–11 days) and rapid platelet engraftment ($>20\,000$ per ul) was $8 \times 10^5$ HSC per kg (see Fig. 7). These results support the approach, thanks to the ability to isolate sufficient numbers of CD34$^+$ Thy$^+$ HSC to enable rapid engraftment while purged of contaminating cancer cells, and T-cells. It would be reasonable in the future to conduct larger, prospective trials of autografts with purified HSC in a two-arm prospectively randomized study to fully address event-free and overall survival.

## Expanding Applications for Hematopoietic Cell Transplantation

As HCT technology and clinical experience and supportive care has improved, and non-myeloablative approaches have been developed, mortality and morbidity from the transplant procedure itself has decreased, and has led to the use of HCT for an expanding number of non-malignant indications.

**Autoimmune disease:** Autoimmune diseases are a heterogeneous group of disorders with genetic and environmental etiologies which occur when there is a breakdown in the immune systems tolerance to normal tissues. The most common autoimmune diseases are rheumatoid arthritis, systemic lupus erythemetosis (SLE) and type I diabetes mellitus. Pathophysiology generally involves persistent manifestations of chronic tissue damage, as cells of the immune system recognize self-antigens on target tissues as foreign. Early experiments demonstrated that bone marrow transplantation could alter the course of autoimmune diseases in animal models.[107] BMT was shown to transfer disease from autoimmune prone animals to unaffected ones, but also to prevent disease if hematopoietic cells were transplanted from unaffected rodents to susceptible ones.[108] Purified HSC from diabetes-resistant donors transplanted to just-diabetic NOD mice have been shown to permanently prevent progression to autoimmune diabetes without GVHD, whereas syngeneic BM or HSC barely delays disease progression.[109] Subsequently, it was observed that in patients receiving HCT for conventional indications, their coexisting autoimmune diseases also improved or were cured.[110] Through international collaborative trials, over 600 patients refractory to standard therapies patients and at high-risk for subsequent mortality have been treated with high dose chemotherapy followed by autologous HCT with their autoimmune disease as the primary transplant indication. The majority of these patients underwent autologous transplantation for severe/refractory multiple sclerosis (MS), systemic sclerosis (scleroderma), rheumatoid arthritis or SLE. Phase I/II trials have demonstrated lasting remissions in some patients with severe autoimmune disorders and are currently moving to randomized trials. Allogeneic transplants have generally been avoided given the risk morbidity of GVHD; however less toxic, non-myeloablative approaches, possibly utilizing purified or HSC or T-cell depleted hematopoietic grafts are now being evaluated.[111]

**Allogeneic hematopoietic cell transplantation for inherited diseases:** Transplantation for sickle cell disease (SCD) has been primarily performed in pediatric patients with clinical features suggestive of poor outcome (e.g. frequent vaso-occlusive episodes) and significant sickle-related morbidity.[112] In multicenter trials of 59 children transplanted with HLA-identical sibling transplantation, the probabilities of survival and disease-free survival were 93% and 84%.[113] Newer approaches to treat SCD with non-myeloablative preparative regimens leading to donor chimerism (as even partial production of normal hemoglobin can significantly effect the clinical course) have been attempted with variable success, complicated by some episodes of graft rejection.[114] Animal model experimentation continues,[115] as does a multi-center collaborative clinical trial, to define an effective non-myeloablative regimen for children with symptomatic SCD. Allogeneic hematopoietic cell transplantation from healthy sibling or HLA-identical donors for otherwise healthy beta-thalassemia major patients (without significant organ damage) is considered the treatment of choice and is the only therapeutic modality for curing this disease, preventing the long-term morbid effects of iron-overload in chronically transfused patients. and has a very high probability of cure with very low morbidity and mortality.[116]

**HSC transplant and gene therapy of HSCs for immunodeficiency:** Severe combined immunodeficiency syndrome (SCID) and other congenital immunodeficiencies are a heterogeneous group of lethal, congenital disorders which result in the inability of T and B lymphocytes to mount an antigen specific response.[117] The first successful human allogeneic HCT was performed in an infant with SCID from an HLA-matched sibling over 3 decades ago, and the recipient remains alive and well. Bone marrow from HLA-matched siblings remains the treatment of choice, and GVHD prophylaxis is often not required possibly to the young age of the donor/recipient pairs and decreased requirement for cytoreductive conditioning.[118] In a small number of cases successful *in utero* transplants have been performed using paternal bone marrow or CD34$^+$ selected cells.[119] More recently, gene therapy trials utilizing genetically modified autologous lymphocytes or HSC in an effort to correct the defect without the need for cytoreduction or risk of GVHD have been performed, with ADA deficiency being the first genetic disease to be treated by gene therapy.[120] In 2000, the first successful correction of X-linked SCID with autologous, gamma-chain transduced CD34$^+$ hematopoietic cells as the sole therapy was performed on several infants, with resulting normalizing of CD3$^+$ cells in the peripheral blood. Nine of eleven recipients had successful reconstitution of normal T and B cell function; however, unfortunately three of the eleven children treated developed a T-cell leukemia when the transgene inserted into an oncogene on chromosome 11.[121] Safer vector technology portends significant benefit for gene therapy of hematopoietic cells for the treatment of ADA deficiency and other immunodeficiency disorders.

## Clinical utility of hematopoietic progenitors

In mouse models, the downstream progenitors of the HSC have been identified. These include the common myeloid progenitor (CMP),[122] granulocyte macrophage progenitor (GMP), the megakaryocyte erythrocyte progenitor (MEP), and the common lymphoid progenitor (CLP), and as discussed above, analogous populations have been identified in the human system.

Clinical applications of hematopoietic progenitor populations have reached proof of concept in animal models and may soon translate to clinical use. Transplantation with purified syngeneic or allogeneic CLP has led to protection from otherwise lethal infection with CMV in murine models.[123] Transplantation with purified syngeneic or fully allogeneic CMP leads to rapid myeloid engraftment and protection from otherwise lethal aspergillosis infection[124] (a common cause of mortality in allogeneic transplant recipients) as well as resistance to pseudomonal infection.[125]

Human CMP have been identified.[126] Clinical trials are planned utilizing allogeneic CMP (expanded *in vitro*) to rapidly generate an effective neutrophil population in otherwise neutropenic patients following chemotherapy or HCT.

# Future Directions for HSC Transplantation

**Expansion of HSC:** There exists a need, and desire to expand HSC; for example, allogeneic cord blood transplants are often not possible due to the limited total CD34$^+$ dose within a particular cord unit, and therefore larger recipients are not candidates or will engraft very slowly. Higher LT-HSC doses correlate with faster engraftment and better long-term survival in both autologous and allogeneic settings.[127] Expansion of LT-HSC should lead to more rapid immune reconstitution, and in the autologous setting, enables expansion free of malignant cells.[128]

Attempts to expand human HSC in culture have primarily focused upon attempts to expand cord blood hematopoietic stem cells, through genetic modifications (such as through the addition of genes to enhance self-renewal (Wnt,[129] beta-catenin), and additionally through three-dimensional stromal cultures which attempt to recapitulate the marrow environment, or with various supportive endothelial populations,[130] and cytokines.[131,132] Most work to date reveals expansion of downstream myelo-erythroid cells, with however, little to no clear expansion of true HSC. Challenges to expansion of HSC include the optimization of culture conditions; identification of the optimal human populations to expand; identification and expansion of human, long-term repopulating cells and concerns that expansion may lead to early senescence or deceased functional capacity.[133] Recently, preclinical studies have suggested novel approaches that may lead to significant *in vivo* engraftment. These include the use of the homeobox gene, Hox-B4, which, when expressed following gene transfer, has led to increased numbers of HSC and ability to engraft adult irradiated hosts.[134]

**HSC transplantation to mediate tolerance for organ allografts:** Use of purified HSC in allotransplantation can result in full donor derived reconstitution without GVHD. Furthermore, animals previously reconstituted with mismatched, allogeneic donor marrow or HSC have been shown to develop lifelong tolerance to completely mismatched donor heart allografts (e.g. donor C57Black hearts into Balb/c recipients).[135] Co-transplantation of other tissues or organs from the same HSC donor should be possible, and permit long-term tolerance induction to replacement cells or organs. This approach has been advanced in murine and primate studies, demonstrating that it is feasible to induce transplantation tolerance in chimeric and nonchimeric hosts using sublethal total lymphoid irradiation (TLI) as a pretransplantation conditioning regimen.[136] Application of this concept has recently led to human trials, in which patients receiving living related kidney donor allografts have subsequently undergone non-myelablative transplantation with marrow or MPB from the same kidney donor. Several recipients have stable or transient chimerism, and as evidence of improved tolerance to the renal allograft, have been able to discontinue immunosuppression, at least transiently.[137]

**HSCs as vehicles for gene therapy:** The HSC is an attractive target for gene therapy, as many genetic diseases are the result of single gene defects affecting the mature hematopoietic cell progeny of affected HSC. Examples include inherited blood disorders such as the various anemias (including beta-thalassemia, Diamond-Blackfan syndrome, globoid cell leukodystrophy, sickle-cell anemia, severe combined immunodeficiency, X-linked lymphoproliferative syndrome, and Wiskott-Aldrich syndrome), inborn errors of metabolism (examples include Hunter's syndrome, Hurler's syndrome, Lesch Nyhan syndrome, and osteopetrosis); in addition to the transduction of HSC to carry genes to supply missing proteins from other tissues. As mentioned above, clinical trials involving transduction of autologous CD34$^+$ progenitors for the treatment of X-linked SCID have demonstrated successful reconstitution of normal lymphocyte function; however this was complicated in some patients by insertional mutagenesis leading to T-cell leukemia.

Genetic modification of hematopoietic stem cells with genes that inhibit replication of human immunodeficiency virus (HIV) could lead to development of T lymphocytes and monocytic cells resistant to HIV infection after transplantation (thus a form of "intracellular vaccination"). Transducing autologous HSC[138] in AIDS patient to carry protective genes such as anti-sense to HIV-*gag*, RevM10 and anti HIV-*tat*,[139] has been pursued in early clinical trials.[140]

Additional potential therapeutic strategies utilizing HSC gene therapy might be aimed at more effective treatment of malignancy, including targeting HSC and their progeny with genes for regulated drug chemotherapy and radiotherapy resistance (such as MDR-1 and BCL-2),[141] to enable temporal resistance to therapies that destroy tumors at the cost of lymphohematopoietic failure. An additional example includes HSC-derived immunotherapy for malignancy, which may enable the generation of targeted graft versus tumor effect without GVHD. In this approach, a fraction of autologous or allogeneic donor HSC or common lymphocyte progenitors would be transduced with alpha-beta T-cell receptor (TCR) specific to known tumor associated antigens (such as Her2Neu in breast cancer, WT-1 or PR-1 in many leukemias) presented in the context of their HLA Class I or II, such that large numbers of antigens specific TCR transgenic T-cells would be developed from TCR transduced HSC or progenitors, and generate a robust and targeted graft versus tumor effect.

## Other HSC mediated regenerative therapies, and the question of HSC plasticity

A dogma of hematopoietic stem cell biology had been that HSC are restricted to the hematolymphoid lineage, with multipotent, self-renewing HSC at the top of the pyramid, committed progenitor cells in the middle, and lineage-restricted mature blood cells at the bottom. Recent years have seen a great deal of debate regarding the "plasticity" of adult HSC, and their ability to differentiate into a variety of non-hematopoietic tissues, with several investigators claiming that HSC are pluripotent, with the ability to "transdifferentiate" into a variety of tissues (review[142,143]), including to tissues of the central nervous system,[144] cardiomyocyte,[145] skeletal muscle,[146] hepatic,[147] pancreas[148] and other organ systems. Transdifferentiation refers to the change of a cell of one tissue lineage into a cell of an entirely distinct lineage, with loss of specific markers and function of the original, and acquisition of markers and function of the transdifferentiated cell type. This "plasticity" of the hematopoietic system, usually described following transfer of marrow-derived stem cells into the injured recipient mice, has led to clinical trials. These trials could be considered premature given the lack of basic understanding of the mechanisms and validity of animal experimentation. Recent clinical trials using unselected or HSC- enriched bone marrow populations have described functional improvement in cardiac function following myocardial infarction and congestive heart failure; however, the mechanisms of improvement remain unclear, with hypothesized etiologies ranging from transdifferentiation or fusion to a paracrine effect from multiple factors which may be secreted by autologous marrow derived cells when localized to the infarct or peri-infarct region of the heart.[149]

Controversial claims of adult stem cell mediated tissue regeneration have since been reevaluated in light of subsequent work, demonstrating that much of what appeared to be transdifferentiation was the results of cell-cell fusion.[150] For example, in a mouse model of hereditary tyrosinemia, the ability of BM cells to regenerate functional hepatocytes was demonstrated to be fusion with hepatocytes in the livers of the recipient mice.[151] Fusion occurs, however, at low frequency, and this implies that such rare events are unlikely to significantly contribute to normal tissue regeneration. Analysis of parabiotic animals sharing a common blood vasculature via tissue anastomoses for up to nine months, in which all the cells of one partner were marked with green fluorescent protein (GFP), while the other partner was unmarked, has revealed no evidence of homeostatic cell replacement from circulating cells in nonhematopoietic tissues, either via cell fusion or transdifferentation, despite significant HSC cross engraftment in these animal pairs.[152] Potential mechanisms and explanations for the observations of adult stem cell plasticity are summarized in Fig. 8. These include transdifferentiation, de-differentiation, heterogenous stem cell sources, pluripotent stem cells, and cell fusion.

In many reports claiming HSC plasticity, rigorously defined, purified true stem cell populations were not administered.[153] In most studies to date, various preparations from hematopoietic tissue have been utilized in *in vivo* studies of stem-cell plasticity. This includes unselected cells from bone marrow and peripheral blood which contain a heterogeneous population of stem cells, which can include HSC, mesenchymal stem cells, multipotent adult progenitior cells (MAPC) and endothelilal precursor cells.

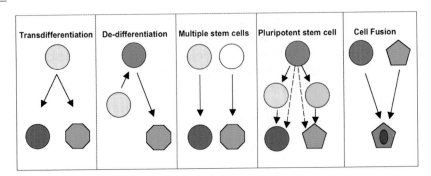

**Figure 8. Adult stem cell plasticity: Potential mechanisms and explanations for observations.** Tissue specific stem cells are represented in yellow or green, pluripotent stem cells in blue, and differentiated cells from "yellow" lineage by red ovals, and of the green lineage by green hexagons. In the first model (Panel A), pluripotent stem cells have the ability to directly differentiate into cell types of different lineages; (Panel B), mature cells dedifferentiate into cells with stem-cell like characteristics and eventually redifferentiate into terminally differentiated cells of a different tissue. In the third model (Panel C), distinct stem cells differentiate, each into its own organ-specific cell; in the fourth (Panel D), pluripotent stem cells differentiate directly, or via multipotent stem cells into different tissue lineages; and finally (Panel E), stem or mature cells from distinct lineages undergo cell fusion (adapted from Wagers and Weissman, 2004).

In order to rigorously prove that HSC transform or differentiate into non-hematopoietic tissues, this should be documented by cell marking, preferably at the prospectively isolated, single-cell level, and the resulting cells must be shown to have become integral to the acquired tissue and have acquired the function of the particular organ, both by expressing tissue specific proteins and showing organ specific function. Such work utilizing c-kit-enriched BM cells, Lin$^-$ c-Kit$^+$ BM cells and c-Kit$^+$ Thy1.1$^{lo}$ Lin$^-$ Sca-1$^+$ long-term reconstituting haematopoietic stem cells failed to regenerate myocardium in a murine infarct model.[154]

Future studies in the preclinical setting, and randomized, double blinded clinical trials should ideally utilize rigorously defined, homogenous populations of hematopoietic or other adult-derived stem cell populations to define mechanisms, the most favorable intervention timing, delivery mechanism and optimal cell populations which may confer a significant clinical benefit.

## REFERENCES

1. Weissman IL. (2000) Translating stem and progenitor cell biology to the clinic: Barriers and opportunities. *Science* **287**(5457):1442–1446.

2. Kondo M, Wagers AJ, Manz MG, *et al.* (2003) Biology of hematopoietic stem cells and progenitors: Implications for clinical application. *Annu Rev Immunol* **21**:759–806. (Review.)

3. Ford CE, Hamerton JL, Barnes DH, Loutit JF. (1956) Cytological identification of radiation-chimaeras. *Nature* **177**:452–454.

4. Till JE, McCulloch EA. (1961) A direct meaurement of the radiation sensitivity of normal mouse bone marrow cells. *Radiat Res* **14**:1419–1430.

5. Becker A, McCulloch E, Till J. (1963) Cytologic demonstration of the clonal nature of spleen colonies derived from transplanted mouse marrow cells. *Nature* **197**:452–454.

6. Wu AM, Sikminovitch L, Till JE, McCulloch EA. (1968) Evidence for a relationship between mouse hematopoietic stem cells and cells forming colonies in culture. *Proc Natl Acad Sci* **59**:1209–1215.

7. Muller-Sieberg CF, Whitlock CA, Weissman IL. (1986) Isolation of two early B lymphocyte progenitors from mouse marrow: A committed pre-pre B cell and a clongenic Thy-1-lo hematopoietic stem cell. *Cell* **44**:653–662.

8. Spangrude GH, Heimfeld S, Weissman IL. (1988) Purification and characterization of mouse hematopoietic stem cells. *Science* **241**:58–62.

9. Ikuta K, Weissman IL. (1992) Evidence that hematopoietic stem cells express mouse c-kit, but do not depend on steel factor for their generation. *Proc Natl Acad Sci* **89**:1502–1506.

10. Smith LG, Weissman IL, Heimfeld S. (1991) Clonal analysis of hematopoietic stem-cell differentiation *in vivo*. *Proc Natl Acad Sci* **88**:2788–2792.

11. Osawa M, Hanada K, Handa H, Nakauchi H. (1996) Long-term lympho-hematopoietic reconstitution by a single CD34-low/negative hematopoietic stem cell. *Science* **273**:242–245.

12. Scharenberg CW, Harkey MA, Torok-Storb B. (2002) The ABCG2 transporter is an efficient Hoechst 33342 efflux pump and is preferentially expressed by immature human hematopoietic progenitors. *Blood* **99**(2): 507–512.

13. Morrison SJ, Wandycz AM, Hemmati HD, *et al.* (1997) Identification of a lineage of multipotent hematopoietic progenitors. *Development* **124**(10): 1929–1939.

14. Matthews W, Jordan CT, Wiegand GW, *et al.* (1991) A receptor tyrosine kinase specific to hematopoietic stem and progenitor cell-enriched populations. *Cell* **65**(7):1143–1152.

15. Christensen JL, Weissman IL. (1999) Flk-2 is a marker in hematopoietic stem cell differentiation: A simple method to isolate long-term stem cells. *Proc Natl Acad Sci* **96**:3120–3125.

16. Kondo M, Wagers AJ, Manz MG, *et al.* (2003) Biology of hematopoietic stem cells and progenitors: Implications for clinical application. *Annu Rev Immunol* **21**:759–806.

17. Morrison SJ, Wandycz AM, Hemmati HD, *et al.* (1997) Identification of a lineage of multipotent hematopoietic progenitors. *Development* **124**:1929–1939.

18. Uchida N, Weissman IL. (1992) Searching for hematopoietic stem cells: Evidence that Thy-1.1$^{lo}$ Lin$^-$ Sca$^-$1$^+$ cells are the only stem cells in C57BL/Ka-Thy-1.1 bone marrow. *J Exp Med* **175**:175–184.

19. Weissman IL, Anderson DJ, Gage F. (2001). Stem and progenitor cells: Origins, phenotypes, lineage commitments, and transdifferentiations. *Annu Rev Cell Dev Biol* **17**:387–403. (Review.)

20. Shizuru JA, Jerabek L, Edwards CT, Weissman IL. (1996) Transplantation of purified hematopoietic stem cells: Requirements for overcoming the barriers of allogeneic engraftment. *Biol Blood Marrow Transplant* **2**:3–14.

21. Metcalf D. (1989) The molecular control of cell division, differentiation, commitment and maturation in haemopoietic cells. *Nature* **339**:27–30.

22. Kondo M, Weissman IL, Akashi K. (1997) Identification of clonogenic common lymphoid progenitors in mouse bone marrow. *Cell* **91**:661–672.

23. Traver D, Akashi K, Manz M, *et al.* (2000) Development of CD8a-positive dendritic cells from a common myeloid progenitor. *Science* **290**:2152–2154.

24. Akashi K, Traver D, Miyamoto T, Weisman IL. (2000) A clonogenic common myeloid progenitor that give rise to all myeloid lineages. *Nature* **404**: 193–197.

25. Na Nakorn T, Miyamoto T, Weissman IL. (2003) Characterization of mouse clonogenic megakaryocyte progenitors. *Proc Natl Acad Sci* **100**:205–210.

26. Weilbaecher K, Weissman I, Blume K, Hemsfeld S. (1991) Culture of phenotypically defined hematopoietic stem cells and other progenitors at limiting dilution on Dexter monolayers. *Blood* **78**:945–952.

27. Hao QL, Thiemann FT, Petersen D, *et al.* (1996) Extended long-term culture reveals a highly quiescent and primitive human hematopoietic progenitor population. *Blood* **88**:3306–3313.

28. Bosma MJ, Carroll AM. (1991)The SCID mouse mutant: Definition, characterization, and potential uses. *Annu Rev Immunol* **9**:323–350.

29. Kemale-Reis S, Dick JE. (1988) Engrafment of immune-deficient mice with human hematopoietic stem cells. *Science* **242**:1706–1709.

30. Bhatia M, Wang JC, Kapp U, *et al.* (1997) Purification of primitive human hematopoietic cells capable of repopulating immune-deficient mice. *Proc Natl Acad Sci* **94**:5320–5325.

31. Conneally E, Cashman J, Petzer A, Eave C. (1997) Expansion *in vitro* of transplantable human cord blood stem cells demonstrated using a quantitative assay of their lympo-myeloid repopulating activity in nonobese diabetic sc/scid mice. *Proc Natl Acad Sci* **94**:9836–9841.

32. McCune JM, Namikawa R, Kaneshiima H, *et al.* (1988) The SCID-hu mouse: Murine model for the analysis of human hematolymphoid differentiation and function. *Science* **241**:1632–1639.

33. Kraft DL, Weissman IL, Waller EK. (1993) Differentiation of CD3-4-8-human fetal thymocytes *in vivo*: Characterization of a CD3-4+8- intermediate. *J Exp Med* **178**(1):265–277.

34. Su L, Kaneshima H, Bonyhadi M, *et al.* (1995) HIV-1-induced thymocyte depletion is associated with indirect cytopathogenicity and infection of progenitor cells *in vivo*. *Immunity* **2**(1):25–36.

35. Meissner EG, Duus KM, Loomis R, *et al.* (2003) HIV-1 replication and pathogenesis in the human thymus. *Curr HIV Res* **1**(3):275–285. (Review.)

36. Goldman JP, Blundell MP, Lopes L, *et al.* (1998) Enhanced human cell engraftment in mice deficient in RAG2 and the common cytokine receptor chain. *Br J Haematol* **103**(2):335.

37. Traggiai E, Chicha L, Mazzucchelli L, *et al.* (2004) Development of a human adaptive immune system in cord blood cell-transplanted mice. *Science* **304**(5667):104–107.

38. Flake AW, Harrison MR, Adzick NS, Zanjani ED. (1986) Transplantation of fetal hematopoietic stem cells in utero: The creation of hematopoietic chimeras. *Science* **233**:776–778.

39. Zanjani ED, Flake AW, Rice H, *et al.* Long-term repopulating ability of xenogeneic transplanted human fetal liver hematopoietic stem cells in sheep. *J Clin Invest* **93**:1051–1055.

40. Civin CI, Strauss LC, Brovall C, *et al.* (1984) Antigenic analysis of hematopoiesis. A hematopoietic progenitor cell surface antigen defined by a monoclonal antibody raised against KG-1a cells. *J Immunol* **133**:157–165.

41. Krause DS, Fackler MJ, Civin CI, May WS. (1996) CD34 structure, biology, and clinical utility. *Blood* **87**:1–13.

42. Fina L, Molgaard HV, Roberson D, *et al.* (1990) Expression of the CD34 gene in vascular endothelial cells. *Blood* **75**:2417–2426.

43. Healy L, May G, Gale K, *et al.* (1995) The stem cell antigen CD34 functions as a regulator of hematopoietic cell adhesion. *Proc Natl Acad Sci* **92**: 12240–12244.

44. Krause DS, Ito T, Fackler MJ, *et al.* Characterization of murine CD34, a marker for hematopoietic progenitor and stem cells. *Blood* **84**(3):691–701.

45. Baum CM, Weissman IL, Tsukamoto AS, *et al.* (1992) Isolation of a candidate human hematopoietic stem-cell population. *Proc Natl Acad Sci* **89**:2804–2808.

46. Larochell A, Vormoor J, Hanenberg H, *et al.* (1996) Identification of primitive human hematopoietic cells capable of repopulating NOD/SCID mouse bone marrow: Implications for gene therapy. *Nat Med* **2**:1329–1343.

47. Guenechea G, Gan OI, Dorrell C, Dick JE. (2001) Distinct classed of human stem cells that differ in proliferative and self-renewal potential. *Nat Immunol* **2**:75–82.

48. Yin AH, Miraglia S, Zanjani Ed, *et al.* (1997) AC133, a novel marker for human hematopoietic stem cell and progenitor cells. *Blood* **90**:5002–5012.

49. Wognum AW, Eaves AC, Thomas TE. (2003) Identification and isolation of hematopoietic stem cells. *Arch Med Res* **34**(6):461–475. (Review.)

50. Gallacher L, Murdoch B, Wu DM, *et al.* (2000) Isolation and characterization of human CD34-Lin- and CD34+Lin- hematopoietic stem cells using cell surface markers AC133 and CD7. *Blood* **95**:2813–2820.

51. Gordon PR, Leimig T, Babarin-Dorner A, *et al.* (2003) Large-scale isolation of CD133+ progenitor cells from G-CSF mobilized peripheral blood stem cells. *Bone Marrow Transplant* **31**(1):17–22.

52. Lang P, Bader P, Schumm M, *et al.* (2004) Transplantation of a combination of CD133+ and CD34+ selected progenitor cells from alternative donors. *Br J Haematol* **124**(1):72–79.

53. Torrente Y, Belicchi M, Sampaolesi M, *et al.* (2004) Human circulating AC133(+) stem cells restore dystrophin expression and ameliorate function in dystrophic skeletal muscle. *J Clin Invest* **114**(2):182–195.

54. Sherwood RI, Christenson JL, Conboy IM, *et al.* (2004) Isolation of adult myogenic progenitors; functional heterogeneity of cells within engrafting skeletal muscles. *Cell* **119**(4): 543–554.

55. Stamm C, Kleine HD, Westphal B, *et al.* (2004) CABG and bone marrow stem cell transplantation after myocardial infarction. *Thorac Cardiovasc Surg* **52**(3):152–158.

56. Balsam LB, Wagers AJ, Christensen JL, *et al.* (2004) Haematopoietic stem cells adopt mature haematopoietic fates in ischaemic myocardium. *Nature* **428**(6983):668–673.

57. Hess DA, Meyerrose TE, Wirthlin L, *et al.* (2004) Functional characterization of highly purified human hematopoietic repopulating cells isolated according to aldehyde dehydrogenase activity. *Blood* **104**(6):1648–1655.

58. Wright DE, Wagers AJ, Gulati AP, *et al.* (2001) Physiological migration of hematopoietic stem and progenitor cells. *Science* **294**(5548):1933–1936.

59. Abkowitz JL, Robinson AE, Kale S, *et al.* (2003) Mobilization of hematopoietic stem cells during homeostasis and after cytokine exposure. *Blood* **102**(4): 1249–1253.

60. Richman CM, Weiner RS, Yankee RA. (1976) Increase in circulating stem cells following chemotherapy in man. *Blood* **45**:1031–1039.

61. Siena S, Bregni M, Brando B, *et al.* (1989) Circulation of CD34+ hematopoietic stem cells in the peripheral blood of high dose cyclophosphamide-treated patients: Enhancemnt by intravenous recombinant human granulocyte-macrophage colony-stimulating factor. *Blood* **74**:1904–1915.

62. Morrison SJ, Wright DE, Weissman IL. (1997) Cyclophosphamide/ granulocyte colony stimulating factor induces hematopoietic stem cells to proliferate prior to moblization. *Proc Natl Acad Sci* **94**:1908–1913.

63. Galy A, Travis M, Cen D, Chen B. (1995) Human T, B, natural killer, and dendritic cells arise from a common bone marrow progenitor cell subset. *Immunity* **3**:459–473.

64. Hao QL, Zhu J, Price MA, *et al.* (2001) Identification of a novel, human multilymphoid progenitor in cord blood. *Blood* **97**:3683–3690.
65. Landsdorp PM, Sutherland HJ, Eaves CJ. (1990) Selective expression of CD45 isoforms on functional subpopulations of CD34+ cells from human bone marrow. *J Exp Med* **172**:363–366.
66. Huang S, Chen Z, Yu JF, *et al.* (1999) Correlation between IL-3 receptor expression and growth potential of human CD34+ hematopoietic cells from different tissues. *Stem Cells* **17**:265–272.
67. Manz MG, Miyamoto T, Akashi K, Weissman IL. (2002) Prospective isolation of human clonogenic common myeloid progenitors. *Proc Natl Acad Sci* **99**: 11872–11877.
68. International Bone Marrow Transplant Registry; 2002 data from the Statistical center of the IBMTR and ABMTR.
69. Applebaum FR, Herzig GP, Ziegler JL, *et al.* (1978) Successful engraftment of cryopreserved autologous bone marrow in patients with malignant lymphoma. *Blood* **52**:85–95.
70. Oudshoorn M, Leeuwen A, van Zanden HGM, vanRood JJ. (1994) Bone marrow donors worldwide a successful exercise in international cooperation. *Bone Marrow Transplant* **14**:3–8.
71. Aversa F, Tabilio A, Velardi A, *et al.* (1998) Treatment of high-risk acute leukemia with T-cell-depleted stem cells from related donors with one fully mismatched HLA haplotype. *N Engl J Med* **339**(17):1186–1193.
72. Statistical Center of the IBMTR and ABMTR, 2004.
73. Antin JH. (1993). Graft-versus-leukemia: No longer an epiphenomenon. *Blood* **82**:2273–2277.
74. Chao NJ. (1997) Graft versus host disease: The viewpoint from the donor T cell. *Biol Blood Marrow Transplant* **3**:1–10.
75. Deeg HJ, Storb R. (1984) Graft versus host disease: Pathophysiological and clinical aspects. *Annu Rev Med* **35**:11–24.
76. Ferrara J, Levy R, Chao N. (1999) Pathophysiologic mechanisms of acute graft vs host disease. *Biol Blood Marrow Transplant* **5**:347–356.
77. Brown JM. (2003) "Fungal Infections after hematopoietic cell transplantation," in *Thomas' Hematopoietic Cell Transplantation*, 3rd ed. Blume: Forman and Appelbaum. pp. 683–397.
78. Martin PF, Hansen FA, Buckner CD, *et al.* (1985) Effects of *in vitro* depletion of T cells in HLA identical allogeneic marrow grafts. *Blood* **66**:664–672.
79. Horowitz MM, *et al.* (1990) Graft-versus-leukemia reactions after bone marrow transplantation. *Blood* **75**:555–562.
80. Maris M, Storb R. (2002) Outpatient allografting in hematologic malignancies and nonmalignant disorders — Applying lessons learned in the canine model to humans. *Cancer Treat Res* **110**:149–75. (Review.)

81. Sykes M, Spitzer TR. (2002) Non-myeloblative induction of mixed hematopoietic chimerism: Application to transplantation tolerance and hematologic malignancies in experimental and clinical studies. *Cancer Treat Res* **110**:79–99. (Review.)
82. Georges GE, Storb R. (2003) Review of "minitransplantation": Nonmyeloablative allogeneic hematopoietic stem cell transplantation. *Int J Hematol* **77**(1):3–14.
83. Dey BR, McAfee S, Colby C, *et al.* (2003) Impact of prophylactic donor leukocyte infusions on mixed chimerism, graft-versus-host disease, and antitumor response in patients with advanced hematologic malignancies treated with nonmyeloablative conditioning and allogeneic bone marrow transplantation. *Biol Blood Marrow Transplant* **9**(5):320–329.
84. Flowers ME, Traina F, Storer B, *et al.* (2005) Serious graft-versus-host disease after hematopoietic cell transplantation following nonmyeloablative conditioning. *Bone Marrow Transplant* **35**(5):535.
85. Socinski MA, Elias A, Schnipper L, *et al.* (1988) GMCSF expands the circulating haematopoietic prgenitor cell compartment in man. *Lancet* **1**:1194.
86. Chao NJ, Schriber JR, Grimes K, *et al.* (1993) GMCSF mobilized peripheral blood progenitor cells accelerate granulocyte and platelet recover after high-dose chemotherapy. *Blood* **81**:2031.
87. Lapidot T, Petit I. (2002) Current understanding of stem cell mobilization: The roles of chemokines, proteolytic enzymes, adhesion molecules, cytokines, and stromal cells. *Exp Hematol* **30**(9):973–981.
88. Flowers ME, Parker PM, Johnston LJ, *et al.* (2002) Comparison of chronic graft-versus-host disease after transplantation of peripheral blood stem cells versus bone marrow in allogeneic recipients: Long-term follow-up of a randomized trial. *Blood* **100**(2):415–419.
89. Morton J, Hutchins C, Durrant S. (2001) Granulocyte-colony-stimulating factor (G-CSF)-primed allogeneic bone marrow: Significantly less graft-versus-host disease and comparable engraftment to G-CSF-mobilized peripheral blood stem cells. *Blood* **98**(12):3186–3191.
90. Ji SQ, Chen HR, Wang HX. (2002) GCSF primed haploidentical marrow transplantation without *ex vivo* T cell depletion: An excellent alternative for high-risk leukemia. *Bone Marrow Transplant* **30**:861–866.
91. Broxmeyer HE, Hangoc G, Cooper S, *et al.* (1992) Growth characteristics and expansion of human umbicard cod blood and estimation of its potential for transplantation in adults. *Proc Natl Acad Sci* **89**:4109–4113.
92. Rocha V, Sanz G, Gluckman E. (2004) Eurocord and European Blood and Marrow Transplant Group. Umbilical cord blood transplantation. *Curr Opin Hematol* **11**(6):375–385.
93. Gluckman E, Broxmeyer HE, Auerbach AD, *et al.* (1989) Hematopoietic reconstitution in a patient with Fanconi's anemia by means of umbilical cord blood from an HLA-identical sibling. *N Engl J Med* **321**:1174–1178.

94. Rocha V, Sanz G, Gluckman E. (2004) Eurocord and European Blood and Marrow Transplant Group. Umbilical cord blood transplantation. *Curr Opin Hematol* **11**(6):375–385.

95. Hwang WY. (2004) Haematopoietic graft engineering. *Ann Acad Med Singapore* **33**(5):551–558. (Review.)

96. Mohr M, Hilgenfeld E, Fietz T, *et al.* (1999) Efficacy and safety of simultaneous immunomagnetic CD34+ cell selection and breast cancer cell purging in peripheral blood progenitor cell samples used for hematopoietic rescue after high-dose therapy. *Clin Cancer Res* **5**(5):1035–1040.

97. Umiel T, Prilutskaya M, Nguyen NH, *et al.* (2000) Breast tumor contamination of peripheral blood stem cell harvests: Increased sensitivity of detection using immunomagnetic enrichment. *J Hematother Stem Cell Res* **9**(6): 895–904.

98. Sharp JG, Kessinger A, Vaughan WP, *et al.* (1992) Detection and clinical signficance fo minimal tumor cell contamination of peripheral stem cell harvests. *Int J Cell Cloning* **10**:92–94.

99. Bakkus MH, Bouko Y, Samson D, *et al.* (2004) Post-transplantation tumour load in bone marrow, as assessed by quantitative ASO-PCR, is a prognostic parameter in multiple myeloma. *Br J Haematol* **126**(5):665–674.

100. Rill DR, Santana VM, Roberts WM, *et al.* (1994) Demonstration that autologous bone marrow transplantation for solid tumors can return a multiplicity of tumorigenic cells. *Blood* **84**(2):380–383.

101. Brenner MK, Rill DR, Moen RC, *et al.* (1994) Gene marking and autologous bone marrow transplantation. *Ann NY Acad Sci* **31**(716):204-227. Review.

102. Deisseroth AB, Zu Z, Claxton D, *et al.* (1994) Genetic marking shows that Ph+ cells present in autologous transplants of chronic myelogenous leukemia (CML) contribute to relapse after autologous bone marrow in CML. *Blood* **83**(10):3068–3076.

103. Sasaki DT, Tichenor EH, Lopez F, *et al.* (1995) Development of a clinically applicable high-speed flow cytometer for the isolation of transplantable human hematopoietic stem cells. *J Hematother* **4**(6):503–514.

104. Negrin RS, Atkinson K, Leemhuis T, *et al.* (2000) Transplantation of highly purified CD34+Thy-1+ hematopoietic stem cells in patients with metastatic breast cancer. *Biol Blood Marrow Transplant* **6**(3):262–271.

105. Stadtmauer EA, O'Neill A, Goldstein LJ, *et al.* (2000) Conventional-dose chemotherapy compared with high-dose chemotherapy plus autologous hematopoietic stem-cell transplantation for metastatic breast cancer. *N Engl J Med* **342**(15):1069–1076.

106. Vose JM, Bierman PJ, Lynch JC, *et al.* (2001) Transplantation of highly purified CD34+Thy-1+ hematopoietic stem cells in patients with recurrent indolent non-Hodgkin's lymphoma. *Biol Blood Marrow Transplant* **7**(12)680–687.

107. Morton JI, Siegel BV. (1964) Transplantation of autoimmune potential. Development of antinuclear antibodies in H-2 histcompatible recipients

of bone marrow from New Zealand black mice. *Proc Natl Acad Sci* **71**:2162–2165.

108. Van Bekkhum DW. (2002) Experimental basis of hematopoietic stem cell transplantation for treatment of autoimmune diseases. *J Leuk Biol* **72**: 609–620.

109. Beilhack GF, Scheffold YC, Weissman IL, *et al.* (2003) Purified allogeneic hematopoietic stem cell transplantation blocks diabetes pathogenesis in NOD mice. *Diabetes* **52**(1):59–68.

110. Marmout ATA, Gratwohl A, Vischer T. (1995) Haematopoietic precursor cell transplants for autoimmune diseases. *Lancet* **345**:978.

111. Oyama Y, Traynor AE, Barr W, Burt RK. (2003) Allogeneic stem cell transplantation for autoimmune diseases: Nonmyeloablative conditioning regimens. *Bone Marrow Transplant* **32**(1):S81–3.

112. Hoppe CC, Walters MC. (2001) Bone marrow transpantation in sickle cell anemia. *Curr Opin Oncol* **13**:85–90.

113. Mentzer WC. (2000) Bone marrow transplantation for hemoglobinopathies. *Curr Opin Hematol* **7**:95–100.

114. Krishnamurti IL, Blazar BR, Wagner JE. (2001) Bone marrow transplantation without myeloablation for sickle cell disease. *N Engl J Med* **344**:68.

115. Ionnone R, Luznik L, Engstron LW, *et al.* (2001) Effects of mixed hematopoietic chimerism in a mouse model of bone marrow transplantaion for sickle cell anemia. *Blood* **97**:3960–3965.

116. Lucarelli G, Clift RA, (2004) "Marrow transplantation in Thalessemia" in *Thomas' Hematopoietic Stem Cell Transplantation*, 3rd ed., pp. 1409–1416.

117. Buckley RH. (2000) Primary immunodeficiency diseases due to defects in lymphocytes. *N Engl J Med* **343**:1313–1323.

118. Small TN. (2000) Hematopoietic stem cell transplantation for severe combined immunodefiency disease. *Immunol Allergy Clin North Am* **20**:207–220.

119. Flake A, Roncarolo M, Puck J, *et al.* (1996) Treatment of X linked severe combined immunodeficiency by in utero tranplantation of paternal bone marrow. *New Engl J Med* **335**:1806–1810.

120. Cavazzana-Calvo M, Hacein-Bey S, Yates F, *et al.* (2001) Gene therapy of severe combined immunodeficiencies. *J Gene Med* **3**:201–206. (Review.)

121. Buckley RH. (2002) Gene therapy for SCID: A complication after remarkable progress. *Lancet* **360**:1185–1186.

122. Akashi K, Traver D, Miyamoto T, Weissman IL. (2000) A clonogenic common myeloid progenitor that gives rise to all myeloid lineages. *Nature* **404**(6774): 193–197.

123. Arber C, BitMansour A, Sparer TE, *et al.* (2003) Common lymphoid progenitors rapidly engraft and protect against lethal murine cytomegalovirus infection after hematopoietic stem cell transplantation. *Blood* **102**(2): 421–428.

124. BitMansour A, Burns SM, Traver D, *et al.* (2002) Myeloid progenitors protect against invasive aspergillosis and Pseudomonas aeruginosa infection following hematopoietic stem cell transplantation. *Blood* **100**(13):4660–4667.

125. Bitmansour A, Cao TM, Chao S, *et al.* (2005) Single infusion of myeloid progenitors reduces death from Aspergillus fumigatus following chemotherapy-induced neutropenia. *Blood* **105**(9):3535–3537.

126. Manz MG, Miyamoto T, Akashi K, Weissman IL. (2002) Prospective isolation of human clonogenic common myeloid progenitors. *Proc Natl Acad Sci USA* **99**:11872–11877.

127. Uchida N, Tsukamoto A, He D, *et al.* (1998) High doses of purified stem cells cause early hematopoietic recovery in syngeneic and allogeneic hosts. *J Clin Invest* **101**(5):961–966.

128. Devine SM, Lazarus HM, Emerson SG. (2003) Clinical application of hematopoietic progenitor cell expansion: Current status and future prospects. *Bone Marrow Transplant* **31**:241–252.

129. Reya T, Duncan AW, Ailles L, *et al.* (2003) A role for Wnt signaling in self-renewal of haematopoietic stem cells. *Nature* **423**(6938):409–414.

130. Chute JP, Muramoto G, Fung F, Oxford C. (2004) Quantitative analysis demonstrates expansion of SCID-repopulating cells and increased engraftment capacity in human cord blood following *ex vivo* culture with human brain endothelial cells. *Stem Cells* **22**:202–215.

131. Balducci E, Azzarello G, Valenti MT, *et al.* (2003) The Impact of progenitor enrichment, serum, and cytokines on the *ex vivo* expansion of mobilized peripheral blood stem cells: A controlled trial. *Stem Cells* **21**:33–40.

132. Peters SO, Kittler EL, Ramshaw HS, Quesenberry PJ. (1996) *Ex vivo* expansion of murine marrow cells with interleukin-3 (IL-3), IL-6, IL-11, and stem cell factor leads to impaired engraftment in irradiated hosts. *Blood* **87**:30–37.

133. Brown JM, Weissman IL. Progress and prospects in hematopoietic stem cell expansion and transplantation. *Exp Hematol* **32**(8):693–695.

134. Krosl J, Austin P, Beslu N, *et al.* (2003) *In vitro* expansion of hematopoietic stem cells by recombinant TATHOXB4 protein. *Nat Med* **9**:1428–1432.

135. Shizuru JA, Weissman IL, Kernoff R, *et al.* (2002) Purified hematopoietic stem cell grafts induce tolerance to alloantigens and can mediate positive and negative T cell selection. *Proc Natl Acad Sci USA* **97**:9555–9560.

136. Strober S, Lowsky RJ, Shizuru JA, *et al.* (2004) Approaches to transplantation tolerance in humans. *Transplantation* **77**(6):932–936. (Review.)

137. Millan MT, Shizuru JA, Hoffmann P, *et al.* (2002) Mixed chimerism and immunosuppressive drug withdrawal after HLA-mismatched kidney and hematopoietic progenitor transplantation. *Transplantation* **73**(9):1386–1391.

138. Su L, Lee R, Bonyhadi M, *et al.* (1997) Hematopoietic stem cell-based gene therapy for acquired immunodeficiency syndrome: Efficient transduction and expression of RevM10 in myeloid cells *in vivo* and *in vitro*. *Blood* **89**(7): 2283–2290.

139. Rosenzweig M, Marks DF, Hempel D, *et al.* (1997) Transduction of CD34+ hematopoietic progenitor cells with an anti tat gene protects T-cell and macrophage progeny from AIDS virus infection. *J Virol* **71**(4):2740–2746.

140. Kohn DB, Bauer G, Rice CR, *et al.* (1999) A clinical trial of retroviral-mediated transfer of a rev-responsive element decoy gene into CD34(+) cells from the bone marrow of human immunodeficiency virus-1-infected children. *Blood* **94**(1):368–371.

141. Domen J, Gandy KL, Weissman IL. (1998) Systemic overexpession of BCL-2 in the hematopoietic system protects transgenic mice from the consequences of lethal irradiation. *Blood* **91**(7):2272–2282.

142. Korbling M, Estrov Z. (2003) Adult stem cells for tissue repair — A new therapeutic concept? *N Engl J Med* **349**: 570–582. (Review.)

143. Grove JE, Bruscia E, Krause DS. (2004) Plasticity of bone marrow-derived stem cells *Stem Cells.* **22**(4):487–500.

144. Brazelton TR, Rossi FM, Keshet GI, Blau HM. (2000) From marrow to brain: Expression of neuronal phenotypes in adult mice. *Science* **290**:1775–1779.

145. Orlic D, Kajstura J, Chimenti S, *et al.* (2001) Bone marrow cells regenerate infarcted myocardium. *Nature* **410**(6829):701–705.

146. Camargo FD, Green R, Capetenaki Yk, *et al.* (2003) Single hematopoietic stem cells generate skeletal muscle through myeloid intermediates. *Nat Med* **9**(12):1520–1527.

147. Austin TW, Lagasse E. (2003) Hepatic regeneration from hematopoietic stem cells. *Mech Dev* **120**(1):131–135. (Review.)

148. Hussain MA, Theise ND. (2004) Stem-cell therapy for diabetes mellitus *Lancet* **364**(9429):203–205.

149. Perin EC, Silva GV. (2004) Stem cell therapy for cardiac diseases. *Curr Opin Hematol* **11** (6):399–403.

150. Alvarez-Dolado M, Pardal R, Garcia-Verdugo JM, *et al.* (2003) Fusion of bone marrow derived cells with Purkinje neurons, cardiomyocytes and hepatocytes. *Nature* **425**:968–973.

151. Vassilopoulos G, Wang PR, Russell DW. Transplanted bone marrow regenerates liver by cell fusion. *Nature* **422**:901–904.

152. Wagers AJ, Sherwood RI, Christenson JL, Weissman IL. (2004) Little evidence for develomental plasticity of adult hematopoietic stem cells. *Science* **297**:2256–2259.

153. Wagers AJ, Weissman IL. (2004) Plasticity of adult stem cells. *Cell* **116**(5): 639–648.

154. Balsam LB, Wagers AJ, Christensen JL, *et al.* (2004) Haematopoietic stem cells adopt mature haematopoietic fates in ischaemic myocardium. *Nature* **428**(6983):668–673.

# 14

# Hematopoietic Stem Cells for Leukemias and other Life Threatening Hematological Disorders

Patrick Tan

## Introduction

In recent years, hematopoietic stem cell transplantation (HSCT) has not only become a reality, but has become an important weapon in the treatment of cancers, organ failures and even in diseases caused by autoimmunity.

HSCT, is a rapidly developing area, drawing on cell biology, molecular biology, virology, immunology, cell quantitation techniques and biochemical engineering. It has potential in many clinical settings and provides new hope to patients who are suffering from conditions known to be unresponsive to conventional treatment with surgery, chemotherapy and radiotherapy.

Of late, the term "hematopoietic stem cell transplantation" has been used to replace "bone marrow transplantation." It is a more precise term and emphasises the "hematopoietic stem cell" as the key element. The hematopoietic stem cells (HSCs) are the earliest, most immature group of cells that possess the greatest potential for self-renewal and long term marrow repopulating capacity. These cells can be harvested from the bone marrow, peripheral blood, umbilical cord and even fetal liver.

The sources of these stem cells may be a compatible family member or an unrelated donor (allogeneic transplantation), an identical twin (syngeneic transplantation) or the patient's own cells previously collected and suitably stored (autologous transplantation).

---

*Correspondence: Haematology and Stem Cell Transplant Centre, Mt Elizabeth Hospital, Singapore.
E-mail: patrick_tan@gleneagles.com.sg

The successful transplantation of bone marrow cells in three patients with congenital immunodeficiency disorders in the late 1960s in human marked the birth of HSCT.[1-3]

# Indications

The number of stem cell transplants done worldwide has progressively increased since its first clinical use. To a large extent, this is related to liberalization of eligibility criteria, particularly with extension of its use to the older age group and broadening of its indications.

Below is a table listing the blood disorders where marrow or stem cell transplants have been done.

# Sources of Stem Cells

The sources of the hematopoietic stem cells may be:

1. A compatible family member or an unrelated donor (allogeneic transplantation)
2. An identical twin (syngeneic transplantation)
3. Patient's own cells previously collected and suitably stored (autologous transplantation)

These stem cells can be harvested from the bone marrow directly, or collected from the blood by apheresis after being mobilized by granulocyte colony stimulating factors or be harvested from the placenta at birth.

**Table 1.** Common Diseases for Stem Cells Transplant

| Congenital | Acquired Malignant | Acquired Non-malignant |
|---|---|---|
| Immunodeficiency disorder | The Acute leukemias | Aplastic anemia |
| Fanconi's anemia | Chronic myeloid leukemia | Paroxysmal nocturnal hemoglobinuria |
| Ostepetrosis | Chronic lymphatic leukemia | Pure red cell aplasia |
| Mucolipidoses | Myelodysplastic syndromes | Autoimmune disorders |
| Mucopolysaccharidoses | Hodgkin's and Non-Hodgkin's lymphomas | Hypereosinophilic syndrome |
| Lysosomal diseases | Multiple myeloma | Amyloidosis |

# A. The Non-malignant Disorders

## Restorative therapy

Restoration of marrow and immunological functions is the basis for allogeneic marrow transplant for severe aplastic anemias,[4] immunodeficiency diseases and the other non-malignant disorders. In the classical allogeneic stem cell transplant, donor cells are introduced to repopulate the recipient's failing marrow but only after high dose chemotherapy which by destroying host immune cells provide the needed immunosuppression to prevent rejection. Sustainable long term engraftment of the donor hemopoietic stem cell is the key factor in marrow and immune functions restoration.

The inability to accurately characterize and isolate the stem cells has led to the use of bulk marrow which contains not only the stem cells but also many other committed cells, including potent immune cells which could mediate immunological damage to the host (graft versus host disease). Selection of donors by matching histocompatibility antigens at major loci may help to reduce the graft-versus-host reaction. Nevertheless, it must also be appreciated that the donors' T lymphocytes play an important role in creating the "immunological space" for sustained engraftment of the donor's HSC.

# B. Leukemias and Other Hematological Malignancies

## (a) Immunotherapy

Hematopoietic stem cells can generate lymphocytes that can react against targets in the recipient. These lymphocytes generated from normal donor stem cells can exert an antitumor effect. In the leukemics, it has been shown that the antileukemic effect by the donor T lymphocytes plays an important role in preventing relapse. It is well known that the chance of acute leukemia relapse is high (60%) after a syngeneic (twin) marrow transplant. The absence of HLA disparity which is essential for T cell recognition and reaction in the twins is associated with little or no graft versus leukemia effect. Such transplant is usually uncomplicated but relapse rate is high. Consistent with this, is the high relapse rate seen when the donor marrow underwent rigorous T-cell depletion processing in an attempt to remove all the T cells.[5–7]

In contrast, much lower relapse rates (10% to 20%) are being reported with matched (related or unrelated) allogeneic marrow transplant despite the use of similar high dose myeloablative regimen. Furthermore, the relapse rate is shown to be significantly lower in patients who had developed some form of graft-versus-host disease than in those who had developed none.

Such observations are consistently seen in many large studies and have led to the conclusion that the graft-versus-leukemia effect of the accompanying T lymphocytes plays a vital role in the prevention of relapse. Nevertheless, the initial donor stem cell engraftment is a prerequisite for the long term survival of donor T cells and graft-versus-leukemia effect. Donor T cells will be rapidly rejected by the host without an initial donor stem cell engraftment.

The classical allogeneic HSC transplant is generally found to be beneficial in young patients (< 50-years old) with the acute leukemias, chronic myeloid leukemias, multiple myelomas, and in certain cases of lymphomas.

## (b) Adoptive immunotherapy

The management of leukemic relapse after an allogeneic bone marrow transplant is a major therapeutic challenge, as palliative therapy with oral cytotoxic drugs will merely alleviate symptoms by controlling blood cell counts and the size of the spleen, but does not have the capacity for cure. A second transplant may be an option in selected cases, but has a high morbidity and mortality.

Since the early 1990s, studies have shown that lymphocytes from the original marrow donor are highly effective in treating patient with chronic leukemias (CML & CLL) and those with myelomas who had post-transplant relapse. Cure rates of greater than 90% have been reported for patients with chronic myeloid leukemias who developed early cytogenetic relapse after transplant. Similar results have been reported with the chronic lymphatic leukemias and the myelomas. Donor lymphocyte infusion (DLI) is, however, found to be less effective in the acute leukemias, largely because of the rapid progression.[6–8]

## Evolvement into Two Forms as an Anti-Tumor Therapy

The classical allogeneic HSC transplant combines the effect of two major "weapons," namely, the "high dose chemotherapy" and "allogeneic cells" when used in treating malignancies.

Over the years, the "high dose chemotherapy arm" has developed in its fullest form in tandem with autologous HSC transplant (rescue). The use of autologous HSC rescue has allowed high doses of chemotherapy to be used to treat patients with tumors refractory to standard dose but sensitive to high doses chemotherapy.

The "allogeneic cellular therapy arm," on the other hand, has evolved into the non-myeloablative HSC transplant, where the use of high dose chemotherapy has given way to the use of immunosuppressive agents as conditioning regimens.

## A. Hematopoietic stem cell rescue

Clinical evidence of dose-response effect in the treatment of high grade lymphoma has led to trials of autologous marrow transplants for the non-Hodgkin's lymphoma in the mid-1980s. The outcomes were promising and autologous bone marrow transplant after high dose chemotherapy became widely accepted as a salvage therapy in patients with relapsed lymphoma. The discovery of the surface molecule CD 34 which is the surrogate marker for the hematopoietic stem cell in the early 1990s allowed easy and rapid evaluation of the stem cell dose in a given collected sample. This has led to rapid increase in the number of autologous stem cell rescue procedures done for many conditions worldwide. The CD 34 molecule does not mark the stem cell but rather the population of cells that contain the stem cells.[9−11]

Hematopoietic stem cell rescue has allowed us to use high doses of chemoradiotherapy to treat resistant tumors. The use of high dose of chemotherapy +/− radiotherapy followed by autologous peripheral stem cell rescue is currently being widely explored in the treatment of some tumors like lymphoma, myelomas, as well as certain cases of leukemias when patients could not find an HLA identical donors. Beneficial outcomes have been reported in selected groups of patients.[9−11]

## B. The non-myeloablative approach: cellular therapy

The classical method also called the "myeloablative allogeneic stem cell transplant" is associated with significant toxicities, which are responsible for considerable transplant-related mortality. In many series, transplant related mortality in the classical allogeneic stem cell transplant may reach up to between 20% and 30%. These toxicities are largely due to the use of high dose chemoradiotherapy prior to transplantation, which was thought to be the main "engine" that leads to the cure of malignancy, and the transplant itself was merely a supportive measure designed to allow the patient to receive myeloablative treatment without experiencing permanent aplasia.

However, it is now clear that the results seen with allogeneic transplantation are partly attributable to an immune effect mediated by donor lymphocytes, recognized as the graft-versus-tumor effect. The success of donor lymphocyte infusions in inducing remissions in patients relapsing following allogeneic transplantation, suggests that long term disease control may be feasible without a high dose conditioning regimen. The emphasis has now switched to attempting to achieve a reduction in transplant-related morbidity and mortality with subsequent improvements in the quality of life.

Cellular therapy is the main element in the non-myeloablative allogeneic stem cell transplant (NASCT). Hematopoietic stem cells as well as immune cells like the T lymphocytes are the main elements used to mediate the "cure." Immunosuppressive agents with little cytotoxic effects are used as conditioning regimens to suppress host immune cells in order to prevent graft rejection, thus allowing donor stem cells and other progenitor cells to proliferate and occupy the marrow.

Restoration of hematopoiesis in patients with severe aplastic anemias and correction of immunodeficiency states can be achieved by allowing long-term engraftment of donor stem cells. Non-myeloablative regimens followed by transfusion of allogeneic progenitor cells have been investigated in a number of centers as a way of harnessing the graft-versus-tumor effect.

One of the most well studied NASCT protocol is the use of fludarabine, and low dose total body irradiation as a conditioning regimen, and cyclosporine A and mycophenolate mofetil as the post-transplant immunosuppressants by Storb *et al.*, Fred Hutchinson Cancer Research Center in Seattle.[12] Many patients with life-threatening malignant and non-malignant diseases have been entered into this protocol and significant numbers are enjoying disease free survival up to 3 years.[12–16]

## (1) The Trojan horse and the army

Hematopoietic stem cell engraftment is a prerequisite for effective immune cell therapy. Allogeneic hematopoietic stem cell (Trojan horse) is required: 1) To maintain the "state of tolerance" for sustained survival of allogeneic immune cells; 2) To ensure continual supply of hematopoietic cells, which also include the immunocompetent cells.

Potent immunotherapeutic affects, presumably mediated by donor T lymphocytes and natural killer cells (the "army") are the key elements in the "fight" against malignancies. These donor immune cells could only work in the presence of the donor's HSC, which maintains the state of tolerance, thus allowing the immune cells to survive a much longer period for sustainable graft-versus-tumor effect.

## (2) The non-myeloablative HSCT and the targeted cell therapy

It is hoped that the incorporation of more effective but less toxic agents in conditioning regimens using the non-myeloablative approach will reduce procedure-related morbidity and mortality and ultimately improve long-term outcomes for patients undergoing allogeneic transplantation. The ideal is achieved when full donor's HSC engraftment could be achieved with zero risk. The donor's HSC is the platform (Trojan horse) for immune cell therapy. HSC by itself will have no major impact on tumors, but will allow sustained survival of the immune cells (T and B lymphocytes, natural killer cells, dendritic cells, etc.), which in due course will have powerful antitumor effects. The ideal method, then, is to use donors' immune cells which have been specifically cloned against the tumors. These cells will have no capabilities of causing graft-versus-host disease, but will mediate strong antitumor effect.

## The Future

There are major scientific, clinical and regulatory hurdles that still need to be overcome to bring the full potential clinical benefits of HSCT and cell therapy to patients. Nevertheless, it seems that the "burning" enthusiasm and enormous effort exhibited by many in the field, both scientists and clinicians, will bring great promise to many future patients.

## REFERENCES

1. Good RA, Meuwissen HJ, Gatti RA, Allen HD. (1969) Successful marrow transplant for correction of immunological deficit in lymphopenic aggammaglobulinemia and treatment of immunologically induced pancytopenia. *Exp Hematol* **19**:4–10.
2. Gatti RA, Meuwissen HJ, Allen HD, *et al.* (1968) Immunological reconstitution of sex-linked lymphopenic immunological deficiency. *Lancet* **2**:1366–1369.
3. Bach FH, Albertini RJ, Anderson JL. (1968) Bone marrow transplantation in a patient with Wiskott-Aldrich syndrome. *Lancet* **2**:1364–1366.

4. Anasetti C, Doney KC, Storb R. (1986) Marrow transplantation for severe aplastic anaemia. Long term outcome in fifty untransfused patients. *Ann Intern Med* **104**:461.

5. Antin JH. (1993) Graft versus leukemia: No longer an epiphenomenon. *Blood* **82**:2273–2277.

6. Kolb HJ, Mittermuller J, Clemm C, *et al.* (1990) Donor leukocyte transfusions for teatment of recurrent chronic myelogenous leukaemia in marrow transplant patients. *Blood* **76**:2462–2465.

7. Bar BM, Schattenbrg A, Mensink EJ, *et al.* (1993) Donor leukocyte infusions for chronic granulocytic leukemia relapsed after allogeneic bone marrow transplantation. *Clin Oncol* **11**:513–519.

8. Barrett AJ. (1995) Strategies to enhance the graft versus malignancy effect in allogeneic transplants. Bone marrow transplantation: Foundations for the 21st century, *Ann NY Acad Sci* **770**:203–212.

9. Herzig GP, Herzig RH. (1990) Current concepts in dose intensity and marrow transplantation. In *Acute Myelogenous Leukemia: Progress and Controversies,* ed. RP Gale, p. 333. New York: Wiley-liss.

10. Appelbaum FR, Deisseroth AB, Graw RG. (1978) Prolonged complete remission following high dose chemotherapy of Burkitt's lymphoma in relapse. *Cancer* **41**:1059–1063.

11. Armitage JO, Jagannath S, Spitzer G. (1986) High dose therapy and autologous marrow transplantation as salvage treatment for patients with diffuse large cell lymphoma. *Eur J Cancer Clin Oncol* **22**:871–877.

12. Storb R, Yu C, Wagner JL. (1997) Stable mixed hematopoietic chimerism in DLA identical littermate dogs given sublethal total body irradiation before and pharmacological immunosuppression after marrow transplantation. *Blood* **89**:3048–3054.

13. Yu C, Storb R, Deeg HJ, *et al.* (1995) Synergism between mycophenolate mofetil and cyclosporine in preventing graft versus host disease in lethally irradiated dogs given DLA- nonidentical unrelated marrow grafts. (Abstract) *Blood* **86**(1):577a.

14. Slavin S, Nagler A, Naparstek E, *et al.* (1998) Non-myeloablative stem cell transplantation and cell therapy as an alternative to conventional bone marrow transplantation with lethal cytoreduction for the treatment of malignant and nonmalignant hematologic diseases. *Blood* **91**:756–763.

15. Sykes M, Preffer F, McAfee S, *et al.* (1999) Mixed lymphohemopoietic chimerism and graft-versus-lymphoma effects after nonmyeloablative therapy and HLA-mismatched bone-marrow transplantation. *Lancet* **353**:1755–1759.

16. McSweeney PA, Storb R. (1999) Mixed chimerism: Preclinical studies and clinical applications (review). *Biol Blood Marrow Transplant* **5**:192–203.

# Differentiation of Human Embryonic Stem Cells to Cardiomyocytes

Chris Denning, Robert Passier and Christine Mummery*

## Introduction

Differentiation of human embryonic stem cells (hESC) to cardiomyocytes is of interest to developmental biologists for the study of human heart development as well as to electrophysiologists for studying the effects of pharmacological agents on the electrical phenotype of normal and aberrant human heart cells. However, the most interest is from cardiac physicians who view hESC as a potential source of cardiomyocytes for cell replacement therapies in cardiac disease. For all applications, hESC would ideally differentiate efficiently to defined cardiac cell types in culture. Understanding the molecular control of normal heart development and its relevance to cardiomyocyte differentiation of mouse embryonic stem cells (mESC) provides a useful basis for inducing a program of cardiogenesis in hESC. These areas are reviewed in the present chapter. A survey of the current methods used in the differentiation of hESC to cardiomyocytes (hES-CMs), is followed by a description of their morphological, phenotypic and electrical properties and a discussion of their potential use in drug screening, as (genetic) disease models and in transplantation to the diseased heart.

## Development of the Mammalian Heart

### Differentiation and early morphogenesis

Heart formation is initiated in vertebrate embryos soon after gastrulation when the three embryonic germ layers, ectoderm, endoderm and mesoderm, are established in the primitive streak. The heart is derived from

---

*Correspondence: Hubrecht Laboratory, Uppsalalaan 8, 3584 CT Utrecht, the Netherlands. Tel.: +31 30 2121800, fax: +31 30 2516464, e-mail: christin@niob.knaw.nl

the mesoderm and is the first definitive organ to form in development. Its morphogenesis, growth and integrated function are essential for the survival of the embryo. Heart forming or cardiac progenitor cells are for the most part localized in the anterior primitive streak. Different populations of precursor cells are distributed within the streak (in relation to the organizing center, or node) in the same anterior-posterior order that they are later found in the tubular heart.[1] As a result, cells furthest from the node end up in the atrium, those nearer the node end up in the ventricle, whilst those nearest the node later form the outflow tract. In addition to precursors in the primitive streak, there are also precursors bilaterally distributed in the epiblast directly adjacent to the streak. As development proceeds, the precursor population of precardiac mesoderm emigrates from the streak in an anterolateral direction, giving rise to the heart-forming fields on either side of the streak. These heart fields harbor not only progenitors of the atrial, ventricular and outflow tract lineages but also endocardial progenitor cells. The axial distribution is maintained as the fields migrate to fuse and form the cardiac crescent. Cells of the cardiac crescent then adopt a definitive cardiac fate in response to cues from the adjacent anterior endoderm.[2] The anterior endoderm, in particular, appears to have an instructive function in cardiogenesis in various species [reviewed by Brand (Ref. 3)]. Ablation of the anterior endoderm in amphibians results in loss of myocardial specification,[4] whilst explants of the posterior, blood-forming mesoderm in chick are reprogrammed to express cardiac instead of blood-restricted marker genes if combined with the anterior endoderm.[5] After the cardiac mesoderm has been specified, it is then directed towards the midline where the heart fields fuse and form a single heart tube. In mutant mouse and zebrafish in which this fails to take place, two tube-like structures form (*cardiac bifida*) which both acquire contractile activity.[6–9] These mutants include several that lack endodermal tissue. Thus, the endoderm is not only important for differentiation of cardiac precursors but is also essential for cardiac mesoderm migration, although it is clear that tube formation also involves a cell autonomous function of the cardiac mesoderm itself.

Once this process of primary cardiac induction is complete, cells are recruited from the lateral plate mesoderm, medial to the primary heart field, to give rise to the secondary, anterior heart field (AHF). These cells contribute to the primitive right ventricle and the outflow tract.[10] Whilst the heart tube is initially almost straight, the ventricular segment at this stage starts to bulge ventrally, flips to the right and begins to form a C-shaped heart. Left-right asymmetry thus becomes evident and cardiac looping morphogenesis has commenced. Through a series of "ballooning" steps and morphogenetic movements, the four-chambered heart eventually forms.[11]

The sequential activation of the transcription factors that results in the formation of nascent- then precardiac mesoderm, and eventually determines cardiac cell fate is likely to be controlled in hESC as in embryos *in vivo*. The signals may be known or novel, in either case emanating from the endoderm, node or acting cell autonomously in the mesoderm. In the following section, some of the most relevant aspects of this molecular control are considered.

## Molecular control of cardiac development

Three families of peptide growth factors have been studied most intensely for their positive and negative effects on cardiogenesis. These are the bone morphogenetic proteins (BMPs), members of the transforming growth factor β superfamily, the wnts and the fibroblast growth factors (FGFs). Members of these three families or their inhibitors are expressed in endoderm. Disrupted expression of ligands, receptors or their downstream target genes has dramatic and distinct effects on cardiac development that are highly conserved between species [reviewed by Olson and Schneider (Ref. 2)]. In general, BMP signaling promotes cardiogenesis in vertebrates[12–17] and is also required to generate mesoderm/cardiac muscle cells from mouse teratocarcinoma stem cells and embryonic stem (ES) cells in culture.[18–19]

Wingless in *Drosophila* and related wnt proteins in vertebrates are involved in cardiac specification although their function is complex. Wnts were initially considered suppressive of heart formation but both induction and inhibition have since been reported [reviewed by Olson and Schneider (Ref. 2)]. Results have not yet been reconciled but may relate to distinct effects of the canonical (acting via β-catenin/GSK3 to repress cardiogenesis) versus non-canonical (acting via PKC/JNK to promote cardiogenesis) signaling pathways, and/or indirect effects in certain model systems (for example, induction, expansion or augmentation of BMP-producing endoderm-like cells). Finally, limited studies in chick and zebrafish have implicated a cardioinductive role for FGFs,[20–22] although in *Drosophila*, FGFs appear to provide positional cues to cells for specification. It is of interest that BMP2 is able to upregulate FGF8 ectopically and that BMP2 and FGF8 probably synergize to drive mesodermal cells into myocardial differentiation. These signaling pathways are essential not only for primary cardiogenesis but are also involved in secondary (AHF) cardiogenesis [reviewed in (Ref. 3)].

Once anterior mesoderm cells have received appropriate signals, such as those described above, they switch on a set of cardiac-restricted transcription factors that interact in combination to control downstream genes in the cardiac pathway. The homeodomain transcription factor Nkx2.5[23] and the T-box protein Tbx5[24,25] are among the earliest markers of the cardiac lineage and are activated shortly after cells have formed the heart fields. Nkx2.5 is thought to be required in mice specifically for left ventricular chamber development,[26] whilst loss of Tbx5 results in severe hypoplasia of both the atrial and left ventricular compartments[27] and may thus be important for the formation of both. Nkx2.5 and Tbx5 associate with members of the GATA family of zinc finger transcription factors and with serum response factor (SRF) to activate cardiac structural genes, such as actin, myosin light chain (MLC), myosin heavy chain (MHC), troponins and desmin. Tbx5 can also cooperate with Nkx2.5 to activate expression of ANF and the junctional protein connexin 40.[27,28] Members of the myocyte enhancer factor 2 (MEF2) family of transcription factors also play key roles in cardiomyocyte differentiation by switching on cardiac muscle structural genes.[29] In addition, association of SRF with a nuclear protein myocardin, activates cardiac specific promoters.[30] HOP, a cardiac homeodomain protein, which affects cardiomyocyte proliferation and differentiation,[31] also associates with SRF and inhibits myocardin transcriptional activity.[32] Thus, multiple complex interactions take place between various transcription factors to control initial differentiation, proliferation and maturation of cardiomyocytes. Apart from their functional role, many of these factors serve as excellent markers of cardiomyocytes in differentiating cultures of hESC and mESC (see Table 1), and can be useful in identifying their degree of maturity and the kinetics with which differentiation is taking place because their normal expression is under tight temporal control. A recent addition to this list is Isl1, a LIM homeodomain transcription factor, which identifies a cardiac progenitor population that proliferates prior to differentiation and contributes the majority of cells to the heart.[33] Unlike skeletal muscle cells, where differentiation and proliferation are mutually exclusive, embryonic cardiomyocytes differentiate and assemble sarcomeres even while they proliferate, although organization is much greater postnatally once proliferation has stopped. The prospect of using Isl1 as a marker for the undifferentiated cardiac progenitor state is exciting and cell sorting of differentiating hESC on the basis of Isl1 expression could allow their further characterization.

## Cardiomyocytes Derived from Mouse Embryonic Stem Cells

Pluripotent embryonal carcinoma (EC) cells derived from teratocarcinoimas as well as ES cells derived from embryos retain the capacity to form derivatives of the three germ layers in culture. *In vitro* differentiation usually requires an initial aggregation to form structures termed "embryoid bodies" (EBs). After a few days of culture under appropriate conditions of cell density, culture medium and serum supplement, cardiomyocytes form between an outer epithelial layer of the EB and basal mesenchymal cells and become readily identifiable by spontaneous contraction [review by Boheler *et al.* (Ref. 34)]. Differentiation may be enhanced by culture supplements or co-culture with endoderm-like cells[35,36] [reviewed by Rathjen and Rathjen (Ref. 37)]. Wnt11, which activates the non-canonical wnt pathway, induced GATA4 and Nkx2.5 expression in P19 EC cells[38] and repressed the canonical pathway although wnt-3a, acting through the canonical pathway, promoted cardiomyogenesis in a P19 subclone.[39] These differences are unresolved. The effects of wnt signalling on cardiomyogenesis of mESC have not been described.

Cardiomyocyte differentiation in murine EBs recapitulates the programed expression of cardiac genes observed in the mouse embryo *in vivo* both in the kinetics and the sequence in which the genes are upregulated. GATA-4 and Nkx2.5 transcripts appear before mRNAs encoding ANF, MLC-2v, α-MHC and β-MHC. Sarcomeric proteins are also established in a manner similar to that seen in normal myocardial development. The electrical properties and phenotype of cardiomyocytes derived from mouse EB cultures have been examined in some detail [reviewed in Boheler *et al.* (Ref. 34)]. The rate of contraction decreases with differentiation and maturation in culture, as in normal mouse development, and their differentiation as such can be divided into three developmental stages: early (pacemaker-like or primary myocardial-like cells); intermediate; and terminal (atrial-, ventricular-, nodal-, His, and Purkinje-like cells).[40] In the early stages, the nascent myofibrils are sparse and irregular but myofibrillar and sarcomeric organization increases with maturation. Functional gap junctions develop between cells and eventually their phenotype resembles that of neonatal rat myocytes. Likewise, the electrophysiological properties of mES-CMs develop with differentiation in a manner reminiscent of their development in the mouse embryo.[34] Fully differentiated mES-CMs are responsive to β-adrenergic stimulation whilst early mES-CMs are not.[41] They also exhibit many features of excitation-contraction coupling found in isolated fetal or neonatal cardiomyocytes.

Several of the features described above in combination with the amenability of mESC to genetic manipulation have made it possible to develop strategies for genetic selection of cardiomyocytes from mixed populations of differentiating cells. Since cell surface antibodies recognizing cardiomyocytes or their precursors are not available, this has been important for initiating studies to transplant pure cell populations of specific cardiac lineages to the adult mouse heart. The first study of this type was carried out using mESC expressing a fusion gene composed of the α-MHC promoter and a neo$^R$ cassette.[42] After selection in G418, the percentage of cardiomyocytes increased from 3–5% to almost 100%. Similar studies have been carried out using the MLC-2v or α-cardiac actin promoter coupled to either neo$^R$ or GFP for selection.[43–45] Unfortunately in the adult heart post-transplantation, survival of mES-CMs is poor (∼5% survival) and many undergo apoptosis.[42] Recently, however, several studies demonstrated that cardiac function significantly improved, after transplantation of mES-CMs in mouse,[46,47] rat,[48] and pig[49] models of myocardial infarction. No evidence of graft rejection, sudden cardiac death or tumor formation was observed up to 12 weeks after transplantation.[47] In humans, an estimated $10^8$–$10^9$ cells lost during myocardial infarction would be required per patient for effective therapy. A useful step towards upscaling has recently been made when $10^9$ mES-CMs were produced by genetic selection in a bioreactor.[50]

## Cardiomyocytes Derived from Human Embryonic Stem Cells

### *In vitro* differentiation of hESC to cardiomyocytes

Several groups have shown that hESC can differentiate to cardiomyocytes in culture. The first report,[51] was published almost 3 years after the first isolation of hESC lines. In this study, hESC from the H9.2 hESC line, were dispersed into small clumps (3–20 cells) using collagenase IV and grown for 7–10 days in suspension in plastic petri dishes. During this stage, cells are aggregated to form structures resembling EBs from mESC although without a distinct outer epithelial cell layer.[52] The EBs were then transferred to 0.1% gelatin-coated culture dishes where they attached. In the EB outgrowths, beating areas were first observed 4 days after plating, i.e. 11–14 days after the start of the differentiation protocol. A maximum number of beating areas was observed 20 days after plating (27–30 days of differentiation),

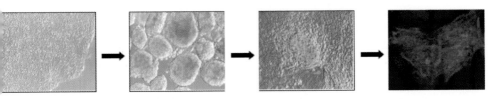

**Figure 1.** Differentiation from undifferentiated hESC to beating embryoid bodies (from left to right), containing α-actinin positive (red; nuclei are blue) cardiomyocytes (right panel).

when 8.1% of 1884 EBs counted was beating. This "spontaneous" differentiation to cardiomyocytes by culturing the hESC as EBs was repeated by other groups (Fig. 1). For example, Xu *et al.*[53] demonstrated beating EBs from the H1, H7, H9, H9.1 and H9.2 hESC lines. Here, however, approximately 70% of the EBs displayed beating areas, after differentiation for 20 days. As early as day 8 of their differentiation protocol (growing in suspension followed by plating on culture dishes), 25% of the EBs were beating. A third group later also demonstrated spontaneous differentiation of cardiomyocytes from hESC lines H1, H7, H9 and H14 with 10–25% of the EBs beating after 30 days of differentiation.[54] Recently, Zeng *et al.*[55] described the differentiation of two other hESC lines BG01 and BG02. Again, following dissociation of hESC into small clumps using collagenase IV, cells were grown for 7 days as EBs in suspension and cultured on adherent plates for another 7 days. Although the objectives of this study were not related to the cardiac field, immunoreactivity was demonstrated for the cardiac marker cardiac troponin I (cTnI).

An alternative method for the derivation of cardiomyocytes from hESC was described by Mummery *et al.*[35,56] Beating areas were observed following co-culture of hESC with a mouse visceral endoderm-like cell-line (END-2). As described above, endoderm plays an important role in the differentiation of cardiogenic precursor cells in the adjacent mesoderm *in vivo*. Earlier, cardiomyocyte differentiation had been observed in co-cultures of END-2 cells with mouse P19 EC cells[57] and mESC.[35] For the derivation of cardiomyocytes from hESC, END-2 cells were seeded on a 12-well plate, mitotically inactivated with mitomycin C and co-cultured with the hESC line, HES-2 (ESI, Singapore). This resulted in beating areas in approximately 35% of the wells after 12 days in co-culture[56] (Fig. 2).

Whilst these various methods appear to be effective, all produce cardiomyocytes at low efficiency and in insufficient numbers to treat adult

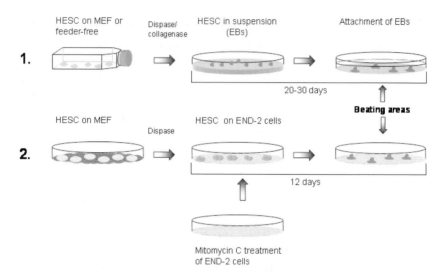

**Figure 2.** Differentiation from undifferentiated hESC towards cardiomyocytes by growing hESC in suspension as embryoid bodies (EBs) (1.), or as co-cultures with mitomycin C-treated END-2 cells (2.). The number of days from the start of differentiation to maximal number of beating areas in culture is indicated for each procedure.

humans. In addition, the risk of teratoma formation from any undifferentiated hESC escaping the differentiation protocol has to be rigorously assessed. As described above, drug selection in combination with a cardiac-specific promoter in mESC derived EBs results in a cell population of nearly 100% cardiomyocyte. Similar methods have not yet been applied effectively in hESC, in part because it has proven difficult to obtain stable, genetically modified clonal cell lines. The best methods to date have relied on physical selection methods. Following manual dissection and dissociation of beating areas from hESC, between 2–70% of the cells stained for cardiac markers, depending on the particular aggregate dissected.[54,56] The only enrichment method described to date used discontinuous Percoll gradient purification (40.5% over 58.5%).[53] This resulted in approximately 4-fold enrichment (70% positively stained for cardiac markers) in a particular cell fraction compared with the initial differentiated cell suspension. FACS-analysis would be a useful approach to quantification; the fact that it has not been widely used for hES-CMs reflects the lack of cardiomyocyte cell surface antibodies and the fact that most (specific) antibody markers are against structural proteins. The striated pattern of the structural proteins is a necessary confirmation of cardiomyocyte identity.

## Improving cardiomyocyte differentiation from hESC

Despite differences in morphology, growth and the molecular pathways controlling self-renewal versus differentiation in mESC and hESC, aspects of the control of cardiomyocyte differentiation of mESC may be conserved in hESC.[34,58] In addition, as described above, P19 EC cell differentiation to the cardiac lineage has been studied for more than 2 decades,[59] and may also identify potential cardiac inducing or stimulating factors for hESC. For instance, both BMP-2 and BMP-4, as described earlier, induce cardiac differentiation in chick explant studies from a tissue that normally does not give rise to cardiac tissue.[14] BMP-2 induces cardiac differentiation in mESC.[60] Retinoic acid and DMSO, enhance or promote cardiac differentiation in mESC or P19 EC cells, respectively.[61-63] Oxytocin, a nonapeptide originally recognized as a female reproductive hormone, was recently found to increase cardiomyocyte differentiation in P19 EC cells.[64] Other factors such as members of the TGFβ, FGF, and wnt families have been discussed above [reviewed in Refs. 34 and 58]. Finally, of a range of factors tested recently, ascorbic acid was striking for its ability to promote cardiomyocyte differentiation.[65]

Several of these potential cardiogenic factors have been tested in hESC. No significant improvement in the efficiency of cardiomyocyte differentiation was observed by adding DMSO and retinoic acid[51,53] or BMP-2.[66] It is not clear whether these factors do not play an important role in cardiac differentiation from hESC; whether differentiation protocols were not optimal; or whether specific co-factors were absent. Variations in concentration, timing and combinatorial effects of potential cardiogenic factors in hESC differentiation may have a crucial effect on the outcome of differentiation protocols. The only factor that has been shown to enhance cardiomyocyte differentiation of hESC significantly is the demethylating agent 5-aza-2′deoxycytidine, which has also been shown to stimulate cardiomyocyte differentiation in human mesenchymal stem cells.[67] Treatment of hESC with 5-aza-2′deoxycytidine resulted in a time- and concentration-dependent effect on cardiomyocyte differentiation, with a maximum 2-fold upregulation in the expression of cardiac α myosin heavy chain, as determined by real time RT-PCR, under optimal conditions observed.[53]

Another aspect warranting consideration, is the presence of fetal calf serum as a culture supplement. All studies on cardiomyocyte differentiation of hESC so far have been done in the presence of serum. Serum contains stimulatory as well as inhibitory factors for cardiomyocyte differentiation. Recently, Sachinidis *et al.*[68] observed a 4.5-fold upregulation in the percentage of the number of beating EBs from mESC after changing to a serum-free differentiation medium. This effect of serum has been more commonly observed in the differentiation of skeletal muscle and neural precursors, hematopoietic stem cells and endothelial cells, but has received relatively little attention for the other lineages.

## Characteristics of Human Embryonic Stem Cell Derived Cardiomyocytes

The previous sections have described the current methods to promote spontaneous and induced differentiation of hESC towards the cardiac lineage. Although functional cardiomyocytes can easily be identified *in vitro* by their beating phenotype, only more detailed investigation can establish the specific cardiac cell types generated, the degree of maturation they achieve compared with *in vivo* cardiac development and whether they possess fully functional excitation-contraction coupling machinery that responds appropriately to pharmacological agents. To realize the scientific and therapeutic potential of hES-CMs comprehensive characterization is therefore required.

### Transcriptional profile of hESC derived cardiomyocytes

Differentiation of hESC to the cardiac lineage creates a gene expression profile (Table 1) reminiscent of both mESC differentiation and the early stages of normal mouse heart development.[69] Analysis of hES-CM RNA and proteins has demonstrated the presence of cardiac transcription factors, including GATA-4, myocyte enhancer factor (MEF-2) and Nkx2 transcription factor related locus 5 (Nkx2.5).[51,53] Correspondingly, structural components of the myofibers are appropriately expressed. These include α-, β- and sarcomeric-myosin heavy chain (MHC); atrial and ventricular forms of myosin light chain (MLC-2a and -2v); tropomyosin; α-actinin; and desmin. Furthermore, hES-CMs fail to react with antibodies from non-cardiac lineages (Table 1).

Antibody reactivity to two members of the troponin complex, cardiac troponin T (cTnT), which binds to tropomyosin, and cTnI, which provides a calcium sensitive molecular switch for the regulation of striated muscle contraction, have been demonstrated. cTnI appears to be truly cardiac specific as antibodies to this protein only react with cells arising from beating and not non-beating regions. In addition, upregulation of atrial natriuretic

**Table 1.** Markers for hES-CM Characterization. Methods of detection are RT-PCR, reverse transcriptase-polymerase chain reaction; IF, immunofluorescence; W, western blot analysis; pharmacological agents are given as agent name (ph.)

| Marker | hESC line(s) | Method(s) of detection | Expression in hES-CMs | References |
|--------|--------------|------------------------|----------------------|------------|
| *Transcription factors* | | | | |
| NK2 transcription factor related locus 5 (Nkx2.5) | H1, H7, H9, H9.1, H9.2 | RT-PCR | + | (51, 53) |
| GATA4 | H1, H7, H9, H9.1, H9.2 | RT-PCR, IF, W | + | (51, 53) |
| Myocyte enhancer factor 2 (Mef-2) | H1, H7, H9, H9.1, H9.2 | IF | + | (53) |
| *Structural elements* | | | | |
| Cardiac tropinin I (cTnI) | H1, H7, H9, H14, H9.1, H9.2 | IF, W, RT-PCR | + | (51, 53, 54, 70) |
| Cardiac tropinin T (cTnT) | H1, H7, H9, H9.1, H9.2 | RT-PCR/IF | + | (51, 53) |
| α-myosin heavy chain (α-MHC) | H1, H7, H9, H9.1, H9.2 | RT-PCR/IF | + | (51, 53) |
| β-myosin heavy chain (β-MHC) | H1, H7, H9, H9.1, H9.2 | IF | + | (53) |
| Sarcomeric myosin heavy chain (sMHC) | H1, H7, H9, H9.1, H9.2, H14 | IF | + | (53, 54) |
| Myosin light chain 2a (MLC-2a) | H9.2, HES-2 | RT-PCR/IF | + | (51, 56) |
| Myosin light chain 2v (MLC-2v) | H9.2, HES-2 | RT-PCR/IF | + | (51, 56) |
| α-actinin | H1, H7, H9, H9.1, H9.2, H14, HES-2 | RT-PCR/IF | + | (51, 53, 54, 56) |
| Tropomyosin | H1, H7, H9, H9.1, H9.2, HES-2 | IF | + | (53, 56) |
| Desmin | H1, H7, H9, H9.1, H9.2 | IF | + | (51, 53) |
| Smooth muscle actin (SMA) | H1, H7, H9, H9.1, H9.2 | IF | + | (53) |
| *Receptors & regulatory elements* | | | | |
| Atrial natriuretic factor (ANF) | H1, H7, H9, H9.1, H9.2, HES-2 | RT-PCR/IF | + | (51, 53, 56) |
| Phospholamban (PLN) | HES-2 | RT-PCR | + | (56) |
| Ryanodine receptor (RyR) | HES-2 | IF | + | (56) |
| Creatine kinase-MB (CK-MB) | H1, H7, H9, H9.1, H9.2 | IF | + | (53) |

**Table 1.** (*Continued*)

| Marker | hESC line(s) | Method(s) of detection | Expression in hES-CMs | References |
|---|---|---|---|---|
| Myoglobin | H1, H7, H9, H9.1, H9.2 | IF | + | (53) |
| α1-adrenoceptors | H1, H7, H9, H9.1, H9.2, HES-2 | IF/phenylephrine (ph.) | + | (51, 53, 54, 56) |
| β1-adrenoceptors | H1, H7, H9, H9.1, H9.2, HES-2 | IF/isoprenaline (ph.) | + | (51, 53, 56) |
| β2-adrenoceptors | H1, H7, H9, H9.1, H9.2 | IF/clenbuterol (ph.) | + | (53) |
| Muscarinic receptors | HES-2 | Carbachol (ph.) | + | (56) |
| Phosphodiesterase | H1, H7, H9, H9.1, H9.2 | IBMX (ph.) | + | (51, 53) |
| Adenylate cyclase | H9.2 | Forskolin (ph.) | + | (51) |
| Ki67 (cell division) | H1, H7, H9, H9.1, H9.2 | IF | + | (53, 70) |
| *Gap junction & adhesion proteins* | | | | |
| Connexin 43 | H9.2, HES-2 | IF | + | (72) (56) |
| Connexin 45 | H9.2 | IF | + | (72) |
| Connexin 40 | H9.2 | IF | + | (72) |
| N-cadherin | H1, H7, H9, H9.1, H9.2 | IF | + | (53) |
| *Ion channels* | | | | |
| L-type $Ca^{2+}$ channel; $I_{ca-L}$ ($\alpha1c$) | H1, H7, H9, H9.1, H9.2, HES-2 | RT-PCR/IF/ diltiazem (ph.) | + | (53, 56) |
| Transient outward $K^+$ channel $I_{TO}$ (*Kv4.3*) | HES-2 | RT-PCR | + | (56) |
| Slow delayed rectifier $K^+$ channel $I_{Ks}$ (*KvLQT1*) | HES-2 | RT-PCR | + | (56) |
| Slow delayed rectifier $K^+$ channel $I_{Kr}$ (*HERG*) | H9, H14 | E4031 (ph.) | + | (54) |
| *Non-cardiac proteins* | | | | |
| Myogenin (skeletal muscle) | H1, H7, H9, H9.1, H9 | IF | − | (53) |
| Nebulin (skeletal) | H9.2 | IF | − | (51) |
| α-fetal protein (AFP; endodermal) | H1, H7, H9, H9.1, H9.2 | IF | − | (53) |
| β-tubulin III (neuronal) | H1, H7, H9, H9.1, H9.2 | IF | − | (53) |

factor (ANF), a hormone that is actively expressed in both atrial and ventricular cardiomyocytes in the developing heart, has also been observed during cardiac differentiation of hESC. Moreover, these cells express creatine kinase-MB (CK-MB) and myoglobin.[53] CK-MB is found to be involved in high energy phosphate transfer and facilitates diffusion of high energy phosphate from the mitochondria to myofibril in myocytes. Myoglobin is a cytosolic oxygen binding protein responsible for the storage and diffusion of oxygen in myocytes. Thus many of the transcription factors, structural proteins and metabolic regulators of cardiac development are found within hES-CMs.

These cardiomyocytes do, however, react with antibodies to smooth muscle actin, a protein found in embryonic and fetal, but not adult cardiomyocytes, suggesting a limited degree of maturation.[53] Adding weight to this notion, single hES-CM, rather than showing the more defined rod shape of mature cells, display numerous different morphologies, including spindle, round, tri- or multi-angular. Sarcomeric immunostaining reveals sarcomeric striations organized in separated bundles, which parallels the pattern seen in human fetal cardiomyocytes and not the highly organized parallel bundles seen in human adult cardiomyocytes.[56]

Ultramicroscopical analysis does, however, demonstrate hES-CM maturation during extended culture. While during the early stages of differentiation (~10–20 days) cardiomyocytes have a large nucleus to cytoplasm ratio with disoriented myofibrils lacking sarcoplasmic pattern distributed throughout the cytoplasm in a random fashion, both numbers and organization of myofibrils increased at later times (~20–50 days).[70] During this time, cells elongated and Z-line assembly from periodically aligned Z-bodies was observed. At the late stages (>50 days), a high degree of sarcomeric organization was observed and discrete A (dark) and I (light) bands could be seen in some sarcomeres. Furthermore, hES-CMs progressively withdraw from the cell cycle during culture. However, although this developmental pattern is reminiscent of mESC differentiation to cardiomyocytes, maturation of hES-CMs proceeds more slowly, is more heterogeneous, and does not reach the level of maturity, typical of adult cardiomyocytes. This was manifested, for example, by the lack of developed T tubule system.[70] Thus, it has been suggested maturation may be aided by the addition of prohypertrophic factors such as cardiotrophin. Alternatively, subjecting the hES-CMs to oscillating mechanical load.[56,71] or culturing them in 3-dimensional matrices may stimulate maturation. Considering that cardiomyocytes in the infarcted heart regress to an embryonic phenotype, it will be interesting to see whether hES-CMs with an embryonic or adult phenotype are more suited to functional and electrical integration following transplantation.

## A functional conduction system: Excitation-contraction coupling machinery

Cardiomyocytes from hESC have a beat rate of $\sim$30 to 130 bpm and respond appropriately to pharmacological agents. Ligand binding to the adrenoceptors (AR) generates a signaling cascade that results in elevated cAMP levels. This activates cAMP-dependent protein kinase (PKA), which then phosphorylates and alters the function of a few cardiac proteins that have key effects on the overall cardiac function. The presence of $\alpha$1- and $\beta$1-ARs in hES-CMs has been demonstrated by immunocytochemical analysis. Pharmacological induction with phenylephrine ($\alpha$1-AR agonist) resulted in dose dependent increase in contraction rate in both human fetal and hES-CMs. Increases in beat rate and amplitude were observed after isoprenaline ($\beta$1-AR agonist) treatment.[51,53,54,56] However, $\beta$2-AR agonists, such as clenbuterol only elicited a response during the late stages of *in vitro* differentiation (day 61 to 72) and not in the early stages (day 22 and 39),[54] consistent with the observation that cardiac contractility to $\beta$-adrenergic stimulation changes during development.[73] Negative chronotropic responses to muscarinic agonists, such as carbachol, have also been observed, again consistent with data from mouse fetal and mES-CMs.[73,74] Alternatively, perturbation of pathways downstream of receptors can also modulate cardiac output. Inhibition of phosphodiesterase (which converts cAMP into 5'-AMP) by isobutyl methylxanthine (IBMX) and activation of adenylate cyclase by forskolin, results in increased or decreased beat rate, respectively.[51]

Other known and unknown factors influence beat rate. Human ES-CMs are exquisitely sensitive to temperature; reduced temperature correlates with decreased beat rate. In EBs, continuous and episodic beat patterns have been observed, with the latter being speculatively attributed to conduction block related to tissue geometry, impaired cell to-cell coupling, reduced cellular excitability or immature $Ca^{2+}$ regulatory system. Human ES-CMs do express adhesion molecules (N-cadherin), and the gap junction proteins connexins 43 and 45, but Cx45 usually only in the early stages of *in vivo* development.[56,72] They are also well coupled by functional gap junctions, as evidenced by visualizing Lucifer Yellow dye spread or $Ca^{2+}$ movement between connected cells. However, visualization of $Ca^{2+}$ ion transients via fura-2 fluorescence revealed the rate to peak concentration in each beat was significantly longer in the hES-CMs.[72]

In heart muscle, the main currents involved in the action potential are influx of $Ca^{2+}$ and $Na^+$ during depolarization (phase 0) and efflux or maintenance of $K^+$ during repolarization and resting potential (phases 1–4). Sodium channels are the major current of depolarization in atrial and ventricular cells, but in pacemaker cells it is $Ca^{2+}$ via L-type calcium channels ($I_{Ca-L}$). Molecularly, $I_{Ca-L}$ channels are expressed in hES-CMs[56] and functionally they are inhibited, hence their beat rate reduced in a dose dependent manner by blockers such as diltiazem.[53] Verapamil also inhibited action potential in human fetal and hES-CMs. By contrast, mES-CMs in the early stages of differentiation are nonresponsive despite the presence of $I_{Ca-L}$ channels. Thus, while there are many similarities between human fetal or hES-CMs and mES-CMS, their $Ca^{2+}$ channel modulation resembles that of the adult mouse.[56]

The sarcoplasmic reticulum is also a major source of $Ca^{2+}$ release in myocytes. Expression of ryanodine receptors (RyR; $Ca^{2+}$ activated $Ca^{2+}$ release) and phospholamban (PLN; regulation of SERCA2a-mediated $Ca^{2+}$ uptake) has been demonstrated in hES-CMs.

Repolarization is initiated by numerous $K^+$ channels. Among these are transient outward current ($I_{TO1}$; encoded by the genes *Kv4.2* and *Kv4.3*), which causes the early rapid repolarization seen in phase 1 of the action potential; and slow activating current ($I_{Ks}$; encoded by *KvLQT1* with the *mink* subunit), which causes repolarization associated with phase 3 of the action potential. Mutations in *KvLQT1* are the cause of long QT syndrome in humans; this is the rapid component of delayed rectification ($I_{Kr}$; encoded by human *eag*-related gene [HERG]), which is involved in all phases of repolarization, but is most active during phase 3. Mutations in *HERG* have been linked to a congenital form of long QT syndrome. Furthermore, these channels are the targets of anti-arrhythmic agents such as E4031.

RNA for *Kv4.3* and *KvLQT1* has been detected by RT-PCR in hES-CMs.[56] However, while *Kv4.3* could be detected in differentiating cells several days before the onset of beating, *KvLQT1* was expressed in undifferentiated hESC but transcripts disappeared during early differentiation and reappeared later. Functionally, application of E-4031 leads to increased duration of phase 3 (terminal repolarization) and triggered early after depolarizations (EADs) based arrhythmias, providing pharmacological evidence that $I_{Kr}$ contributes to repolarization in hES-CMs.[54] Delayed after depolarizations (DADs) typically occur during $Ca^{2+}$ overload, such as produced by injury or digitonin toxicity and were observed to occur spontaneously, possibly a result of microelectrode impalement or spontaneous $Ca^{2+}$ release.

Interestingly, forced electrical stimulation at increasing frequencies resulted in action potential shortening adaptation in ventricular-like cardiomyocytes. This physiological response leads to systolic shortening at high heart rates, thereby maintaining diastolic time for ventricular filling, as shown in the ventricular myocardium of human embryos.[54]

Based on electrophysiological characteristics of the action potential, such as resting potential, upstroke velocity, amplitude and duration, it is possible to assign cell type (pacemaker, atrial, ventricular, nodal) to hES-CMs in culture. Upstroke velocities, a measure of the rate of depolarization, were particularly low in the ventricular-like cells (average 8 V/s,[56] 5–30 V/s[54]), which is comparable to cultured fetal ventricular cardiomyocytes, but very different to the rapid upstroke of adult ventricular cells ($\sim$150–350 V/s).[54] Similarly, the relatively positive resting potential of atrial- and ventricular-like hES-CMs (approximately $-40$ to $-50$ mV) is comparable to the early stages of fetal development. While some studies describe up to 85% of hES-CMs being ventricular-like,[56] others report a lesser predominance of this cell type.[54,72] This may relate to the time of analysis since the cardiomyocyte composition of mES-derived embryoid bodies changes during culture duration.[40] It may also reflect differences in the hESC lines or methods of differentiation used, or that the analysis by Mummery *et al.*[56] used dissociated hES-CMs rather than EBs.

Therefore, it is clear that hESC can be differentiated towards cardiomyocytes that appropriately respond to different stimuli, and this infers functional expression of many of the components required for excitation-coupling. However, while maturation of hES-CMs does occur during prolonged culture, currently these cells fail to attain the characteristics of adult cardiomyocytes. It will be important to assess novel methods to stimulate maturation so *in vitro* produced cardiomyocytes with embryonic or adult characteristics are at the disposal of the scientific and clinical community.

# Modeling Cardiac Diseases in Human Embryonic Stem Cells: Duchenne Muscular Dystrophy and Channelopathies as Examples

Duchenne muscular dystrophy (DMD) is a fatal X-linked genetic disorder affecting $\sim$1 in 3000 new-born males. Developing and progressing *in utero*, the condition presents in patients by the age of 2–5 years when they experience walking difficulties. They are wheelchair-bound by the early teens and death due to cardiac or respiratory failure usually ensues by the late teens/early twenties.[75] Recently, the technique of noninvasive ultrasonic tissue characterization has demonstrated cardiac abnormalities in all the patients, even in the youngest (4-years old) included in the study.[76]

The genetic lesion involved has been mapped to the dystrophin gene, which encodes multiple alternatively-spliced RNA transcripts[75] that are expressed either globally or in a tissue specific manner (e.g. in muscle, neural, retinal and gastric tissue). The functional significance of improper dystrophin production was originally thought to perturb only a structural role of actin cytoskeleton linkage to the cytoplasmic surface of the sarcolemma; with failing membrane integrity, intracellular calcium levels rise, resulting in cell death. However, it is now clear that dystrophin also has more complex functions, such as signaling, via a large multi-protein complex, the dystrophin-associated protein complex (DAPC).[77] This complex appears to interact with proteins associated with multiple signaling pathways, including calmodulin, Grb2 and NOS.

Overall, however, the molecular and cellular consequences of the absence of dystropin are poorly understood. In part, this is due to mouse-human differences in the genetics of the disorder, with mice requiring ablation of at least two genes before the disease starts to parallel the human condition.[78] Another factor is the inability to biopsy certain material from patients, such as heart or neural tissue, during the development of the condition.

DMD is one example of a number of diseases, for which there are no adequate mouse models: mice bearing these mutations do not have a phenotype but cardiac physiology in the mouse differs so significantly from humans. This is not surprising. Another example is a set of diseases known as "channelopathies." Genetic alterations in several ion channels have been associated with heritable cardiac arrhythmias that can lead to sudden death. The underlying mechanism is distortion of the stable balance between inward and outward currents during the generation of action potentials. Many different mutations in relatively few independent ion channels are thought to be involved in these channelopathies. One underlying dysfunctional mechanism is the Long-QT syndrome. LQTS is a cardiovascular disorder characterized by abnormal cardiac repolarization. Two forms of inherited LQTS are known: the Romano-Ward syndrome (autosomal dominant) and the Jervell and Lange Nielsen (leading to deafness, autosomal recessive). Mutations in several ion channels are involved: KCNQ1, HERG, SCN5a, KCNE1, KCNJ2 and KCNE2.[79] Mutations in SCN5a, a $Na^+$ channel,[79,80] has been associated with Brugada syndrome. Several $K^+$ channels are involved, most predominantly KCNQ1 (42%) and HERG (45%).[79] Together, KCNQ1 and HERG account for 87% of the hereditary mutations now identified as linked with cardiac arrest. Coordinated down-regulation of KCNQ1 and KCNE1 expression in hypertrophied hearts can affect the normal current densities.[81] Individuals predisposed to ventricular tachyarrythmias caused by unstable repolarization may have "silent" forms of congenital LQTS, which are not manifest until drug exposure or environmental cues affect the repolarization. Detailed functional analyses of these channels, wt and mutant variants, in combination with screening of pharmacological compounds in a "stable" (i.e. genetically constant/defined) human cellular background may contribute to answering questions on how best to "prevent" and control arrhythmias. Exploitation of the unique opportunities of hESC to generate *in vitro* models in hES-CMs by coupling gene-targeting and directed differentiation, may lead to a better understanding of cardiac disease and possible treatment strategies.

## Concluding Remarks

In this chapter, we have discussed how hESC differentiate to cardiomyocytes, and have reviewed avenues of research currently being followed to improve the efficiency, and upscale production and specificity, based on information derived from the study of normal development and protocols effective in mouse embryonic stem cells. Whilst these approaches will undoubtedly contribute significantly to, and probably solve, current culture problems, a major unanswered question is whether the resulting cardiomyocytes will improve cardiac function following transplantation to a diseased human heart. In some cases, this would involve replacement of lost "contraction power," but in others, it may involve a source of controlled pacemaker activity to improve cardiac electrical function. In either case, extensive animal experiments, firstly in rodents then in animals with cardiac physiology more reminiscent of humans (such as pigs), will be necessary before the value and safety of hESC become clear.

## REFERENCES

1. Garcia-Martinez V, Schoenwolf GC. (1993) Primitive-streak origin of the cardiovascular system in avian embryos. *Dev Biol* **159**:706–719.
2. Olson EN, Schneider MD. (2003) Sizing up the heart: development redux in disease. *Genes Dev* **17**:1937–1956.

3. Brand T. (2003) Heart development: molecular insights into cardiac specification and early morphogenesis. *Dev Biol* **258**:1–19.
4. Nascone N, Mercola M. (1995) An inductive role for the endoderm in Xenopus cardiogenesis. *Development* **121**:515–523.
5. Schultheiss TM, Xydas S, Lassar AB. (1995) Induction of avian cardiac myogenesis by anterior endoderm. *Development* **121**:4203–4214.
6. Molkentin JD, Lin Q, Duncan SA, et al. (1997) Requirement of the transcription factor GATA4 for heart tube formation and ventral morphogenesis. *Genes Dev* **11**:1061–1072.
7. Roebroek AJ, Umans L, Pauli IG, et al. (1998) Failure of ventral closure and axial rotation in embryos lacking the proprotein convertase Furin. *Development* **125**:4863–4876.
8. Saga Y, Miyagawa-Tomita S, Takagi A, et al. (1999) MesP1 is expressed in the heart precursor cells and required for the formation of a single heart tube. *Development* **126**:3437–3447.
9. Reiter JF, Alexander J, Rodaway A, et al. (1999) Gata5 is required for the development of the heart and endoderm in zebrafish. *Genes Dev* **13**: 2983–2995.
10. Kelly RG, Buckingham ME. (2002) The anterior heart-forming field: Voyage to the arterial pole of the heart. *Trends Genet* **18**:210–216.
11. Christoffels VM, Habets PE, Franco D, et al. (2000) Chamber formation and morphogenesis in the developing mammalian heart. *Dev Biol* **223**: 266–278.
12. Zaffran S, Frasch M. (2002) Early signals in cardiac development. *Circ Res* **91**:457–469.
13. Schneider MD, Gaussin V, Lyons KM. (2003) Tempting fate: BMP signals for cardiac morphogenesis. *Cytokine Growth Factor Rev* **14**:1–4.
14. Schultheiss TM, Burch JB, Lassar AB. (1997) A role for bone morphogenetic proteins in the induction of cardiac myogenesis. *Genes Dev* **11**:451–462.
15. Shi Y, Katsev S, Cai C, et al. (2000) BMP signaling is required for heart formation in vertebrates. *Dev Biol* **224**:226–237.
16. Krishnan P, King MW, Neff AW, et al. (2001) Human truncated Smad 6 (Smad 6s) inhibits the BMP pathway in Xenopus laevis. *Dev Growth Differ* **43**: 115–132.
17. Gaussin V, Van de PT, Mishina Y, et al. (2002) Endocardial cushion and myocardial defects after cardiac myocyte-specific conditional deletion of the bone morphogenetic protein receptor ALK3. *Proc Natl Acad Sci USA* **99**: 2878–2883.
18. Monzen K, Shiojima I, Hiroi Y, et al. (1999) Bone morphogenetic proteins induce cardiomyocyte differentiation through the mitogen-activated protein kinase TAK1 and cardiac transcription factors Csx/Nkx-2.5 and GATA-4. *Mol Cell Biol* **19**:7096–7105.

19. Johansson BM, Wiles MV. (1995) Evidence for involvement of activin A and bone morphogenetic protein 4 in mammalian mesoderm and hematopoietic development. *Mol Cell Biol* **15**:141–151.

20. Lough J, Barron M, Brogley M, *et al.* (1996) Combined BMP-2 and FGF-4, but neither factor alone, induces cardiogenesis in non-precardiac embryonic mesoderm. *Dev Biol* **178**:198–202.

21. Barron M, Gao M, Lough J. (2000) Requirement for BMP and FGF signaling during cardiogenic induction in non-precardiac mesoderm is specific, transient, and cooperative. *Dev Dyn* **218**:383–393.

22. Alsan BH, Schultheiss TM. (2002) Regulation of avian cardiogenesis by Fgf8 signaling. *Development* **129**:1935–1943.

23. Lints TJ, Parsons LM, Hartley L, *et al.* (1993) Nkx-2.5: A novel murine homeobox gene expressed in early heart progenitor cells and their myogenic descendants. *Development* **119**:969.

24. Bruneau BG, Logan M, Davis N, *et al.* (1999) Chamber-specific cardiac expression of Tbx5 and heart defects in Holt-Oram syndrome. *Dev Biol* **211**:100–108.

25. Horb ME, Thomsen GH. (1999) Tbx5 is essential for heart development. *Development* **126**:1739–1751.

26. Yamagishi H, Yamagishi C, Nakagawa O, *et al.* (2001) The combinatorial activities of Nkx2.5 and dHAND are essential for cardiac ventricle formation. *Dev Biol* **239**:190–203.

27. Bruneau BG, Nemer G, Schmitt JP, *et al.* (2001) A murine model of Holt-Oram syndrome defines roles of the T-box transcription factor Tbx5 in cardiogenesis and disease. *Cell* **106**:709–721.

28. Habets PE, Moorman AF, Clout DE, *et al.* (2002) Cooperative action of Tbx2 and Nkx2.5 inhibits ANF expression in the atrioventricular canal: Implications for cardiac chamber formation. *Genes Dev* **16**:1234–1246.

29. Black BL, Olson EN. (1998) Transcriptional control of muscle development by myocyte enhancer factor-2 (MEF2) proteins. *Annu Rev Cell Dev Biol* **14**:167–196.

30. Wang D, Chang PS, Wang Z, *et al.* (2001) Activation of cardiac gene expression by myocardin, a transcriptional cofactor for serum response factor. *Cell* **105**:851–862.

31. Shin CH, Liu ZP, Passier R, *et al.* (2002) Modulation of cardiac growth and development by HOP, an unusual homeodomain protein. *Cell* **110**:725–735.

32. Wang D, Passier R, Liu ZP, *et al.* (2002) Regulation of cardiac growth and development by SRF and its cofactors. *Cold Spring Harb Symp Quant Biol* **67**:97–105.

33. Cai CL, Liang X, Shi Y, *et al.* (2003) Isl1 identifies a cardiac progenitor population that proliferates prior to differentiation and contributes a majority of cells to the heart. *Dev Cell* **5**:877–889.

34. Boheler KR, Czyz J, Tweedie D, *et al.* (2002) Differentiation of pluripotent embryonic stem cells into cardiomyocytes. *Circ Res* **91**:189–201.
35. Mummery C, Ward D, van den Brink CE, *et al.* (2002) Cardiomyocyte differentiation of mouse and human embryonic stem cells. *J Anat* **200**: 233–242.
36. Rathjen J, Lake JA, Bettess MD, *et al.* (1999) Formation of a primitive ectoderm like cell population, EPL cells, from ES cells in response to biologically derived factors. *J Cell Sci* **112** (Pt 5):601–612.
37. Rathjen J, Rathjen PD. (2001) Mouse ES cells: Experimental exploitation of pluripotent differentiation potential. *Curr Opin Genet Dev* **11**:587–594.
38. Pandur P, Lasche M, Eisenberg LM, *et al.* (2002) Wnt-1l activation of a non-canonical Wnt signalling pathway is required for cardiogenesis. *Nature* **418**:636–641.
39. Nakamura T, Sano M, Songyang Z, *et al.* (2003) Wnt- and beta-catenin-dependent pathway for mammalian cardiac myogenesis. *Proc Natl Acad Sci USA* **100**:5834–5839.
40. Hescheler J, Fleischmann BK, Lentini S, *et al.* (1997) Embryonic stem cells: A model to study structural and functional properties in cardiomyogeneis. *Cardiovasc Res* **36**:149–162.
41. Maltsev VA, Ji GJ, Wobus AM, *et al.* (1999) Establishment of beta-adrenergic modulation of L-type $Ca^{2+}$ current in the early stages of cardiomyocyte development. *Circ Res* **84**:136–145.
42. Klug MG, Soonpaa MH, Koh GY, *et al.* (1996) Genetically selected cardiomy-ocytes from differentiating embronic stem cells from stable intracardiac grafts. *J Clin Invest* **98**:216–224.
43. Meyer N, Jaconi M, Landopoulou A, *et al.* (2000) A fluorescent reporter gene as a marker for ventricular specification in ES-derived cardiac cells. *FEBS Lett* **478**:151–158.
44. Muller M, Fleischmann BK, Selbert S, *et al.* (2000) Selection of ventricular-like cardiomyocytes from ES cells *in vitro. FASEB J* **14**:2540–2548.
45. Kolossov E, Fleischmann BK, Liu Q, *et al.* (1998) Functional characteristics of ES cell-derived cardiac precursor cells identified by tissue-specific expression of the green fluorescent protein. *J Cell Biol* **143**:2045–2056.
46. Yang Y, Min JY, Rana JS, *et al.* (2002) VEGF enhances functional improvement of postinfarcted hearts by transplantation of ESC-differentiated cells. *J Appl Physiol* **93**:1140–1151.
47. Hodgson DM, Behfar A, Zingman LV, *et al.* (2004) Stable benefit of embryonic stem cell therapy in myocardial infarction. *Am J Physiol Heart Circ Physiol* **287**:H471–H479.
48. Min JY, Yang Y, Converso KL, *et al.* (2002) Transplantation of embryonic stem cells improves cardiac function in postinfarcted rats. *J Appl Physiol* **92**: 288–296.

49. Min JY, Sullivan MF, Yang Y, *et al.* (2002) Significant improvement of heart function by cotransplantation of human mesenchymal stem cells and fetal cardiomyocytes in postinfarcted pigs. *Ann Thorac Surg* **74**:1568–1575.

50. Zandstra PW, Bauwens C, Yin T, *et al.* (2003) Scalable production of embryonic stem cell-derived cardiomyocytes. *Tissue Eng* **9**:767–778.

51. Kehat I, Kenyagin-Karsenti D, Snir M, *et al.* (2001) Human embryonic stem cells can differentiate into myocytes with structural and functional properties of cardiomyocytes. *J Clin Invest* **108**:407–414.

52. Pera MF. (2001) Human pluripotent stem cells: a progress report. *Curr Opin Genet Dev* **11**:595–599.

53. Xu C, Police S, Rao N, *et al.* (2002) Characterization and enrichment of cardiomyocytes derived from human embryonic stem cells. *Circ Res* **91**:501–508.

54. He JQ, Ma Y, Lee Y, *et al.* (2003) Human embryonic stem cells develop into multiple types of cardiac myocytes: Action potential characterization. *Circ Res* **93**:32–39.

55. Zeng X, Miura T, Luo Y, *et al.* (2004) Properties of pluripotent human embryonic stem cells BG01 and BG02. *Stem Cells* **22**:292–312.

56. Mummery C, Ward-van Oostwaard D, Doevendans P, *et al.* (2003) Differentiation of human embryonic stem cells to cardiomyocytes: Role of coculture with visceral endoderm-like cells. *Circulation* **107**:2733–2740.

57. van den Eijnden-van Raaij AJ, van Achterberg TA, van der Kruijssen CM, *et al.* (1991) Differentiation of aggregated murine P19 embryonal carcinoma cells is induced by a novel visceral endoderm-specific FGF-like factor and inhibited by activin A. *Mech Dev* **33**:157–165.

58. Sachinidis A, Fleischmann BK, Kolossov E, *et al.* (2003) Cardiac specific differentiation of mouse embryonic stem cells. *Cardiovasc Res* **58**:278–291.

59. van der Heyden MA, Defize LH. (2003) Twenty one years of P19 cells: What an embryonal carcinoma cell line taught us about cardiomyocyte differentiation. *Cardiovasc Res* **58**:292–302.

60. Behfar A, Zingman LV, Hodgson DM, *et al.* (2002) Stem cell differentiation requires a paracrine pathway in the heart. *FASEB J* **16**:1558–1566.

61. Wobus AM, Kaomei G, Shan J, *et al.* (1997) Retinoic acid accelerates embryonic stem cell-derived cardiac differentiation, and enhances development of ventricular cardiomyocytes. *J Mol Cell Cardiol* **29**:1525–1539.

62. McBurney MW, Jones-Villeneuve EM, Edwards MK, *et al.* (1982) Control of muscle and neuronal differentiation in a cultured embryonal carcinoma cell line. *Nature* **299**:165–167.

63. Ventura C, Maioli M. (2000) Opioid peptide gene expression primes cardiogenesis in embryonal pluripotent stem cells. *Circ Res* **87**:189–194.

64. Paquin J, Danalache BA, Jankowski M, *et al.* (2002) Oxytocin induces differentiation of P19 embryonic stem cells to cardiomyocytes. *Proc Natl Acad Sci USA* **99**:9550–9555.

65. Takahashi T, Lord B, Schulze PC, *et al.* (2003) Ascorbic acid enhances differentiation of embryonic stem cells into cardiac myocytes. *Circulation* **107**: 1912–1916.

66. Pera MF, Andrade J, Houssami S, *et al.* (2004) Regulation of human embryonic stem cell differentiation by BMP-2 and its antagonist noggin. *J Cell Sci* **117**:1269–1280.

67. Makino S, Fukuda K, Miyoshi S, *et al.* (1999) Cardiomyocytes can be generated from marrow stromal cells *in vitro*. *J Clin Invest* **103**:697–705.

68. Sachinidis A, Gissel C, Nierhoff D, *et al.* (2003) Identification of plateled-derived growth factor-BB as cardiogenesis-inducing factor in mouse embryonic stem cells under serum-free conditions. *Cell Physiol Biochem* **13**:423–429.

69. Fijnvandraat AC, van Ginneken AC, Schumacher CA, *et al.* (2003) Cardiomyocytes purified from differentiated embryonic stem cells exhibit characteristics of early chamber myocardium. *J Mol Cell Cardiol* **35**: 1461–1472.

70. Snir M, Kehat I, Gepstein A, *et al.* (2003) Assessment of the ultrastrucutral and proliferative properties of human embryonic stem cell-derived cardiomyocytes. *Am J Physiol Heart Circ Physiol* **285**:H2355–H2363.

71. Caspi O, Gepstein L. (2004) Potential applications of human embryonic stem cell-derived cardiomyocytes. *Ann N Y Acad Sci* **1015**:285–298.

72. Kehat I, Gepstein A, Spira A, *et al.* (2002) High-resolution electrophysiological assessment of human embryonic stem cell-derived cardiomyocytes: A novel *in vitro* model for the study of conduction. *Circ Res* **91**:659–661.

73. Wobus AM, Wallukat G, Hescheler J. (1991) Pluripotent mouse embryonic stem cells are able to differentiate into cardiomyocytes expressing chronotropic responses to adrenergic and cholinergic agents and $Ca^{2+}$ channel blockers. *Differentiation* **48**:173–182.

74. An RH, Davies MP, Doevendans PA, *et al.* (1996) Developmental changes in beta-adrenergic modulation of L-type $Ca^{2+}$ channels in embryonic mouse heart. *Circ Res* **78**:371–378.

75. Blake DJ, Weir A, Newey SE, *et al.* (2002) Function and genetics of dystrophin and dystrophin-related proteins in muscle. *Physiol Rev* **82**: 291–329.

76. Giglio V, Pasceri V, Messano L, *et al.* (2003) Ultrasound tissue characterization detects preclinical myocardial structural changes in children affected by Duchenne muscular dystrophy. *J Am Coll Cardiol* **42**:309–316.

77. Rando TA. (2001) The dystrophin-glycoprotein complex, cellular signaling, and the regulation of cell survival in the muscular dystrophies. *Muscle Nerve* **24**:1575–1594.

78. Grady RM, Teng H, Nichol MC, *et al.* (1997) Skeletal and cardiac myopathies in mice lacking utrophin and dystrophin: A model for Duchenne muscular dystrophy. *Cell* **90**:729–738.

79. Splawski I, Shen J, Timothy KW, *et al.* (2000) Spectrum of mutations in long-QT syndrome genes KVLQT1, HERG, SCN5A, KCNE1, and KCNE2. *Circulation* **102**:1178–1185.

80. Wilde AA, Antzelevitch C, Borggrefe M, *et al.* (2002) Proposed diagnostic criteria for the Brugada syndrome: consensus report. *Circulation* **106**: 2514–2519.

81. Ramakers C, Vos MA, Doevendans PA, *et al.* (2003) Coordinated down-regulation of KCNQ1 and KCNE1 expression contributes to reduction of I(Ks) in canine hypertrophied hearts. *Cardiovasc Res* **57**:486–496.

# 16

# Differentiation of Human Embryonic, Fetal and Adult Stem Cells to Islet Cells of the Pancreas

Enrique Roche and Bernat Soria*

## Introduction

Stem cells display the ability of robust self renewal by symmetric cell divisions, but at the same time under certain conditions these cells can differentiate in a committed lineage by asymmetric cell fate. Taken together these abilities of stem cells represent a new challenge in Regenerative Medicine and offer unlimited sources of tissues to treat degenerative diseases caused by defective functioning of a specific cell type.

In this context, the main objective of Regenerative Medicine is to generate customized tissues in order to restore the lost function in the organism, ideally in the absence of immune rejection. The possibility of *in vitro* regeneration of entire organs is still very unlikely, although obtaining specific cell types seems to be a more realistic goal. Therefore, those pathologies originated by cellular dysfunction would, in principle, be candidates to benefit from cell replacement strategies. This includes neurodegenerative diseases, cardiovascular alterations and osteoarticular pathologies, besides others.

Diabetes, one of the most prevalent diseases in the industrialized world, can be included into this group of pathologies.[1] The disease is caused by a malfunctioning of pancreatic β-cells, which are the unique cells in the adult, responsible for insulin production and secretion. This peptidic hormone is instrumental in nutrient homeostasis and the absence of other compensatory hormones obliges diabetic people to depend for survival on daily injections of the hormone from exogenous sources. Unfortunately, insulin injections cannot mimic time- and dose-precise endogenous insulin

---

*Correspondence: Instituto de Bioingenieria, Universidad Miguel Hernandez, 03550-San Juan, Alicante, Spain. Tel.: 34-96 591 9531, Fax: 34-96-591 9546, e-mail: bernat.soria@umh.es

release by the β-cell and this is why patients develop diabetic complications, such as neuropathy, nephropathy, retinopathy and diverse cardiovascular disorders.[1]

The first successful attempt in Regenerative Medicine in the treatment of diabetes has been the transplantation of islets of Langerhans isolated from cadaveric donors.[2] The transplantation protocol, initiated 25 years ago, has recently been improved by the group of Shapiro in Edmonton, Canada, and it presents several advantages compared to the classical double transplant of kidney and whole pancreas. First of all, only the structures containing insulin (islets) are transplanted, thereby reducing the volume of the organ that has to be manipulated, optimizing the surgical procedure and reducing immune response. A second advantage of the Edmonton protocol is the immunosuppressive drug regime that has less deleterious effects on islet survival.

More than 400 patients in the world have benefited from islet transplantation and some of them have not had the need for daily insulin injection in the last 4 years. No major adverse effects and diabetic complications were observed in the follow-up of these patients. However, due to scarcity of donor material and the low yield of islets obtained during the isolation procedure (less than 50% of the content), not all diabetic people can have access to this new surgery protocol.

Stem cells would become a key alternative for cell replacement protocols in diabetes. In addition, the use of emerging technologies, such as nuclear transfer, oocyte parthenogenesis and cell reprogramming may be the next revolution in Regenerative Medicine, allowing the production of customized cells to treat specific diseases. Once these techniques become available, the generation of reliable insulin-producing tissues will be a reality. At present, only a few approaches have been used to obtain insulin-producing cells from embryonic, fetal and adult stem cells. This chapter intends to highlight the key points of published material in this field in order to demonstrate the possibilities that stem cells offer in the context of Regenerative Medicine for a future cure of diabetes.

# Differentiation of Islet Cells from Mouse and Human Embryonic Stem Cells

## Mouse

Differentiation protocols involve removal of replicating embryonic stem cells (ESCs) from monolayers in the presence of leukemia inhibitory factor

(LIF) or feeder layers, allowing culture in suspension in the absence of LIF or feeders. Under these conditions cells spontaneously tend to aggregate and form spheroid structures known as embryoid bodies (EBs).[3] These seem to be instrumental in initiating cell differentiation programs through operating molecular mechanisms that at present are still unknown. Many cell types displaying gene markers from ectoderm, mesoderm and endoderm arise from EBs when they are forced to grow on adherent monolayers.

Many authors, including us, have reported the spontaneous differentiation of mouse embryonic stem cells (mESCs) to insulin-producing cells. In those studies, insulin gene expression, together with expression of other pancreatic hormones, was observed in EBs as early as 10–23 days.[4–6] However, the fraction of insulin positive cells in the subsequent outgrowth phase is less than 0.1% of the total cells present in the culture. This observation suggested improvements in two fronts in order to obtain insulin-producing cells:

a) Increase the percentage of insulin-positive cells in total culture.
b) Eliminate contaminating cells present in the culture in order to obtain a pure population of insulin-positive cells.

With these ideas in mind, we were the first to develop a protocol that demonstrates that ESC can differentiate *in vitro* to insulin-containing cells.[7] The elimination of cells that do not express insulin was performed by using a double-selection gating system, and its efficiency was previously reported when cardiomyocytes were obtained.[8] The construct used for transfection contained a hygromycin resistance gene under the control of the constitutive promoter of the phosphoglycerate kinase gene, thereby allowing the selection of single clonal ESCs during the expansion stage. The other element was formed by a neomycin selection cassette under the control of the regulatory regions of the human insulin gene. Therefore, cells resistant to neomycin are expressing the insulin gene at the same time. This strategy allowed the selection of insulin-producing lineages from others that are eliminated when neomycin was added to the culture medium.[9]

To increase the percentage of insulin-positive cells in the culture, we combined the selection strategy with a final maturation process that was performed during the outgrowth culture phase. Three determinants, used also in subsequent protocols by others, were instrumental in this strategy:

a) Incubation was carried out in bacterial Petri dishes, allowing cells to form aggregates that mimic islet cell-to-cell interactions and favoring subsequent manipulations for animal transplantation.

b) Lowering of glucose concentration from 25 mM (typical of Dubecco's Modified Eagle's Medium: DMEM) to 5 mM in order to normalize expression levels of glucose-regulated genes, which are linked to correct glucose sensing and hormone secretion.

c) Addition of nicotinamide which acts as a differentiation factor, as it has been previously described in human fetal pancreatic cells.[10]

Several insulin-producing clones were obtained. Insulin was produced in low amounts in the majority of the cells (less than 0.1 ng/$\mu$g protein) and only few clones (around 1%) displayed high amounts of insulin (0.1–0.2 ng/$\mu$g protein). Two clones had significant amounts of insulin suitable for further maturation and transplantation protocols (0.5 and 0.8 ng/$\mu$g protein respectively). The possible cause of this variability will be discussed later on in this chapter. These results suggest redefinition of new DNA constructs for cell selection and subsequent changes during the maturation phase in order to increase the yield of insulin production.

Streptozotocin diabetic mice transplanted with the best insulin-containing cells recovered to normoglycemia compared with sham-diabetic animals. Furthermore, *in vivo* glucose tolerance tests showed an euglycemic response in transplanted animals, very similar to controls, that was completely absent in sham-diabetic mice. The circulating levels of the hormone in transplanted animals were around 50% of the insulin measured in control animals and less than 5% in streptozotocin-diabetic mice. These results suggest that the isolated bioengineered cells are not exactly differentiated $\beta$-cells, but they have a certain ability to mimic $\beta$-cell function which needs further improvements.

Based on this work, we have recently developed a new protocol using current knowledge of key transcription factors in $\beta$-cell development.[11] Gating technology incorporated the use the Nkx6.1 gene promoter controlling the resistance to neomycin. Nkx6.1 is a transcription factor involved in the expansion of $\beta$-cell precursors.[12] An additional change comprised the presence of differentiation factors already in the EB phase, yielding 20% of cells positive for insulin. The selection with neomycin produced a pure population of cells co-expressing insulin, Pdx1, Nkx6.1, as well as genes coding for key proteins involved in the glucose-sensing process such as glucokinase, GLUT-2 and Sur-1. The differentiation factors include nicotinamide, anti-sonic hedgehog or coculture with pancreatic buds. All these factors, and in particular coculture with pancreatic rudiments, were very effective during the maturation phase, yielding cells that responded to increasing concentrations of extracellular glucose. Subsequent transplantation into the kidney capsule restored normoglycemia for 3 weeks in streptozotocin-diabetic mice and reverted to hyperglycemia once the graft was removed. Altogether, this new strategy offers reproducible amounts of intracellular insulin content which can be improved by adding differentiation factors to the medium or using new maturation protocols. Comparison between both protocols is shown in Table 1.

Another set of protocols used another directed differentiation strategy without any further selection (Table 2). These protocols have been subsequently developed based on the idea that nestin-positive cells could be precursors of neuronal and endocrine pancreatic cells. Although this is a matter of scientific debate, nestin-expressing cells have been detected in

**Table 1.** Directed Differentiation and Selection Strategies to Obtain Insulin-producing Cells from mESCs

|  | Soria *et al.*[7] | León-Quinto *et al.*[11] |
|---|---|---|
| *Cell type* | R1<br>Transfected with:<br>Ins-neo$^r$/pGK-hygro$^r$ | D3<br>Transfected with:<br>Nkx6.1-neo$^r$/pGK-hygro$^r$ |
| *Expansion phase* | +LIF<br>+hygromycin (clonal selection)<br>10% FBS | +LIF<br>+hygromycin (clonal selection)<br>15% FBS |
| *Differentiation phase* | EBs 8-10d<br>10% FBS | EBs 7d<br>+Differentiation factors*<br>3% FBS |
| *Maturation phase* | Plating 5-8d<br>+neomycin<br>Cluster formation 14d<br>+10 mM nicotinamide<br>+5 mM glucose<br>10% FBS | Plating 7d<br>+Differentiation factors*<br>10% FBS<br>Selection 20–30d<br>+neomycin<br>15% FBS |
| *Insulin content before transplantation* | 0.1–0.8 ng/µg protein | 0.2 ng/µg protein |

*Differentiation factors: 10 mM nicotinamide or anti-sonic hedgehog or coculture with pancreatic rudiments.

FBS: fetal bovine serum; hygro$^r$: hygromycin resistence gene; Ins: regulatory regions of human insulin gene; neo$^r$: neomycin resistance gene; Nkx6.1: Nkx6.1 gene promoter; pGK: phosphoglycerate kinase promoter.

**Table 2.** Directed Differentiation Strategies to Obtain Insulin-producing Cells from mESCs and hESCs

| | Lumelsky et al.[15] | Hori et al.[16] | Moritoh et al.[19] | Blyszczuk et al.[17] | Miyazaki et al.[18] | Segev et al.[21] |
|---|---|---|---|---|---|---|
| *Cell type* | E 14.5 E 14.1 B5 | JM1 ROSA | EB3 Ins2-neo[r] | R1 CMV-Pax4 | EB3 Pdx1-Tet off | H9.2 |
| *Stage 1* | Expansion 2-3d +LIF | Expansion 2d +LIF | Expansion +LIF | Expansion +LIF | Expansion 2–3d +LIF | Expansion on MEF* |
| *Stage 2* | EBs 4d | EBs | EBs 4-5d | EBs 5d | EBs 4-5d | EBs 7d |
| *Stage 3* | Nestin+ 6-7d +ITSFn Serum free | Nestin+ 6d +ITSFn Serum free | Outgrow 2d 10% FCS 4-7d +ITSFn Serum free | Nestin+ 4d | Outgrow 1 – 2d Nestin+ 3-6d +ITSFn Serum free | Plating 7d +ITSFn |
| *Stage 4* | 6d +N2 +B27 +bFGF | 6d +N2 +B27 +bFGF | 6-8d +N2 +B27 +KGF +EGF +bFGF | 8d +N2 +B27 +NIC 20% FBS | 6–8d +MHM +bFGF +EGF +KGF | Expansion +N2 +B27 +bFGF |
| *Stage 5* | 6d +N2 +B27 +NIC | 6d +N2 +B27 +NIC +10 µM LY294002 | | 8d +N2 +B27 +NIC Serum free | MHM −bFGF −EGF −KGF | +N2 +B27 +NIC Low glucose |
| *Insulin content* | 0.145 ng/µg prot | 0.39 ng/µg prot | | 0.45 ng/µg prot | 0.05 ng/µg prot | |

bFGF: basic fibroblast growth factor; EGF: epidermal growth factor; FBS: fetal bovine serum; KGF: keratinocyte growth factor; MEF*: mitotically inactivated mouse embryonic fibroblasts; neo[r]: neomycin resistance gene; NIC: nicotinamide.

rat and human pancreatic islets.[13] However, cell-lineage tracing strategies suggest that nestin-positive progenitors contribute to the formation of the microvasculature of the islet.[14] The first protocol was developed by the group of McKay[15] and comprised 5 stages:

1) Expansion of ESCs in the presence of LIF.
2) EB formation in the absence of LIF.
3) Selection of nestin-positive cells in the outgrowth phase by plating the EBs in serum-free ITSFn medium (DMEM/F12 containing insulin, transferrin, selenium and fibronectin).
4) Expansion of nestin-positive cells in serum-free N2 medium [DMEM/F12 containing B27 supplement, insulin, transferrin, selenium, progesterone and putrescine supplemented with bFGF (basic Fibroblast Growth Factor)] or FGF2.
5) Expression of insulin, performed by incubation of the cells in clusters with the previously described N2 medium, but in the absence of bFGF and in the presence of nicotinamide.

Subsequent protocols have added some changes to this general scheme:

— Use of cell growth inhibitors in the last stage, such as the phosphoinositide 3-kinase inhibitor LY294002.[16]
— Achieving constitutive or regulated expression of key transcription factors in β-cell function, such as Pdx1 or Pax4.[17,18]
— Changing the "cocktail" of growth factors added during the expansion of nestin-positive cells.[18]
— Combining directed differentiation with gating selection strategies.[18,19]

All the protocols yielded cells positive for insulin in different amounts that were able to correct hyperglycemia in transplanted animals in some studies.

## Human

Human embryonic stem cells (hESCs) do not respond to human LIF and even when cultured on feeder layers of mitotically inactivated fibroblasts, they initiate differentiation processes. Therefore, it is more difficult to maintain them in a pluripotent state in culture. Despite this problem, some groups have able to obtain insulin-positive cells from hESCs. The approaches used are an extension of those developed in the mouse model: spontaneous differentiation[20] and directed differentiation.[21]

Spontaneous differentiation performed in one study successfully yielded insulin-positive cells, but intracellular hormone content or C-peptide presence was not reported.[20] A subsequent report introduced a directed differentiation strategy based on the selection of nestin-precursors developed in mice.[21] The method was improved by lowering the glucose concentration in a final step of the differentiation protocol (Table 2). In fact, the combination of low glucose concentration, presence of nicotinamide and formation of clusters, as previously described in mESCs,[7] has been shown to be critical determinant in improving insulin content and secretion in cells derived from hESCs. However, the authors of this publication claimed that the final cells were immature islet-like aggregates. Their conclusions were based on the following observations: low responsiveness to glucose and co-expression of insulin with other pancreatic endocrine hormones, such as glucagon and somatostatin.[22] Although improvements in this protocol are necessary, it is important to point out that hESCs can be bioengineered *in vitro* to produce insulin-secreting cells and thus future strategies to treat diabetes can be envisaged.

Altogether, the approaches from mESCs or hESCs need specific improvements in order to set up a definitive protocol. Optimization of characterization criteria will be necessary in the near future in order to compare results between different laboratories. For instance, insulin staining and radioimmuno-detection are inadequate, because the hormone can be taken up from the culture medium into the cells, especially when they enter into apoptosis.[23] Insulin is used in several steps of protocols based on the selection of nestin-positive cells and the serum added to the medium is another source of exogenous insulin. This observation therefore suggests that C-peptide detection must be used as a probe for endogenous insulin production.

In addition, since some ectoderm-derived tissues have the ability to express insulin genes, bioengineering protocols may also have to consider the presence of ectodermal derived cells positive for insulin. Indeed, pancreatic β-cells derive from the endoderm, but they share many phenotypic characteristics with some neuroectoderm-derived cells. These include:

a) Cell excitatory machinery, such as SUR1, Kir6.2, $K_{ATP}$ channels and voltage dependent L-type $Ca^{2+}$ channels.[24]
b) Proteins of the secretory machinery as well as exocytotic mechanisms.[25–27]
c) Glucose-sensing machinery, including the glucose transporter GLUT-2 and glucokinase.[28,29]

d) Transcription factors such as Neurogenin3, Beta2, Pax4, Pax6, Pdx1, Nkx2.2 and Nkx6.1.[30–34]

e) The insulin itself that has been detected transiently during embryogenesis in ectoderm-derived tissues such as neuronal tube and neuronal crest of transgenic mice.[35] In this context, rodents possess two non-allelic related genes identified as insulin I and insulin II. Insulin I gene expression is typical of pancreatic β-cells, while insulin II is expressed in yolk sac, developing brain and also in pancreatic β-cells.[36]

This last point is extremely important because insulin-positive tissues derived from the ectoderm and endoderm display some key differences in the mechanisms involved in insulin gene regulation, the amount of hormone synthesized, and the mechanisms of processing and secretion. Contrary to rodents, chickens and humans have a single insulin gene, but it can be also expressed in neuroectoderm-derived tissues.[37,38] Ectodermal insulin is synthesized and quickly secreted as unprocessed proinsulin due to the absence of endopeptidase PC2. In addition, the amount of ectoderm-derived insulin produced is thousand times much lower than the insulin produced by the endocrine pancreas. Furthermore, the physiological roles of both peptides are different: ectodermal insulin acts as a growth factor during embryogenesis while endoderm-derived insulin is instrumental in nutrient homeostasis in the adult organism.[39–41]

All these data indicate that the insulin gene can be expressed in ectodermal derived tissues *in vivo* and that this possibility may also be considered in bioengineering protocols. Figure 1 summarizes the strategies followed so far in our laboratory. These are not exclusive of other possibilities not yet published, such as the use of 3-D structures resembling extracellular matrix, use of conditioned media, cell reprogramming with cell extracts, etc. Taken altogether, these observations suggest the possibility of obtaining ESC-lineages committed to endoderm, detection of insulin I expression at least in mESCs and development of additional tests to assess the presence of functional groups of proteins, which include the nutrient sensing machinery, the exocytotic apparatus and the pathway responsible for hormone synthesis, processing and storage. At present, the final cells obtained in the published protocols have partially addressed these points. Therefore, fulfilling these criteria is necessary to mimic β-cell function and to produce human-derived candidate cells for transplantation trials.

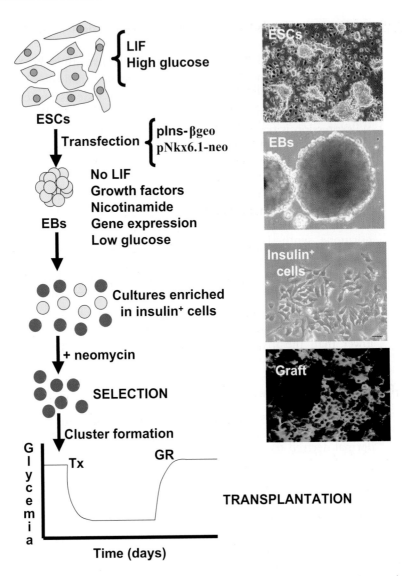

**Figure 1.** Summary of strategies used in our laboratory to obtain insulin-producing cells from embryonic stem cells. Abbreviations used: GR: graft removal; Tx: transplant. Pictures on the right correspond to D3-ESCs and derived cells. ESC: embryonic stem cells; EB: embryoid bodies; Insulin+ cells after differentiation and selection; Graft: insulin-positive (green) cells in the removed graft.

## Differentiation of Islet Cells from Human Fetal Stem Cells

Not much work has been performed with human fetal stem cells (hFSCs) compared with ESCs. In addition, many functional aspects of endocrine fetal pancreas are still poorly characterized. The main characteristic of fetal immature β-cells is the unresponsiveness to extracellular glucose concentrations. However, these cells secrete insulin when exposed to amino acids *in vitro*.[42] As mentioned before, co-expression of insulin with other pancreatic endocrine hormones is another feature of immature islets.[22] Furthermore, the amount of islet clusters obtained is lower than in an adult human pancreas. Altogether, these observations indicate that strategies for expansion and further differentiation in mature structures are necessary.

In this context several factors have been tried to expand and differentiate human fetal pancreatic islets, such as fibrin extracellular matrices, three-dimensional structures that favor cell-to-cell interactions, nicotinamide, hepatic growth factor, activin A and betacellulin.[10,43–45] In order to design directed differentiation and proliferation strategies, it is important to identify the nature of the precursor(s) cell(s) present in endocrine fetal pancreatic tissue. It has been reported that nestin-positive cells could be involved,[46] but this is still a matter of debate.[47,48]

## Differentiation of Islet Cells from Human Adult Stem Cells

Adult stem cells (ASCs) represent a good alternative for cell replacement protocols, mainly because these cells may be obtained from the patient, thus will not produce immune rejection once implanted into a recipient. However, the low proliferative capacity *in vitro* and the commitment to specific cell lineages could constrain their clinical application. With respect to the pancreas, we have to consider two types of ASCs: pancreatic and extra-pancreatic stem cells.

### Pancreatic stem cells

Endocrine pancreas, as well as the nervous system and heart, are organs with a very limited proliferative capacity compared to epidermis of the skin, intestinal epithelium, muscle, liver (after hepatectomy) and the very well studied bone marrow tissue (responsible for blood cell generation and osteoarticular turnover). Nevertheless, an interesting finding is that the tissues with no proliferation, such as the nervous system, bear progenitor stem cells. This is a key finding, because the replication and differentiation of these cells could be stimulated *in vivo* or *in vitro* prior to recipient implantation. Unfortunately, the presence of stem cells in pancreatic tissue still remains an elusive question.

Several reports indicate that islet neogenesis from pancreatic duct cells play an important role in the recovery of endocrine mass after partial pancreatectomy. Furthermore, islet-like cell structures or cultivated human islets buds (CHIBs) have been obtained from the density gradient fraction rich in duct cells during islets fractionation.[49] To achieve this, cells need to be expanded in serum free ITS medium (insulin + transferrin + selenium) containing nicotinamide and keratinocyte growth factor. After several days of culture, confluent cells were overlayed with Matrigel, which favors the formation of tridimensional structures positive for insulin and glucagon. However, the amount of insulin produced and the β-cell mass obtained are still very low for transplantation protocols, suggesting the need to search for protocols to promote cell expansion and enrich the hormone content.

These data seem to favor the existence of a progenitor or stem-like cell in the pancreas, located in ducts that can be bioengineered *in vitro* to produce insulin-positive cells. If these cells are implicated *in vivo* in the endocrine pancreas, turnover is still a matter of debate. In this context, Melton has provided evidence that new β-cells arise from replication of pre-existing differentiated β-cells, questioning the existence of a stem cell population in pancreatic tissue.[50] The experimental design is based on genetic lineage tracing of insulin gene expression cells. It is difficult to argue against the rationale and results of these elegant experiments. Even more interesting, when the same gradient density fraction was devoid from pancreatic β-cells by exposure to alloxan or streptozotocin (that selectively attack insulin-producing cells) no CHIBs were obtained (unpublished results by several groups, including ours).

Since regeneration depends on the size of the insult, it may be concluded that β-cell replication does not exclude islet neogenesis in the rodent after subtotal pancreatectomy (90% removal) or the participation of pancreatic progenitors from islets, provided that those unipotent progenitors are capable of expressing insulin. In addition, it is still unknown if the endocrine pancreatic stem cells respond equally to different stimuli, including mechanical (i.e. pancreatectomy) or chemical (i.e. streptozotocin administration) insults. Altogether, the data indicate that complementary approaches should be adopted to address this relevant question for pancreas regeneration.

Therefore, the main conclusion from all these studies is that the potential pancreatic stem cell has not been isolated nor identified so far. Some laboratories are claiming that nestin-positive cells within the duct could be this potential cell. But nestin is a protein of intermediary filaments present in neurons derived from the ectoderm, while endocrine pancreatic cells are derived from the endoderm. In addition, recent evidence supports the participation of nestin-positive cells in vascular intra-islet structures.[14]

Demonstration and subsequent characterization of the putative pancreatic stem-progenitor cell would be a key finding in the field, opening new possibilities for autotransplantation. Replacement protocols using pancreas from cadaveric donors would require the setting up of *in vitro* expansion and differentiation protocols. In addition, increasing the insulin production of ductal tissue opens new possibilities on the use of cadaveric pancreatic tissue, autotransplantation and pancreas regeneration. However, more research is required to exploit this area.

Although very preliminary, the possibility of using de-differentiated exocrine tissue to obtain islet-like structures has been suggested. Under certain circumstances exocrine tissue may transdifferentiate into liver tissue, a close derivative of the endoderm. The proposal would be that exocrine tissue may lose the differentiation stage to a progenitor that may be further engineered to endocrine tissue.

## Extra-pancreatic stem cells

Bone marrow has been considered as a stem cell reservoir in the adult organism, containing cells with multipotentiality beyond blood and bone precursors. Mesenchymal cells from the bone marrow, side population cells and the recently described MAPCs (multipotent adult progenitor cells)[51] display a multipotentiality much higher than expected. MAPCs from humans, primates and rodents proliferate actively and may be induced to differentiate into cells positive for ectoderm, mesoderm and endoderm markers. However, the production of insulin-containing cells from these cells has not been reported.

A subsequent report added insulin-secreting cells to the list of cells that can transdifferentiate from bone marrow stem cells.[52] However, this finding has not been supported by other laboratories.[53,54] There is no clear explanation for these contradictory findings, because transdifferentiation has been in great debate recently due to the unexpected capacity of certain ASCs to fuse with differentiated cell types and acquire the recipient phenotype. However, bone marrow stem cells can contribute to neovasculature by becoming endothelial cells.[55] In addition, signals derived from the vasculature play an instrumental role in early pancreas development.[56] In this context, new blood vessels should have the ability to induce pancreas regeneration in streptozotocin-diabetic mice.[55] The existence of a pancreatic pluripotent stem cell responding to a variety of extracellular factors, including those derived from intra-islet endothelium, cannot be excluded.

Peripheral blood may also contain pluripotential progenitors. Recent reports have described the presence of monocyte-derived pluripotent stem cells.[57–59] A subset of progenitor cells of monocyte origin (PCMO) differentiates into endodermal-derived tissues, such as hepatocytes and insulin-containing cells.[57,58] Albeit very preliminary, these observations open the possibility of using patient peripheral blood cells to generate insulin-producing cells for replacement protocols.

Gene therapy protocols have shown that insulin can be expressed in other cell types besides β-cells. However, the expression of the insulin gene in ectopic tissues is not a sufficient requisite, because the hormone needs to be correctly processed and properly secreted. Interesting results have been observed in hepatic tissue which, as well as β-cell, originates from endoderm. The expression of certain transcription factors essential for β-cell function such as Pdx1 or NeuroD in hepatic cells resulted in the induction of the insulin gene.[60,61] Fetal liver stem cells have shown to be easily engineered to express a β-cell phenotype.[62] All these studies offer the potential use of liver biopsies to generate functional islet-like structures for cell therapy of diabetes and represent an alternative source of ASCs for autotransplantation.

Although important improvements have been performed in this area, several issues need to be still resolved with respect to the use of pancreatic precursors or ASCs from extra-pancreatic tissues. It is clear that adult β-cells have a limited capacity to expand and differentiate and pancreatic ductal tissue has inadequate insulin content to restore hyperglycemia in transplantation protocols. Therefore, the search for strategies to increase β-cell proliferation *in vitro* or insulin production from ductal tissue will be instrumental in taking more advantage of pancreases from cadaveric donors. On the other hand, transdifferentiation from extra-pancreatic tissues requires further study to clarify certain caveats, including genetic reprogramming mediated by either cell fusion events or extracellular signals generated after implantation in specific niches.

## Conclusions

A reasonable number of reports clearly demonstrate that ESCs, FSCs and ASCs could be, in different degrees, a potential source of cells for the treatment of diabetes in Regenerative Medicine (Fig. 2). However, optimization of protocols and standardization of characterization criteria are the main gaps to fill before the establishment of a definitive strategy. Laboratory experience from the different groups is important in order to generate key information for a final step-by-step protocol to obtain insulin-producing cells from ESCs, FSCs or ASCs. The final cell product has to mimic as much as possible the phenotype and function of a mature β-cell in order to assure appropriate cell replacement and restoration of the lost function in the organism.

However, this is just the beginning because transplantation of bioengineered cells is likely to pose new challenges that scientists would have to face.

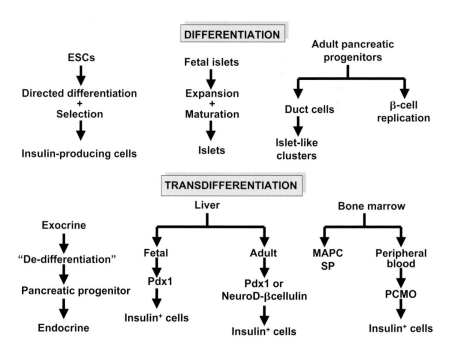

**Figure 2.** Scheme representing the different strategies to obtain insulin-producing cells from ESCs and ASCs. Abbreviations used: MAPC: multipotent adult progenitor cell; SP: side population.

Some of them are:

— Immune rejection.
— Survival of the implant, taking in account anti-necrotic and anti-apoptotic mechanisms.
— Site of implantation, to assure correct nutrition and oxygen supply.
— Biosafety mechanisms, to eliminate the graft in case of non-function or tumor formation.

In conclusion, we are beginning to know the potential of stem cells and their possible applications to cure diabetes.[63,64] It is possible that new β-cells will be created in the near future by using all those emerging technologies. To achieve this goal we need to improve on the actual protocols and further investigate the basic biology of stem cells.

## REFERENCES

1. DeFronzo RA, Ferrannini E, Keen H, Zimmet P. (2004) *International Textbook of Diabetes Mellitus*, 3rd ed. Chichester, UK: John Wiley and Sons.
2. Shapiro AMJ, Lakey JRT, Ryan EA, *et al.* (2000) Islet transplantation in seven patients with type 1 diabetes mellitus using a corticoid-free immunosuppressive regime. *N Eng J Med* **343**: 230–238.
3. Smith AG. (1991) Culture and differentiation of embryonic stem cells. *J Tissue Culture Methods* **13**: 89–94.
4. Soria B, Skoudy A, Martín F. (2001) From stem cells to beta cells: New strategies in cell therapy of diabetes mellitus. *Diabetologia* **44**: 407–415.
5. Shiroi A, Yoshikawa M, Yokota H, *et al.* (2002) Identification of insulin-producing cells derived from embryonic stem cells by zinc-chelating dithizone. *Stem Cells* **20**: 284–292.
6. Kahan BW, Jacobson LM, Hullett DA, *et al.* (2003) Pancreatic precursors and differentiated islet cell types from murine embryonic stem cells. An *in vitro* model to study islet differentiation. *Diabetes* **52**: 2016–2024.
7. Soria B, Roche E, Berná G, *et al.* (2000) Insulin-secreting cells derived from embryonic item cells normalize glicemia in streptozotocin-induced diabetic mice. *Diabetes* **49**: 157–162.
8. Klug MG, Soonpa MH, Koh GY, Field LJ. (1996) Genetically selected cardiomyocytes from differentiating embryonic stem cells form stable intracardiac grafts. *J Clin Invest* **98**: 216–224.
9. Roche E, Burcin MM, Esser S, *et al.* (2003) The use of gating technology in bio-engineering insulin-secreting cells from embryonic stem cells. *Cytotechnology* **41**: 145–151.

10. Otonkoski T, Beattie GM, Mally MI, *et al.* (1993) Nicotinamide is a potent inducer of endocrine differentiation in cultured human fetal pancreatic cells. *J Clin Invest* **92**: 1459–1466.

11. León-Quinto T, Jones J, Skoudy A, *et al.* (2004) *In vitro* directed differentiation of mouse embryonic stem cells into insulin-producing cells. *Diabetologia* **47**: 1442–1451.

12. Chakrabarti SK, Mirmira RG. (2003) Transcription factors direct the development and function of pancreatic β cells. *Trends Endocrinol Metab* **14**: 78–84.

13. Zulewski H, Abraham EJ, Gerlach MJ, *et al.* (2001) Multipotential nestin-positive stem cells isolated from adult pancreatic islets differentiate *ex vivo* into pancreatic endocrine, exocrine and hepatic phenotypes. *Diabetes* **50**: 521–533.

14. Treutelaar MK, Skidmore JM, Dias-Leme CL, *et al.* (2003) Nestin-lineage cells contribute to the microvasculature but not endocrine cells of the islet. *Diabetes* **52**: 2503–2512.

15. Lumelsky N, Blondel O, Laeng P, *et al.* (2001) Differentiation of embryonic stem cells to insulin-secreting structures similar to pancreatic islets. *Science* **292**: 1389–1394.

16. Hori Y, Rulifson IC, Tsai BC, *et al.* (2002) Growth inhibitors promote differentiation of insulin-producing tissue from embryonic stem cells. *Proc Natl Acad Sci USA* **99**: 16105–16110.

17. Blyszczuk P, Czyz J, Kania G, *et al.* (2003) Expression of Pax4 in embryonic stem cells promotes differentiation of nestin-positive progenitor and insulin-producing cells. *Proc Natl Acad Sci USA* **100**: 998–1003.

18. Miyazaki S, Yamato E, Miyazaki J. (2004) Regulated expression of Pdx-1 promotes *in vitro* differentiation of insulin-producing cells from embryonic stem cells. *Diabetes* **53**: 1030–1037.

19. Moritoh Y, Yamato E, Yasui Y, *et al.* (2003) Analysis of insulin-producing cells during *in vitro* differentiation from feeder-free embryonic stem cells. *Diabetes* **52**: 1163–1168.

20. Assady S, Maor G, Amit M, *et al.* (2001) Insulin production by human embryonic stem cells. *Diabetes* **50**: 1691–1697.

21. Segev H, Fishman B, Ziskind A, *et al.* (2004) Differentiation of human embryonic stem cells into insulin-producing clusters. *Stem Cells* **22**: 265–274.

22. Polak M, Bouchareb-Banaei L, Scharfmann R, Czernichow P. (2000) Early pattern of differentiation in the human pancreas. *Diabetes* **49**: 225–232.

23. Rajagopal J, Anderson WJ, Kume S, *et al.* (2003) Insulin staining of ES cell progeny from insulin uptake. *Science* **299**: 363.

24. Dumm-Meynell AA, Rawson NE, Levin BE. (1998) Distribution and phenotype of neurons containing the ATP-sensitive $K^+$ channel in rat brain. *Brain Res* **814**: 41–54.

25. Lang J. (1999) Molecular mechanisms and regulation of insulin exocytosis as a paradigm of endocrine secretion. *Eur J Biochem* **259**: 3–17.
26. Gerber SH, Südhof TC. (2002) Molecular determinants of regulated exocytosis. *Diabetes* **51** (**Suppl. 1**): S3–S11.
27. Rorsman P, Renström E. (2003) Insulin granule dynamics in pancreatic beta cells. *Diabetologia* **46**: 1029–1045.
28. Yang XJ, Kow LM, Funabashi T, Mobbs CHV. (1999) Hypothalamic glucose sensor. Similarities to and differences from pancreatic β-cell mechanisms. *Diabetes* **48**: 1767–1772.
29. Schuit FC, Huypens P, Heimberg H, Pipeleers DG. (2001) Glucose sensing in pancreatic beta-cells: A model for the study of other glucose-regulated cells in gut, pancreas, and hypothalamus. *Diabetes* **50**: 1–11.
30. Sommer L, Ma Q, Anderson DJ. (1996) Neurogenins, a novel family of atonal-related bHLH transcription factors, are putative mammalian neuronal determination genes that reveal progenitor cell heterogeneity in the developing CNS and PNS. *Mol Cell Neurosci* **8**: 221–241.
31. Chu K, Nemoz-Gaillard E, Tsai MJ. (2001) BETA2 and pancreatic islet development. *Recent Prog Horm Res* **56**: 23–46.
32. Lemke G. (1993) Transcriptional regulation of the development of neurons and glia. *Curr Opin Neurobiol* **3**: 703–708.
33. Pérez-Villamil B, Schwartz PT, Vallejo M. (1999) The pancreatic homeodomain transcription factor IDX1/IPF1 is expressed in neuronal cells during brain development. *Endocrinology* **140**: 3857–3860.
34. Pattyn A, Vallstedt A, Días JM, *et al.* (2003) Complementary roles for Nkx6 and Nkx2 class proteins in the establishment of motoneuron identity in the hindbrain. *Development* **130**: 4149–4159.
35. Alpert S, Hanahan D, Teitelman G. (1988) Hybrid insulin genes reveal a developmental lineage for pancreatic endocrine cells and imply a relationship with neurons. *Cell* **53**: 295–308.
36. Melloul D, Marshak S, Cerasi E. (2002) Regulation of insulin gene transcription. *Diabetologia* **45**: 309–326.
37. Pérez-Villamil B, de la Rosa EJ, Morales AV, de Pablo F. (1994) Developmentally regulated expression of the preproinsulin gene in the chicken embryo during grastrulation and neurulation. *Endocrinology* **135**: 2342–2350.
38. Hernández-Sánchez C, Mansilla A, de la Rosa EJ, *et al.* (2003) Upstream AUGs in embryonic proinsulin mRNA control its low translation level. *EMBO J* **22**: 5582–5592.
39. Vicario-Abejón C, Yusta-Boyo MJ, Fernández-Moreno C, de Pablo F. (2003) Locally born olfactory bulb stem cells proliferate in response to insulin-related factors and require endogenous insulin-like growth factor-I for differentiation into neurons and glia. *J Neurosci* **23**: 895–906.

40. Hernández-Sánchez C, Rubio E, Serna J, de la Rosa EJ, de Pablo F. (2002) Unprocessed proinsulin promotes cell survival during neurulation in the chick embryo. *Diabetes* **51**: 770–777.

41. de la Rosa EJ, de Pablo F. (2000) Cell death in early neural development: Beyond the neurotrophic theory. *Trends Neurosci* **23**: 454–458.

42. Nielsen JH. (1985) Growth and function of the pancreatic β-cell *in vitro*: Effects of glucose, hormones and serum factors on mouse, rat and human pancreatic islets in organ culture. *Acta Endocrinol* **108**: 1–40.

43. Beattie GM, Rubin JS, Mally MI, *et al.* (1996) Regulation of proliferation and differentiation of human fetal pancreatic islet cells by extracellular matrix, hepatocyte growth factor and cell–cell contact. *Diabetes* **45**: 1223–1228.

44. Beattie GM, Montgomery AM, Lopez AD, *et al.* (2002) A novel approach to increase human islet cell mass while preserving β-cell function. *Diabetes* **51**: 3435–3439.

45. Demeterco C, Beattie GM, Dib SA, *et al.* (2000) A role for activin A and betacellulin in human fetal pancreatic cell differentiation and growth. *J Clin Endocrinol Metab* **85**: 3892–3897.

46. Huang H, Tang X. (2003) Phenotypic determination and characterization of nestin-positive precursors derived from human fetal pancreas. *Lab Invest* **83**: 539–547.

47. Humphrey RK, Bucay N, Beattie GM, *et al.* (2003) Characterization and isolation of promoter-defined nestin-positive cells from the human fetal pancreas. *Diabetes* **52**: 2519–2525.

48. Gao R, Ustinov J, Pulkkinen M-A, *et al.* (2003) Characterization of endocrine progenitor cells and critical factors for their differentiation in human adult pancreatic cell culture. *Diabetes* **52**: 2007–2015.

49. Bonner-Weir S, Taneja M, Weir GC, *et al.* (2000) *In vitro* cultivation of human islets from expanded ductal tissue. *Proc Natl Acad Sci USA* **97**: 7999–8004.

50. Dor Y, Brown J, Martínez OI, Melton DA. (2004) Adult pancreatic β-cells are formed by self-duplication rather than stem-cell differentiation. *Nature* **429**: 41–46.

51. Jiang Y, Jahagirdar BN, Reinhardt RL, *et al.* (2002) Pluripotency of mesenchymal stem cells derived from adult marrow. *Nature* **418**: 41–49.

52. Ianus A, Holz GG, Theise ND, Hussain MA. (2003) *In vivo* derivation of glucose-competent pancreatic endocrine cells from bone marrow without evidence of cell fusion. *J Clin Invest* **111**: 843–850.

53. Choi JB, Uchino H, Azuma K, *et al.* (2003) Little evidence of transdifferentiation of bone marrow-derived cells into pancreatic beta cells. *Diabetologia* **46**: 1366–1374.

54. Lechner A, Yang Y-G, Blacken RA, *et al.* (2004) No evidence for significant transdifferentiation of bone marrow into pancreatic β-cells *in vivo*. *Diabetes* **53**: 616–623.

55. Hess D, Li L, Martin M, *et al.* (2003) Bone marrow-derived stem cells initiate pancreatic regeneration. *Nat Biotech* **21**: 763–770.
56. Kim SK, Hebrok, M. (2001) Intercellular signals regulating pancreas development and function. *Genes Dev* **15**: 11–127.
57. Ruhnke M, *et al.* (2005) *Gastroenterology* (in press).
58. Ruhnke M, *et al.* (2005) *Transplantation* (in press).
59. Zhao Y, Glesne D, Huberman E. (2003) A human peripheral blood monocyte-derived subset acts as pluripotent stem cell. *Proc Natl Acad Sci USA* **100**: 2426–2431.
60. Ferber S, Halkin A, Cohen H, *et al.* (2000) Pancreatic and duodenal homeobox gene 1 induces expression of insulin genes in liver and ameliorates streptozotocin-induced hyperglycemia. *Nat Med* **6**: 568–572.
61. Kojima H, Fujimiya M, Matsumura K, *et al.* (2003) NeuroD-betacellulin gene therapy induces islet neogenesis in the liver and reverses diabetes in mice. *Nat Med* **9**: 596–603.
62. Zalzman M, Gupta S, Giri RK, *et al.* (2003) Reversal of hyperglycemia in mice by using human expandable insulin-producing cells differentiated from fetal liver progenitor cells. *Proc Natl Acad Sci USA* **100**: 7253–7258.
63. Colman A. (2004) Making new beta cells from stem cells. *Semin Cell Dev Biol* **15**: 337–345.
64. Roche E, Soria B. (2004) Generation of new islets from stem cells. *Cell Biochem Biophys* **40** (**Suppl**): 113–123.

# 17

# Progress in Islet Transplantation

A.M. James Shapiro* and Sulaiman Nanji

## Introduction

Phenomenal progress has occurred within the recent five years in cellular replacement therapy for selected patients with type 1 diabetes. Islet transplantation replaces cells destroyed by autoimmune destruction with cells derived from cadaveric donor pancreas organs, but recipients need to take potent anti-rejection drugs to prevent allograft destruction. Islet transplantation can provide a new lease of life for patients with unstable type 1 diabetes, and many are rendered completely free of exogenous insulin injection therapy with excellent glucose control. The treatment has generated enormous enthusiasm for the approach, particularly within the patient community. Unfortunately, there are not sufficient organ donors to supply anywhere close to the potential need. Furthermore, the side effects and potential risks of the anti-rejection therapies needed to sustain graft function can only be justified in those few patients with the severest forms of brittle diabetes. The process involved with islet extraction from the whole organ pancreas is complex, unreliable and can only be done in specialized centers. Not all isolations result in a clinical transplant, and since patients usually need two successful islet preparations to render them insulin free, upwards of four to six donor pancreas organs may be processed in order to provide the two good preparations needed to treat a single patient. Furthermore, many cells are destroyed during the isolation process, during islet storage in culture and many islets fail to engraft once transplanted into the recipient. Therefore, while islet transplantation today has generated huge hope and promise, it is clear that islet replacement therapy in its current form cannot provide a practical cure for type 1 diabetes. While islet transplantation provides proof of concept that cellular replacement therapy

*Correspondence: Clinical Research Chair in Transplantation (CIHR/Wyeth), University of Alberta, 2000 College Plaza, 8215 112th Street, Edmonton AB Canada T6G 208. Tel: (780) 407 7330, fax: (780) 407 6933, e-mail: shapiro@islet.ca

in diabetes is feasible, alternate sources of regulated, insulin-secreting cells are needed if this therapy is to become more of a mainstay treatment. Xenotransplantation using pig islets as a donor source is evolving slowly in preclinical models and early clinical trials in children with diabetes in Mexico have met with controversy. Methods to proliferate isolated islets in culture using combination growth factors offers one practical approach that will likely deliver first in the clinic. Ultimately stem cell derived, insulin-secreting tissues will render islet transplantation and a need to isolate islets from cadavers redundant.

# Background

Diabetes mellitus afflicts over 200 million people worldwide, representing the third most common disease and fourth leading cause of death in North America.[1] There are two principal categories of this disease: type 1 refers to those with autoimmune pancreatic islet destruction and typically manifests in childhood or early adulthood. There are 30 000 new type 1 patients diagnosed with diabetes annually in North America;[2] type 2 diabetes typically affects obese individuals with peripheral insulin resistance and reduced insulin secretion. The incidence of type 2 diabetes is increasing rapidly, and accounts for the major impact of this disease. Type 2 diabetes is lifestyle-related, and recent data has established that intervention through combined diet and physical activity with modest weight reduction can substantially reduce the incidence of the disease by up to 40% to 60% over 4 years in those able to comply with therapy.[3] The overall cost burden of diabetes and its secondary complications to the global society comprises 9% to 15% of healthcare expenses in developed countries, and the global cost has been estimated to accrue at over \$153 billion (US) on an annual basis. There is no known cure for type 1 diabetes, and mainstay treatment consists of chronic insulin injection. While exogenous insulin therapy has dramatically reduced mortality from diabetes, patients often succumb to the long-term sequelae of diabetic angiopathy, either in the form of nephropathy, neuropathy or retinopathy. Maintaining rigorous glycemic control with intensive insulin therapy has been shown to delay and sometimes prevent the progression of these complications,[4] but patients are at risk of severe and sometimes fatal hypoglycemic events.[5,6] Although insulin pumps and implantable insulin-secreting devices are a promising approach to improved glucose homeostasis, the development of reliable and accurate glucose sensor technology has been a limiting factor. A more physiologic approach to correct the diabetic state is the transplantation of insulin-producing tissue.

Transplantation of insulin secreting tissue through beta cell replacement therapy may be accomplished today either by vascularized whole pancreas transplantation or by islet transplantation. Over 20 000 whole pancreas transplants have been performed worldwide, and while this is an invasive and potentially risky surgical procedure, it can very effectively restore normal endogenous insulin secretion, maintain long-term glucose homeostasis, and improve quality of life.[7] Accrued evidence has established that some secondary complications of diabetes can be stabilized and even reversed by successful pancreas transplantation.[8] Simultaneous pancreas and kidney transplantation is presently considered the standard of care for selected patients with type 1 diabetes with end-stage renal failure, with a success rate of greater than 90% at one year.[8] Recent modifications in surgical technique, with portal venous drainage and enteric exocrine drainage have led to more physiological glucose control, and reduced morbidity compared with the previous bladder-drainage technique. Furthermore, avoidance or rapid tapering of steroid therapy early post transplant has improved wound healing, reduced the risks of surgical site infection, and improved metabolic control. Nonetheless, the procedure remains invasive and associated with the potential risk of mortality. The risk-benefit of whole pancreas transplantation has been debated intensively, and Venstrom *et al.* have claimed that patient survival is compromised after pancreas alone or pancreas after kidney transplants, compared with patients awaiting this procedure in the USA.[9] Others have questioned the analysis of that data, and the issue remains controversial. Furthermore, in view of the risks associated with surgery and long-term immunosuppressive drug therapy, pancreas transplantation is largely reserved for patients with diabetes with clinically significant complications, where the severity of their disease justifies accepting the risks of the procedure and immunosuppression. Therefore, with the exception of rare patients with severe, labile forms of diabetes, pancreas transplantation is not a practical option for young patients with diabetes who have not yet developed complications.

A promising alternative is the transplantation of islet cells isolated from donor pancreases and embolized into the recipient liver via the portal vein (Fig. 1). Compared with pancreas transplantation, islet transplantation is technically far simpler, has a low morbidity, and offers the opportunity for storage of the islet graft in tissue culture prior to infusion. Moreover, the fact that islets can be kept in culture provides a unique opportunity to immunologically manipulate the islet graft, as well as optimize recipient conditioning prior to transplantation, thereby

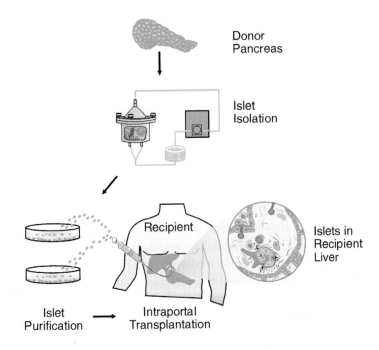

**Figure 1.** Processes involved with islet extraction from the donor cadaveric pancreas.

facilitating tolerance induction. The low morbidity of the procedure and the potential for tolerance induction make islet transplantation a promising strategy for correcting diabetes in young patients, including children,[10] prior to the establishment of secondary complications.

Dr Paul Lacy and colleagues were the first to show that chemical diabetes could be cured in mice and rats through islet transplantation in the late 1960s.[11,12] Drs John Najarian and David Sutherland began the first series of islet autografts and allografts at the University of Minnesota in 1977. Initial clinical attempts with islet allografts were disappointing, but islet auto-transplantation was often effective in rendering patients insulin free.[13,14] One of these early patients remains insulin independent with excellent glucose control to this day, almost 19 years after her transplant.

Recent progress in the science of islet isolation has increased the efficiency of the process, and has had a major impact in enhancing the consistency and quality of highly purified islet preparations for safe transplantation into patients.[15,16] Delivery of collagenase enzymes injected down the pancreatic duct led to cleavage of islets from their acinar-islet interface more effectively than any method described previously, but it still led to significant islet destruction through inadvertent islet enzyme penetration.[17] However, the process did permit the successful isolation of islets from the human pancreas.[18] The approach was recently refined further by Lakey *et al.* to allow precise control of the temperature and perfusion pressure.[19]

A major advance came with the introduction of a semi-automated dissociation chamber and process originally developed by Dr Camillo Ricordi *et al.* in 1988, modifications of which have now become the universal standard for successful high yield large animal and human islet isolation.[20] The chamber is now universally known as the Ricordi Chamber. The collagenase-distended pancreas is placed inside a stainless steel chamber containing glass marbles and a 500 μm mesh screen and mechanically dissociated by gentle agitation, with tissue samples evaluated sequentially to determine the end-point before liberated islets become fragmented by over-digestion. This novel approach minimizes trauma to the islets in a continuous digestion process with the collection of free islets as they are liberated from the digestion chamber. Since the introduction of this technique, many laboratories around the world have utilized this system for the isolation of human islets.

Large-scale purification of human islets of suitable quality for safe transplantation into the human portal vein was enhanced considerably by the introduction of an automated refrigerated centrifuge system (COBE 2991) by Lake *et al.* which permitted rapid large volume Ficoll gradient processing in a closed system 600 ml bag.[21] This system reduced the exposure time of islets to potentially toxic Ficoll, and typically renders a final islet preparation of 2–4 ml of tissue for safe embolization into the portal vein.

A major limitation to successful pancreatic digestion has been the source, quality and variability in collagenase activity and contaminants in various enzyme blends. A class of purified enzyme blend (Liberase™, Roche Pharmaceuticals, Indianapolis, USA) was developed that contained collagenase I, II, thermolysine, clostripain and clostridial neutral protease and had a low endotoxin activity. This provided more consistent and enhanced islet yield compared with crude collagenase, but still remains the weakest link in the entire process, as the enzymes do not remain stable during storage.[22] Despite the key advances in collagenase quality, intraductal enzyme delivery, automated dissociation and purification outlined above, inconsistency remains in the overall success of the islet isolation procedure, which may reflect variability in donor-related factors (donor inotropic need, duration of cardiac arrest, hyperglycemia, age and obesity in the donor, in addition to the skills of the local procurement team).[23]

## Evolution in Outcomes in Clinical Islet Transplantation

Transplantation of isolated pancreatic islets is an appealing approach to the treatment of insulin-dependent diabetes mellitus. However, the perennial hope that such an approach would result in long-term freedom from the need for exogenous insulin, with stabilization of the secondary complications of diabetes, has been slow to materialize in practice (Fig. 2).

A total of over 447 attempts to treat type 1 diabetes with islet allografts were reported to the Islet Transplant Registry between 1974 and 2000, the vast majority of which occurred within the most recent decade.[24] Mainstay immunosuppression was largely based on the combination of glucocorticoids, cyclosporine and azathioprine, with anti-lymphocyte serum induction.[25] The majority of these grafts were combined islet-kidney transplants, since it was felt inappropriate to initiate new immunosuppression in islet-alone recipients who would not have otherwise required therapy to sustain another solid organ kidney or liver graft. Under these protocols, fewer than 10% of patients were able to discontinue insulin therapy for longer than one year, although 28% had sustained C-peptide secretion at one year post transplant.[26] These disappointing results contrasted with the success of islet autografts, and partial success of islet allografts in

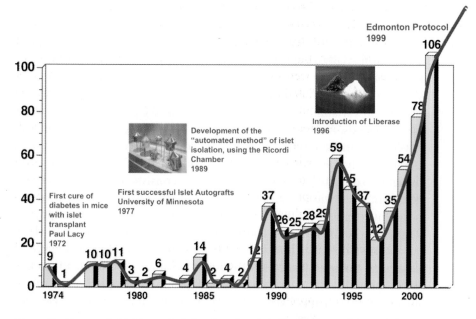

**Figure 2.**  Progress in islet transplantation from bench to the clinic over the past 30 years.

non-diabetic pancreatectomized recipients where glucocorticoid-free immunosuppression was combined with unpurified islet preparations.[27]

A key question remained unanswered — were the previous poor results of islet allografts in type 1 diabetic recipients a result of poor control of alloimmune pathways, or did they reflect recurrence of autoimmune diabetes? Insulin independence was only rarely achievable under glucocorticoid and cyclosporine-based immunosuppression. C-peptide secretion diminished to zero over time in most cases, suggesting islet graft loss from acute rejection or possible recurrence of autoimmune diabetes. Results of whole pancreas transplantation indicated that stable graft function was achievable over time, even with lower dose maintenance immunosuppression, suggesting that prevention of autoimmune destruction might be more readily achieved than prevention of alloimmune rejection. Autoimmune recurrence after whole pancreas transplantation only appears to be a challenge when no immunosuppression is given, as occurred in the unique situation of a living-donor hemi-pancreas transplant between identical twins, where autoimmune recurrence led to graft loss within two months.[28]

Results of islet transplantation improved in the mid-1990s under cyclosporine, glucocorticoid and azathioprine immunosuppression, together with anti-IL2 receptor induction and antioxidants. Combined data from the Giessen and Geneva (GRAGIL consortium) groups suggested that up to 50% of patients had detectable C-peptide secretion (indicating graft insulin production), but only 20% of these patients remained insulin free at one year.[29,30] Islets were cultured for a mean of two days, and mean islet implant mass was 9000 IE/kg, derived from single donors in half of the cases. Two of 10 patients achieved insulin independence after single-donor islet infusions, but it took 6 to 8 months to achieve independence, and both were recipients of shipped islets from a central islet isolation site.[29]

The University of Milan subsequently reported experience in type 1 diabetic islet after kidney recipients using antilymphocyte serum together with cyclosporine, azathioprine and corticosteroids, and compared outcomes with an alternative regimen using antithymocyte globulin, cyclosporine, mycophenolate mofetil and metformin.[31] Rates of insulin independence were enhanced from 33% to 59% with the elimination of prednisone, and addition of mycophenolate and metformin. Over 50% of patients maintained insulin independence beyond one year, likely as a result of more effective immunosuppression coupled with anti-inflammatory, less diabetogenic and improved insulin action with the newer protocol.

## The Edmonton Protocol

A new protocol was developed in Edmonton, Canada in 1999 that radically changed the face of clinical islet transplantation. The first seven patients treated under the Edmonton Protocol all achieved and maintained insulin independence beyond one year, demonstrating for the first time that islet transplantation could achieve insulin independence with rates similar to whole pancreas transplantation, but without a need for a major inter-ventional surgery.[32] The success of this protocol has been attributed to a number of key modifications from previous clinical trials (Fig. 3). Of note, patients were given an adequate number of high-grade islets prepared from an average of two donor organs. Furthermore, a more potent but less diabetogenic, steroid-free anti-rejection therapy was developed using sirolimus, low-dose tacrolimus and an anti-interleukin-2 receptor mono-clonal antibody (anti-IL-2R mAb).

Since the release of the early Edmonton results, considerably more expe-rience has been accrued both in Edmonton and at other centers worldwide. At the University of Alberta, a total of 70 patients have now received islet-alone transplants. (Tables 1, 2 and 3) Most patients continue to require two islet infusions in order to provide adequate engraft mass (approximately 12 000 IE/kg islet mass, based on the recipient body weight), but approx-imately 10% become insulin-free after just one islet infusion. Of patients undergoing completed islet transplants, 82% remain insulin-free by the

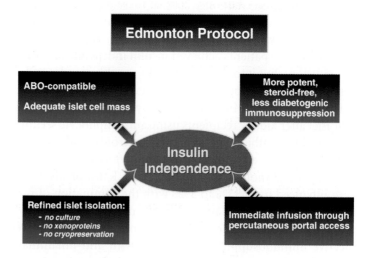

**Figure 3.** Key components and concepts behind the Edmonton Protocol.

**Table 1.** Product Release Criteria Prior to Clinical Islet Transplantation

| | |
|---|---|
| Islet count | ≥ 250 000 IE |
| Islet mass per kg | ≥ 4000 IE/kg (based on recipient weight) |
| Tissue packed cell volume | ≤ 5.0cc |
| Islet Viability | ≥ 70% |
| Gram stain on final prep | Negative |
| Endotoxin content | < 5 EU/kg (based on recipient weight) |

**Table 2.** Indications for Clinical Islet Transplantation

1. Islet-alone transplantation

   a. Type 1 diabetes (C-peptide negative) for more than 5 years
   b. Evidence for compliance with intensive and optimal insulin and monitoring regimen
   c. Evidence for failure of optimal insulin therapy, with severe hypoglycemia, hypoglycemic unawareness or glycemic instability, as measured by:
      i. Hypo Score > 900
      ii. Lability Index > 400

2. Islet-after kidney transplantation

   a. Type 1 diabetes and stable kidney allograft function
   b. Immunosuppression with "islet-friendly," steroid free, sirolimus/tacrolimus based therapy

3. Other

   a. Type 1 diabetes and stable other solid organ allograft function (e.g. heart, lung, liver, etc.)
   b. Immunosuppression with "islet-friendly," steroid free, sirolimus/tacrolimus based therapy

end of one year. There is some fall off in insulin independence, with 70% remaining insulin-free at two years and 50% free at three years post transplant. Most patients who return to insulin continue to secrete endogenous insulin (and C-peptide) in sufficient amounts to continue to stabilize risk of hypoglycemic reactions or of glycemic lability, and 88% of patients continue to demonstrate islet function as long as five years post transplant. Islet transplantation has proven to be remarkably successful in stabilizing glucose control to a degree that is vastly superior to even intensive insulin therapy, and patients typically demonstrate normalization of HbA1C.[33] As a result of this high level of success, a number of Provinces in Canada

**Table 3.** Current Contraindications to Clinical Islet Transplantation

| | |
|---|---|
| 1. | Severe co-existing, uncorrectable cardiac disease |
| 2. | Active alcohol or substance abuse |
| 3. | Psychiatric disorder if it makes the subject non-compliant with therapy |
| 4. | History of non-adherence to prescribed regimens |
| 5. | Active infection, including hepatitis C, hepatitis B, HIV |
| 6. | Any history of or current malignancies except squamous or basal skin cancer |
| 7. | Age less than 18 or greater than 65 years |
| 8. | Creatinine clearance $< 65\,\mathrm{mL/min/1.73\,m^2}$ |
| 9. | Serum creatinine $> 150\,\mu\mathrm{mol/L}(1.7\,\mathrm{mg/dL})$ |
| 10. | Macroalbuminuria (urinary albumin excretion rate $> 300\,\mathrm{mg/24\,h}$) |
| 11. | Positive pregnancy test, intent for future pregnancy or failure to follow effective contraceptive measures, or presently breast feeding |

now regard islet transplantation as a "non-experimental" alternative standard of care for selected patients with unstable forms of type 1 diabetes. An international multicentre trial of the Edmonton Protocol was recently completed by the Immune Tolerance Network in nine sites, and this study demonstrated that the original Edmonton findings could be replicated at times to a very high level of success, depending on the experience of the site.[34] Worldwide, there have now been over 400 patients treated since 1999, and increasing momentum and focus on the remaining challenges of islet isolation, alternative insulin-secreting regulated sources, better immunosuppression with less side effects, and the possibility of immunological tolerance continue to drive the field forward.

## Recent Advances since the Edmonton Protocol

Over the past five years since the introduction of the Edmonton Protocol, extensive progress has continued, and notably there are now more patients transplanted with islets since the year 2000 than the total number in the preceding 30 years. This expanded experience has not only confirmed the initial Edmonton findings, but has highlighted important limitations that must be overcome if islet transplantation is to be more broadly applied as a potential cure for diabetes. The results of islet after kidney transplants appear to match the success of islet alone transplants under sirolimus-based immunosuppression. The Milan group has recently shown that long-term islet graft function (with persistent C-peptide secretion) can not only prolong the half-life of a kidney transplant, but is associated with a reduced incidence of diabetic vascular complications leading to significantly enhanced patient survival.[35]

Notable progress includes the introduction of the perfluorodecalin (PFC) "two-layer" method for pancreas protection during transportation and rescue of marginal donors, successful single-donor islet transplants from obese, non-heart-beating donors, and the routine use of islet culture rather than immediate transplantation to further improve the purity, practicality and safety of the procedure.[36] The risk of acute bleeding following percutaneous transhepatic access to the portal vein has been reduced substantially by physical and mechanical ablation of the catheter tract using combinations of coils and thrombostatic agents. The use of a bag rather than a syringe for islet delivery has further improved the sterility and safety of the procedure.[37]

Dr Bernhard Hering and colleagues at the University of Minnesota have recently achieved a remarkably high level of success with single donor islet infusions using refinements in pancreas shipment, islet processing and recipient immune suppression.[38] It is not certain whether these refinements make the biggest contribution to the high level of single donor success, or whether it relates more to selection of optimal organ donors, short cold ischemic shipment times and selection of only optimal, low weight, insulin sensitive recipients.[39]

Recent attention has been focused on the loss of viable islets not only during the isolation and purification process,[40] but also when embolized into the portal vein of the recipient liver. Based on metabolic tests in post-transplant recipients, it is estimated that only 25% to 50% of the implanted islet mass actually engrafts in the patient.[33] Recently, Dr Olle Korsgren and colleagues in Sweden have shown that human islets exposed to ABO-compatible blood triggers an "instant blood mediated inflammatory reaction" (IBMIR), characterized by activation of platelets and the coagulation and complement systems, leading to islet damage by clot formation and leukocyte infiltration.[41] Further investigation into the mechanisms of this phenomenon revealed that tissue factor and thrombin play critical roles in mediating IBMIR, indicating that strategies to block binding of these factors may have considerable therapeutic potential in islet transplantation. Furthermore, in recent years, several experimental strategies have been developed to enhance islet engraftment. For instance, anti-inflammatory treatment with TNF-alpha-receptor antibody in a marginal mass islet model in mice, as well as antioxidant therapy with nicotinamide, vitamin D3, pentoxiphylline or cholesterol lowering agents pravastatin or simvastatin, have all demonstrated positive impact in the pre-clinical setting, and suggest a potential role in future clinical trials designed to improve islet engraftment.

## Immediate Next Steps for Improvement in Islet Transplant

Islet transplantation has clearly come of age, but refinements are needed to move this therapy to a point where it can be considered an effective alternative to insulin for the broad population of desperate patients with type 1 diabetes. Key areas of focus include:

a) Optimization and more biologically relevant monitoring of the manufacture of purified collagenase enzyme blends to improve the reliability of islet isolation. Use of HPLC profiles to measure peaks of type 1 and type 2 collagenase may be helpful, and is beginning to provide an opportunity to blend an exact collagenase activity to adjust for different digestion characteristics of a particular human pancreas.

b) Development of real-time, predictive potency assays for *in vitro* assessment of islets before proceeding with transplantation. *In vitro* islet stimulation in high glucose to measure the insulin stimulation provides only a crude index of function, and refinements under development include dynamic islet perifusion, oxygen consumption rate analysis, islet ATP consumption, and calcium flux oscillations.

c) Use of a variety of growth factors (e.g. hepatocyte growth factor, epidermal growth factor, gastrin, Exendin-4, GLP-1, INGAP) to promote islet proliferation and expansion during culture, or systemically in the recipient to promote expansion either of transplanted islets or possibly of endogenous islets within the native pancreas.

d) Further refinements in anti-rejection therapy with the goal of preventing allograft rejection or autoimmune recurrence without non-immunosuppressant side effects that are frequently encountered in islet recipients currently.

## Recent Progress in Recipient Immunosuppression

While the Edmonton immunosuppressive protocol represents a major step forward, the medications used in this protocol are associated with significant side effects that unfavorably alter the risk benefit ratio of islet transplantation. The therapy is currently unsuitable for the majority of patients with type 1 diabetes that are adequately stabilized on insulin. Approximately 10% of patients with type 1 diabetes, however, have severe forms of glycemic lability or recurrent hypoglycemia, and islet-alone transplantation has therefore been restricted to this target population initially. If the myriad of nonimmune side effects associated with current immunosuppressive therapy could be avoided, islet transplantation could be more broadly applied in diabetes. The side effects of current anti-rejection therapies are due in large part to the reagents' imprecise mechanisms of action, as most clinically approved immunosuppressive maintenance drugs target signaling pathways with a near ubiquitous cellular distribution.

Calcineurin inhibitors have numerous unwanted side effects, including nephrotoxicity, diabetes, hypertension, impaired lipid metabolism, and hirsuitism. Even when drug levels are kept low, significant side effects may develop. This is particularly true in the diabetic patient population where renal function may already be impaired. The incidence of chronic renal failure in nonrenal transplant recipients is an astonishing 16.5%.[42] Similarly, the Edmonton group has reported that patients with underlying impaired renal function can experience accelerated nephrotoxicities even when low doses of tacrolimus are used.[33] This underscores the appealing and essential nature of a calcineurin inhibitor-free immunosuppressive regimen, particularly for islet transplantation.

Nonimmune toxicities are also observed with sirolimus in islet recipients. These include mouth ulceration, peripheral edema, dyslipidemia, weight loss, leukopenia, ovarian cysts and anemia. While not directly nephrotoxic itself, sirolimus appears to potentiate calcineurin-inhibitor associated nephrotoxicity, and this may be particularly troublesome with increased proteinuria and rise in creatinine when sirolimus is used at high dose together with tacrolimus when there is underlying diabetic nephropathy. While alterations in immunosuppressive regimens alone will not solve the issue of having an adequate supply of islets for transplantation, the negative impact of tacrolimus and sirolimus on islet engraftment and function after transplantation should not be underestimated. This point is underscored by the recent observation that the incidence of new onset immunosuppressive-related diabetes mellitus approaches 30% in the first two years following renal transplantation, when tacrolimus is used as the primary immunosuppressive agent. Unfortunately, dose reduction seldom reverses this condition. Recent data from the kidney transplant literature suggests that sirolimus may also exacerbate the diabetogenicity of tacrolimus. Furthermore, as a powerful non-selective anti-proliferative agent, *high*-dose sirolimus may impair early islet revascularization and engraftment, and may further impede the transdifferentiation of ductal-derived CK19 positive elements within the islet graft that have the potential capacity for islet expansion after transplantation. The antiangiogenic effects of sirolimus were clearly demonstrated in tumor metastasis in growth model in mice, where sirolimus was linked to markedly reduced production of vascular endothelial growth factor (VEGF).[43] By improving islet engraftment and function, the initial development of tacrolimus and steroid-free with *low*-dose sirolimus immunosuppression, and ultimately tacrolimus and sirolimus free regimens, is predicted to reduce the islet mass required to achieve euglycemic insulin-independence and thus facilitate successful transplantation using only single donors.

An exciting array of potentially "islet-friendly" alternative antirejection therapies are now becoming available that will likely further change the face of islet transplantation in the next few years. If the risk of nonimmune related side effects could be further reduced, then patients could be treated earlier in the course of type 1 diabetes, perhaps even at the time of diagnosis or including children with the disease. New agents include:

a) Non-depletional T-cell therapies, including hOKT3$\gamma_1$-Ala-Ala used in the Hering trial.[38]

b) Depletional T-cell therapies, including Alemtuzumab (Campath-1H), an anti-CD52 monoclonal antibody that depletes lymphocytes for prolonged periods after transplantation. An anti-CD3 diphtheria-conjugated immunotoxin has shown considerable promise in primate kidney and islet transplant models and is being developed for clinical trials presently. This agent was most effective when combined with 15-deoxyspergualin (DSG).[44–46]

c) Costimulation blockade therapies. T cells require two signals in order to trigger an immune response — Signal 1 being engagement of the T-cell receptor with allograft derived antigens, and Signal 2 being a number of costimulatory molecules that increase the avidity of binding between T and B cells, and strongly promote intracellular T-cell signaling events. Blocking costimulation while leaving T-cell receptor-antigen engagement unaltered effectively renders the alloreactive T-cell population anergic, forcing them to apoptose.[47–49] Two critical costimulatory pathways include CD28:B7 and CD40:CD40L, and the blockade of these pathways with CTLA4-Ig or CD40L mAb, respectively, has been shown to promote long-term allograft survival in a variety of small and large animal transplantation models. These promising experimental results led to Phase 1 clinical trials of CD40L blockade using a humanized antibody called hu5C8. Unfortunately, the use of this agent produced unexpected thromboembolic complications that resulted in one patient mortality.[50,51] Subsequent reports demonstrated that both platelets and endothelium express high levels of CD40L,[52,53] leading researchers to consider blocking the CD40 epitope on antigen presenting cells, thereby avoiding concerns of cross-reactivity associated with CD40L blockade.[54] Recently, this approach has demonstrated prolongation of renal allografts without evidence of thromboembolic complications.[55]

Furthermore, development of LEA29Y, a second generation CTLA4-Ig with an approximately 10-fold more potency *in vitro* has demonstrated significant benefit in primate models of islet transplantation when combined with sirolimus and an anti-IL-2R mAb.[56] Presently, a multicenter Phase III clinical trial of LEA29Y is underway in renal transplant recipients, and preliminary results suggest low rates of acute rejection, excellent graft function and minimal side effects. A trial of LEA29Y in clinical islet transplantation is currently underway at the Edmonton and Emory sites.

d) Agents that alter trafficking and recruitment of lymphocytes. Chemokines play a critical role in allograft rejection by orchestrating lymphocyte migration and activation. FTY-720, a sphingosine 1-phosphate agonist, inhibits lymphocyte trafficking and is effective in preventing allograft rejection both in preclinical and clinical trials. Interest in exploring FTY720 in islet transplantation is based on recent promising data of this compound in experimental and clinical transplantation. It has been shown to prevent allograft rejection in several rodent models of allotransplantation, and more recently in primate renal transplantation.[57] With particular relevance to islet transplantation, FTY720 has been shown to potently inhibit autoimmune diabetes and recurrent disease in NOD mice, and this agent may enhance insulin action without any diabetogenic side effects in mouse and primate models. Moreover, studies at the University of Minnesota and Miami in nonhuman primate islet transplantation have demonstrated that the combination of FTY720 and RAD (Everolimus) is effective in maintenance of immunosuppression following basiliximab induction therapy.[58] Based on this study, as well as strong preliminary data in mouse models of islet allograft rejection, and in autoimmune diabetes,[59,60] clinical trials using FTY720 in islet transplantation are imminent.

It remains to be seen, as stem cell technologies evolve, whether recipient immunosuppression will be required. Certainly, if autologous stem cells with insulin secretional capacity are developed, immunosuppression will not be necessary unless these cells express target epitopes that are susceptible to autoimmune destruction. If allogeneic stem cells are developed for clinical use, then either immunosuppressant strategies as outlined above, or tolerance promoting techniques may be required to sustain graft function.

## Tolerance Induction

The possibility of achieving a permanent state of unresponsiveness to an allograft without the need for chronic immunosuppression remains an important focus in transplantation research. It has been suggested that islet transplantation could serve as a primary testing ground for novel tolerance protocols, since a lack of efficacy would result in a patient's return to insulin therapy rather than potential death in the case of failure of a life-sustaining heart or liver transplant. Moreover, the fact that islets can survive in culture provides the opportunity to not only to immunologically manipulate the graft during the cultured state but also optimize recipient conditioning prior to transplantation. Although islet transplantation offers a unique opportunity to test new tolerance strategies for clinical application, it may prove to be a challenging model because of the need to overcome both alloimmune and autoimmune barriers, and different mechanistic approaches may ultimately be required to achieve this.

Tolerance has been achieved experimentally in small animal inbred strain models using two key approaches — mixed chimerism through myeloablation and donor bone marrow transplantation, or through peripheral mechanisms, including deletion, anergy and regulation of T cells. Attempts to induce tolerance in the clinic through mixed chimerism have been successful, with rare reports of bone marrow transplant recipients with established donor chimerism that have been able to accept a renal transplant from the same donor without further immunosuppression.[61] However, serious concerns regarding toxicities with recipient pre-conditioning and the risk of graft-versus-host disease have been major barriers to more widespread clinical applications of this approach thus far. For this approach to become more practical and safe, techniques to avoid toxic preconditioning of the recipient will be needed.

In islet transplantation, strategies to induce mixed chimerism may be of particular interest since it has the potential of preventing autoimmune recurrence by restoring self-tolerance through the bone marrow transplant, in addition to achieving permanent islet allograft acceptance. However, until less toxic preconditioning strategies are developed, it is not justifiable to impose the current risks associated with bone marrow transplantation in a patient whose disease is controlled with insulin therapy.

## Conclusions

Hope is on the horizon for patients with type 1 diabetes (Fig. 4). The advent of successful outcomes with cellular replacement therapy in diabetes has renewed enthusiasm that a limitless source of insulin secreting tissue will be found, and that cellular transplantation will replace chronic insulin therapy in due course. Islet transplantation in its current form can provide superb glucose control and can completely normalize glycated HbA1C in recipients. Such glycemic control is already far superior to what can be

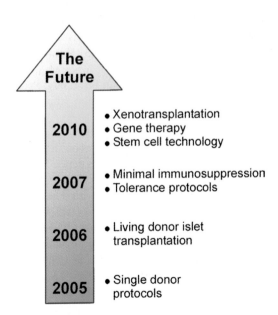

**Figure 4.** Future directions in cellular replacement therapy for diabetes.

achieved with injected insulin or pump therapies. The current drawbacks of islet transplant therapy are the need for chronic immunosuppression with its attendant side-effects and risks, and the fact that the cadaveric islet donor pool is so scarce that it cannot possibly supply sufficient cells to meet even a small fraction of the real need in type 1 diabetes. A limitless cell source would also open up the opportunity to test therapies in type 2 diabetes where most of the disease burden arises. It is anticipated that outcomes of clinical islet transplantation will further improve as current research opportunities begin to be translated to the clinic. The potential for islet graft proliferation in culture or *in vivo* after transplantation, using combinations of potent growth factors, is an area that is progressing rapidly. Progress in safer, less toxic immunosuppression is also occurring at a rapid pace, and as agents such as LEA29Y, FTY720 and CD3 immunotoxin emerge in early clinical trials, it is anticipated that such approaches will dramatically enhance the tolerability and durability of islet replacement therapy.

# REFERENCES

1. Boyle JP, Honeycutt AA, Narayan KM, *et al.* (2001) Projection of diabetes burden through 2050: Impact of changing demography and disease prevalence in the U.S. *Diabetes Care* **24**:1936–1940.
2. Libman I, Songer T, Laporte R. (1993) How many people in the U.S. have IDDM? *Diabetes Care* **16**:841–842.
3. Williamson DF, Vinicor F, Bowman BA. (2004) Primary prevention of type 2 diabetes mellitus by lifestyle intervention: Implications for health policy. *Ann Intern Med* **140**:951–957.
4. The Diabetes Control and Complications Trial Research Group. (1993) The effect of intensive treatment of diabetes on the development and progression of long-term complications in insulin dependent diabetes mellitus. *N Engl J Med* **329**:977–986.
5. Adverse events and their association with treatment regimens in the diabetes control and complications trial. (1995) *Diabetes Care* **18**(11):1415–1427.
6. The Diabetes Control and Complications Trial Research Group. (1997) Hypoglycemia in the diabetes control and complications trial. *Diabetes* **46**:271–286.
7. Gross CR, Limwattananon C, Matthees BJ. (1998) Quality of life after pancreas transplantation: A review. *Clin Transplant* **12**:351–361.
8. Sutherland DE, Gruessner RW, Dunn DL, *et al.* (2001) Lessons learned from more than 1,000 pancreas transplants at a single institution. *Ann Surg* **233**:463–501.

9. Venstrom JM, McBride MA, Rother KI, *et al.* (2003) Survival after pancreas transplantation in patients with diabetes and preserved kidney function. *JAMA* **290**:2817–2823.

10. Hathout E, Lakey J, Shapiro J. (2003) Islet transplant: An option for childhood diabetes? *Arch Dis Child* **88**:591–594.

11. Ballinger W, Lacy P. (1972) Transplantation of intact pancreatic islets in rats. *Surgery* **72**:175–186.

12. Lacy P, Kostianovsky M. (1967) Method for the isolation of intact islets of Langerhans from the rat pancreas. *Diabetes* **16**:35–39.

13. Najarian JS, Sutherland DE, Matas AJ, *et al.* (1977) Human islet transplantation: A preliminary report. *Transplant Proc* **9**:233–236.

14. Najarian JS, Sutherland DE, Matas AJ, Goetz FC. (1979) Human islet autotransplantation following pancreatectomy. *Transplant Proc* **11**:336–340.

15. Horaguchi A, Merrell RC. (1981) Preparation of viable islet cells from dogs by a new method. *Diabetes* **30**:455–458.

16. Noel J, Rabinovitch A, Olson L, *et al.* (1982) A method for large-scale, high-yield isolation of canine pancreatic islets of Langerhans. *Metabolism* **31**:184–187.

17. van Suylichem PT, Wolters GH, van Schilfgaarde R. (1992) Peri-insular presence of collagenase during islet isolation procedures. *J Surg Res* **53**:502–509.

18. Gray DW, McShane P, Grant A, Morris PJ. (1984) A method for isolation of islets of Langerhans from the human pancreas. *Diabetes* **33**:1055–1061.

19. Lakey JR, Warnock GL, Shapiro AM, *et al.* (1999) Intraductal collagenase delivery into the human pancreas using syringe loading or controlled perfusion. *Cell Transplant* **8**:285–292.

20. Ricordi C, Lacy PE, Scharp DW. (1989) Automated islet isolation from human pancreas. *Diabetes* **38 Suppl** 1:140–142.

21. Lake SP, Bassett PD, Larkins A, *et al.* (1989) Large-scale purification of human islets utilizing discontinuous albumin gradient on IBM 2991 cell separator. *Diabetes* **38 Suppl** 1:143–145.

22. Linetsky E, Bottino R, Lehmann R, *et al.* (1997) Improved human islet isolation using a new enzyme blend, liberase. *Diabetes* **46**:1120–1123.

23. Lakey JR, Warnock GL, Rajotte RV, *et al.* (1996) Variables in organ donors that affect the recovery of human islets of Langerhans. *Transplantation* **61**:1047–1053.

24. Brendel M, Hering B, Schulz A, Bretzel R. (1999) International Islet Transplant Registry Report. University of Giessen, Germany; Report No.: Newsletter No. 8.

25. Boker A, Rothenberg L, Hernandez C, *et al.* (2001) Human islet transplantation: update. *World J Surg* **25**:481–486.

26. Bretzel RG, Brandhorst D, Brandhorst H, *et al.* (1999) Improved survival of intraportal pancreatic islet cell allografts in patients with type-1 diabetes mellitus by refined peritransplant management. *J Mol Med* **77**:140–143.

27. Ricordi C, Tzakis AG, Carroll PB, *et al.* (1992) Human islet isolation and allotransplantation in 22 consecutive cases. *Transplantation* **53**:407–414.

28. Sibley RK, Sutherland DE, Goetz F, Michael AF. (1985) Recurrent diabetes mellitus in the pancreas iso- and allograft. A light and electron microscopic and immunohistochemical analysis of four cases. *Lab Invest* **53**:132–144.

29. Oberholtzer J, Benhamou P, Toso C, *et al.* (2001) Human islet transplantation network for the treatment of type 1 diabetes: First (1999–2000) data from the Swiss-French GRAGIL Consortium. *Am J Transplant* **1**:182.

30. Oberholzer J, Triponez F, Mage R, *et al.* (2000) Human islet transplantation: Lessons from 13 autologous and 13 allogeneic transplantations. *Transplantation* **69**:1115–1123.

31. Maffi P, Bertuzzi F, Guiducci D, *et al.* (2001) Per and peri-operative management influences the clinical outcome of islet transplantation. *Am J Transplant* **1**:181.

32. Shapiro AM, Lakey JR, Ryan EA, *et al.* (2000) Islet transplantation in seven patients with type 1 diabetes mellitus using a glucocorticoid-free immunosuppressive regimen. *N Engl J Med* **343**:230–238.

33. Ryan EA, Lakey JR, Paty BW, *et al.* (2002) Successful islet transplantation: Continued insulin reserve provides long-term glycemic control. *Diabetes* **51**:2148–2157.

34. Shapiro AM, Ricordi C, Hering B. (2003) Edmonton's islet success has indeed been replicated elsewhere. *Lancet* **362**:1242.

35. Fiorina P, Folli F, Zerbini G, *et al.* (2003) Islet transplantation is associated with improvement of renal function among uremic patients with type I diabetes mellitus and kidney transplants. *J Am Soc Nephrol* **14**:2150–2158.

36. Kuroda Y, Fujino Y, Morita A, *et al.* (1992) Oxygenation of the human pancreas during preservation by a two-layer (University of Wisconsin solution/perfluorochemical) cold-storage method. *Transplantation* **54**:561–562.

37. Baidal D, Froud T, Ferreira J, *et al.* (2003) The Bag method for islet cell infusion. *Cell Transplantation* **12**:809–813.

38. Hering BJ, Kandaswamy R, Harmon JV, *et al.* (2004) Transplantation of cultured islets from two-layer preserved pancreases in type 1 diabetes with anti-CD3 antibody. *Am J Transplant* **4**:390–401.

39. Shapiro AM, Ricordi C. (2004) Unraveling the secrets of single donor success in islet transplantation. *Am J Transplant* **4**:295–298.

40. Paraskevas S, Maysinger D, Wang R, *et al.* (2000) Cell loss in isolated human islets occurs by apoptosis. *Pancreas* **20**:270–276.

41. Bennet W, Groth CG, Larsson R, *et al.* (2000) Isolated human islets trigger an instant blood mediated inflammatory reaction: implications for intraportal islet transplantation as a treatment for patients with type 1 diabetes. *Ups J Med Sci* **105**:125–133.

42. Ojo AO, Held PJ, Port FK, *et al.* (2003) Chronic renal failure after transplantation of a nonrenal organ. *N Engl J Med* **349**:931–940.
43. Guba M, von Breitenbuch P, Steinbauer M, *et al.* (2002) Rapamycin inhibits primary and metastatic tumor growth by antiangiogenesis: Involvement of vascular endothelial growth factor. *Nat Med* **8**:128–135.
44. Contreras JL, Wang PX, Eckhoff DE, *et al.* (1998) Peritransplant tolerance induction with anti-CD3-immunotoxin: A matter of proinflammatory cytokine control. *Transplantation* **65**:1159–1169.
45. Thomas JM, Contreras JL, Jiang XL, *et al.* (1999) Peritransplant tolerance induction in macaques: Early events reflecting the unique synergy between immunotoxin and deoxyspergualin. *Transplantation* **68**:1660–1673.
46. Contreras JL, Eckhoff DE, Cartner S, *et al.* (2000) Long-term functional islet mass and metabolic function after xenoislet transplantation in primates. *Transplantation* **69**:195–201.
47. Harding FA, McArthur JG, Gross JA, *et al.* (1992) CD28-mediated signalling co-stimulates murine T cells and prevents induction of anergy in T-cell clones. *Nature* **356**:607–609.
48. Lenschow DJ, Walunas TL, Bluestone JA. (1996) CD28/B7 system of T cell costimulation. *Annu Rev Immunol* **14**:233–258.
49. Schwartz RH. (1990) A cell culture model for T lymphocyte clonal anergy. *Science* **248**:1349–1356.
50. Knosalla C, Gollackner B, Cooper DK. (2002) Anti-CD154 monoclonal antibody and thromboembolism revisited. *Transplantation* **74**:416–417.
51. Buhler L, Alwayn IP, Appel JZ, 3rd, *et al.* (2001) Anti-CD154 monoclonal antibody and thromboembolism. *Transplantation* **71**:491.
52. Garlichs CD, Geis T, Goppelt-Struebe M, *et al.* (2002) Induction of cyclooxygenase-2 and enhanced release of prostaglandin E(2) and I(2) in human endothelial cells by engagement of CD40. *Atherosclerosis* **163**:9–16.
53. Szabolcs MJ, Cannon PJ, Thienel U, *et al.* (2000) Analysis of CD154 and CD40 expression in native coronary atherosclerosis and transplant associated coronary artery disease. *Virchows Arch* **437**:149–159.
54. Larsen CP, Pearson TC. (1997) The CD40 pathway in allograft rejection, acceptance, and tolerance. *Curr Opin Immunol* **9**:641–647.
55. Pearson TC, Trambley J, Odom K, *et al.* (2002) Anti-CD40 therapy extends renal allograft survival in rhesus macaques. *Transplantation* **74**:933–940.
56. Adams AB, Shirasugi N, Durham MM, *et al.* (2002) Calcineurin inhibitor-free CD28 blockade-based protocol protects allogeneic islets in nonhuman primates. *Diabetes* **51**:265–270.
57. Schuurman HJ, Menninger K, Audet M, *et al.* (2002) Oral efficacy of the new immunomodulator FTY720 in cynomolgus monkey kidney allotransplantation, given alone or in combination with cyclosporine or RAD. *Transplantation* **74**:951–960.

58. Wijkstrom M, Kenyon NS, Kirchhof N, *et al.* (2004) Islet allograft survival in nonhuman primates immunosuppressed with basiliximab, RAD, and FTY720. *Transplantation* **77**:827–835.
59. Maki T, Gottschalk R, Monaco AP. (2002) Prevention of autoimmune diabetes by FTY720 in nonobese diabetic mice. *Transplantation* **74**:1684–1686.
60. Fu F, Hu S, Deleo J, *et al.* (2002) Long-term islet graft survival in streptozotocin- and autoimmune-induced diabetes models by immunosuppressive and potential insulinotropic agent FTY720. *Transplantation* **73**: 1425–1430.
61. Spitzer TR, Delmonico F, Tolkoff-Rubin N, *et al.* (1999) Combined histocompatibility leukocyte antigen-matched donor bone marrow and renal transplantation for multiple myeloma with end stage renal disease: The induction of allograft tolerance through mixed lymphohematopoietic chimerism. *Transplantation* **68**:480–484.

# 18

# Differentiating Human Stem Cells to Neurons

## Su-Chun Zhang

## Introduction

Stem cells can replenish themselves in the long-term while producing progenies that are different from themselves. Stem cells from many sources possess such traits. Stem cells isolated from the inner cell mass (ICM) of a pre-implantation blastocyst,[1,2] known as embryonic stem cells (ESCs), can generate all cell lineages that make up an organism. Stem cells isolated from an organ or a tissue generally have restricted potential to produce certain cell types that constitute the organ or tissue. However, precursor cells cultured from many tissues, especially after prolonged expansion, exhibit broader potential and produce cell types that do not belong to the tissue of origin. One example is the bone marrow precursor cells that can produce non-hematopoietic cell types, including neural cells.[3] Thus, non-neural stem cells, as well as neural stem cells, may produce neurons.

Neurons are the building blocks of the nervous system. Like cells in non-neural tissues, neurons are born in a highly ordered process from their precursors, neuroepithelia or neural stem cells, with the first birth of projection neurons followed by interneurons and glial cells. Unlike many other cell types, a neuron in one region differs from that in another place even though they may carry the same transmitters. The temporal sequence of neuronal birth and their topographical locations form the basis of the first order of neuronal connections. Environmental (e.g. glia) and target signals subsequently promote functional maturation and shape neuronal circuitry. These stereotypic as well as regulatory rules likely apply to the differentiation of neurons from stem cells in a culture Petri dish and influence the

*Correspondence: Departments of Anatomy and Neurology. The Stem Cell Research Program. The Waisman Center. The WiCell Institute, University of Wisconsin, Madison, WI, USA. Email: zhang@waisman.wisc.edu.

application of stem cell-produced neurons in cell therapy. For production of neurons from human stem cells, it may be necessary to explore alternatives to these fundamental rules learned from animal studies.

# The Birth of a Neuron

## What is a neuron?

A neuron is a signaling cell in the nervous system with highly specialized processes, axons and dendrites, that receive, integrate, and transduce stereotypic electrochemical signals from one cell to another. Unlike many other cell types in the body, every neuron sits in a unique position of the brain and spinal cord and makes connections (synapses) with neurons or non-neuronal cells (e.g. muscles) in other areas to establish highly ordered communication systems. The positional information and the initial targets of a neuron are endowed during the birth of a neuron in early embryonic development. The establishment of functional connectivity is influenced by activity.

## Specification of a neuron

Birth of a neuron involves several sequential steps orchestrated by signaling events.[4] The initial step is the specification of neuroepithelia from the embryonic ectoderm — a process known as neural induction. This step takes place during the third week of human embryonic development. By the end of the third week, a sheet of columnar neuroepithelia, referred to as the neural plate, has formed and has begun to fold and form the neural tube. The mechanism of neuroectoderm specification, inferred largely from studies using xenopus, chick, and other low vertebrates, remains a topic of scientific debate. The main division is whether neural induction occurs simply by removal of inhibitory factors, such as bone morphogenetic proteins (BMPs), the so-called "default pathway," or requires active inductive signaling such as by fibroblast growth factors (FGFs).[5] It is likely that both instructive and inhibitory signalings are necessary. FGF may instruct a "pro" neural state at an early stage, and BMP antagonists may subsequently stabilize the neural identity.

## Patterning of the neural plate and neural tube

The process of neural plate and neural tube formation does not occur homogenously and simultaneously. Ectodermal cells (or perhaps epiblasts) in the head region are first specified to the neuroectodermal fate. The rostral (forebrain) neuroepithelia gradually extend caudally to form the entire neural plate. Hence, there is a temporal sequence of neural plate formation from the rostral to caudal regions. Meanwhile, the neural plate begins to fold and fuse dorsally at the future neck area at the end of the third week, which extends both rostrally and caudally to form a complete neural tube by the end of the fourth week. Thus, neuroepithelial cells are temporally and spatially different from each other at the time when the neural plate and neural tube are formed. Precursor cells in each subdivision along the rostro-caudal and dorsoventral axes, by exposure to a unique set of morphogens such as Wnt, FGF, BMP, retinoic acid (RA), and sonic hedgehog (SHH)[4,6,7] at specific concentrations, are fated to subtypes of neurons and glial cells. Therefore, neuroepithelial or progenitor cells generated from stem cells in a Petri dish may differ from each other depending on the morphogens used. While neuroectoderm specification and neural tube formation occurs within hours or 1–2 days in most experimental animals, this process takes place over a 2-week period in humans, stretching from the beginning of the third gestation week to the end of the fourth week. Consequently, the time window for applying morphogens into human stem cell cultures for directed neural differentiation becomes a critical issue.

## Differentiation and maturation of neurons

Once the neuroepithelial cells are specified, they differentiate into neurons and glia largely in a cell autonomous manner.[8] In general, neurons are born first, followed by glial cells. Differentiation of glial cells will in turn promote the maturation of neurons such as synaptogenesis. Glial cells in the adult nervous system may also be crucial for maintaining neural progenitors and guiding neuronal differentiation.

If the basic principles of neuronal specification learned from animal studies, as outlined above, hold true for human cells *in vitro*, the key to the generation of neuronal subtypes from human stem cells is the specification of region-specific progenitors. This process is largely completed if the stem/progenitors are taken from neural tissues. For stem cells from an early embryo or non-neural tissues, neuroectodermal specification is a prerequisite step, during which application of a specific set of morphogens at a particular time is crucial for neuronal subtype specification.

## How to Define a Neuron in a Petri Dish?

The first and most important criterion for defining a cell, especially a neural cell, is its location in the body, which is lost in stem cell-differentiated

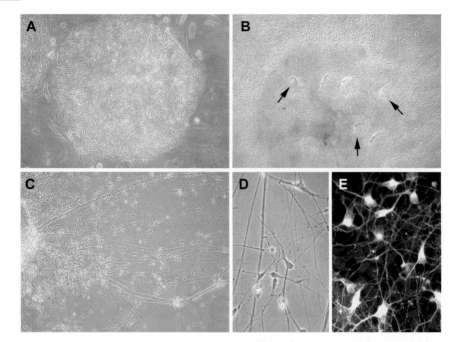

**Figure 1.** Neural differentiation from hESCs. hESCs (**A**), after 2 weeks of differentiation in a chemically defined culture condition,[59] became columnar neuroepithelia and organized into neural tube-like rosettes (arrows) in the colony center, with a ring of non-neural flat cells in the surrounding (**B**). The neuroepithelial cells, after enrichment, differentiated into predominantly neuronal cells in additional 3 weeks (**C**) with distinguishable axons and dendrites (**D**). The neuroepithelial cells, after treatment with FGF8 and SHH, differentiated into a large proportion of dopaminergic neurons that were positively stained for tyrosine hydroxylase in additional 3 weeks (**E**).

cultures. Hence, determination of a cell type requires a combination of markers at morphological, genetic, and functional levels.[9,10] For a cell to be a neuron, the first and foremost feature is its unique process-bearing morphology. Neuronal processes (dendrites and axons) are uniquely different from processes of any other cell types, including those of their cousins astrocytes and oligodendrocytes. These processes are not simple protrusions or extensions of the cell body. In a mature neuron, axons and dendrites are easily distinguishable, with a fine, less-branched axon and thicker, branched dendrites (Figs. 1 C–E). Under serum-free culture conditions, processes of primate astrocytes may appear similar to neurites. Expression of neuronal genes and gene products may help define a neuron, particularly neuronal subtypes. As discussed above, specification and determination of a neuron depends upon the sequentially regulated expression of a set of genes. Therefore, a combination of the presence of one set of genes and the absence of another set is often required to determine a neuronal type, in addition to morphological criteria. Cells under unusual culture environments or stress conditions may aberrantly express many genes, including neural genes. Thus, the mere expression of some neuronal genes without a characteristic neuronal morphology is not sufficient to define a neuron. The absence of non-neural genes is equally important. Care should also be paid to the subcellular localization of the neuronal markers as aberrantly induced gene expression often has atypical subcellular localization. Under most circumstances, the demonstration of a simple stereotypic neuronal electrochemical property, such as a typical resting potential and sodium-gated action potential, in addition to the aforementioned morphological and biochemical criteria, is necessary for defining a neuron produced in a Petri dish, particularly when the presumptive neurons are produced from non-neural cells.

# Differentiation of Human Neural Stem Cells to Neurons

## Establishment of neural stem cells

Human neural stem/progenitor cells have been established from fetal brain and spinal cord tissues, mostly from gestation weeks 8 to 20.[11–17] Similar cultures have also been established from biopsied or postmortem adult brain tissues,[18,19] confirming the notion that stem/progenitor cells exist in adult human brain. These stem/progenitor cells are expanded in the form of "neurospheres" in the presence of mitogens, including epidermal growth factor (EGF), fibroblast growth factor 2 (FGF2), and leukemia inhibitory factor (LIF). The human neural stem/progenitors isolated from the forebrain can be expanded *in vitro* for up to 40 passages, with a doubling time of approximately 2 days, particularly in the presence of LIF.[12,20] Therefore, neural stem/progenitors, at least based on current technology, can be maintained *in vitro* for a reasonably long period.

## *In vitro* differentiation

Neural stem cells can differentiate into neurons, astrocytes, and oligodendrocytes. Based on this simple definition, the above human neural stem/progenitor cells have been shown to be able to give rise to the three general cell types, although no clonal analysis yet confirms the stemness of these precursor cells. The major differentiated progeny is astrocytes.

Neuronal population varies considerably, depending on when and where the neural stem/progenitor cells are derived. Within the neuronal population, the vast majority are interneurons, with few or no large projection neurons. Oligodendrocytes are usually a minority. Attempts to differentiate expanded human neural progenitors to a large proportion of oligodendrocytes have not yet borne fruit, in stark contrast to rodent neural progenitors. The lack of projection neuron and oligodendrocyte differentiation from brain-derived neural stem/progenitor cells conforms to the developmental principle that the large-projection neurons and oligodendrocytes are generally specified at the very early stage of neural development.[21,22]

Neural stem/progenitor cells isolated from human brain tissues and expanded in culture appear to retain some intrinsic properties of the progenitor cells *in vivo*. Precursor cells isolated from the forebrain can be expanded much more extensively than those from the brain stem and spinal cord at the same developmental stage. The neuronal differentiation potential of forebrain progenitors is also significantly higher than those from the brain stem and the spinal cord. The neuronal types produced by the progenitors largely correspond to those that are supposed to exist in the brain regions where the progenitors are isolated. Thus, progenitors isolated from the forebrain and striatum produce predominantly GABA and glutaminergic neurons but rarely generate dopaminergic neurons.[23] These characteristics of neural progenitors reflect the need for more neurons to be produced over a longer period during forebrain development as compared with the hindbrain and spinal cord.

The differentiation potential of neural progenitors also changes with time, again similar to their *in vivo* properties. Neural progenitors, regardless of their origin, produce more neurons and fewer astrocytes in early passages than late passages. Differentiation of neuronal subtypes also changes with time. Primary cultures of fetal midbrain tissues generate dopaminergic neurons, whereas progenitors expanded from the same region hardly produce such projection neurons.[23] Thus, the hierarchy of the birth of neuronal subtypes is largely cell autonomous, *in vivo* or *in vitro*. This raises a question whether expanded neural progenitors have the capacity to generate projection neuronal subtypes, such as midbrain dopaminergic neurons and spinal cord motor neurons, for cell replacement in neural degenerative disorders. This question is also relevant to the capacity of endogenous neural progenitors to replace such neuronal types in adult life.

Technically, current approaches for maintaining human neural stem/progenitor cells *in vitro* are not sufficient to prohibit stem cells from differentiation. Cells in neurospheres are mostly committed progenitors with scarce true neural stem cells, contrasting the current methods for maintaining embryonic stem cells. A recent study using mouse neural stem cells as a model system indicates that factors secreted by vascular endothelial cells help to maintain the neural stem cells for a longer period to produce projection neurons.[24] Whether a similar strategy may be employed for maintaining human neural progenitors remains to be seen.

*In vitro* differentiation of human neural progenitors to neurons is straightforward. It is achieved by plating the progenitors on a laminin substrate in serum-free media using DMEM/F12 or Neurobasal medium as base media. Addition of neurotrophic factor cocktails increases or biases the neuronal differentiation, most likely through enhancing the survival of differentiated neurons.[25] Little evidence suggests that trophic factors specify subtype neuronal fate in the differentiation culture system. Neuronal identities are defined essentially by their typical morphology and expression of neuronal markers at the immunocytochemical level. Few reports provide electrophysiological evidence demonstrating whether the neurons differentiated from *in vitro* expanded neural progenitors possess the same or similar functional properties of neurons in the brain and spinal cord.

## *In vivo* differentiation

The *in vivo* differentiation potential is assessed by implanting the human neural stem/progenitors into rodent brains. For proof-of principle, neural progenitors are generally transplanted into the embryonic or neonatal brains,[26,27] or the subventricular zone/rostral migratory stream of adult rodent brains.[28] In both environments, neurogenic signals are present. Under these conditions, grafted human neural progenitors, identified later by either genetic labeling or using human specific markers, differentiate into neurons that are morphologically, and in some cases neurochemically, similar to endogenous neurons in the particular brain regions.[26,27] This "site-specific" differentiation suggests that human neural progenitors can respond to the environmental signals for differentiation. However, such "site-specific" differentiation should be interpreted with caution. It may simply be the coincidence of neuronal differentiation of the forebrain neural progenitors in the forebrain environment. Different subpopulations of "forebrain" neural progenitors selectively survive and/or home to certain brain regions such as olfactory bulb, hippocampus, striatum, the usual incorporation sites of the above transplant paradigms, and mature to the cells that are essentially the same as their origin.

Human neural progenitors grafted into adult non-neurogenic brain regions usually differentiate in a cell autonomous manner. Forebrain neural progenitors, grafted into the rat striatum, differentiate into neurons that exhibit properties of striatal neuronal phenotypes, such as expression of GABA and/or DARPP32.[29] In the Huntington's background, these transplanted human neural progenitors appear to migrate more extensively than in normal animals and differentiate into both neurons and astrocytes.[30] Under the ischemia environment, grafted human neural progenitor cells migrate preferentially toward the lesion site and differentiate predominantly into neurons, as identified by an immature neuronal marker βIII-tubulin.[31] Thus, the brain environment may influence the differentiation and/or survival of human neural progenitors. It is also observed that the human neurons extend diffusing processes, suggesting that the differentiation of human neurons may be at least partly independent of the mouse environment since grafted rat neurons usually do not extend excessively long axons under this condition.[27,29] In addition, a significant proportion of cells become astrocytes.[28,30]

Transplant studies using human neural stem/progenitor cells to date indicate that they are capable of producing neurons and that the majority of differentiated neurons are late born cells or interneurons.[32] These observations conform to the principles of neural lineage development learned from animal studies and are consistent with *in vitro* results. One exception is a recent report on the differentiation of forebrain progenitors into large projection neurons, including spinal cord motor neurons.[33] This occurs after the cortical progenitor cells are primed with laminin and FGF2 and transfected with associated adenoviral vectors (for cell labeling purpose) before transplantation into normal unlesioned rat brain and spinal cord. Virally labeled cholinergic neurons have been observed in raphi nuclei bilaterally in the hindbrain that are millimeters away from the single transplant site as well as the large motor neurons in the ventral horn of the spinal cord. These results are extraordinary because forebrain human neural progenitors, isolated after brain formation, usually do not produce spinal cord motor neurons that are specified at the earliest stage of neural development and because the efficient differentiation occurs in the normal uninjured adult brain environment. This extraordinary outcome may be attributed to the altered cell behaviors after excessive *in vitro* expansion (up to 85 passages)

in the presence of EGF and FGF2, as the same forebrain human neural progenitors used in that report usually become senescent by 40 passages even in
the presence of LIF.[20] This could also be due to technical issues in identifying
grafted cells.

## Turning Non-Neural Human Stem Cells to Neurons

Stem cells, isolated from non-neural tissues, have been reported to produce
cells that express some neural genes. Most of these studies examine stem
cells isolated from rodent tissues, including hematopoietic, bone marrow, skin, and fat tissues. Controversies abound over whether tissue stem
cells can differentiate across major embryonic germ layer lineages and
whether some of the phenomena are attributed to technical flaws such
as cell fusion and transformation.[34–37] Passionate debates are still being
exchanged through publications in "high-profile" journals. In reality, one
should be able to judge easily the scientific implication and clinical potential of these stem cells simply based on the definition of a stem cell: a cell
capable of long-term self-renewal and robust multi-lineage differentiation
with functional outcomes.

### *In vitro* neural differentiation

Non-neural human stem cells with neural differentiation potential reported
thus far are mainly cells isolated from bone marrow, umbilical cord blood,
and dermal tissues. The initial report involves the use of somewhat peculiar
culture conditions such as the addition of DMSO.[38] More recent approaches
involve the expansion of non-neural (mesenchymal, bone marrow, umbilical cord blood, and dermal) precursors in EGF and/or FGF2 before differentiation in a neural medium.[39–44] The rationale appears that EGF and FGF2
selectively promote the proliferation of precursors with neural tendency
although both mitogens are potent stimulators of non-neural precursors.
Process-bearing cells have been induced and these cells have been stained
immunohistochemically for multiple markers that are normally expressed
by neurons. Most of the neural marker-expressing cells do not exhibit a
neuronal morphology. The processes are either spikes or short extensions
of cytoplasm with no resemblance to dendrites or axons. The immunostaining for multiple diverse neuronal proteins exhibits uniform diffuse staining pattern without characteristic patterns and subcellular localizations,
suggestive of aberrant gene expression. Chandran and colleagues employ

a more sophisticated approach to induce neuronal differentiation from precursors expanded from dermal tissues by co-culturing with astrocytes.[44] Some neural marker-bearing cells do exhibit typical neuronal morphology. Given the potential presence of neural crest derivatives in the dermal cell preparation,[45] a firm conclusion of transdifferentiation from skin to neuron cannot be drawn even although such transdifferentiation would occur within the same embryonic ectoderm.

## *In vivo* neural differentiation

The neural differentiation potential of non-neural stem cells in the brains of patients is usually determined by colabeling of the donor marker (usually Y-chromosome probes) with neuronal markers, if a female patient had received cell transplantation from male donors. Although the major proportion of the few donor-derived cells is found to exhibit glial phenotypes, colabeling of Y-chromosome and neuronal markers such as NeuN is found in patients who received bone marrow transplant, after the possible cell fusion-caused false positive stainining has been carefully excluded.[46] These observations suggest that bone marrow cells may differentiate into neurons. Experimentally, transplanted human cells can be readily identified by human-specific markers or pre-labeled with dyes or genetic markers. Human mesenchymal stem cells, transplanted into normal rodent brains, can differentiate into neural cells although most donor cells in the brain appear to be astrocytes.[47,48] Purified bone marrow stem cells, transplanted into the ischemic area of the mouse brain, have been found to express neuronal markers such as βIII-tubulin, neurofilament proteins, and neuron-specific enolase.[49] However, the neural marker-bearing bone marrow cells do not exhibit neuronal morphology and do not form connections with recipient neurons despite some functional recovery. Apparently, the functional recovery does not have anything to do with the putative neural differentiation. Similar observations have been made with umbilical cord blood stem cells that are transplanted into rodent brains.[41,43] Snyder and colleagues have recently found a rather widespread presence of donor-derived cells in the brain of a pediatric patient who had received umbilical cord blood transplantation. Many of the donor cells adopt the microglial phenotype with expression of macrophage markers but not a single donor-derived cell in the brain colabels with neuronal markers (Evan Snyder, personal communication). This observation appears to argue against the capacity of umbilical cord blood stem cells to transdifferentiate into neurons, given the relatively conducive recipient brain environment and the intimate contact of grafted cells with the environment.

Cross lineage differentiation is, and will likely continue to be, a major scientific endeavor. From the application standpoint, functional demonstration of these neuronally transdifferentiated cells is eagerly awaited. To date, no evidence has been provided that transdifferentiated human neurons are functional or simply exhibit typical eletrophysiological properties seen in neurons isolated from brain tissues. Although transplantation of non-neural stem cells results in behavioral changes in some neurological animal models, evidence is not yet available to demonstrate that the behavioral change is the outcome of transdifferentiation into neurons.

# Development of Human Embryonic Stem Cells to Neurons

## Human embryonic stem cells (hESCs)

Like their mouse counterparts, hESCs are derived from the inner cell mass of a pre-implantation embryo.[50] Unlike mouse ESCs that can be maintained in chemically defined culture conditions with the presence of LIF, hESCs requires unknown factors produced by embryonic mouse fibroblast for self-renewal. hESCs differ from their mouse counterparts in many aspects, including the growth rate, growth requirement, expression of cell surface molecules, and differentiation potential such as the capability to produce trophoblasts.[51] Accordingly, neural differentiation from hESCs requires special considerations.[52]

## *In vitro* neural differentiation of hESCs

### *Differentiation of neuroepithelia from hESCs*

Stem cells or progenitors isolated from embryonic brain tissues are specified or committed to certain subtype neural lineages with a specific positional identity. Consequently, differentiation of these stem/progenitor cells is largely cell autonomous. ESCs, however, are naïve cells capable of differentiation into cells of all three embryonic germ layers. Therefore, the first step of neural differentiation from hESCs is the specification of the neuroectodermal fate. Little is known how embryonic epiblasts are specified to the endoderm, mesoderm, and ectoderm fate. The initial step of ESC differentiation is rather random, i.e. by removal of self-renewing signals (e.g. mouse embryonic fibroblasts) and by promoting cell-cell interactions through aggregation of ESCs in suspension. This random step is then influenced by using special culture conditions including the use of specific morphogens or signaling molecules.[52]

Two major strategies have been developed for differentiating hESCs into neuroepithelial cells. One is to co-culture hESCs with stromal cells, mostly cell lines, which is initially reported for neural differentiation from mESCs.[53] Neural differentiation is achieved through direct contact of hESCs with stromal cells. Conditioned media from the stroma cells do not appear effective. The advantage of this approach is technically simple. It is particularly attractive for achieving dopaminergic neuronal differentiation,[54] since many stromal cell lines tested so far preferentially promote dopaminergic differentiation (see below). The drawback of this approach is the introduction of yet another unknown component and potentially risky factors as most cell lines are tumorigenic.

The more common strategy is the use of morphogens to promote neural differentiation or to enhance the proliferation of hESC-derived neuroepithelial cells. Spontaneously differentiated neural progenitors in hESC cultures have been stimulated to divide by addition of FGF2 so that neurosphere-like clusters of neural progenitors can be obtained.[55] This approach relies on over-growth of hESCs and subsequent random differentiation and therefore is not directed neural differentiation. The most commonly used mESC differentiation protocol, treatment of ESC aggregates with retinoic acid (RA), yields reasonable neural differentiation from hESCs.[56] However, RA appears to have much less potent effect on hESCs than mESCs in neural differentiation, even after a 10 times increase in dosage. One important note is that RA treatment results in the differentiation of neural cells with a caudal (hindbrain and spinal cord) fate. We have demonstrated that treatment of hESC-derived neuroepithelial cells with RA (0.1–1 uM) potently induces Hox gene expression with elimination of rostral gene expression.[57] Thus, neural progenitors differentiated in the presence of RA are likely restricted to a caudal fate. Along this note, the use of B27, a common practice in differentiation of neural stem/progenitor cells, may very well result in the differentiation of caudal neural fate.[52] For this reason, two forms of B27 supplement are now on the market, one of which does not contain RA.

The default model of neural induction implies that the embryonic ectoderm will become neuroectoderm if the signaling of TGFβ family such as BMP is inhibited.[5] Addition of noggin, a BMP antagonist, results in a robust differentiation of neuroepithelia-like cells from hESCs.[58] This observation seems to suggest that neuroectodermal specification from hESCs also follows a default model. However, these noggin-treated precursor cells are somewhat different from neuroepithelial cells, as they can be maintained in the same way as ESCs and express some of the differentiated neural markers. We have found that treatment of hESCs with noggin increases the proportion of neuroepithelial cells but does not affect the initial step of neural specification (Pankratz and Zhang, unpublished observation).

A rival model of neuroectoderm specification is active induction by FGF family members.[5] Based on this theory, we have established a chemically defined culture system for neuroectodermal differentiation from hESCs.[59] This system involves aggregation of hESCs for 4–6 days, followed by adherent colony culture in the presence of FGF2. Under this condition, columnar neuroepithelial cells are observable after 10 days of differentiation from hESCs. By 14–16 days of differentiation, these columnar neuroepithelial cells form neural tube-like structures in every colony (Figs. 1A and 1B). The striking feature of this culture system is its resemblance to *in vivo* neuroectoderm development since neural tube-like rosettes form after 2 weeks of hESC differentiation. The neural tube begins to form after 3 weeks of human embryo development and hESCs are derived from a 6-day-old embryo. The columnar epithelial cells express neural markers such as Pax6 and Sox1, confirming the neural identity. More importantly, the hESC-derived neuroepithelial cells display a forebrain phenotype by expression of a forebrain marker Otx2 but not caudal markers such as hox genes.[57] Again, it mirrors normal neural development as neural plate cells initially carry a forebrain identity. This would suggest that the neuroepithelial cells generated in this way may be plastic for further differentiation into neurons and glial cells with more caudal positional identities (see below), contrary to the caudally restricted neural cells using RA. The neural differentiation is very robust, as 70–90% of the differentiated progenies are neuroepithelial cells, which can be easily enriched through differential adhesion.[59] The neural differentiation with formation of neural tube-like rosettes has been reproduced by many other groups using other hESC lines.[54,60]

## Neuronal subtype differentiation from hESCs

Neuronal differentiation of hESCs is a common phenomenon. Developmentally, neurons are born first, followed by glia. Therefore, hESC-derived neuroepithelial cells usually give rise to mostly neurons in the first few weeks of differentiation with few glial cells.[59] Key to application of hESCs in science and clinics is the directed differentiation of neuronal subtypes.[61] Considerable effort has been made in directing neuronal subtypes, particularly those with potential clinical applications. Information begins to emerge that dopaminergic neurons with midbrain characters can be generated from hESCs in a similar way as mESCs. Studer and colleagues have demonstrated that hESCs differentiate into neuroepithelial cells that form neural tube-like rosettes when co-cultured with stroma cells and further differentiate into dopamine neurons in the presence of FGF8 and SHH. These hESC-produced dopamine neurons possess many of dopaminergic neuron features, including the ability to secrete dopamine in an activity-dependent manner.[54] A similar co-culture approach is also applied by other groups to promote dopaminergic neuron differentiation from other hESC lines, albeit with lower efficiency.[62]

The above approach offers a simple and relatively efficient way of producing dopaminergic neurons from hESCs. However, co-culture with stroma cell lines, which are often tumorigenic, introduces unknown and potential adverse factors into the system, which is not ideal from both the standpoint of basic understanding of neural development and clinical application of the hESC products. We have thus established a chemically defined culture method for directing hESC-derived neuroepithelial cells to dopaminergic neurons. As discussed above, neuroepithelial cells, generated based on our method,[59] initially carry a forebrain identity. Treatment of these neuroepithelial cells with FGF8 and SHH induces a midbrain progenitor identity prior to their differentiation into dopaminergic neurons (Fig. 1E).[63] We have further discovered that patterning the neuroepithelial cells with FGF8 prior to, but not after, formation of neural tube-like rosettes by the Sox1 + neuroepithelial cells is essential for an efficient induction of dopaminergic neurons with a midbrain identity. This contrasts with the classical approach used for differentiating mESCs into dopaminergic neurons in which ESC-derived neuroepithelial cells are expanded with FGF2 and then induced with FGF8 and SHH.[64] This finding is paralleled by another observation that treatment of early but not late neuroepithelial cells with a caudalizing reagent retinoic acid is required for differentiating hESCs into spinal cord motor neurons.[57] Neuroepithelial cells in the neural tube-like rosettes, after expansion in the presence of FGF2, no longer produce spinal cord motor neurons in the presence of RA and SHH. This is because the transcriptional codes that are required for motor neuron specification are no longer induced by the late RA treatment.

Thus, coupling of the intrinsic program of the precursor cells and the morphogens applied in a culture dish is the key to the production of neuronal subtypes.

Our finding that specification of the early born projection neurons such as midbrain dopaminergic neurons and spinal motor neurons may appear, contradict what we have believed that cell lineage differentiation is generally a linear process. In fact, neural progenitors along the rostral-caudal and dorsal-ventral axes differ from each other at the time when the neural plate and neural tube forms, presumably because they are induced by different sets of morphogens almost at the same time. In that regard, our observations provide direct evidence that the identity of the hESC-derived neuroepithelial cells is the direct consequence of the morphogens the precursor cells encounter during the process of neuroectodermal induction. Once the progenitors are defined, the subsequent differentiation process is largely cell autonomous. From an application perspective, our findings will likely provide a generalized strategy for producing neuronal subtypes with specific positional and transmitter phenotypes.

## *In vivo* neural differentiation of hESCs

*In vivo* neural differentiation of hESCs is at the stage of proof-of-concept studies. Almost all the studies to date involve transplantation of hESCs or their neural derivatives into neurogenic environment, e.g. embryonic or neonatal brain and spinal cord. Direct injection of hESCs into the early chick embryos at the stage of neuroectoderm development results in differentiation of the pluripotent hESCs into neuroepithelial cells which organize into neural tube-like rosettes, and to lesser degree, βIII-tubulin + neurons. However, hESCs in contact with non-neural tissues, differentiate into non-neural lineages,[65] suggesting that the fate choice of hESCs is largely influenced by the environmental signals. The easy accessibility of chick embryos offers a unique tool for evaluating the differentiation potential of hESCs.

hESC-derived neural progenitors, following transplantation into the neonatal mouse brain, migrate into a widespread brain regions and differentiate into neurons and glial cells.[55,59] Neuronal differentiation occurs mainly in the neurongenic brain regions at the time of transplantation. Neuronal differentiation appears to be region-specific, at least based on morphological criteria. Neurons, identified by human specific markers in the cerebral cortex, generally line up with host neurons. hESC-derived neurons located in the subcortical regions are mostly multipolar. Glial differentiation is observed chiefly in the white matter areas. No teratoma formation was observed in either of the studies, possibly due to the relatively pure population of donor cells and the short post-transplant period (4–12 weeks).

To further characterize the behaviors of *in vitro* generated human neuroepithelial cells *in vivo*, we have followed their fate after transplantation into the brain ventricles of neonatal immune-compromised SCID mice. Within 3 weeks following transplantation, grafted cells continue to divide, as identified by immunostaining with Ki67, and form clusters of nestin + neuroepithelial cells. By three months, the vast majority of human cells have migrated into the brain parenchyma and express an early neuronal marker, βIII-tubulin. More mature neuronal markers such as NeuN and neurofilament proteins have been observed after survival for longer than three months. Furthermore, differentiated neurons appear to acquire transmitter phenotypes that correspond to host neurons in the regions. Importantly, neurons differentiated from the grafted neuroepithelial cells form synapses with host neurons, as confirmed by immuno-electron microscopy.[32] These observations suggest that differentiation of hESC-derived neuroepithelial cells is partly dependent upon the intrinsic program and partly on the environmental signals the grafted cells encounter. The proliferation profile of grafted cells is essentially absent 9 months after transplantation, except in neurogenic regions such as the olfactory bulb. These findings strongly suggest that hESCs can provide a robust and potentially safe source of neurons for replacement therapy.

## Stem Cell-Produced Human Neurons — From Benchtop to Bedside

Both neural and non-neural stem cells can produce neurons, at least under experimental settings. Whether and how the stem cell-produced neurons will be exploited for repairing the brain and spinal cord, will likely be dictated by the ability and capacity of stem cells in generating particular neuronal subtypes, the ability of the stem cell-produced neurons to integrate into the existing neuronal circuitry in a functional manner, and issues surrounding clinical practice such as immunological compatibility and safety measures. Neural stem cells can be isolated and scaled up to a reasonable amount under clean environment, thus providing a safe source of neural cells. The main limiting factor is the restricted ability to produce neuronal subtypes for cell replacement purposes. Given their neural origin, neural stem cells and their derivatives, including glial cells, can be modified to deliver biomolecules to promote neuronal survival and replacing missing molecules.

The major advantage of non-neural stem cells is that many of these stem cells can be isolated from the patient's own body for autologous transplant. Despite indication of possible neuronal differentiation under certain circumstances, which is scientifically entertaining, the applicability of non-neural stem cells in neuronal replacement is significantly limited by the poor neural differentiation potential and inability to produce neurons with specific positional and transmitter identities. If the lack of neural differentiation potential is not cell autonomous, a strategy needs to be developed for robust differentiation of neuronal subtypes with functional outcomes.

The ability and robustness of hESCs to produce functional neuronal subtypes, along with the readily available culture system for expanding them to a large quantity, is so far unparalleled by any other types of stem cells. Several issues have hindered immediate application of hESC derivatives. Because neurons are generated from hESCs entirely in an artificial culture environment, they often do not retain the phenotypes when placed in the totally new, brain environment. This phenomenon likely also applies to neurons produced from non-neural stem cells in a Petri dish. More sophisticated strategy needs to be devised to adapt the *in vitro* produced neurons to the brain environment. Safety is a major concern given the inherent ability of ESCs for multilineage differentiation. The key to overcome this issue is a robust directed differentiation and means of purifying target cell populations and/or removing undifferentiated stem cells. Strategies are being devised on both fronts. Neuroepithelial cells can be highly enriched [56,59,63] and these cells can be robustly differentiated into neuronal subtypes.[54,57] Genetic as well as epigenetic approaches have been developed for removing undifferentiated mESCs[66] and hESCs.[67] Finally, immunological incompatibility needs to be dealt with.[61] These issues are mostly solvable. Therefore, application of hESC-derived neural cells in patients is achievable.

# Acknowledgment

Studies in my laboratory have been supported by the NIH-NCRR (RR016588), NIH-NINDS (NS045926, NS046587), the Michael J. Fox Foundation, the National ALS Association, and the Myelin Project.

# REFERENCES

1. Martin GR. (1981) Isolation of a pluripotent cell line from early mouse embryos cultured in medium conditioned by teratocarcinoma stem cells. *Proc Natl Acad Sci USA* **78**: 7634–7638.
2. Evans MJ, Kaufman MH. (1981) Establishment in culture of pluripotential cells from mouse embryos. *Nature* **292**: 154–156.
3. Jiang Y, Jahagirdar BN, Reinhardt RL, *et al.* (2002) Pluripotency of mesenchymal stem cells derived from adult marrow. *Nature* **418**: 41–49.
4. Jessell TM. (2000) Neuronal specification in the spinal cord: Inductive signals and transcriptional codes. *Nat Rev Genet* **1**: 20–29.
5. Wilson SI, Edlund T. (2001) Neural induction: Toward a unifying mechanism. *Nat Neurosci* **4** (**Suppl**): 1161–1168.
6. Marquardt T, Pfaff SL. (2001) Cracking the transcriptional code for cell specification in the neural tube. *Cell* **106**: 651–654.
7. Osterfield M, Kirschner MW, Flanagan JG. (2003) Graded positional information: Interpretation for both fate and guidance. *Cell* **113**: 425–428.
8. Edlund T, Jessell TM. (1999) Progression from extrinsic to intrinsic signaling in cell fate specification: A view from the nervous system. *Cell* **96**: 211–224.
9. Zhang SC. (2001) Defining glial cells during CNS development. *Nat Rev Neurosci* **2**: 840–843.
10. Svendsen CN, Bhattacharyya A, Tai YT. (2001) Neurons from stem cells: Preventing an identity crisis. *Nat Rev Neurosci* **2**: 831–834.
11. Svendsen CN, Clarke DJ, Rosser AE, Dunnett SB. (1996) Survival and differentiation of rat and human epidermal growth factor-responsive precursor cells following grafting into the lesioned adult central nervous system. *Exp Neurol* **137**: 376–388.
12. Carpenter MK, Cui X, Hu ZY, *et al.* (1999) *In vitro* expansion of a multipotent population of human neural progenitor cells. *Exp Neurol* **158**: 265–278.
13. Flax JD, Aurora S, Yang C, *et al.* (1998) Engraftable human neural stem cells respond to developmental cues, replace neurons, and express foreign genes. *Nat Biotechnol* **16**: 1033–1039.
14. Zhang SC, Ge B, Duncan ID. (2000) Tracing human oligodendroglial development *in vitro*. *J Neurosci Res* **59**: 421–429.
15. Vescovi AL, Parati EA, Gritti A, *et al.* (1999) Isolation and cloning of multipotential stem cells from the embryonic human CNS and establishment of transplantable human neural stem cell lines by epigenetic stimulation. *Exp Neurol* **156**: 71–83.
16. Uchida N, Buck DW, He D, *et al.* (2000) Direct isolation of human central nervous system stem cells. *Proc Natl Acad Sci USA* **97**: 14720–14725.
17. Buc-Caron MH. (1995) Neuroepithelial progenitor cells explanted from human fetal brain proliferate and differentiate *in vitro*. *Neurobiol Dis* **2**: 37–47.

18. Nunes MC, Roy NS, Keyoung HM, *et al.* (2003) Identification and isolation of multipotential neural progenitor cells from the subcortical white matter of the adult human brain. *Nat Med* **9**: 439–447.

19. Kukekov VG, Laywell ED, Suslov O, *et al.* (1999) Multipotent stem/progenitor cells with similar properties arise from two neurogenic regions of adult human brain. *Exp Neurol* **156**: 333–344.

20. Wright LS, Li J, Caldwell MA, Wallace K, Johnson JA, Svendsen CN. (2003) Gene expression in human neural stem cells: Effects of leukemia inhibitory factor. *J Neurochem* **86**: 179–195.

21. Rowitch DH. (2004) Glial specification in the vertebrate neural tube. *Nat Rev Neurosci* **5**: 409–419.

22. Zhou Q, Anderson DJ. (2002) The bHLH transcription factors OLIG2 and OLIG1 couple neuronal and glial subtype specification. *Cell* **109**: 61–73.

23. Ostenfeld T, Joly E, Tai YT, *et al.* (2002) Regional specification of rodent and human neurospheres. *Brain Res Dev Brain Res* **134**: 43–55.

24. Shen Q, Goderie SK, Jin L, *et al.* (2004) Endothelial cells stimulate self-renewal and expand neurogenesis of neural stem cells. *Science* **304**: 1338–1340.

25. Caldwell MA, He X, Wilkie N, *et al.* (2001) Growth factors regulate the survival and fate of cells derived from human neurospheres. *Nat Biotechnol* **19**: 475–479.

26. Brustle O, Choudhary K, Karram K, *et al.* (1998) Chimeric brains generated by intraventricular transplantation of fetal human brain cells into embryonic rats. *Nat Biotechnol* **16**: 1040–1044.

27. Englund U, Bjorklund A, Wictorin K. (2002a) Migration patterns and phenotypic differentiation of long-term expanded human neural progenitor cells after transplantation into the adult rat brain. *Brain Res Dev Brain Res* **134**: 123–141.

28. Englund U, Fricker-Gates RA, Lundberg C, Bjorklund A, Wictorin K. (2002b) Transplantation of human neural progenitor cells into the neonatal rat brain: Extensive migration and differentiation with long-distance axonal projections. *Exp Neurol* **173**: 1–21.

29. Armstrong RJ, Watts C, Svendsen CN, Dunnett SB, Rosser AE. (2000) Survival, neuronal differentiation, and fiber outgrowth of propagated human neural precursor grafts in an animal model of Huntington's disease. *Cell Transplant* **9**: 55–64.

30. McBride JL, Behrstock SP, Chen EY, *et al.* (2004) Human neural stem cell transplants improve motor function in a rat model of Huntington's disease. *J Comp Neurol* **475**: 211–219.

31. Kelly S, Bliss TM, Shah AK, *et al.* (2004) Transplanted human fetal neural stem cells survive, migrate, and differentiate in ischemic rat cerebral cortex. *Proc Natl Acad Sci USA* **101**: 11839–11844.

32. Guillaume DL, Zhang SC. (2003) Transplantation to replace neurons. In: *Neural Stem Cells: Development and Transplantation.* ed. J.E. Bottenstein, pp. 299–328. Totawa, NJ: Humana Press.

33. Wu P, Tarasenko YI, Gu Y, Huang LY, Coggeshall RE, Yu Y. (2002) Region-specific generation of cholinergic neurons from fetal human neural stem cells grafted in adult rat. *Nat Neurosci* **5**: 1271–1278.

34. Anderson DJ, Gage FH, Weissman IL. (2001) Can stem cells cross lineage boundaries? *Nat Med* **7**: 393–395.

35. Blau HM, Brazelton TR, Weimann JM. (2001) The evolving concept of a stem cell: Entity or function? *Cell* **105**: 829–841.

36. Wagers AJ, Weissman IL. (2004) Plasticity of adult stem cells. *Cell* **116**: 639–648.

37. Raff M. (2003) Adult stem cell plasticity: Fact or artifact? *Annu Rev Cell Dev Biol* **19**: 1–22.

38. Woodbury D, Schwarz EJ, Prockop DJ, Black IB. (2000) Adult rat and human bone marrow stromal cells differentiate into neurons. *J Neurosci Res* **61**: 364–370.

39. Hermann A, Gastl R, Liebau S, *et al.* (2004) Efficient generation of neural stem cell-like cells from adult human bone marrow stromal cells. *J Cell Sci* **117**: 4411–4422.

40. Willing AE, Lixian J, Milliken M, *et al.* (2003) Intravenous versus intrastriatal cord blood administration in a rodent model of stroke. *J Neurosci Res* **73**: 296–307.

41. Zigova T, Song S, Willing AE, *et al.* (2002) Human umbilical cord blood cells express neural antigens after transplantation into the developing rat brain. *Cell Transplant* **11**: 265–274.

42. Sanchez-Ramos JR. (2002) Neural cells derived from adult bone marrow and umbilical cord blood. *J Neurosci Res* **69**: 880–893.

43. McGuckin CP, Forraz N, Allouard Q, Pettengell R. (2004) Umbilical cord blood stem cells can expand hematopoietic and neuroglial progenitors *in vitro. Exp Cell Res* **295**: 350–359.

44. Joannides A, Gaughwin P, Schwiening C, *et al.* (2004) Efficient generation of neural precursors from adult human skin: Astrocytes promote neurogenesis from skin-derived stem cells. *Lancet* **364**: 172–178.

45. Fernandes KJ, McKenzie IA, Mill P, *et al.* (2004) A dermal niche for multipotent adult skin-derived precursor cells. *Nat Cell Biol* **6**: 1082–1093.

46. Cogle CR, Yachnis AT, Laywell ED, *et al.* (2004) Bone marrow transdifferentiation in brain after transplantation: A retrospective study. *Lancet* **363**: 1432–1437.

47. Azizi SA, Stokes D, Augelli BJ, Digirolamo C, Prockop DJ. (1998) Engraftment and migration of human bone marrow stromal cells implanted in the brains of albino rats — Similarities to astrocyte grafts. *Proc Natl Acad Sci USA* **95**: 3908–3913.

48. Kopen GC, Prockop DJ, Phinney DG. (1999) Marrow stromal cells migrate throughout forebrain and cerebellum, and they differentiate into astrocytes after injection into neonatal mouse brains. *Proc Natl Acad Sci USA* **96**: 10711–10716.

49. Zhao LR, Duan WM, Reyes M, *et al.* (2002) Human bone marrow stem cells exhibit neural phenotypes and ameliorate neurological deficits after grafting into the ischemic brain of rats. *Exp Neurol* **174**: 11–20.

50. Thomson JA, Itskovitz-Eldor J, Shapiro SS, *et al.* (1998) Embryonic stem cell lines derived from human blastocysts. *Science* **282**: 1145–1147.

51. Thomson JA, Odorico JS. (2000) Human embryonic stem cell and embryonic germ cell lines. *Trends Biotechnol* **18**: 53–57.

52. Du ZW, Zhang SC. (2004) Neural differentiation from embryonic stem cells: Which way? *Stem Cell & Development* **13**: 372–381.

53. Kawasaki H, Mizuseki K, Sasai Y. (2002) Selective neural induction from ES cells by stromal cell-derived inducing activity and its potential therapeutic application in Parkinson's disease. *Methods Mol Biol* **185**: 217–227.

54. Perrier AL, Tabar V, Barberi T, Rubio ME, *et al.* (2004) Derivation of midbrain dopamine neurons from human embryonic stem cells. *Proc Natl Acad Sci USA* **101**: 12543–12548.

55. Reubinoff BE, Itsykson P, Turetsky T, *et al.* (2001) Neural progenitors from human embryonic stem cells. *Nat Biotechnol* **19**: 1134–1140.

56. Carpenter MK, Inokuma MS, Denham J, Mujtaba T, Chiu CP, Rao MS. (2001) Enrichment of neurons and neural precursors from human embryonic stem cells. *Exp Neurol* **172**: 383–397.

57. Li XJ, Du ZW, Zarnowska ED, *et al.* (2005) Specification of motoneurons from human embryonic stem cells. *Nat Biotechnol* **23**: 215–221.

58. Pera MF, Andrade J, Houssami S, *et al.* (2004) Regulation of human embryonic stem cell differentiation by BMP-2 and its antagonist noggin. *J Cell Sci* **117**: 1269–1280.

59. Zhang SC, Wernig M, Duncan ID, Brustle O, Thomson JA. (2001) *In vitro* differentiation of transplantable neural precursors from human embryonic stem cells. *Nat Biotechnol* **19**: 1129–1133.

60. Schulz TC, Palmarini GM, Noggle SA, *et al.* (2003) Directed neuronal differentiation of human embryonic stem cells. *BMC Neurosci* **4**: 27.

61. Zhang SC. (2003) Embryonic stem cells for neural replacement therapy: Prospects and challenges. *J Hematother Stem Cell Res* **12**: 625–634.

62. Buytaert-Hoefen KA, Alvarez E, Freed CR. (2004) Generation of tyrosine hydroxylase positive neurons from human embryonic stem cells after coculture with cellular substrates and exposure to GDNF. *Stem Cells* **22**: 669–674.

63. Yan YP, Yang DL, Zarnowska ED, *et al.* (2005) Directed differentiation of dopaminergic neuronal subtypes from human embryonic stem cells. *Stem Cells* (in press).

64. Lee SH, Lumelsky N, Studer L, Auerbach JM, McKay RD. (2000) Efficient generation of midbrain and hindbrain neurons from mouse embryonic stem cells. *Nat Biotechnol* **18**: 675–679.

65. Goldstein RS, Drukker M, Reubinoff BE, Benvenisty N. (2002) Integration and differentiation of human embryonic stem cells transplanted to the chick embryo. *Dev Dyn* **225**: 80–86.

66. Bieberich E, Silva J, Wang G, Krishnamurthy K, Condie BG. (2004) Selective apoptosis of pluripotent mouse and human stem cells by novel ceramide analogues prevents teratoma formation and enriches for neural precursors in ES cell-derived neural transplants. *J Cell Biol* **167**: 723–734.

67. Zwaka TP, Thomson JA. (2003) Homologous recombination in human embryonic stem cells. *Nat Biotechnol* **21**: 319–321.

# 19

# Liver Stem Cells

Malcolm R. Alison*, Pamela Vig, Francesco Russo,
Eunice Amofah and Stuart J. Forbes

## Introduction

The liver is normally proliferatively quiescent, but hepatocyte loss through partial hepatectomy, uncomplicated by virus infection or inflammation, invokes a rapid regenerative response from all cell types in the liver to perfectly restore liver mass. Moreover, hepatocyte transplants in animals have shown that a certain proportion of hepatocytes in fetal and adult liver can clonally expand, suggesting that hepatoblasts/hepatocytes are themselves the functional stem cells of the liver. More severe liver injury can activate a facultative stem cell compartment located within the intrahepatic biliary tree, giving rise to cords of bipotential transit amplifying cells (oval cells/hepatic progenitor cells), that can ultimately differentiate into hepatocytes and biliary epithelial cells. A third population of stem cells with hepatic potential resides in the bone marrow; these hematopoietic stem cells may contribute to the albeit low renewal rate of hepatocytes, but can make a more significant contribution to regeneration under a very strong positive selection pressure. In some instances, transdifferentiation of bone marrow cells to hepatocytes does not seem to occur, and instead cell fusion between myeloid or monocytic lineages and damaged hepatocytes appears to be the mechanism by which the hematopoietic genome becomes reprogrammed. There is also evidence that there may be cells in the bone marrow possessing markers of hepatic commitment, being chemoattracted to the liver as a result of organ damage. Bone marrow may also harbor cells with

*Correspondence: Department of Diabetes and Metabolic Medicine, St Mary's School of Medicine & Dentistry, Royal London Hospital, Whitechapel, London E1 1BB. Tel.: (44) 207 377 7000 X3308, Fax: (44) 207 377 7636, e-mail: m.alison@qmul.ac.uk

fibrogenic potential, contributing significantly to end-stage liver fibrosis. In common with other tissues, there is persuasive evidence that in the liver, stem cells are the founder cells of primary malignancies.

So, perhaps borne out of necessity from the plethora of potentially cell-damaging xenobiotics that assail the liver, plus a myriad of other cellular

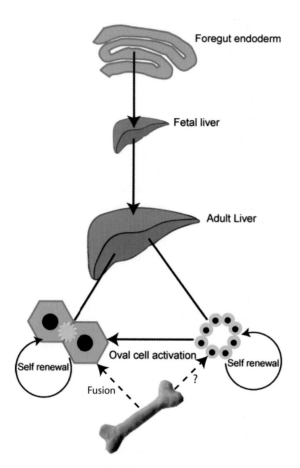

**Figure 1.** Current understanding of the origin and interrelationships of the cells involved in liver development and regeneration. Fetal liver contains bipotential hepatoblasts capable of differentiating into hepatocytes and cholangiocytes. These cells are capable of self-renewal after loss, but when hepatocyte renewal is compromised, bipotential oval cells are activated from canal of Hering cells (potential stem cell niche) to take over the burden of regenerative growth. The bone marrow also harbors cells with liver potential, but the factors determining their durable ingress are poorly understood: in the Fah null mouse, fusion between bone marrow-derived myelomonocytic cells and enzyme deficient hepatocytes occurs. Evidence that bone marrow can transdifferentiate into biliary cells is weak.

insults, e.g. hepatotropic viruses, the mammalian liver can invoke not just one, but at least three apparently distinct cell lineages to contribute to regenerative growth after damage (Fig. 1). This chapter summarizes recent advances in our understanding of these three cell systems.

## Hepatoblasts/Hepatocytes

In response to parenchymal cell loss the hepatocytes are the cells that normally restore the liver mass, rapidly re-entering the cell cycle from the $G_o$ phase (Fig. 2). However, even after a two-thirds partial hepatectomy, the remaining cells have only to cycle on average 1.4 times to restore preoperative cell number. This seemingly modest response leads to the incorrect assumption that hepatocytes had only limited division potential, and thus were not true stem cells. A crucial property that defines a stem cell is its ability to give rise to a large family of descendants, i.e. be clonogenic,[1] and importantly, at least some hepatocytes can do this. Hepatocyte transplantation models have shown that the transplanted cells are capable of significant clonal expansion within the diseased livers of experimental animals and probably humans (see below).

## Fetal liver

In the diseased human liver there may not be the substantial selective growth advantage for transplanted cells that is operative in many of the rodent

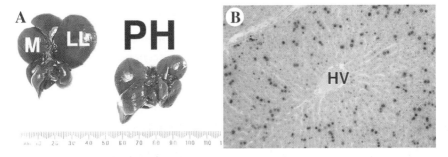

**Figure 2.** **A**. Partial hepatectomy (PH) in a rat involves resection of the left lateral (LL) and median (M) lobes comprising 2/3 of the liver mass and the preoperative mass is restored within 10 days. **B**. 24 hours after PH many hepatocytes are in DNA synthesis as indicated by bromodeoxyuridine labeling, but note the relative absence of labeling around the hepatic vein (HV) at this time, though within 48 hours after PH almost all hepatocytes will have traversed the cell cycle at least once.

models, and it therefore becomes of interest to determine if it is possible to enrich for true stem cells that would continue to expand in the recipient liver in the absence of a major growth stimulus. The fetal rat liver is potentially a rich source of bipotential stem cells for liver transplantation[2–4] and has been the focus of considerable attention. For example, Kubota and Reid[5] have described a population of bipotential progenitors from ED13 fetal rat liver that lacked expression of MHC class I and had modest ICAM-1 expression, features that may allow hepatoblasts to escape from the immune system when transplanted into an MHC-incompatible host. These cells were clonogenic in culture with some descendants expressing phenotypic markers of hepatocytes ($\alpha$-fetoprotein [AFP] and albumin) and others of cholangiocytes (cytokeratin [CK]-19).

In a similar vein, Shafritz and colleagues[6] have demonstrated that fetal liver epithelial progenitors (FLEPs) from ED14 rats are more clonogenic than normal adult hepatocytes; when wild-type FLEPs are injected into recently hepatectomized syngeneic dipeptidyl peptidase (DPPIV-) deficient F344 rats, they proliferate for at least 6 months and constitute 7% of the recipient liver at this time compared with colonization of only 0.06% of the liver by wild-type adult hepatocytes. DPPIV is an exopeptidase expressed in the bile canalicular surface of hepatocytes in contrast to diffuse cytoplasmic expression in bile duct epithelia. Thus, DPPIV-positive hepatocytes are readily detected in the recipient mutant liver either by enzyme histochemistry or immunohistochemistry (IHC). Much greater colonization of the mutant liver was observed when the recipient rats were given prior administration of the DNA-binding pyrrolizidine alkaloid retrorsine (usually two injections of 30 mg/kg each, two weeks apart); at 6 months after transplantation, 60–80% of the recipient liver was occupied by DPPIV + hepatocytes.[7] Likewise, when these FLEPs were transduced with lentiviral vectors expressing green fluorescent protein (GFP) under the control of the albumin promoter, the colonizing cells co-expressed DPPIV and GFP.[8]

Bipotential progenitors have been isolated from fetal mouse liver in a number of studies. Tanimizu *et al.*[9] have selected such cells on the basis of expression of Dlk, a type1 membrane protein that has 6 EGF-like repeats in its extracellular domain, while others have isolated cells capable of large-scale repopulation in the setting of subacute liver failure in uPA/SCID or uPA/Rag-2 transgenic mice.[10,11] Clonogenic cells have also been isolated by Suzuki *et al.*[12] from fetal mouse liver (ED13.5) on the basis of expressing the integrins $\alpha6$ (CD49f) and $\beta1$ (CD29), but not c-kit, CD45 or Ter119 (erythroid precursor antigen). Designated hepatic colony forming units in culture (H-CFU-C), this sorting achieved a 35-fold enrichment of H-CFU-C over total fetal liver cells. Further selection based on c-Met-positivity enriched for H-CFU-C and these cells produced both hepatocytes (albumin-positive) and biliary cells (cytokeratin-19-positive) in culture;[13] EGFP-marked cells from these clonally-derived H-CFU-C also produced hepatocytes and biliary cells when injected into mice, and more surprisingly were found to apparently differentiate into pancreatic ducts and acini and duodenal mucosal cells when injected directly into these organs. Similar CD49[+], c-Met[+], c-Kit[−], CD45[−] hepatic progenitors have been isolated by Hoppo *et al.*,[14] who found that their maturation required direct cell-cell contact with Thy-1[+] cells of probable mesenchymal origin that variously expressed α-SMA and desmin.

## Adult liver

Many studies have examined the transplantation potential of adult hepatocytes in the DPPIV- mutant rat, combining retrorsine treatment with a mitogenic stimulus, such as partial hepatectomy or triiodothyronine (T3), leading to rapid replacement of DPPIV- cells by DPPIV+ donor cells;[15,16] even in the absence of a mitogenic stimulus, near total replacement by donor cells occurs within 12 months.[17]

Intriguingly, when retrorsine-treated adult rats are given a two-thirds partial hepatectomy, regeneration is accomplished by the activation, expansion and differentiation of so-called small hepatocyte-like progenitors (SHPCs).[18] These cells showed phenotypic traits of fetal hepatoblasts, oval cells and fully differentiated hepatocytes, but they were morphologically and phenotypically distinct from all three. Cytochrome (CYP) P450 enzymes have a pivotal role in hepatocyte biology, but typically these cell clusters lacked CYP enzymes that are usually readily induced by retrorsine, and this probably accounted for their resistance to the anti-proliferative effects of retrorsine. When such cells (H4-positive) were isolated, established in short-term culture and then transplanted into syngeneic rats, they gave rise to differentiated hepatocytes as shown by expression of albumin and transferrin, but lack of AFP.[19]

The clonogenic potential of transplanted adult hepatocytes has been very impressively shown in the Fah null mouse, a model of hereditary type 1 tyrosinemia, where there exists a profoundly strong positive selection pressure on the transplanted wild-type cells, since Fah-deficient mice will die as neonates unless rescued by 2-(2-nitro-4-trifluoro-methylbenzoyl)-1, 3-cyclohexanedione (NTBC), a compound that prevents the accumulation of toxic metabolites in the tyrosine catabolic pathway. When $10^4$ normal hepatocytes from congenic male wild-type mice were intrasplenically injected into mutant female mice, these cells rapidly colonized the mutant liver.[20] Moreover, serial transplantations from the colonized livers to other Fah null livers indicated that at least 69 doublings would have been necessary from the original hepatocytes for six rounds of liver repopulation. This estimate is likely to be a minimal figure, since it assumes that all injected hepatocytes migrate to the liver from the spleen and take part equally in the cycles of regeneration. In fact, probably at best only 15% of intrasplenically transplanted hepatocytes migrate to the liver, and if all these participated equally in repopulation, a minimum of 86 doublings would be required for 6 serial transplants. This figure may be even higher if not all the cells that migrated to the liver actually took part in repopulation, and the authors suggested that there might be a sub-population of hepatocyte stem cells designated as "regenerative transplantable hepatocytes" (RTHs). We could speculate that perhaps these RTHs are analogous to the SHPCs described by Gordon *et al.*?[18] The Fah null mouse can also be rescued by pancreatic cells; though most Fah-deficient mice withdrawn from the NTBC treatment and transplanted with pancreatic cells will die, a small proportion do survive with 50–90% replacement of the diseased liver by pancreatic cell-derived hepatocytes.[21] Since animals fed a copper-deficient diet undergo pancreatic exocrine cell atrophy and refeeding induces the surviving ducts to give rise to hepatocytes,[22] it was surprising that pancreatic cell suspensions enriched for pancreatic ducts were poorer than unfractionated pancreatic cells at reconstituting the diseased Fah-deficient liver with functional hepatocytes.

Other cell lines with hepatocyte potential can also be isolated from adult rat liver;[23] these cells were AFP-, albumin- and CK19-negative, but after co-culture with hepatic stellate cells, they expressed albumin, transferrin and α-1 antitrypsin. It is well known that the WB-F344 rat liver diploid epithelial cell line readily differentiates into hepatocytes when transplanted into syngeneic rats, but biliary differentiation (CK19+, BDS7+, gamma glutamyl transpeptidase [GGT]+) can be induced by *in vitro* culture on Matrigel.[24] Tateno and Yoshizato[25] have defined the conditions for the long-term expansion of cells isolated from adult Fisher 344 rats. Fujikawa *et al.*[26] have, like Suzuki and colleagues,[12,13] identified a population of CD49f+,

CD29$^+$, c-kit$^-$, CD45$^-$ and Ter119$^-$ cells, this time in adult mouse liver, that were bipotential and clonogenic *in vitro*.

## Clinical application

Hepatic stem cells from whatever source, may be therapeutically useful for treating a variety of diseases that affect the liver. This would include a number of genetic diseases that produce liver disease such as Wilson's disease (copper accumulation), Crigler Najjar syndrome (lacking bilirubin conjugation activity) and tyrosinemia, as well as cases where there is extrahepatic expression of the disease, e.g. Factor IX deficiency. In terms of therapeutic potential, we have already noted the rescue of the Fah null mouse and the DPPIV-negative rat by hepatocytes, and there are other examples. Transplantation of adult rat hepatocytes has been effective in normalizing bilirubin levels and improving bilirubin conjugation activity in Gunn rats (a model of Crigler-Najjar syndrome).[27,28] These cells were reversibly immortalized and transduced with the bilirubin-uridine 5′-diphosphoglucuronate glucuronsyltransferase gene (*ugt1a1*) and engraftment was improved by prior irradiation and partial hepatectomy of the recipient rats. Likewise, an infusion of isolated hepatocytes through the portal vein equivalent to 5% of the parenchymal mass to a patient with Crigler Najjar syndrome, achieved a medium-term reduction in serum bilirubin and increased bilirubin conjugate levels in the bile.[29] Hepatocyte transplantation has also been successful in the treatment of human glycogen storage disease type 1a,[30] but not in the treatment of severe ornithine transcarbamylase deficiency, where rejection of the transplanted cells was thought to be the reason for only temporary (11 days) relief.[31] There has also been a drive to propagate human hepatocytes in a xenogenic setting to study the likes of drug metabolism and infectious disease. Indeed, immunodeficient uPA/recombinant activation gene-2 (Rag-2) mice support the growth of human hepatocytes[32] and remain permissive for human hepatitis B virus infection. Similarly, Alb-uPA/SCID mice have been used to study hepatitis C virus infection *in vivo*, as human cell transplantation results in the development of large persistent nodules of human hepatocytes.[33]

# Biliary Epithelial Cells

## Rodent studies

When either massive damage is inflicted upon the liver or regeneration after damage is compromised, a *potential* stem cell compartment located within the smallest branches of the intrahepatic biliary tree is activated. This so-called "oval cell" or "ductular reaction" amplifies the biliary population before these cells differentiate into either hepatocytes or cholangiocytes.[34-38] Careful studies in rats indicate that oval cells are predominantly derived from the canal of Hering, and thus this is the location of a stem cell niche.[39] In rats and mice, the canals of Hering barely extend beyond the limiting plate, but the resultant oval cell proliferation can result in arborizing ducts that express AFP (Fig. 3A), which stretch to the midzonal areas (Fig. 3B) before these cells differentiate into hepatocytes (Fig. 3C). A wide range of markers has been used to identify ovals cells, including GGT and glutathione-S-transferase (GST-P) activity along with a host of monoclonal antibodies raised against cytoskeletal proteins and unknown surface antigens.[40-42] Moreover, antigens traditionally associated with hematopoietic cells can also be expressed by oval cells, including c-kit, flt-3, Thy-1 and CD34.[43-46] This similarity has given support to the notion that at least some hepatic oval cells are directly derived from a precursor of bone marrow origin, particularly when the biliary tree is damaged,[47] though other studies indicate that many/most oval cells are derived from the direct intrahepatic proliferation of cells already located within the biliary tree.[35] Regarding the contribution of bone marrow (see below), Hatch *et al.*[48] have suggested that only when the rat liver is severely damaged do hepatocytes upregulate the chemokine stromal-derived factor-1 alpha (SDF-1α), and both oval cells are activated and bone marrow cells recruited through SDF-1α/CXCR4 interactions.

In 1996, Goodell *et al.*[49] reported on a new method for the isolation of hematopoietic stem cells (HSCs) based on the ability of the HSCs to efflux a fluorescent dye. Like the activity of the P-glycoprotein (encoded by the *mdr1* gene), this activity was verapamil-sensitive. Cells subjected to Hoechst 33342 dye staining that actively efflux the Hoechst dye appeared as a distinct population of cells on the side of a flow cytometry profile, hence the name the "side population" (SP) was given to these cells.[50] Numerous studies now point to the fact that the SP phenotype of HSCs in mice and humans is largely determined by the expression of a protein known as the ABCG2 transporter (ATP-binding cassette [ABC] subfamily G member 2, also known as BCRP1).[51] Perhaps not surprisingly, we now have reports

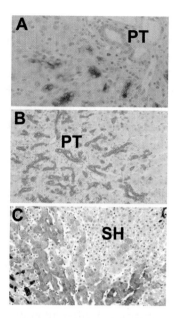

**Figure 3.**  Oval cell behavior in the rat liver treated by the AAF/PH protocol. (**A**) AFP expression (IHC staining) is typically observed in the migrating oval cells; note absence of staining in the interlobular duct in the portal tract (PT). (**B**) The oval cell response can be visualized by CK8 immunostaining, with cords of cells emanating from the portal tract (PT). (**C**) At later times the cords of oval cells differentiate into small hepatocytes (SH) but with a notable lack of CYP immunoexpression (brown staining). Note occasional residual oval cell ductules expressing CK19 (purple staining).

of the upregulation of several ABC transport proteins in damaged human liver, particularly in regenerating ductules[52] and of ABCG2/BCRP1 in rat oval cells,[53] adding support to the belief that ABC transporter proteins are intimately involved in the biology of stem/progenitor cells in many tissues. However, is the expression of these ABC transporters in liver ductules a genuine marker of stem cells/progenitors or merely a reflection of the protective role these proteins undoubtedly perform within the biliary tree against toxic bile constituents?[54] The correct answer is probably the latter, since it is highly unlikely that all the reactive ductular cells are stem cells or even progenitors, and many are imminently going to differentiate into hepatocytes and cholangiocytes. Nevertheless, the expression of ABC transporters adds incrementally to the battery of already established markers for this stem cell response. Wulf *et al.*[55] isolated both CD45$^+$ and CD45$^-$ SP cells from mouse liver, claiming that such cells could differentiate into both hepatocytes and biliary epithelia in a model of hepatic damage.

Most models of oval cell activation have employed potential carcino-gens to inhibit hepatocyte replication in the face of a regenerative stim-ulus, though replicative senescence associated with a fatty liver can also activate oval cells.[56] In the rat, protocols have included administering 2-acetylaminofluorene (2-AAF) prior to a two-thirds partial hepatectomy (AAF/PH) or a necrogenic dose of carbon tetrachloride, feeding a choline deficient diet supplemented with ethionine (CDE), or simply treating ani-mals with the likes of 3′-methyl-diaminobenzidine (3′-Me-DAB), galac-tosamine or furan. Cells derived by such procedures are clearly not relevant to human studies, but Sell and co-workers[57,58] have demonstrated that it is possible to derive bipotential liver progenitor cells (LPCs) from rat livers without using mutagenic chemicals. Allyl alcohol causes periportal necro-sis, and the resultant oval cells can be isolated and propagated, producing a number of clonally-derived cell lines, capable of at least 100 doublings in the presence of a feeder layer of lethally irradiated fibroblasts. These cell lines were able to differentiate into hepatocytes after removal of the feeder cells; supplementation with either a combination of oncostatin M and dex-amethasone or bFGF promoted differentiation. Recent observations have raised the possibility that the feeder cells themselves may be the origin of these cells![59]

We have little idea of the precise cues that promote the activation of oval cells, but blockade of the sympathetic nervous system enhances oval cell accumulation.[60] In terms of therapeutic potential, one of the most impres-sive demonstrations of the hepatocytic potential of oval cells has come from experiments in which oval cells were isolated from Long Evans Cinnamon (LEC) rats (a model of Wilson's disease); transduced *ex vivo* with a reporter gene, β-galactosidase, and then transplanted into LEC/Nagase analbumine-mic double mutant rats, these cells differentiated into albumin-expressing hepatocytes.[61] Oval cells can also act as transplantable hepatocyte pro-genitors in the Fah knockout mouse; furthermore, the fact that oval cells generated in bone marrow transplanted wild-type mice can repopulate Fah knockouts, but lack markers of the original bone marrow donor, strongly suggests that oval cells *do not* originate from bone marrow in this particular model.[62]

## Human studies

Oval cells also occur in human liver, but a recent consensus paper[63] rec-ommends that the term "oval cells" be discontinued as the organization

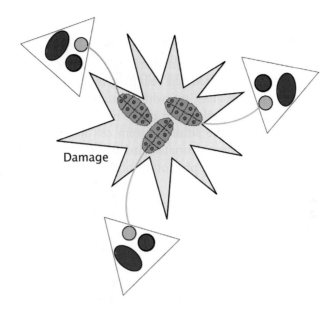

Damage

**Figure 4.** The canals of Hering (green) extend from the portal areas into the proximate third of the hepatic lobule in the human liver (see Ref. 64) and major hepatocyte damage activates the lining cells to divide and some differentiate into hepatocytes.

of the biliary tree in humans is unlike that found in the commonly used rodent models. In normal human liver, terms such as "cholangiocytes" or "progenitor cells" are preferred, while in diseased states, "oval cell reactions" should be replaced by "ductular reactions." Three-dimensional reconstructions of serial sections of human liver immunostained for cytokeratin-19 have shown that the smallest biliary ducts, the canals of Hering, unlike those in rodents, normally extend into the proximate third of the lobule,[64] envisaging that these canals react to massive liver damage (akin to a trip-wire), proliferating and then differentiating into hepatocytes (Fig. 4). The magnitude of ductular reactions in human liver rise with increasing severity of liver disease[65] and this ductular reaction is widely accepted to be a stem cell response rather than a ductular metaplasia of "damaged" hepatocytes. Notwithstanding, Falkowski *et al.*[66] still felt compelled to formally dispel such a notion in human liver, showing that "cholestatic" hepatocytes were very largely *not* associated with ductular reactions and moreover, that in cirrhosis most (94%) intraseptal buds of hepatocytes *were* associated with ductular reactions.

## Oval cells may be pluripotential

Two studies have suggested that if oval cells are transplanted to extrahepatic locations, they may transdifferentiate to a phenotype appropriate for that site. Deng *et al.*[67] transplanted oval cells from mouse liver into early postnatal brain, observing apparent transdifferentiation into neurons, microglia and astrocytes, though cell fusion could not be discounted (see below). On the other hand, if the adult rat stem cell-like liver line (WB F344) is co-cultured with rodent neonatal cardiac cells, then the WB F344 cells will adopt a cardiomyocyte phenotype and function, with no cell fusion occurring.[68]

## Bone Marrow

Some oval cells/hepatocytes were first revealed to be derived from circulating bone marrow cells in the rat (Table 1a). Petersen *et al.*[47] followed the fate of syngeneic male bone marrow cells transplanted into lethally irradiated female recipient animals whose livers were subsequently injured by a regime of 2-acetylaminofluorene (which blocks hepatocyte regeneration) and carbon tetrachloride (which causes hepatocyte necrosis) designed to cause oval cell activation. Y chromosome-positive oval cells were found at 9 days after liver injury and some Y chromosome-positive hepatocytes were seen at 13 days when oval cells were differentiating into hepatocytes. Additional evidence for hepatic engraftment of bone marrow cells was forthcoming from a rat whole liver transplant model. Lewis rats expressing the MHC class II antigen L21-6 were recipients of livers from Brown Norway rats that were negative for L21-6. Subsequently, ductular reactions in the transplants contained both L21-6-negative and L21-6-positive cells, indicating that some oval cells were of *in-situ* derivation and some were of recipient origin, presumably from circulating bone marrow cells.

Using a similar gender mismatch bone marrow transplantation approach in mice to track the fate of bone marrow cells, Theise *et al.*[71] found that over a 6 month period, 1–2% of hepatocytes in the murine liver may be derived from bone marrow in the absence of any obvious liver damage, suggesting that bone marrow contributes to normal "wear and tear" renewal (Table 1b). On the other hand, others have found no evidence that bone marrow contributes to murine hepatocyte renewal, even in the chronically damaged liver.[80]

**Table 1a.** Hepatocyte Differentiation of Hematopoietic Cells in Rats

| Authors | Procedure | Injury | Evidence | Hematopoietic Contribution to Hepatocyte Population (%) | Comments |
|---|---|---|---|---|---|
| 47 | Male BMTx to females Male wild-type to DPPIV-null | 2-AAF/CCl$_4$ | Y+ cells in female DPPIV+ hepatocytes in DPPIV- liver | 0.16 | Also noted Y+ oval cells |
| 69 | Strain-mismatch liver transplantation | Organ rejection | C3 antigen not detected in cells integrated into hepatic plates | NS | No positive evidence that cells were hepatocytes |
| 70 | CD45 mismatch BMTx | Retrorsine and CCl$_4$ | Donor MHC antigens (IHC) | None | Hepatocyte hypertrophy responsible for restoration of liver mass |

Abbreviations: 2-AAF/CCl$_4$, acetylaminofluorene followed by carbon tetrachloride; BMTx, bone marrow transplant; DPPIV, dipeptidyl peptidase IV; IHC, Immunohistochemistry; NS, not stated.

**Table 1b.** Hepatocyte Differentiation of Hematopoietic Cells in Mice

| Authors | Procedure | Injury | Evidence | Hematopoietic Contribution to Hepatocyte Population (%) | Comments |
|---|---|---|---|---|---|
| 71 | Male BMTx to female mice | None | Y+/albumin mRNA+ | Up to 2.2% | |
| 72 | Male BMTx to female Fah null mice | Liver failure | Y+/Fah+ hepatocytes | 30–50 | |
| 73 | Male BMTx to female Fah null mice | Liver failure | Y+/Fah+ hepatocytes | >30 | Without liver failure got same initial engraftment but no clonal expansion |
| 74 | Male Bcl-2 BMTx to female mice | 8 × Jo2 antibody | Y+/Bcl-2+/CK 8, 18, 19+ | 0.8 | Each Jo2 injection destroyed 50% liver |
| 75 | GFP BMTx | 2/3 PH | GFP+ hepatocytes | None | Major contribution of GFP+ cells to endothelia and Kupffer cells |
| 76 | Female BMTx to male Fah null mice | Liver failure | Genotype of Fah+ cells. Fah+/Y+ cells | ~ 50 | Fah+ cells had mixture of donor and recipient genotype and were Y+: fusion occurring |
| 77 | Male BMTx to female Fah null mice | Liver failure | Genotype of Fah+ cells | NS | Mixed genotype (as above) = cell fusion |

**Table 1b.** (*Continued*)

| | | | | |
|---|---|---|---|---|
| 78 | Cre recombinase+ BMTx to conditional Cre reporter mice | None | LacZ+ hepatocytes, albumin+, with bile canaliculi | <0.01 | Cell fusion occurring. Also fusion of BM with Purkinje cells and cardiomyocytes |
| 79 | GFP+ BMTx | $CCl_4$-induced cirrhosis | GFP cells in cords, mostly Liv2 + | ~ 25 | Periportal location of GFP+ cells. Rise in serum albumin. |
| 80 | Marked (EGFP or LacZ) BMTx | 3 models of chronic liver injury | Y+ or GFP+ or LacZ+ | None | |
| 55 | Male BMTx (SP) to females | DDC injury | Y+/ZK31+ | 0.2–1.3 | Also contribution to oval cells |
| 81 | GFP+/Y+ BMTx to females | DEN injury | Y+/GFP+/AFP+ | Infrequent clusters | |
| 82 | Cre reporter BMTx to Cre recombinase recipients | None | CK+/Y+ | 0.05 | No fusion observed |
| 83 | GMPs and BMMs to male Fah null mice | Liver failure | Fah+/Y+ | ~1 cluster/$10^5$ hepatocytes | All result from fusion of myelomonocytic cells with Fah-/- hepatocytes |
| 84 | LysM-Cre BMTx to Fah-/-/R26R | Liver failure | β-gal/+Fah+ | NS | All result from fusion of myelomonocytic cells with Fah-/- hepatocytes |

Abbreviations: AFP, alpha fetoprotein; BMMs, bone marrow-derived macrophages; BMTx, bone marrow transplant; CK, cytokeratin; DDC, 3,5-diethoxycarbonyl-1, 4-dihydrocollidine; EGFP, enhanced green fluorescent protein; Fah, fumarylacetoacetate hydrolase; GFP, green fluorescent protein; GMPs, granulocyte/macrophage progenitors; LysM, lysozyme M (expressed by myelomonocytic lineage); NS, not stated; SP, side population.

In two contemporaneous papers, Alison *et al.*[85] and Theise *et al.*[86] demonstrated that hepatocytes can also be derived from bone marrow cell populations in humans (see Table 1c). The degree of hepatic engraftment of HSCs into human liver was highly variable, most likely related to the severity of parenchymal damage, with up to 40% of hepatocytes and cholangiocytes derived from bone marrow in a liver transplant recipient with recurrent hepatitis.[86] Subsequent human investigations with G-CSF mobilized peripheral blood CD34+ stem cells have shown that these cells are also apparently able to transdifferentiate into hepatocytes, with 4–7% of the hepatocytes in female livers being Y chromosome-positive after the cell transplant from male donors.[88] However, it is worth noting that some other studies, examining the contribution of recipient cells to liver allografts in humans, have failed to register any real engraftment in the allografted liver[89–91] (see Table 1c).

## The Fah story

The most convincing "proof of principle" demonstration of the potential therapeutic utility of bone marrow came from the curing of mice with a metabolic liver disease.[72] Female mice deficient in the enzyme fumarylacetoacetate hydrolase (*fah-/-*, a model of fatal hereditary tyrosinemia type 1), a key component of the tyrosine catabolic pathway, can be rescued biochemically by a million whole bone marrow cells that are wild-type for Fah. Moreover, only purified HSCs (c-kit$^{high}$Thy$^{low}$Lin$^-$Sca-1$^+$) were capable of this functional repopulation, with as few as 50 of these cells being capable of hepatic engraftment when hematopoiesis was supported by $2 \times 10^5$ *fah-/-* congenic adult female bone marrow cells. The salient point to arise from this powerful demonstration of the therapeutic potential of bone marrow cells was that though the initial engraftment was low, approximately one bone marrow cell for every million indigenous hepatocytes, the strong selection pressure exerted thereafter on the engrafted bone marrow cells resulted in their clonal expansion to eventually occupy almost half the liver. This positive selection was achieved by cycles of withdrawal of NTBC, the compound that blocks the breakdown of tyrosine to hepatotoxic fumarylacetoacetate (FAA) and maleylacetoacetate in the Fah deficient mice, so protecting against liver failure. In the absence of NTBC, FAA accumulates and destroys the hepatocytes; thus, the ensuing regenerative stimulus promotes the growth of the engrafted cells. Furthermore, in the absence of NTBC, no clonal expansion was seen.[73]

**Table 1c.** Hepatocyte Differentiation of Hematopoietic Cells in Man

| Authors | Procedure | Injury | Evidence | Hematopoietic Contribution to Hepatocyte Population (%) | Comments |
|---|---|---|---|---|---|
| 85 | Male BMTx to female Female allograft in male | Variable | Y+/CK8+ cells | 1–2 | Some small clusters |
| 86 | Male BMTx to female Female allograft in male | Variable | Y+/CAM5.2+ cells | 4 – 43* | *recurrent HCV |
| 87 | Allografted liver | Variable | Genotype chimerism | NS | Cholangiocyte chimerism also seen |
| 88 | Male PBSC Tx to female | Variable | Y+/CAM5.2+ cells | 4–7 | Also Y+ cells in skin and gastrointestinal tract |
| 89 | Sex-mismatched liver transplant | Variable | X and Y, CD45 | None | Followed up to 12 years |
| 90 | Sex-mismatched liver transplant | Variable | Genotype chimerism | 0.62 | Most Y+ cells in liver were macrophages |
| 91 | Female allograft in male | Variable | Y+ hepatocytes | Infrequent or non-existent | Followed for up to 16 years |

Abbreviations: BMTx, bone marrow transplant; CK, cytokeratin; PBSC, peripheral blood stem cell.

Further study of this model has now shown that the new healthy liver cells in the *fah-/-* mouse contain chromosomes from both the recipient and donor cells, with the donor hematopoietic cell nuclei being reprogrammed when they fused with the unhealthy *fah-/-* hepatocyte nuclei to become functional hepatocytes.[76,77] In one experiment, a million donor bone marrow cells (*fah+/+*) from Fanconi anemia group C (*fancc-/-*) homozygous mutant mice were serially transplanted into lethally irradiated *fah-/-* recipients.[76] The usual repopulation (~50%) of the mutant liver by Fah-positive hepatocytes was noted, but Southern blot analysis of the purified repopulating cells revealed that they were heterozygous for alleles (*fah+/+*; *fancc-/-*) that were unique to the donor marrow — fusion with host liver cells must have occurred. To confirm this conclusion, in a second experiment *fah+/+* bone marrow from ROSA26 female mice was transplanted into male *fah* knockout mice. Cytogenetic analysis of the LacZ-positive, sorted bone marrow-derived hepatocytes revealed that most, if not all, had a Y chromosome, thus confirming fusion. Before bone marrow transplant, most host hepatocytes had a karyotype of either 40,XY or 80,XXYY, but after transplantation with *fah+/+* bone marrow, the commonest karyotype of Fah-positive heptocytes was either 80,XXXY, suggesting fusion between a diploid female donor cell and a diploid male host cell; or 120,XXXXYY, suggesting fusion between a female donor diploid blood cell and a tetraploid male host hepatocyte. However, a substantial proportion of bone marrow-derived hepatocytes were aneuploid, suggesting fusion had created a genetic instability with the hybrid cells randomly shedding chromosomes. The fusogenic partners in the Fah setting have now been identified and are not the HSCs themselves, but are bone marrow-derived macrophages (BMMs) and granulocyte/macrophage progenitors (GMPs).[83,84] In one study, T and B cells were found not to be involved, and direct intrasplenic transplantation of either BMMs or GMPs, without lethal irradiation or hematopoietic reconstitution, leads to robust replacement of the Fah null liver with Fah[+] hepatocytes.[83] Likewise, Camargo and colleagues[84] came to similar conclusions from a series of experiments, including a Cre/lox strategy, involving transplanting bone marrow from mice that expressed Cre recombinase only in the myelomonocytic lineage to Fah null mice that had a floxed β-galactosidase reporter gene; nodules that were positive for both β-galactosidase and Fah indicated that probably Kupffer cells were the fusogenic partners for the Fah-deficient hepatocytes.

## Other models

Fusion of bone marrow cells has also been found to occur in the normal mouse, not only with hepatocytes, but also with Purkinje cells and cardiomyocytes.[78] These were very elegant studies *in vivo* and *in vitro* in which a reporter gene was activated only when cells fused. However, unlike the Fah null mouse, no selection pressure (liver damage) was operative, and even after 10 months only 9–59 fused cells/$5.5 \times 10^5$ hepatocytes were found: importantly, they also found evidence that with time either donor genes had been inactivated or eliminated, again suggestive of genetic instability in heterokaryons. On the other hand, data from Krause and colleagues[82] suggest that under normal physiological circumstances, true transdifferentiation rather than cell fusion prevails. In their study, lethally irradiated female mice that ubiquitously expressed Cre recombinase were the recipients of male bone marrow that would only express EGFP if fusion occurred; 2–3 months after transplantation, 0.05% of 36,000 hepatocytes were Y chromosome positive but none expressed EGFP.

Mouse bone marrow-derived hepatocytes can also be expanded selectively if they are engineered to overexpress Bcl-2, and then the indigenous cells are targeted for destruction by an anti-Fas antibody.[74] One could also add that if fusion was responsible for all these observations made in the liver, then clearly these hybrids have a selective growth advantage, turning unhealthy hepatocytes into metabolically competent hepatocytes and would not negate the therapeutic potential of bone marrow cells in the liver. Expressing a similar sentiment, Blau[92] has suggested that if cell fusion was responsible for the apparent reprogramming of certain adult cells, then there is something "exciting" about rescuing damaged cells through fusion, with, for example, bone marrow-derived cells providing a healthy and entire genetic complement, even one that has been manipulated for gene therapy. As in the human arena, not all murine studies are in anything like complete accord; e.g. Terai *et al.*[79] report an impressive 25% contribution of bone marrow to the parenchyma after $CCL_4$, but Kanazawa and Verma[80] failed to find any evidence for bone marrow engraftment in 3 models of chronic liver injury, including $CCL_4$ (see Table 1b). On the other hand, many studies have testified as to the ability of human cord blood cells to transdifferentiate into hepatocytes in the liver of the immunodeficient mouse (Table 1d), albeit at a low level.

**Table 1d.** Hepatocyte Differentiation of Human Umbilical Cord Blood (h.UCB) Cells in Immunodeficient Mice

| Authors | Procedure | Injury | Evidence | Hematopoietic Contribution to Hepatocyte Population (%) | Comments |
|---|---|---|---|---|---|
| 93 | CD34+/-, C1qRp+ h.UCB to NOD/SCID mice | None | h.albumin (RT-PCR) c-Met (IHC) | 0.05–0.1 | Illustrated cells were commonly binucleate — fusion? |
| 94 | CD34+ or CD45+ h.UCB to NOD/SCID/BMG-mice | None | h.albumin (RT-PCR) HepPar1 (IHC) | 1–2 | FISH with human and murine centromeric probes found no fusion |
| 95 | h.UCB to NOD/SCID mice | None | HepPar1 (IHC) h.DNA sequences (FISH) | NS | No evidence for fusion |
| 96 | CD34+ h.UCB to NOD/SCID and NOD/SCID/BMG- | CCl$_4$ | h.Alu sequences h.albumin mRNA | 1–10 are human in liver, but only 1 in 20 express albumin | rhHGF increased h.albumin expression |
| 97 | h.UCB to SCID mice | 2-AAF/PH | h. X chromosome (FISH) h.albumin (IHC) | 0.1–1 | Claimed albumin+ cells in clusters (data not shown) |
| 98 | h.UCB (CD34+) to NOD/SCID | CCl$_4$ | h.albumin | ~50 – 175/1.5 × 10$^6$ | Occasional clusters. Human cells adjacent to SDF-1+ bile ducts. HGF+SDF induced lamellipodia on CD34+ cells |

Abbreviations: 2-AAF/PH, acetylaminofluorene followed by partial hepatectomy; FISH, fluorescence *in situ* hybridization; h.UCB, human umbilical cord blood; IHC, immunohistochemistry; NOD/SCID/BMG-, non-obese/severe combined immunodeficient/β2 microglobulin negative; NS, not stated; PBSC, peripheral blood stem cell; RT-PCR, reverse transcription-polymerase chain reaction; SP, side population.

## Bone marrow cell homing

While it seems logical to believe that parenchymal damage is a stimulus to hepatic engraftment by HSCs, the molecules that mediate this homing reaction to the liver are not well understood. One obvious mechanism is that cells in the liver express the stem cell chemoattractant "stromal derived factor-1" (SDF-1), for which HSCs have the appropriate receptor known as CXCR4.[99] Hatch *et al.*[48] have persuasive evidence that SDF-1 is involved in oval cell activation, speculating furthermore, that this chemokine may secondarily recruit bone marrow to the injured liver. More definitive proof was provided by Kollet *et al.*[98] who observed increased SDF-1 expression (particularly in biliary epithelia) after parenchymal damage, and concomitant with such damage was increased HGF production, a cytokine that was very effective in promoting protrusion formation and CXCR4 upregulation in human CD34+ hematopoietic progenitors. In a similar vein, Ratajczak *et al.*[100] provide evidence that already committed liver progenitors (CK19+, $\alpha$FP+) expressing CXCR4 "hide-out" in human and murine bone marrow, becoming mobilized consequent to liver damage. HSCs apparently primed to become hepatocytes (c-Met+, $\alpha$FP+) also exist in rat bone marrow and can be terminally differentiated by a combination of HGF and EGF.[101] The influence of hepatocytes on HSCs can be quite dramatic, Jang *et al.*[102] co-cultured injured liver cells with highly purified HSCs (separated by a trans-well membrane) and, over a very short time, many HSCs lost expression of CD45 but gained expression of albumin, liver-enriched transcription factors and markers of hepatocyte terminal differentiation. Remarkably, some former HSCs became binucleated and even tetraploid, ploidy changes characteristic of hepatocytes.

## Stem Cells and Liver Disease

Liver fibrosis and liver cancer are two very major causes of human morbidity and mortality, and both appear to have their foundations in stem cells. In the mouse liver, transplanted bone marrow cells can acquire the phenotype of quiescent stellate cells and after liver injury can become activated into $\alpha$SMA-expressing cells.[103] In human cirrhosis, by examining sex-mismatched allografts, Forbes *et al.* have shown that up to 40% of myofibroblasts are of bone marrow origin.[104]

Stem cell biology and cancer are inextricably linked. In continually renewing tissues such as the intestinal mucosa and epidermis, where a steady flux of cells occurs from the stem cell zone to the terminally differentiated cells that are imminently to be lost, it is widely accepted that cancer is a disease of stem cells, since these are the only cells that persist in the tissue for a sufficient length of time to acquire the requisite number of genetic changes for neoplastic development. In the liver, the identity of the founder cells for the two major primary tumors, hepatocellular carcinoma (HCC) and cholangiocarcinoma, is more problematic. The reason for this is that no such unidirectional flux occurs in the liver; moreover, the existence of bipotential hepatic progenitor cells (HPCs), often called oval cells, along with hepatocytes endowed with longevity and long-term repopulating potential, suggests there may be more than one type of carcinogen target cell. Irrespective of which target cell is involved, what is clear is that cell proliferation at the time of carcinogen exposure is pivotal for "fixation" of the genotoxic injury into a heritable form. Taking this view, Sell has opined that in models of experimental hepatocarcinogenesis as a whole, there may be at least four distinct cell lineages susceptible to neoplastic transformation.[105] This supposition is based on the fact that there is considerable heterogeneity in the proliferative responses that ensue after injury in the many different models of hepatocarcinogenesis. Thus, hepatocytes are implicated in some models of HCC; direct injury to the biliary epithelium implicates unipotent cholangiocytes in some models of cholangiocarcinoma; while HPC/oval cell activation accompanies very many instances of liver damage irrespective of etiology, making such cells very likely carcinogen targets. A fourth cell type that might be susceptible to neoplastic transformation is the so-called non-descript periductular cell that responds to periportal injury; the suggestion that such a cell may be of bone marrow origin would be experimentally verifiable in the context of a sex-mismatch bone marrow transplantation (see above) and the appropriate carcinogenic regimen. In the mouse, an origin of HCC from bone marrow has been discounted in a model of chemical hepatocarcinogenesis.[81]

The direct involvement of hepatocytes in hepatocarcinogenesis has been clearly established in rats. Gournay *et al.*[106] found that some preneoplastic foci (expressing gamma glutamyl transpeptidase and the placental form of glutathione-S-transferase) were directly descended from hepatocytes. This was achieved by stably labeling hepatocytes at one day after a 2/3 PH with β-galactosidase, using a recombinant retroviral vector containing the β-galactosidase gene; subsequent feeding with

2-acetylaminofluorene lead to foci, some of which were composed of β-galactosidase-expressing cells. Using the same labeling protocol, Bralet *et al.*[107] observed that 18% of hepatocytes expressed β-galactosidase at the completion of regeneration after a 2/3 PH; subsequent chronic treatment with diethylnitrosamine (DEN) resulted in many HCCs, of which 17.7% of the tumors expressed β-galactosidase, leading to the conclusion that a random clonal origin of HCC from mature hepatocytes was operative in the model.

As discussed above, there is now compelling evidence that oval cells/HPCs are at the very least bipotent, capable of giving rise to both hepatocytes and cholangiocytes. The fact that oval cell activation (ductular cell reaction) precedes the development of HCC in almost all models of hepatocarcinogenesis and invariably accompanies chronic liver damage in humans, makes it almost certain that the mature hepatocyte is not the cell of origin of all HCCs; indeed, perhaps only a small minority of HCCs are derived from the mature hepatocyte. The fact that oval cells/HPCs can be infected with HBV is also consistent with a possible histogenesis of HCC from such cells.[108] An origin of HCC from HPCs is often inferred from the fact that many tumors contain an admixture of mature cells and cells phenotypically similar to HPCs. This would include small oval-shaped cells expressing OV-6, CK7 and 19, and chromogranin-A, along with cells with a phenotype intermediate between HPCs and the more mature malignant hepatocytes.[109] Cells with an HPC phenotype have also been noted in a relatively rare subset of hepatic malignancies, where there are clearly two major components, an HCC component and a cholangiocarcinoma component, again suggestive of an origin from a bipotential progenitor.[110] Direct evidence of a role for oval cells in the histogenesis of HCC can be obtained experimentally; Dumble *et al.*[111] isolated oval cells from p53 null mice and when the cells were transplanted into athymic nude mice they produced HCCs.

## Conclusions

This chapter has highlighted recent progress in identifying cells with hepatic potential in rodents and humans. The chronic shortage of livers for orthotopic liver transplantation has provided a strong impetus for the search for alternative sources of cellular therapy, in particular hepatocyte or even bone marrow transplants. The clinical need for healthy functioning hepatocytes is very clear, not only for the correction of inherited metabolic liver disease, but also for acute liver failure, hepatocellular carcinoma, cirrhosis, bioartificial liver support, hepatotropic viral studies and drug toxicity testing. Hepatocyte transplants have been moderately successful in humans, but clearly there is a need to enrich for cells with potent clonogenic potential; studies in rodents have gone some way to the identification of such cells in these species. Clearly, the key to the success of hepatocyte transplants will be to selectively enhance the growth of the transplanted cells, but unfortunately treatments such as retrorsine (used with conspicuous success in rats) are not clinically acceptable.

Turning to the vexacious subject of bone marrow stem cells as a source of hepatocytes, then the jury is still very much out. Impressive functional repopulation has been achieved in the Fah-deficient mouse liver, though this has been achieved through cell fusion between bone marrow-derived myelomonocytic cells and deficient hepatocytes. If deficient cells are reprogrammed in this way, then that *per se* is not necessarily a bad thing, but if the formed heterokaryons are genetically unstable, then this has significant pathological implication. It also begs the question whether this process is going on all the time in healthy individuals without experimental manipulation such as irradiation and bone marrow transplantation? Like hepatocyte transplants, the key to success will be the selective amplification of bone marrow-derived hepatocytes; studies in the Fah-deficient mouse indicate that the level of initial engraftment is independent of damage, though subsequent clonal expansion is very much dependent on the induction of liver failure. On the other hand, other models of chronic liver injury have failed to detect bone marrow-derived hepatocytes; such conundrums will exercise investigators in this exciting field over the next few years.

## REFERENCES

1. Alison MR, Poulsom R, Forbes S, Wright NA. (2002) An introduction to stem cells. *J Pathol* **197**:419–423.
2. Shafritz DA, Dabeva MD. (2002) Liver stem cells and model systems for liver repopulation. *J Hepatol* **36**:552–564.
3. Fiegel HC, Lioznov MV, Cortes-Dericks L, *et al.* (2003) Liver-specific gene expression in cultured human hematopoietic stem cells. *Stem Cells* **21**:98–104.
4. Fiegel HC, Park JJ, Lioznov MV, *et al.* (2003) Characterization of cell types during rat liver development. *Hepatology* **37**:148–154.
5. Kubota H, Reid LM. (2000) Clonogenic hepatoblasts, common precursors for hepatocytic and biliary lineages, are lacking classical major histocompatibility complex class I antigen. *Proc Natl Acad Sci USA* **97**:12132–12137.

6. Sandhu JS, Petkov PM, Dabeva MD, Shafritz DA. (2001) Stem cell properties and repopulation of the rat liver by fetal liver epithelial progenitor cells. *Am J Pathol* **159**:1323–1334.

7. Dabeva MD, Petkov PM, Sandhu J, *et al.* (2000) Proliferation and differentiation of fetal liver epithelial progenitor cells after transplantation into adult rat liver. *Am J Pathol* **156**:2017–2031.

8. Oertel M, Rosencrantz R, Chen YQ, *et al.* (2003) Repopulation of rat liver by fetal hepatoblasts and adult hepatocytes transduced ex vivo with lentiviral vectors. *Hepatology* **37**:994–1005.

9. Tanimizu N, Nishikawa M, Saito H, *et al.* (2003) Isolation of hepatoblasts based on the expression of Dlk/Pref-1. *J Cell Sci* **116**:1775–1786.

10. Strick-Marchand H, Morosan S, Charneau P, *et al.* (2004) Bipotential mouse embryonic liver stem cell lines contribute to liver regeneration and differentiate as bile ducts and hepatocytes. *Proc Natl Acad Sci USA* **101**:8360–8365.

11. Cantz T, Zuckerman DM, Burda MR, *et al.* (2003) Quantitative gene expression analysis reveals transition of fetal liver progenitor cells to mature hepatocytes after transplantation in uPA/RAG-2 mice. *Am J Pathol* **162**:37–45.

12. Suzuki A, Zheng Y, Kondo R, *et al.* (2000) Flow-cytometric separation and enrichment of hepatic progenitor cells in the developing mouse liver. *Hepatology* **32**:1230–1239.

13. Suzuki A, Zheng YW, Kaneko S, *et al.* (2002) Clonal identification and characterization of self-renewing pluripotent stem cells in the developing liver. *J Cell Biol* **156**:173–184.

14. Hoppo T, Fujii H, Hirose T, *et al.* (2004) Thy1-positive mesenchymal cells promote the maturation of CD49f-positive hepatic progenitor cells in the mouse fetal liver. *Hepatology* **39**:1362–1370.

15. Laconi E, Oren R, Mukhopadhyay DK, *et al.* (1998) Long-term, near-total liver replacement by transplantation of isolated hepatocytes in rats treated with retrorsine. *Am J Pathol* **153**:319–329.

16. Oren R, Dabeva MD, Karnezis AN, *et al.* (1999) Role of thyroid hormone in stimulating liver repopulation in the rat by transplanted hepatocytes. *Hepatology* **30**:903–913.

17. Laconi S, Pillai S, Porcu PP, *et al.* (2001) Massive liver replacement by transplanted hepatocytes in the absence of exogenous growth stimuli in rats treated with retrorsine. *Am J Pathol* **158**:771–777.

18. Gordon GJ, Coleman WB, Hixson DC, Grisham JW. (2000) Liver regeneration in rats with retrorsine-induced hepatocellular injury proceeds through a novel cellular response. *Am J Pathol* **156**:607–619.

19. Gordon GJ, Butz GM, Grisham JW, Coleman WB. (2002) Isolation, short-term culture, and transplantation of small hepatocyte-like progenitor cells from retrorsine exposed rats. *Transplantation* **73**:1236–1243.

20. Overturf K, Al-Dhalimy M, Ou CN, *et al.* (1997) Serial transplantation reveals the stem-cell-like regenerative potential of adult mouse hepatocytes. *Am J Pathol* **151**:1273–1280.

21. Wang X, Al-Dhalimy M, Lagasse E, *et al.* (2001) Liver repopulation and correction of metabolic liver disease by transplanted adult mouse pancreatic cells. *Am J Pathol* **158**:571–579.

22. Rao MS, Reddy JK. (1995) Hepatic transdifferentiation in the pancreas. *Semin Cell Biol* **6**:151–156.

23. Nagai H, Terada K, Watanabe G, *et al.* (2002) Differentiation of liver epithelial (stem-like) cells into hepatocytes induced by coculture with hepatic stellate cells. *Biochem Biophys Res Commun* **293**:1420–1425.

24. Couchie D, Holic N, Chobert MN, *et al.* (2002) In vitro differentiation of WB-F344 rat liver epithelial cells into the biliary lineage. *Differentiation* **69**:209–215.

25. Tateno C, Yoshizato K. (1996) Long-term cultivation of adult rat hepatocytes that undergo multiple cell divisions and express normal parenchymal phenotypes. *Am J Pathol* **148**:383–392.

26. Fujikawa T, Hirose T, Fujii H, *et al.* (2003) Purification of adult hepatic progenitor cells using green fluorescent protein (GFP)-transgenic mice and fluorescence-activated cell sorting. *J Hepatol* **39**:162–170.

27. Tada K, Roy-Chowdhury N, Prasad V, *et al.* (1998) Long-term amelioration of bilirubin glucuronidation defect in Gunn rats by transplanting genetically modified immortalized autologous hepatocytes. *Cell Transplant* **7**:607–616.

28. Guha C, Parashar B, Deb NJ, *et al.* (2002) Normal hepatocytes correct serum bilirubin after repopulation of Gunn rat liver subjected to irradiation/partial resection. *Hepatology* **36**:354–362.

29. Fox IJ, Chowdhury JR, Kaufman SS, *et al.* (1998) Treatment of the Crigler-Najjar syndrome type I with hepatocyte transplantation. *N Engl J Med* **338**:1422–1426.

30. Muraca M, Gerunda G, Neri D, *et al.* (2002) Hepatocyte transplantation as a treatment for glycogen storage disease type 1a. *Lancet* **359**:317–318.

31. Horslen SP, McCowan TC, Goertzen TC, *et al.* (2003) Isolated hepatocyte transplantation in an infant with a severe urea cycle disorder. *Pediatrics* **111**:1262–1267.

32. Dandri M, Burda MR, Torok E, *et al.* (2001) Repopulation of mouse liver with human hepatocytes and in vivo infection with hepatitis B virus. *Hepatology* **33**:981–988.

33. Mercer DF, Schiller DE, Elliott JF, *et al.* (2001) Hepatitis C virus replication in mice with chimeric human livers. *Nat Med* **7**:927–933.

34. Alison MR, Vig P, Russo FF, *et al.* (2004) Hepatic stem cells: From inside and outside the liver? *Cell Prolif* **37**:1–21.

35. Alison MR, Golding M, Sarraf CE, *et al.* (1996) Liver damage in the rat induces hepatocyte stem cells from biliary epithelial cells. *Gastroenterology* **110**:1182–1190.

36. Alison MR, Golding M, Lalani E-N, *et al.* (1997) Wholesale hepatocytic differentiation in the rat from ductular oval cells, the progeny of biliary stem cells. *J. Hepatol.* **26**:343–352.

37. Alison MR, Golding M, Sarraf CE. (1997) Liver stem cells: When the going gets tough they get going. *Int J Exp Path* **78**:365–381.

38. Alison MR, Golding M, Lalani E-N, Sarraf CE. (1998) Wound healing in the liver with particular reference to stem cells. *Phil Trans R Soc Lond B* **353**:1–18.

39. Paku S, Schnur J, Nagy P, Thorgeirsson SS. (2001) Origin and structural evolution of the early proliferating oval cells in rat liver. *Am J Pathol* **158**:1313–1323.

40. Alison MR. (2003) Characterization of the differentiation capacity of rat-derived hepatic stem cells. *Seminars in Liver Disease* **23**:325–336.

41. Hixson DC, Brown J, McBride AC, Affigne S. (2000) Differentiation status of rat ductal cells and ethionine-induced hepatic carcinomas defined with surface-reactive monoclonal antibodies. *Exp Mol Pathol* **68**:152–169.

42. Hixson DC, Chapman L, McBride A, *et al.* (1997) Antigenic phenotypes common to rat oval cells, primary hepatocellular carcinomas and developing bile ducts. *Carcinogenesis* **18**:1169–1175.

43. Baumann U, Crosby HA, Ramani P, *et al.* (1999) Expression of the stem cell factor receptor c-kit in normal and diseased pediatric liver: Identification of a human hepatic progenitor cell? *Hepatology* **30**:112–117.

44. Petersen BE, Goff JP, Greenberger JS, Michalopoulos GK. (1998) Hepatic oval cells express the hematopoietic stem cell marker Thy-1 in the rat. *Hepatology* **27**:433–445.

45. Lemmer ER, Shepard EG, Blakolmer K, *et al.* (1998) Isolation from human fetal liver of cells co-expressing CD34 haematopoietic stem cell and CAM 5.2 pancytokeratin markers. *J Hepatol* **29**:450–454.

46. Omori N, Omori M, Evarts RP, *et al.* (1997) Partial cloning of rat CD34 cDNA and expression during stem cell-dependent liver regeneration in the adult rat. *Hepatology* **26**:720–727.

47. Petersen BE, Bowen WC, Patrene KD, *et al.* (1999) Bone marrow as a potential source of hepatic oval cells. *Science* **284**:1168–1170.

48. Hatch HM, Zheng D, Jorgensen ML, Petersen BE. (2002) SDF-1alpha/CXCR4: A mechanism for hepatic oval cell activation and bone marrow stem cell recruitment to the injured liver of rats. *Cloning Stem Cells* **4**:339–351.

49. Goodell MA, Brose K, Paradis G, *et al.* (1996) Isolation and functional properties of murine hematopoietic stem cells that are replicating in vivo. *J Exp Med* **183**:1797–1806.

50. Alison MR. (2003) Tissue-based stem cells: ABC transporter proteins take centre stage. *J Pathol* **200**:547–550.
51. Scharenberg CW, Harkey MA, Torok-Storb B. (2002) The ABCG2 transporter is an efficient Hoechst 33342 efflux pump and is preferentially expressed by immature human hematopoietic progenitors. *Blood* **99**:507–512.
52. Ros JE, Libbrecht L, Geuken M, *et al.* (2003) High expression of MDR1, MRP1, and MRP3 in the hepatic progenitor cell compartment and hepatocytes in severe human liver disease. *J Pathol* **200**:553–560.
53. Shimano K, Satake M, Okaya A, *et al.* (2003) Hepatic oval cells have the side population phenotype defined by expression of ATP-binding cassette transporter ABCG2/BCRP1. *Am J Pathol* **163**:3–9.
54. Scheffer GL, Kool M, de Haas M, *et al.* (2002) Tissue distribution and induction of human multidrug resistant protein 3. *Lab Invest* **82**:193–201.
55. Wulf GG, Luo KL, Jackson KA, *et al.* (2003) Cells of the hepatic side population contribute to liver regeneration and can be replenished with bone marrow stem cells. *Haematologica* **88**:368–378.
56. Yang S, Koteish A, Lin H, *et al.* (2004) Oval cells compensate for damage and replicative senescence of mature hepatocytes in mice with fatty liver disease. *Hepatology* **39**:403–411.
57. Yin L, Lynch D, Sell S. (1999) Participation of different cell types in the restitutive response of the rat liver to periportal injury induced by allyl alcohol. *J Hepatol* **31**:497–507.
58. Yin L, Sun M, Ilic Z, *et al.* (2002) Derivation, characterization, and phenotypic variation of hepatic progenitor cell lines isolated from adult rats. *Hepatology* **35**:315–324.
59. Leffert HL, Sell S. (2004) Unexpected artifacts raise new caution in stem cell culture research. *Hepatology* **39**:258.
60. Oben JA, Roskams T, Yang S, *et al.* (2003) Sympathetic nervous system inhibition increases hepatic progenitors and reduces liver injury. *Hepatology* **38**:664–673.
61. Yasui O, Miura N, Terada K, *et al.* (1997) Isolation of oval cells from Long-Evans Cinnamon rats and their transformation into hepatocytes in vivo in the rat liver. *Hepatology* **25**:329–334.
62. Wang X, Foster M, Al-Dhalimy M, *et al.* (2003) The origin and liver repopulating capacity of murine oval cells. *Proc Natl Acad Sci USA* **100 Suppl 1** 11881–11888.
63. Roskams TA, Theise ND, Balabaud C, *et al.* (2004) Nomenclature of the finer branches of the biliary tree: Canals, ductules, and ductular reactions in human livers. *Hepatology* **39**:1739–1745.
64. Theise ND, Saxena R, Portmann BC, *et al.* (1999) The canals of Hering and hepatic stem cells in humans. *Hepatology* **30**:1425–1433.

65. Lowes KN, Brennan BA, Yeoh GC, Olynyk JK. (1999) Oval cell numbers in human chronic liver diseases are directly related to disease severity. *Am J Pathol* **154**:537–541.

66. Falkowski O, An HJ, Ianus IA, *et al.* (2003) Regeneration of hepatocyte "buds" in cirrhosis from intrabiliary stem cells. *J Hepatol* **39**:357–364.

67. Deng J, Steindler DA, Laywell ED, Petersen BE. (2003) Neural transdifferentiation potential of hepatic oval cells in the neonatal mouse brain. *Exp Neurol* **182**:373–382.

68. Muller-Borer BJ, Cascio WE, Anderson PA, *et al.* (2004) Adult-derived liver stem cells acquire a cardiomyocyte structural and functional phenotype ex vivo. *Am J Pathol* **165**:135–145.

69. Avital I, Inderbitzin D, Aoki T, *et al.* (2001) Isolation, characterization, and transplantation of bone marrow-derived hepatocyte stem cells. *Biochem Biophys Res Commun* **288**:156–164.

70. Dahlke MH, Popp FC, Bahlmann FH, *et al.* (2003) Liver regeneration in a retrorsine/CCl₄-induced acute liver failure model: Do bone marrow-derived cells contribute? *J Hepatol* **39**:365–373.

71. Theise ND, Badve S, Saxena R, *et al.* (2000) Derivation of hepatocytes from bone marrow cells in mice after radiation induced myeloablation. *Hepatology* **31**:235–240.

72. Lagasse E, Connors H, Al-Dhalimy M, *et al.* (2000) Purified hematopoietic stem cells can differentiate into hepatocytes in vivo. *Nature Med* **6**:1229–1234.

73. Wang X, Montini E, Al Dhalimy M, *et al.* (2002) Kinetics of liver repopulation after bone marrow transplantation. *Am J Pathol* **161**:565–574.

74. Mallet VO, Mitchell C, Mezey E, *et al.* (2002) Bone marrow transplantation in mice leads to a minor population of hepatocytes that can be selectively amplified in vivo. *Hepatology* **35**:799–804.

75. Fujii H, Hirose T, Oe S, *et al.* (2002) Contribution of bone marrow cells to liver regeneration after partial hepatectomy in mice. *J Hepatol* **36**:653–659.

76. Wang X, Willenbring H, Akkari Y, *et al.* (2003) Cell fusion is the principal source of bone-marrow-derived hepatocytes. *Nature* **422**:897–901.

77. Vassilopoulos G, Wang PR, Russell DW. (2003) Transplanted bone marrow regenerates liver by cell fusion. *Nature* **422**:901–904.

78. Alvarez-Dolado M, Pardal R, Garcia-Verdugo JM, *et al.* (2003) Fusion of bone-marrow-derived cells with Purkinje neurons, cardiomyocytes and hepatocytes. *Nature* **425**:968–973.

79. Terai S, Sakaida I, Yamamoto N, *et al.* (2003) An in vivo model for monitoring trans-differentiation of bone marrow cells into functional hepatocytes. *J Biochem* (Tokyo) **134**:551–558.

80. Kanazawa Y, Verma IM. (2003) Little evidence of bone marrow-derived hepatocytes in the replacement of injured liver. *Proc Natl Acad Sci USA* **100** (1):11850–11853.

81. Ishikawa H, Nakao K, Matsumoto K, *et al.* (2004) Bone marrow engraftment in a rodent model of chemical carcinogenesis but no role in the histogenesis of hepatocellular carcinoma. *Gut* **53**:884–889.

82. Harris RG, Herzog EL, Bruscia EM, *et al.* (2004) Lack of a fusion requirement for development of bone marrow-derived epithelia. *Science* **305**:90–93.

83. Willenbring H, Bailey AS, Foster M, *et al.* (2004) Myelomonocytic cells are sufficient for therapeutic cell fusion in liver. *Nat Med* **10**:744–748.

84. Camargo FD, Finegold M, Goodell MA. (2004) Hematopoietic myelomono-cytic cells are the major source of hepatocyte fusion partners. *J Clin Invest* **113**:1266–1270.

85. Alison MR, Poulsom R, Jeffery R, *et al.* (2000) Hepatocytes from non-hepatic adult stem cells. *Nature* **406**:257.

86. Theise ND, Nimmakalu M, Gardner R, *et al.* (2000) Liver from bone marrow in humans. *Hepatology* **32**:11–16.

87. Kleeberger W, Rothamel T, Glockner S, *et al.* (2002) High frequency of epithe-lial chimerism in liver transplants demonstrated by microdissection and STR-analysis. *Hepatology* **35**:110–116.

88. Korbling M, Katz RL, Khanna A, *et al.* (2002) Hepatocytes and epithelial cells of donor origin in recipients of peripheral-blood stem cells. *N Engl J Med* **346**:738–746.

89. Fogt F, Beyser KH, Poremba C, *et al.* (2002) Recipient-derived hepatocytes in liver transplants: A rare event in sex-mismatched transplants. *Hepatology* **36**:173–176.

90. Ng IO, Chan KL, Shek WH, *et al.* (2003) High frequency of chimerism in transplanted livers. *Hepatology* **38**:989–998.

91. Wu T, Cieply K, Nalesnik MA, *et al.* (2003) Minimal evidence of transdiffer-entiation from recipient bone marrow to parenchymal cells in regenerating and long-surviving human allografts. *Am J Transplant* **3**:1173–1181.

92. Blau HM. (2002) A twist of fate. *Nature* **419**:437.

93. Danet GH, Luongo JL, Butler G, *et al.* (2002) C1qRp defines a new human stem cell population with hematopoietic and hepatic potential. *Proc Natl Acad Sci USA* **99**:10441–10445.

94. Ishikawa F, Drake CJ, Yang S, *et al.* (2003) Transplanted human cord blood cells give rise to hepatocytes in engrafted mice. *Ann N Y Acad Sci* **996**:174–185.

95. Newsome PN, Johannessen I, Boyle S, *et al.* (2003) Human cord blood-derived cells can differentiate into hepatocytes in the mouse liver with no evidence of cellular fusion. *Gastroenterology* **124**:1891–1900.

96. Wang X, Ge S, McNamara G, *et al.* (2003) Albumin-expressing hepatocyte-like cells develop in the livers of immune-deficient mice that received transplants of highly purified human hematopoietic stem cells. *Blood* **101**:4201–4208.
97. Kakinuma S, Tanaka Y, Chinzei R, *et al.* (2003) Human umbilical cord blood as a source of transplantable hepatic progenitor cells. *Stem Cells* **21**:217–227.
98. Kollet O, Shivtiel S, Chen YQ, *et al.* (2003) HGF, SDF-1, and MMP-9 are involved in stress-induced human CD34+ stem cell recruitment to the liver. *J Clin Invest* **112**:160–169.
99. Whetton AD, Graham GJ. (1999) Homing and mobilization in the stem cell niche. *Trends Cell Biol* **9**:233–238.
100. Ratajczak MZ, Kucia M, Reca R, *et al.* (2004) Stem cell plasticity revisited: CXCR4-positive cells expressing mRNA for early muscle, liver and neural cells "hide out" in the bone marrow. *Leukemia* **18**:29–40.
101. Miyazaki M, Akiyama I, Sakaguchi M, *et al.* (2002) Improved conditions to induce hepatocytes from rat bone marrow cells in culture. *Biochem Biophys Res Commun* **298**:24–30.
102. Jang YY, Collector MI, Baylin SB, *et al.* (2004) Hematopoietic stem cells convert into liver cells within days without fusion. *Nat Cell Biol* **6**:532–539.
103. Baba S, Fujii H, Hirose T, *et al.* (2004) Commitment of bone marrow cells to hepatic stellate cells in mouse. *J Hepatol* **40**:255–260.
104. Forbes SJ, Russo FP, Rey V, *et al.* (2004) A significant proportion of myofibroblasts are of bone marrow origin in human liver fibrosis. *Gastroenterology* **126**:955–963.
105. Sell S. (2003) Mouse models to study the interaction of risk factors for human liver cancer. *Cancer Res* **63**:7553–7562.
106. Gournay J, Auvigne I, Pichard V, *et al.* (2002) In vivo cell lineage analysis during chemical hepatocarcinogenesis in rats using retroviral-mediated gene transfer: Evidence for dedifferentiation of mature hepatocytes. *Lab Invest* **82**:781–788.
107. Bralet MP, Pichard V, Ferry N. (2002) Demonstration of direct lineage between hepatocytes and hepatocellular carcinoma in diethylnitrosamine-treated rats. *Hepatology* **36**:623–630.
108. Hsia CC, Thorgeirsson SS, Tabor E. (1994) Expression of hepatitis B surface and core antigens and transforming growth factor-alpha in "oval cells" of the liver in patients with hepatocellular carcinoma. *J Med Virol* **43**:216–221.
109. Libbrecht L, Roskams T. (2002) Hepatic progenitor cells in human liver diseases. *Semin Cell Dev Biol* **13**:389–396.
110. Theise ND, Yao JL, Harada K, *et al.* (2003) Hepatic "stem cell" malignancies in adults: Four cases. *Histopathology* **43**:263–271.
111. Dumble ML, Croager EJ, Yeoh GC, Quail EA. (2002) Generation and characterization of p53 null transformed hepatic progenitor cells: Oval cells give rise to hepatocellular carcinoma. *Carcinogenesis* **23**:435–445.

# 20

# Stem Cells of the Eye

Leonard P. K. Ang and Donald T. H. Tan*

## Introduction

Two prerequisites are essential for good vision. The first is a clear optical focusing system. The main components of this (the cornea, lens and intraocular fluids) are designed to bring visual images to a focus on the retina, with the cornea contributing most of the refractive power of the eye. The second is an intact neural system, which detects and transmits images in a coherent fashion from the retina, through an intricate series of neural pathways, to the visual cortex, where they are perceived and interpreted. Ocular stem cell research has concentrated primarily on these two major components of the ocular system that are essential for maintaining clear vision: the ocular surface (comprising the cornea, limbus and conjunctiva) and the retina.

## Ocular Surface Stem Cells

The greatest advances in ocular stem cell biology and treatment have been in the area of ocular surface stem cells. The ocular surface is a complex biological continuum responsible for maintenance of corneal clarity and elaboration of a stable tear film for clear vision, as well as protection of the eye against microbial and mechanical insults. The ocular surface epithelium comprises corneal, limbal and conjunctival epithelium (Fig. 1). The cornea is a highly specialized tissue designed to provide the eye with a clear optical surface for vision. The corneal epithelium consists of a stratified squamous non-keratinizing epithelium that is approximately 5 layers thick. The limbus is a 1.5 to 2 mm wide area that straddles the cornea and bulbar conjunctiva, extending 8–10 layers in thickness.

*Correspondence: Department of Ophthalmology, National University of Singapore, 5 Lower Kent Ridge Road, Singapore 119074. Email: snecdt@pacific.net.sg

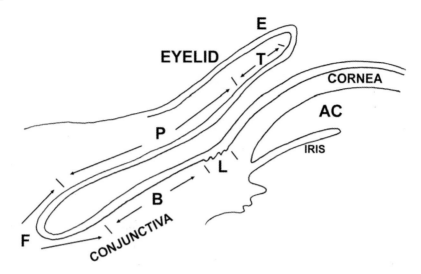

**Figure 1.** Schematic diagram showing the ocular surface comprising the cornea, limbus and conjunctiva. E: epidermis; T: transitional zone between the palpebral conjunctiva and epidermis (mucocutaneous junction); P: palpebral conjunctiva; F: fornix conjunctival; B: bulbar conjunctiva; L: limbus; AC: anterior chamber.

The conjunctiva extends from the corneal limbus up to the mucocutaneous junction at the lid margin. It is divided anatomically into: 1) bulbar conjunctiva, which lies adjacent to the limbus; 2) forniceal conjunctiva, located in the recess created by the lids and the globe; and 3) palpebral conjunctiva, which lines the inner aspect of the eyelids, and terminates at the mucocutanous junction of the lid margin. These regions are lined by 2 to 5 layers of nonkeratinized, stratified columnar epithelium. Clusters of goblet cells are interspersed among the nongoblet epithelial cells, and are scattered throughout the conjunctiva.

Adult corneal and conjunctival stem cells represent the earliest progenitor cells responsible for the homeostasis and regeneration of the ocular surface. The ocular surface is an ideal region to study epithelial stem cell biology, because of the unique spatial arrangement of stem cells and transient amplifying cells. Many intrinsic and extrinsic factors modulate the rate of proliferation and differentiation of stem cells to ensure the maintenance of a steady-state population of ocular surface cells.

## Limbal Stem Cells

The corneal epithelium is under constant cell-turnover, where loss of the terminally differentiated cells located superficially results in replacement

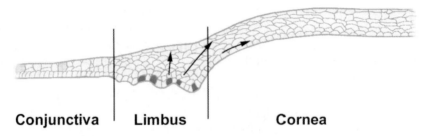

**Conjunctiva    Limbus                    Cornea**

**Figure 2.** The limbal basal epithelial cells (shaded) are believed to contain corneal stem cells. These cells divide to form transient amplifying cells which migrate centrally to occupy the basal layer of the cornea (indicated by arrows). Subsequent cellular divisions result in post-mitotic cells which occupy the suprabasal layers. Progressive differentiation of PMCs result in terminally differentiated cells in the superficial layers.

of the cells by the basal epithelial cells.[1–3] Current evidence now supports the concept that corneal epithelial cells arise from specific progenitor cells located in the basal cell layer of the limbus, which are involved in renewal and regeneration of the corneal epithelium (Fig. 2).[1–6] The limbal stem cells divide to form transient amplifying cells, which migrate superficially to occupy the suprabasal location of the limbus, and centrally to form the basal layer of the corneal epithelium. These transient amplifying cells differentiate to form "post-mitotic cells," which in turn proceed along the path of differentiation to end up as "terminally differentiated cells." These migrate superficially and take on the final phenotypic characteristics of the tissue. The post-mitotic cells and terminally differentiation cells are incapable of cell division.

The initial notion that epithelial cells of the limbus are involved in the regeneration of epithelial cells of the cornea was proposed by Davanger and Evensen in 1971.[7] They observed that in heavily pigmented eyes, pigmented epithelial lines migrated from the limbal region to the central cornea in the healing of corneal epithelial defects. The current evidence supporting the concept of the limbal location of corneal stem cells is summarized below.

Limbal basal epithelial cells contain the least differentiated cells of the corneal epithelium. Schermer *et al.*[8] demonstrated the presence of a 64 kDa keratin, termed K3, among differentiated corneal epithelial cells. This cornea-specific keratin K3 was expressed in the differentiated cells in the suprabasal layer of the limbus, and throughout the corneal epithelium. K3 was essentially absent among limbal basal cells, suggesting that the limbal basal cells represented a more primitive, non-differentiated subpopulation of cells that did not express this cornea-differentiation-related cytokeratin. Further support was provided by Kurpakus *et al.*,[9] who demonstrated that the cornea-specific keratin K12, expressed in the suprabasal cells of the limbus and throughout the entire corneal epithelium, was absent from the limbal basal cells.[9]

To date, no molecular markers that are specific for stem cells have been identified. This has significantly limited our capacity to study the characteristics and behavior of these cells. Taking advantage of the slow-cycling characteristic of stem cells, an indirect method of labeling stem cells was developed.[10] Continuous administration of tritiated thymidine ($^3$H-TdR) for a prolonged period labels all dividing cells. Cells engaged in DNA replication, i.e. in the S-phase of the cell cycle, incorporate the $^3$H-TdR into their DNA. Once labeled, this is followed by a prolonged "chase" period, where slow–cycling cells retain the isotope for an extended period of time.[10] These are termed "label-retaining" cells, and are believed to represent stem cells. Using this technique, Cotsarelis *et al.* provided supporting evidence that the limbus was the site of corneal stem cells.[11] They demonstrated that $^3$H-TdR was incorporated for long time intervals in the limbal basal cells.

This small subpopulation of limbal basal epithelial cells that are normally slow-cycling during the resting state, have been shown to have a significantly higher reserve capacity and proliferative response to wounding and stimulation by tumor promoters, as compared with the peripheral cornea or central cornea.[4,6,11] Lavker *et al.*[6] and Cotsarelis *et al.*[11] showed that stimulation by phorbol myristate (a tumor promoter), resulted in a drastic increase in the labeling index in the limbal basal epithelium. The proliferative response of the limbal cells was maintained over a prolonged period, demonstrating a significant proliferative reserve. No such cells were present in the central corneal epithelium. These kinetic studies provide evidence of a population of label-retaining cells, present in the limbus, that exhibit growth properties that are consistent with that of stem cells.

Stem cells have the greatest growth potential in *in vitro* culture conditions, and regions that are enriched in stem cells demonstrate the highest proportion of colony forming cells. It has been suggested that stem cells are able to undergo at least 120–160 generations.[12,13] In addition, limbal epithelial cells possess a higher proliferative capacity *in vitro* than the central and peripheral corneal cells.[14–19]

## Structural Features of the Limbus that Contribute to the Stem Cell Microenvironment

The features of the limbus are well suited for harboring and protecting the stem cells of the cornea. Stem cells in the body are usually located in the deeper tissue layers, presumably for protection — limbal epithelium is approximately 8–10 cell layers thick, as opposed to the corneal epithelium which usually comprises 5 cell layers. The limbus also tends to be heavily pigmented, especially in pigmented races, and it is believed that this serves to protect the basal cells from the carcinogenic effects of ultraviolet radiation.[11,21] In addition, the palisades of Vogt have an undulating epithelial-stromal junction, which provides greater adhesion properties, thereby rendering the epithelium resistant to shearing forces. These folds also greatly increase the surface area of the basal cells. The stromal component of the limbus is well innervated, and is supplied by a rich vascular network. This allows the modulation and regulation of limbal stem cell growth and proliferation through various cytokine-mediated and neural-mediated pathways, thus establishing the potential for a regulated stromal microenvironment (stem cell niche) which may be very important in the regulation of stem cell activity.

## Corneal Epithelial Healing

Thoft[22] first proposed the X, Y, Z hypothesis of corneal epithelial maintenance (Fig. 3). He suggested that the maintenance of the corneal epithelium could be viewed as a result of three separate, independent mechanisms. X represented the proliferation of basal epithelial cells, Y represented the proliferation and centripetal migration of peripheral cells, and Z referred to the epithelial cell loss from the surface. Corneal epithelial maintenance, which involved a balance of these processes, was defined by the equation: $X + Y = Z$. It is estimated that the corneal epithelium is constantly renewed every 7 to 10 days. Following corneal injury with resultant epithelial cell loss, the regenerative mechanisms designed to replace the corneal epithelium are set into motion, with resultant centripetal movement of the cells from the periphery to the central area. Other investigators have also demonstrated this migration of epithelial cells from the peripheral cornea and limbus.[15,23,24]

Corneal epithelial defects, irrespective of the nature of injury, results in a fairly consistent pattern of re-epithelization.[17,25] Three to six convex leading

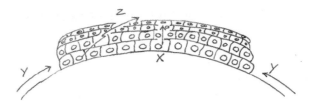

**Figure 3.**  The X, Y and Z hypothesis of corneal epithelial maintenance.
X = proliferation of basal cells
Y = centripetal movement of cells
Z = cell loss from the surface.

fronts of migrating epithelial sheets develop along the circumference of the defect and progress towards the center, where the advancing fronts of the epithelium meet. These tongues of tissue eventually merge imperceptibly to repopulate the entire surface.[25] Lavker *et al.*[21] suggested that the driving force in the centripetal migration of the corneal epithelial cells was not in the limbal population of cells "forcing" its way towards the central cornea, but that as the preferential desquamation of central corneal epithelial cells occurred, the peripheral cells were "drawn inwards" to help repopulate the defect.

# Conjunctival Stem Cells

Conjunctival and corneal epithelial cells belong to two distinct cell lineages, and it is currently believed that corneal and conjunctival epithelia arise from different stem cell populations.[26] Conjunctival epithelium consists of keratinocytes and goblet cells. Wei *et al.*[18] showed that both the non-goblet epithelial cells and goblet cell populations of the conjunctival epithelium arose from a common bipotent progenitor cell.[18] Single cells that were *ex vivo* expanded in culture gave rise to mixed populations of non-goblet epithelial cells and goblet cells. This could only have happened if a bipotent progenitor cell was responsible for the production of both cell lines. Pellegrini *et al.*[16] also showed that goblet cells were found in cultures of transient amplifying cells, occurring late in the life of a single conjunctival clone. They also found that the differentiation to goblet cells occurred at fairly specific time points in the life span of these transient amplifying cells, suggesting that the decision for the conjunctival keratinocytes to differentiate into goblet cells was dependent upon an intrinsic "cell doubling clock."[16] As such, conjunctival epithelial stem cells are probably bipotent, and these cells undergo intrinsic divergence pathways leading to nongoblet and goblet cell populations.

## Site of Conjunctival Stem Cells

Despite the importance of the conjunctiva in maintaining the homeostasis of the ocular surface, relatively little is known about the nature and location of conjunctival stem cells. The forniceal conjunctiva has been shown to be the site that is enriched in conjunctival stem cells in rabbit and mice models. A well-established feature of stem cells is its slow-cycling nature. The forniceal epithelium was found to contain a significantly higher proportion of slow-cycling label-retaining cells, as compared with the bulbar and palpebral conjunctiva (Fig. 4).[27] Following stimulation with a tumor promoter or wounding, the forniceal basal cells underwent a significantly greater proliferative response that was sustained over a longer period of time, as compared with the other regions.

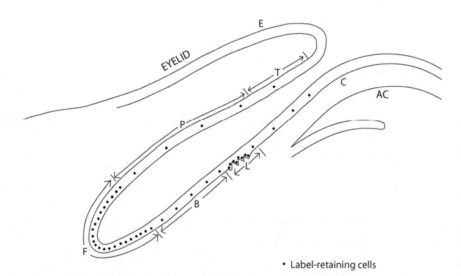

• Label-retaining cells

**Figure 4.** Schematic diagram showing the relative densities of label-retaining cells in the palpebral, forniceal and bulbar conjunctiva in the mice model. The highest concentration is noted in the forniceal conjunctiva, which is believed to be the site enriched in conjunctival stem cells. It also represents the site with the highest density of goblet cells. E: epidermis; T: transitional zone (muco-cutaneous junction); P: palpebral conjunctiva; F: fornix conjunctiva; B: bulbar conjunctiva; L: limbus; C: cornea.

The forniceal region thus possesses a stable, slow-cycling population of progenitor cells, with a significant proliferative reserve that can be recruited to undergo numerous cell divisions following various perturbations of the ocular surface. Additional evidence of the fornix as the site enriched in conjunctival stem cells comes from *in vitro* culture experiments, which takes advantage of the higher growth capacity of stem cells.[17] Cells from the forniceal conjunctiva were shown to have a greater growth potential in culture than the bulbar and palpebral regions.

Although the conjunctival fornix appears to be the site that contains the greatest proportion of stem cells, it is likely that pockets of conjunctival stem cells may also exist throughout the conjunctival epithelium. This could explain the observation by other investigators who analyzed the *in vitro* proliferative capacities of the different conjunctival regions, and suggested that stem cells may instead be uniformly distributed over the bulbar and for-niceal conjunctiva.[16] Some studies have suggested that the mucocutaneous junction at the lid margin may also contribute to conjunctival epithelial regeneration.[28]

Several features of the fornix make it advantageous for stem cells to be clustered in this region. The fornix, located well within the upper and lower recesses created by the closely apposed eyelid and globe, is further away from the external environment than the other regions, in a well protected region of the ocular surface. The conjunctival stem cells are therefore protected from extrinsic insult and injury. The stroma of the fornix is composed of a network of collagen and elastic fibres, which protects the epithelial cells from shearing and mechanical forces. The fornix is also the most richly vascularized and innervated region of the conjunctiva. This allows the stem cells to respond promptly to various stimuli through cytokine or neural mediated mechanisms. These features are important in maintaining the homeostatic environment of the ocular surface.

# Epithelial-Stromal Interaction and the Role of the Microenvironment in Protecting the Stem Cell Niche

Factors responsible for maintaining the "stemness" of stem cells have yet to be determined. It is believed that both intrinsic factors (inherent to the cell), and extrinsic factors (environmental factors surrounding the cell) are involved in the regulation and maintenance of the stem cell niche.[29,30]

The maintenance of "stemness" by extrinsic factors was suggested by Schofield.[30] Stem cells were believed to exist in a microenvironment, or "niche," that promotes the maintenance of the stem cells in its undifferen-tiated state. Several factors that may be involved in the stromal-epithelial interaction that contributes to the microenvironment are described below.

It has been shown that limbal basal cells express higher levels of epider-mal growth factor receptor (EGFR) levels as compared with basal cells of the central cornea. Cells that were more mature and differentiated were found to express lower levels of EGFR.[31] It is postulated that the presence of high levels of EGF receptors might allow these cells to be rapidly stim-ulated by growth factors to undergo cell division during development and following wounding. Limbal basal cells have also been found to express the intermediate filaments, cytokeratin 19 and vimentin.[32,33] They also express $\alpha_6\beta_4$-integrin, metallothionein, AE1, and transferrin receptor.[33] The unique phenotype of these cells differs from that of the surrounding basal cells. Intermediate filaments are involved in the maintenance of the cytoarchi-tecture of cells, and may play a role in the anchorage of these cells to the underlying tissues.

Several other proteins have been found to be present in higher concentrations in the basal cells of the limbal epithelium than in the basal cells of the central corneal epithelium. These include pro-teins that are metabolic enzymes, such as Na-K-ATPase,[1] cytochrome oxidase,[34] and carbonic anhydrase.[35] The regional differences in the con-centration and distribution of these proteins between the corneal and limbal epithelium, suggest an inherent difference in the physiologic and metabolic characteristics of these cells. They may also contribute to the microenvironment that is involved in the maintenance and regulation of stem cells.

The basement membrane of the limbus has different characteristics from that of the central cornea. The basement membrane of the central corneal epithelium possesses a protein identified by monoclonal antibody AE27, which is present at low levels in the limbal area.[36] Conversely, collagen type IV is present in abundance in the basement membrane of the limbus, and is absent in the cornea. The basement membrane of human corneal and conjunctival epithelium can be divided into at least three domains: the conjunctival basement membrane (type IV collagen-positive, AE27-weak); the limbal basement membrane (type IV collagen-positive, AE27-strong); and corneal basement membrane (type IV collagen-negative, AE27-strong). The basement membrane heterogeneity may play a functional role in reg-ulating keratin expression and other aspects of differentiation of corneal epithelium. These properties, together with the anchoring fibrils of the limbus, might enhance the adhesion properties of the basal cells to the underlying stroma.

Ocular surface epithelial cells have been shown to express a myriad of cytokines.[37-39] Stromal-epithelial interactions are believed to be extremely important in supporting normal corneal function. Intercellular communications must therefore occur in a highly coordinated manner between the corneal stromal and epithelial cells. This is particularly crucial during early development, homeostasis, and wound healing. It has been shown that various growth factors, such as transforming growth factor-α (TGF-α), platelet-derived growth factor B (PDGF-B) and interleukin-1β (IL-1β) are synthesized by epithelial cells, while their receptors are present among the stromal fibroblasts.[37] The best characterized stromal to epithelial interaction in the cornea is mediated by the paracrine mediators, hepatocyte growth factor (HGF) and keratinocyte growth factor (KGF).[40] It was found that KGF transcripts and proteins were expressed mostly by limbal fibroblasts, whereas HGF transcripts and proteins were expressed by corneal fibroblasts.[38,39] Because KGF has been known to play an important role in wound healing, the uniquely high expression of KGF may play an important role in the regulation of the proliferation, motility, and differentiation of epithelial stem cell division in wound healing. These findings support the hypothesis that limbal stem cells are under the regulation of the microenvironment, and that the epithelial-stromal interaction of growth factors and cytokines presumably plays a role in stem cell homeostasis and regulation.

## Ocular Surface Disease Arising from Stem Cell Deficiency

Limbal stem cell deficiency can be caused by a variety of hereditary or acquired disorders.[1,3] Possible inherited disorders in which limbal stem cells may be congenitally absent or dysfunctional are shown in Table 1. Limbal stem cells may also be deficient or destroyed by various acquired conditions. Acquired disorders form the majority of cases seen in the clinical setting.

Limbal stem cell deficiency results in abnormal healing and epithelization of the cornea. It is characterized by persistent or recurrent epithelial defects, ulceration, corneal vascularization, stromal inflammation and scarring, and conjunctivalization (conjunctival epithelial ingrowth), with resultant loss of the clear demarcation between corneal and conjunctival epithelium at the limbal region.[1-3] Chronic instability of the corneal epithelium and chronic ulceration may lead to progressive melting of the cornea with the risk of

**Table 1.** Causes of Limbal Stem Cell Deficiency

**Acquired Conditions**
Stevens-Johnson syndrome
Chemical injury
Ocular cicatricial pemphigoid
Contact lens induced keratopathy
Multiple surgeries or cryotherapies to the limbal region
Neurotrophic keratopathy
Peripheral ulcerative keratitis

**Inherited Disorders**
Aniridia keratitis
Keratitis associated with multiple endocrine deficiency

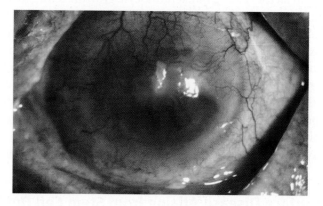

**Figure 5.** Limbal stem cell deficiency in a patient with Stevens-Johnson syndrome. There is loss of transparency of the cornea, stromal scarring, corneal vascularization and conjunctivalization.

perforation. The pathognomonic feature of limbal stem cell deficiency is the presence of conjunctival epithelial ingrowth of the corneal surface.[3] Biomicroscopic examination of the cornea reveals a dull and irregular reflex of the cornea, with vascularization and loss of transparency (Fig. 5). Impression cytology confirms the presence of conjunctival goblet cells on the surface.[41]

Limbal deficiency may be localized (partial) or diffuse (complete).[5,20,25] In localized limbal stem cell deficiency, some sectors of the limbal and corneal epithelium are normal, and conjunctivalization is restricted to the regions devoid of healthy epithelium. In small, localized areas of limbal deficiency, the disease may remain subclinical with no apparent manifestations, as the proliferative reserve of the adjacent healthy limbal tissue may be sufficient to repopulate the corneal surface.

In ocular surface disorders, the diagnosis of the presence of limbal deficiency is crucial as these patients are poor candidates for conventional corneal transplantation alone. Conventional penetrating keratoplasty does not replace the corneal stem cells, and hence the corneal graft is subject to similar complications, with a high risk of graft rejection. In addition, these eyes are often vascularized, with chronic stromal inflammation, and may be accompanied by lid margin irregularities and keratinization. Conventional penetrating keratoplasty is therefore prone to failure.

# Treatment of Severe Ocular Surface Diseases and Stem Cell Deficiency

## Limbal stem cell transplantation

With the widespread acceptance of the limbus as the site of corneal stem cells, limbal transplantation was introduced as a definitive means of replacing the diseased eye with corneal stem cells.[3,42–46] Limbal autograft transplantation was first described in detail by Kenyon and Tseng.[42] The success of limbal transplantation in regenerating the corneal epithelial surface, as compared with conjunctival transplantation, was demonstrated by Tsai *et al.*,[44] where limbal transplantations were shown to be more effective in restoring the corneal epithelial surface than conjunctival grafts.[44] Limbal autograft transplantation is essentially limited to unilateral cases, or bilateral cases with localized limbal deficiency, where sufficient residual healthy limbal tissue is available for harvesting.[47]

For unilateral limbal stem cell disease, the procedure involves lamellar removal of two limbal segments (usually superior and inferior) from the healthy contralateral eye, each segment with a maximal limit of approximately 3 clock hours in size, and transplantation of these segments to the limbal deficient eye, after a complete superficial keratectomy and conjunctival peritomy to remove unhealthy diseased surface epithelium (Fig. 6). For patients with bilateral diffuse disease, limbal allografts may be obtained from living-related donors or cadaveric donors.[48] In living-related limbal allografting, often from a HLA-matched sibling or parent, donor limbal retrieval is limited to two segments. In cases where cadaveric donor limbus

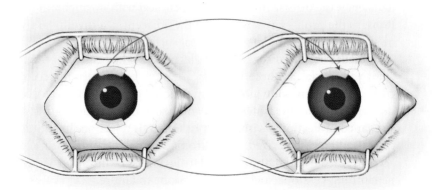

**Figure 6.** Limbal autograft transplantation.

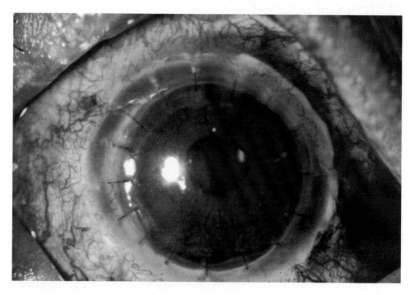

**Figure 7.** A 360° annular limbal allograft transplantation combined with corneal transplantation was performed in this patient with Stevens-Johnson syndrome.

is used, the entire 360° of limbus can be transplanted, either as an intact annular ring, or in several contiguous segments (Fig. 7).

Eyes with limbal deficiency are vascularized and inflamed, and may be associated with lid irregularities and tear film disturbance. This presents a hostile environment for any form of transplantation.[49] The presence of chronic stromal inflammation jeopardizes the success of limbal transplantation.[50] The use of either HLA-matched living related or non-related cadaveric allogenic grafts raises the possibility of graft rejection, and these patients would require aggressive long-term systemic immuno-suppression with cyclosporine, FK506 or mycophenolate mofetil.[51] Despite several studies which reported good early success rates with limbal allograft transplantation, subsequent reviews suggest that medium to long term graft survival rates are more disappointing, with approximately 50% of these grafts failing within 3 to 5 years.[46,52,53]

Limbal transplantation may be combined with penetrating keratoplasty performed either at the same setting, or as a staged procedure. In many of these patients, there is an associated pathology of the conjunctiva and lids. As such, other surgical procedures may be required to reconstruct the ocular surface. These surgical procedures include: lamellar keratoplasty, conjunctival engraftment, forniceal reconstruction and reconstruction of cicatrizing lid disease.

A recent modality for treating ocular surface disorders is the use of human amniotic membranes. These membranes have been used in limbal stem cell deficiency and in the reconstruction of the ocular surface.[52,53] Amniotic membranes possess several characteristics that promote wound healing. It is postulated that the amniotic epithelium produces various growth factors that encourage healing. The basement membrane facilitates the migration of epithelial cells, and reinforces the adhesion of these cells, thereby promoting wound closure.[54,55] Apart from promoting epithelial healing, amniotic membranes have been shown to have anti-inflammatory properties that help reduce the duration and severity of inflammation.

# Bioengineered Ocular Surface Equivalents

## Cultivated limbal stem cell transplantation

Limbal autograft surgery overcomes the problem of immunologic rejection but may only be used for patients with unilateral limbal stem cell deficiency. Because fairly large segments are required, this places the donor eye at risk and may eventually result in surgically-induced limbal stem cell deficiency in the donor eye. The use of autologous cultivated limbal stem cell transplantation has been employed to overcome this problem.[56–59]

The *ex vivo* expansion of limbal epithelial stem cells *in vitro*, followed by transplantation, provides a new modality for the treatment of limbal stem cell deficiency.[56–59] For this procedure, only a small limbal biopsy is required, which minimizes potential damage to the healthy contralateral donor eye. This is then cultivated on various substrates, such as human amniotic membranes or fibrin-based substrates, which results in a composite graft tissue that is then transplanted onto the diseased eyes. Although the long-term results and safety of this procedure have yet to be determined, reasonable success of up to 1 year of follow-up has been achieved.[56–59]

Previous investigators have demonstrated that these amniotic membrane cultures preferentially preserved and expanded limbal epithelial stem cells that retained their *in vivo* properties of being slow-cycling, label-retaining, and undifferentiated.[59] This novel technique proves to be a promising therapeutic option for patients with unilateral or bilateral ocular surface disease, as only small amounts of tissue are required for the expansion of cells, which minimizes iatrogenic injury to the donor eye. The use of these bioengineered corneal surface tissues with a complement of stem cells may thus provide a safer and more effective treatment option.

## Serum-free derived conjunctival tissue-equivalents for transplantation

The use of serum-containing media, often combined with a lethally-treated 3T3 feeder layer, remains the most widely used method for cultivating ocular surface stem cells. The use of animal serum and feeder cells for cultivating cells for transplantation poses the risk of transmission of zoonotic infection and xenograft rejection. The use of a serum-free culture system is therefore significantly advantageous, as it removes the need for animal material in the culture process. Serum-free medium has been shown to support the *in vitro* and *in vivo* proliferation of conjunctival epithelial cells, and has been used in developing conjunctival tissue-equivalents. These bioengineered conjunctival equivalents may be used in the treatment of various ocular surface disorders that require conjunctival excision and replacement, and are particularly useful for extensive or bilateral conjunctival disorders.

The use of bioengineered tissue replacements represents the future for replacement and regeneration of various tissues and organs. Its use in the treatment of ocular surface disease has been promising. In addition, the use of autologous cultivated stem cells allows the patient's own cells to be cryopreserved, so that repeat grafts may be constructed from these cells in the case that the first graft fails.

## Stem Cells of the Retina

The retina is the innermost neural layer of eye, consisting of an inner neurosensory layer and an outer pigmented layer, the retinal pigment epithelium (RPE). The pigmented ciliary body and the peripheral margin of the retina in the adult mammalian eyes are believed to harbor neural progenitors that display stem cell properties and have the capacity to give rise to retinal neurons.[62,63] They have been shown to be multipotential and have the ability to self-renew.[62,63] Interestingly enough, despite our conventional understanding that neural cells are highly specialized and highly differentiated cells, some degree of plasticity has been demonstrated — investigators have shown that human neural stem cells possessed the potential to differentiate into retinal cells, if given the appropriate stimulus.[63]

In comparison with ocular surface stem cell transplantation, the field of retinal stem cell transplantation is in its infancy. Preliminary studies attempted in animals have shown some promising results. Retinal stem cells transplanted into the subretinal space of rats were able to survive and differentiate into cells of photoreceptor lineage.[64,65] Others have attempted transplanting stem cell-derived neural or glial precursors, and showed that they could undergo differentiation and survive for long periods.

## Potential Sources of Ocular Stem Cells

Most of the current work on ocular stem cells has focused primarily on adult stem cells, and research has progressed to the reasonably advanced stage of adult stem cell expansion and transplantation. Embryonic stem (ES) cells represent the most primitive of all progenitor cells. These cells are pluripotent, and their daughter cells are able to develop into all the tissues and organs in the body. Harnessing the immense potential of these cells would have tremendous impact, as this would allow whole organs to be engineered. The biggest hurdle at present is to direct these primitive cells along specific cell lines. Some headway has been made in this area. Investigators have shown that embryonic stem cell-derived neural progenitors are able to acquire retinal phenotypes. Other sources of multipotent stem cells that have been studied for use in the eye include bone marrow derived stem cells and mesenchymal stem cells. The challenge is to be able to induce these to differentiate into specific cell lineages under strict conditions *in vitro*.

## Embryonic Stem Cells and Applications

hESC derivatives are likely to play an important role in the future of regenerative medicine. The RPE plays an important role in maintaining the normal function and survival of the photoreceptors in the neurosensory retina. Degeneration of the RPE with age is thought to play a critical role in the pathogenesis of age-related macular degeneration. Previous animal studies have shown that degenerated RPE cells may be replaced successfully by transplanting donor RPE cells, thereby rescuing the host photoreceptors.[66] Pigmented epithelial cells have been derived from ESCs of the Cynomologous monkey and provided similar protection of the photoreceptors when transplanted into the subretinal space of rats.[67]

Klimanskaya *et al.*[68] demonstrated the derivation of putative RPE cells from hESCs, using a xenobiotic-free culture system. Gene expression profiling of these cells showed a higher similarity to primary RPE tissue than existing human RPE cell lines. With the further development of therapeutic cloning or the creation of banks of homozygous human leucocyte antigen hESCs using parthenogenesis, cell lines derived from hESCs could overcome the problem of immune rejection and may have important clinical applications in the future. These novel treatment options may represent the only hope for sufferers of retinal degenerative disorders, where no treatment currently exists.

## Conclusions

Our knowledge and understanding of ocular surface stem cells has led to improvements in the management of ocular stem cell deficiency. Conjunctivalization, recurrent corneal epithelial defects, vascularization and inflammation are hallmarks of patients with severe limbal stem cell deficiency. Conventional penetrating keratoplasty has been shown to have uniformly poor results in view of the hostile milieu of these diseased eyes, and our understanding of limbal stem cell biology explains the poor outcome of conventional penetrating keratoplasty in the management of ocular surface diseases. Limbal stem cell transplantation has revolutionized the treatment of these difficult cases, by replacing the depleted stem cell population, but the long-term viability of allograft transplantation remains in question.

Although much of the ocular stem cell research has been focused on understanding the physiology and homeostasis of these cells, understanding the "niche" microenvironment where stem cells reside is equally important.

Much remains to be learnt about the structural and biochemical components of the stem cell niche, and the regulatory mechanisms involved in the differentiation of stem cells. The development of cellular markers will also greatly improve our understanding of stem cells and their behavior in normal and diseased states, and an enhanced understanding of the complex interactions leading to stem cell survival and maintenance will allow us to develop new and more successful treatment options for ocular disorders.

# REFERENCES

1. Dua HS, Azuara-Blanco A. (2000) Limbal stem cells of the corneal epithelium. *Surv Ophthalmol* **44**:415–425.
2. Thoft RA, Wiley LA, Sundarraj N. (1989) The multipotential cells of the limbus. *Eye* **3**(Pt 2):109–113.
3. Tseng SC. (1989) Concept and application of limbal stem cells. *Eye* **3**(Pt 2): 141–157.
4. Lehrer MS, Sun TT, Lavker RM. (1998) Strategies of epithelial repair: Modulation of stem cell and transit amplifying cell proliferation. *J Cell Sci* **111**(Pt 19): 2867–2875.
5. Dua HS. (1995) Stem cells of the ocular surface: Scientific principles and clinical applications. *Br J Ophthalmol* **79**:968–969.
6. Lavker RM, Wei ZG, Sun TT. (1998) Phorbol ester preferentially stimulates mouse fornical conjunctival and limbal epithelial cells to proliferate in vivo. *Invest Ophthalmol Vis Sci* **39**:301–307.
7. Davanger M, Evensen A. (1971) Role of the pericorneal papillary structure in renewal of corneal epithelium. *Nature* **229**:560–561.
8. Schermer A, Galvin S, Sun TT. (1986) Differentiation-related expression of a major 64 K corneal keratin in vivo and in culture suggests limbal location of corneal epithelial stem cells. *J Cell Biol* **103**:49–62.
9. Kurpakus MA, Maniaci MT, Esco M. (1994) Expression of keratins K12, K4 and K14 during development of ocular surface epithelium. *Curr Eye Res* **13**:805–814.
10. Bickenbach JR, Mackenzie IC. (1984) Identification and localization of label-retaining cells in hamster epithelia. *J Invest Dermatol* **82**:618–622.
11. Cotsarelis G, Cheng SZ, Dong G, et al. (1989) Existence of slow-cycling limbal epithelial basal cells that can be preferentially stimulated to proliferate: Implications on epithelial stem cells. *Cell* **57**:201–209.
12. Barrandon Y, Green H. (1985) Cell size as a determinant of the clone-forming ability of human keratinocytes. *Proc Natl Acad Sci USA* **82**:5390–5394.
13. Barrandon Y, Green H. (1987) Three clonal types of keratinocyte with different capacities for multiplication. *Proc Natl Acad Sci USA* **84**:2302–2306.

14. Ebato B, Friend J, Thoft RA. (1988) Comparison of limbal and peripheral human corneal epithelium in tissue culture. *Invest Ophthalmol Vis Sci* **29**:1533–1537.

15. Buck RC. (1979) Cell migration in repair of mouse corneal epithelium. *Invest Ophthalmol Vis Sci* **18**:767–784.

16. Pellegrini G, Golisano O, Paterna P, *et al.* (1999) Location and clonal analysis of stem cells and their differentiated progeny in the human ocular surface. *J Cell Biol* **145**:769–782.

17. Wei ZG, Wu RL, Lavker RM, Sun TT. (1993) In vitro growth and differentiation of rabbit bulbar, fornix, and palpebral conjunctival epithelia. Implications on conjunctival epithelial transdifferentiation and stem cells. *Invest Ophthalmol Vis Sci* **34**:1814–1828.

18. Wei ZG, Lin T, Sun TT, Lavker RM. (1997) Clonal analysis of the in vivo differentiation potential of keratinocytes. *Invest Ophthalmol Vis Sci* **38**:753–761.

19. Kruse FE, Tseng SC. (1993) Growth factors modulate clonal growth and differentiation of cultured rabbit limbal and corneal epithelium. *Invest Ophthalmol Vis Sci* **34**:1963–1976.

20. Chen JJ, Tseng SC. (1990) Corneal epithelial wound healing in partial limbal deficiency. *Invest Ophthalmol Vis Sci* **31**:1301–1314.

21. Lavker RM, Dong G, Cheng SZ, *et al.* (1991) Relative proliferative rates of limbal and corneal epithelia. Implications of corneal epithelial migration, circadian rhythm, and suprabasally located DNA-synthesizing keratinocytes. *Invest Ophthalmol Vis Sci* **32**:1864–1875.

22. Thoft RA, Friend J. (1983) The X, Y, Z hypothesis of corneal epithelial maintenance. *Invest Ophthalmol Vis Sci* **24**:1442–1443.

23. Buck RC. (1985) Measurement of centripetal migration of normal corneal epithelial cells in the mouse. *Invest Ophthalmol Vis Sci* **26**:1296–1299.

24. Haddad A. (2000) Renewal of the rabbit corneal epithelium as investigated by autoradiography after intravitreal injection of 3H-thymidine. *Cornea* **19**:378–383.

25. Dua HS, Forrester JV. (1987) Clinical patterns of corneal epithelial wound healing. *Am J Ophthalmol* **104**:481–489.

26. Wei ZG, Sun TT, Lavker RM. (1996) Rabbit conjunctival and corneal epithelial cells belong to two separate lineages. *Invest Ophthalmol Vis Sci* **37**:523–533.

27. Wei ZG, Cotsarelis G, Sun TT, Lavker RM. (1995) Label-retaining cells are preferentially located in fornical epithelium: Implications on conjunctival epithelial homeostasis. *Invest Ophthalmol Vis Sci* **36**:236–246.

28. Wirtschafter JD, Ketcham JM, Weinstock RJ, *et al.* (1999) Mucocutaneous junction as the major source of replacement palpebral conjunctival epithelial cells. *Invest Ophthalmol Vis Sci* **40**:3138–3146.

29. Zieske JD. (1994) Perpetuation of stem cells in the eye. *Eye* **8**(Pt 2):163–169.

30. Schofield R. (1983) The stem cell system. *Biomed Pharmacother* **37**:375–380.
31. Steuhl KP, Thiel HJ. (1987) Histochemical and morphological study of the regenerating corneal epithelium after limbus-to-limbus denudation. *Graefes Arch Clin Exp Ophthalmol* **225**:53–58.
32. Zieske JD, Wasson M. (1993) Regional variation in distribution of EGF receptor in developing and adult corneal epithelium. *J Cell Sci* **106**(Pt 1):145–152.
33. Kasper M. (1992) Patterns of cytokeratins and vimentin in guinea pig and mouse eye tissue: Evidence for regional variations in intermediate filament expression in limbal epithelium. *Acta Histochem* **93**:319–332.
34. Lauweryns B, van den Oord JJ, Missotten L. (1993) The transitional zone between limbus and peripheral cornea. An immunohistochemical study. *Invest Ophthalmol Vis Sci* **34**:1991–1999.
35. Hayashi K, Kenyon KR. (1988) Increased cytochrome oxidase activity in alkali-burned corneas. *Curr Eye Res* **7**:131–138.
36. Kolega J, Manabe M, Sun TT. (1989) Basement membrane heterogeneity and variation in corneal epithelial differentiation. *Differentiation* **42**:54–63.
37. Li DQ, Tseng SC. (1996) Differential regulation of cytokine and receptor transcript expression in human corneal and limbal fibroblasts by epidermal growth factor, transforming growth factor-alpha, platelet-derived growth factor B, and interleukin-1 beta. *Invest Ophthalmol Vis Sci* **37**:2068–2080.
38. Wilson SE, He YG, Weng J, *et al.* (1994) Effect of epidermal growth factor, hepatocyte growth factor, and keratinocyte growth factor, on proliferation, motility and differentiation of human corneal epithelial cells. *Exp Eye Res* **59**:665–678.
39. Wilson SE, Liu JJ, Mohan RR. (1999) Stromal-epithelial interactions in the cornea. *Prog Retin Eye Res* **18**:293–309.
40. Li DQ, Tseng SC. (1997) Differential regulation of keratinocyte growth factor and hepatocyte growth factor/scatter factor by different cytokines in human corneal and limbal fibroblasts. *J Cell Physiol* **172**:361–372.
41. Sridhar MS, Vemuganti GK, Bansal AK, Rao GN. (2001) Impression cytology-proven corneal stem cell deficiency in patients after surgeries involving the limbus. *Cornea* **20**:145–148.
42. Kenyon KR, Tseng SC. (1989) Limbal autograft transplantation for ocular surface disorders. *Ophthalmology* **96**:709–722.
43. Coster DJ, Aggarwal RK, Williams KA. (1995) Surgical management of ocular surface disorders using conjunctival and stem cell allografts. *Br J Ophthalmol* **79**:977–982.
44. Tsai RJ, Sun TT, Tseng SC. (1990) Comparison of limbal and conjunctival autograft transplantation in corneal surface reconstruction in rabbits. *Ophthalmology* **97**:446–455.

45. Tsubota K. (1997) Corneal epithelial stem-cell transplantation. *Lancet* **349**:1556.
46. Tsubota K, Satake Y, Kaido M, *et al.* (1999) Treatment of severe ocular-surface disorders with corneal epithelial stem-cell transplantation. *N Engl J Med* **340**:1697–1703.
47. Frucht-Pery J, Siganos CS, Solomon A, *et al.* (1998) Limbal cell autograft transplantation for severe ocular surface disorders. *Graefes Arch Clin Exp Ophthalmol* **236**:582–587.
48. Dua HS, Azuara-Blanco A. (1999) Allo-limbal transplantation in patients with limbal stem cell deficiency. *Br J Ophthalmol* **83**:414–419.
49. Shimazaki J, Maruyama F, Shimmura S, *et al.* (2001) Immunologic rejection of the central graft after limbal allograft transplantation combined with penetrating keratoplasty. *Cornea* **20**:149–152.
50. Tsai RJ, Tseng SC. (1995) Effect of stromal inflammation on the outcome of limbal transplantation for corneal surface reconstruction. *Cornea* **14**: 439–449.
51. Swift GJ, Aggarwal RK, Davis GJ, *et al.* (1996) Survival of rabbit limbal stem cell allografts. *Transplantation* **62**:568–574.
52. Ilari L, Daya SM. (2002) Long-term outcomes of keratolimbal allograft for the treatment of severe ocular surface disorders. *Ophthalmology* **109**(7): 1278–1284.
53. Solomon A, Ellies P, Anderson DF, *et al.* (2002) Long-term outcome of keratolimbal allograft with or without penetrating keratoplasty for total limbal stem cell deficiency. *Ophthalmology* **109**(6):1159–1166.
54. Meller D, Pires RT, Mack RJ, *et al.* (2000) Amniotic membrane transplantation for acute chemical or thermal burns. *Ophthalmology* **107**:980–989.
55. Tseng SC, Prabhasawat P, Barton K, *et al.* (1998) Amniotic membrane transplantation with or without limbal allografts for corneal surface reconstruction in patients with limbal stem cell deficiency. *Arch Ophthalmol* **116**:431–441.
56. Koizumi N, Inatomi T, Suzuki T, *et al.* (2001) Cultivated corneal epithelial stem cell transplantation in ocular surface disorders. *Ophthalmology* **108**:1569–1574.
57. Pellegrini G, Traverso CE, Franzi AT, *et al.* (1997) Long-term restoration of damaged corneal surfaces with autologous cultivated corneal epithelium. *Lancet* **349**:990–993.
58. Tsai RJ, Li LM, Chen JK. (2000) Reconstruction of damaged corneas by transplantation of autologous limbal epithelial cells. *N Engl J Med* **343**:86–93.
59. Meller D, Pires RT, Tseng SC. (2002) Ex vivo preservation and expansion of human limbal epithelial stem cells on amniotic membrane cultures. *Br J Ophthalmol* **86**:463–471.

60. Ang LP, Tan DT, Beuerman R, *et al.* (2004) The development of a conjunctival epithelial equivalent with improved proliferative properties using a multistep serum-free culture system. *Invest Ophthalmol Vis Sci* **45**(6):1789–1795.

61. Tan DT, Ang LP, Beuerman R. (2004) Reconstruction of the ocular surface by transplantation of a serum-free derived cultivated conjunctival epithelial equivalent. *Transplantation* **77**(11):1729–1734.

62. Ahmad I, Dooley CM, Thoreson WB, *et al.* (1999) In vitro analysis of a mammalian retinal progenitor that gives rise to neurons and glia. *Brain Res* **831**:1–10.

63. Reh TA, Levine EM. (1998) Multipotential stem cells and progenitors in the vertebrate retina. *J Neurobiol* **36**:206–220.

64. Chacko DM, Rogers JA, Turner JE, Ahmad I. (2000) Survival and differentiation of cultured retinal progenitors transplanted in the subretinal space of the rat. *Biochem Biophys Res Commun* **268**:842–846.

65. Kurimoto Y, Shibuki H, Kaneko Y, *et al.* (2001) Transplantation of adult rat hippocampus-derived neural stem cells into retina injured by transient ischemia. *Neurosci Lett* **306**:57–60.

66. Lund RD, Adamson P, Suave Y, *et al.* (2001) Subretinal transplantation of genetically modified human cell lines attenuates loss of visual function in dystrophic rats. *Proc Natl Acad Sci* **98**:9942–9947.

67. Haruta M, Sasai Y, Kawasaki H, *et al.* (2004) In vitro and in vivo characterization of pigment epithelia cells differentiation from primate embryonic stem cells. *Invest Ophthalmol Vis Sci* **45**:1020–1025.

68. Klimanskaya I, Hipp J, Rezai KA, *et al.* (2004) Derivation and comparative assessment of retinal pigment epithelium from human embryonic stem cells using transcriptomics. *Cloning Stem Cells* **6**(3):217–245.

# 21

# Bone Repair and Adult Stem Cells

Dietmar Werner Hutmacher* and Suman Lal Chirammal Sugunan

## Introduction

Continued advances in medical understanding herald the prospect of increased longevity for mankind. However, with this increased longevity comes susceptibility to disease, including those of the skeletal system. Next to blood, bone remains the second most transplanted tissue in the 21st century. Current treatment regimes for the reconstruction or enhancement of function of damaged bone rely on autogenous or allogenic tissue grafts. Autogenous bone, by providing an osteoconductive matrix, growth factors and osteogenic cells which form the critical elements of bone repair, remains the graft material of choice despite the associated donor site morbidity inflicted on the patient, the potential of infection at the bone harvest site, painful surgical procedure, and the risk of injuring surrounding structures. Surgeons are increasingly facing patient groups which have a limited supply of transplantable autologous bone due to multiple operations, poor bone quality, and patient groups where limitations of large bone grafts exist, especially children. The continuing clinical need for improvement of existing treatments for bone disorders, ranging from congenital deformity to reconstructive surgery for tumors and trauma repair, has resulted in the search for novel approaches to skeletal reconstruction.

The most common concept underlying tissue engineering is to combine a scaffold-matrix, living cells, and/or biologically active molecules together to form a "tissue engineered construct (TEC)" to promote the repair and regeneration of tissues. Bone tissue engineering aims to heal bone defects with autologous cells and tissue without the donor site morbidity and expense associated with harvesting autogenous bone. One such

*Correspondence: National University of Singapore, Engineering Drive 1, Singapore 119260.
Tel.: 65-6874-1036, Fax: 65-6872 3069, E-mail: biedwh@nus.edu.sg

strategy involves seeding autologous osteogenic cells *in vitro* throughout a biodegradable scaffold to create a scaffold-cell hybrid (TEC). This involves the isolation of a suitable cell population and expansion to a clinically relevant size *ex vivo*, seeding the expanded population onto an appropriate three-dimensional (3D) scaffold which may be impregnated with appropriate growth factors and other chemical cues to enhance tissue ingrowth, followed by further culture in a bioreactor and/or place *in vivo* at the requisite tissue regeneration site. The scaffold is expected to support cell colonization, migration, growth and differentiation, and to guide the development of the required tissue or to act as a drug delivery device.

Based on this background, novel approaches to skeletal reconstruction will result in orthopedic, plastic and reconstructive surgery undergoing a paradigm transformation by adding bone engineering to their standard treatment concepts. The field of regenerative medicine is driven from a clinical point of view and brings together tissue engineers with backgrounds in biology, materials science and biomedical engineering.[1] The aim of these interdisciplinary groups is directed towards long-term repair and replacement of failing human tissues and organs. This chapter aims to discuss the advances in the use of adult mesenchymal stem cells (MSCs) in scaffold-based bone tissue engineering. The isolation, characterization and *in vitro* differentiation of mesenchymal stem cells and their use along with a scaffold in the making of a bone tissue engineered construct will be discussed.

## Bone Structure and Osteogenesis

A good understanding of the process of osteogenesis and the structure of bone is required before an attempt can be made to mimic the complexities of cellular microenvironment in bone tissue engineering. Bone is a highly specialized form of dense connective tissue which performs a multitude of mechanical, chemical and hematological functions. Being a dynamic and adaptable tissue, bone undergoes subtle continuous remodeling in order to conform to its functions and other mechanical demands. It is different from other connective tissues like muscle, with its hardness being attributed to the deposition of complex mineral substances, calcium hydroxyapatite composed of calcium, phosphorus, sodium, magnesium, fluoride and other ions in trace amounts, within a soft organic matrix known as collagen. Bone also serves as an important reservoir for calcium, which can be drawn upon when required for special metabolic activities. Apart from protective

**Table 1.** Mechanical Properties of Compact and Spongy Bone[6]

| Property | Compact Bone | Spongy Bone |
|---|---|---|
| Compressive strength (MPa) | 100–230 | 2–12 |
| Flexural, tensile strength (MPa) | 50–150 | 10–20 |
| Strain to failure (%) | 1–3 | 5–7 |
| Fracture toughness (MPam$^{1/2}$) | 2–12 | – |
| Young's modulus (GPa) | 7–30 | 0.5–0.05 |

functions, the skeleton acts as a supporting framework for the soft tissues of the body, not only by providing rigidity for the body as a whole, but also by forming a series of mechanical levers to which are attached muscles and ligaments, allowing full mobility. Also, bone tissue serves as a nidus for hemopoietic activities of red marrow and also acts as a storage for energy as lipids stored in the yellow bone marrow.[2]

Bone is a composite material, whose components are primarily collagen and hydroxyapatite, but whose complex structure contains a wealth of mechanically relevant details (Table 1). Collagen possesses a Young's modulus of 1–2 GPa and an ultimate tensile strength of 50–1000 MPa when compared with hydroxyapatite, which has a Youngs modulus of ~130 GPa and an ultimate tensile strength of ~100 MPa.[3] Bone is a composite in several senses, i.e. a porous material, a polymer-ceramic mixture, a lamellar material and a fibre-matrix material. Its mechanical properties will therefore depend on each of these aspects of composition and structure.[4] Macroscopically, cancellous bone has a loosely organized porous matrix where collagen fibrils form concentric lamellae, while compact or cortical bone does not have any spaces or hollows in the bone matrix (Fig. 1).

The major cells forming bone are the osteoblasts, osteoclasts, osteocytes, and the bone lining cells. Osteoblasts regulate the mineralization of the bone matrix and its subsequent synthesis of collagen and other bone proteins. The osteocyte is a mature osteoblast and has several thin processes, which extend from the lacunae into the canaliculi, enabling them to be key cells in several biologic processes, including the regulation of response of bone to the mechanical environment. The bone lining cells are inactive and little is known about their function, while the osteoclasts are large motile, multinucleated cells located on the bone surface participating in bone resorption. While the osteoclast is assumed to be formed by the fusion of mononuclear cells derived from hematopoetic stem cells (HSCs) in the marrow, the bone forming cells, the osteoblast and the osteocyte are derived

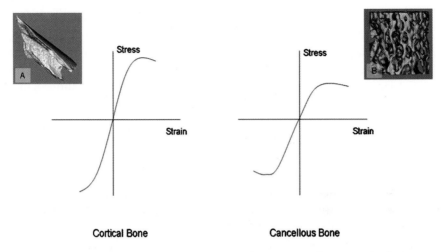

**Figure 1.** Typical stress strain curves for cortical and cancellous bone, to different scales.[5] Micro CT images of cancellous (**A**) and cortical bone (**B**) are shown.

from the mesenchymal stem cells (MSCs) in the bone marrow, otherwise known as osteoprogenitor cells.[7,8]

The osteoprogenitor cells are pluripotent MSCs, which, when appropriately stimulated by growth factors, divide into cells that differentiate into osteoblasts, the process being primarily regulated by the transcription factors Cbfa1 and Osterix. Osteogenesis, or the process of bone formation, is initiated by the uncommitted adult pluripotent MSCs in a regulated sequence of events mediated by a variety of growth factors and cytokines and involves a complex cascade of molecular and morphogenetic processes leading to precisely organized multicellular structures. The MSCs in the bone marrow, hence, are vital to the repair and remodeling mechanisms seen in bone.

## Mesenchymal Stem Cells for Bone Tissue Engineering

Stem and progenitor cells from adult tissues represent an important promise in the therapy of several pathological conditions. Stem cells have the ability to self-replicate for long periods or, in the case of adult stem cells, maintain their differentiation potential throughout the life span of the organism. Progenitor cells are derived from stem cells; they retain the differentiation potential and high proliferation capability, but they have lost the self-replication property. From a classical view, adult stem cells are restricted to differentiating along the lineage pathways of their own tissue. Similarly, progenitor cells are considered committed to the cell phenotypes of their tissues of origin. This concept of lineage restriction has been challenged by experimental evidence over the past few years. Indeed, most stem and progenitor cell types display an amazing plasticity, which is the property of cells to differentiate into phenotypes not restricted to the tissues and, in some cases, not to the germ layers from which they are derived.[9] The definition of a "stem cell" being still elusive, caution should be exercised when using the terminology in the context of regenerative medicine. The current source of cells used in tissue engineering strategies are better labeled precursor/progenitor cells, although there have been experimental evidences demonstrating the plasticity of these cells to differentiate into multiple lineages.

Adult stem/precursor cells were first identified in tissues, characterized by a high rate of cell turnover, such as the bone marrow, and have been well characterized in relation to stem cells originating from other tissues. MSCs reside in close contact with the hematopoetic progenitors in the bone marrow cavity. Recently, MSCs have also been isolated from the periosteum, trabecular bone, adipose tissue, synovium, and deciduous teeth. Adipose tissue has also been shown to contain multipotent stem cells, which have the capacity to differentiate into cells of connective tissue lineages, including bone, fat, cartilage and muscle, in the presence of lineage specific growth factors.[10,11] Osteoprogenitor cells have also been isolated from skeletal muscle in mice and humans.[12] Being less invasive procedures, muscle biopsy and liposuction are hence attractive alternatives in cell-based tissue engineering strategies.

## Osteogenic progenitors from bone marrow-derived mesenchymal stem cells (BMSCs)

The induction of bone formation was first described when the bone marrow derived fibroblastoid cell populations were transferred to ectopic sites delineating the osteogenic properties of BMSCs.[13] It has been proposed that all highly specialized types of hard tissues, including cortical and trabecular bone, tendons, ligaments and different kinds of cartilage as well as stromal microenvironment supporting and regulating hematopoesis, originate from a common type of early mesenchymal progenitor cell.[14–17] The osteogenic precursor cells are presumed to originate from the stromal stem cells possessing self renewing potential; the developmental pathway that the mesenchymal stem cell pursues to differentiate into and osteoblast is still under intense investigation.

A sufficient and reproducible supply of autogenous cells is of paramount importance for bone engineering. Well differentiated cells are limited by the expansion characteristics, and their limited potential for (trans)differentiation. Their accessibility and ease to manipulate *in vitro* has made BMSCs natural candidates for bone tissue engineering as they form the stem cell source for osteoprogenitors and osteoblasts in the bone environment. However, the unambiguous identification of BMSCs is presently hampered by the lack of specific cellular markers.

## Phenotype and isolation of BMSCs

Bone marrow harbors at least two distinct populations of adult stem cells, the HSCs and marrow stromal stem cells or BMSCs. The clonogenic stromal progenitor cells which later came to be described as mesenchymal stem cells were first described as rapidly adherent, non-phagocytic clonogenic cells capable of extended proliferation *in vitro*.[18,19] Assays of colony forming units (CFU-F) from aspirates of human bone marrow yields colony numbers between $1-20/10^5$ mononuclear cells plated.[20,21] The CFU-F in adult human BM is feeder cell-independent.[22] A combination of plastic adherence and *in vitro* culture was first used to establish stromal cell cultures and the contaminating hematopoetic cells were subsequently removed by negative selection using antibodies to CD45, CD34, and CD11b.[23] The stromal cells obtained were shown to exhibit osteogenic, adipose, and chondrogenic differentiation potential *in vitro* and expressed Sca-1, CD29, CD44, and CD106, but not markers of hematopoetic or vascular endothelial cells. Stromal progenitors in human fetal BM express CD34 but are distinguished from the majority of hematopoetic progenitors by their lack of CD38 and HLA-DR.

Subsequent studies demonstrated that stromal progenitors were restricted to a subpopulation of $CD34^+CD38^-HLA-DR^-$ that lacked expression of CD50.[24] The STRO-1 antibody, which is non-reactive with the hematopoetic progenitors, results in a 10- to 20-fold enrichment of CFU-F in fresh aspirates of human bone among the heterogeneous population of adult human bone marrow mononuclear cells. In accordance with the previously described properties of the unfractionated $STRO-1^+$ population, $STRO-1^{bright}$ $VCAM-1^+$ cells assayed at the clonal level, exhibited differentiation into cells with the characteristics of adipose, cartilage and bone cells *in vitro* and formed human bone tissue following transplantation into immunodeficient SCID mice.[24]

Collectively, these data strongly suggest that primitive stromal precursors, including putative stromal stem cells with the capacity for differentiation into multiple mesenchymal lineages, are restricted to the STRO-1$^+$ fraction in adult human bone marrow.[25,26] MSC enrichment could also be accomplished using several markers, including Thy-1, CD49a, CD10, Muc18/CD146, and in accordance with their response to growth factors, antibodies to receptors for PDGF and EGF.[27,28] Additional antibodies that identify human bone marrow have been described but have not yet been able to be verified by other groups.[29]

Other strategies for stem cell isolation use positive selection, e.g. Thy-1$^+$ and c-Kit$^+$ cells, after lineage-depletion or CD34-enrichment. Although the utility of CD34 as a stem cell marker for human cells is well established, there is evidence for the existence of a very primitive population of CD34$^-$ stem cells. Data suggest that the frequency of the CD34$^-$ HSCs is highest early in ontogeny, and CD34$^-$ HSC can generate CD34$^+$ cells *in vivo* and may even have the capacity for non-hematopoetic differentiation. This supports the idea that CD34$^-$ cells may represent developmental precursors of CD34$^+$ HSCs and non-hematopoetic stem cells.[26,28] Studies have shown that human bone marrow MSC can be directly selected by virtue of expression of CD49a, the a1-integrin subunit of the very late antigen (VLA)-1, which is a receptor for collagen and laminin.[30] This population is CD49a$^+$ CD45$^{med/low}$ and differentiates into several mesodermal directions. All CFU-F found in human bone marrow are included within the CD49a$^+$ CD45$^{med/low}$ fraction. In contrast to the above MSCs that are apparently restricted to mesodermal lineages, Jiang *et al.*[31] described the isolation and characterization of cells copurifying with MSCs, termed mesenchymal adult progenitor cells (MAPCs), that are by far more plastic than previously ascribed to mesenchymal cells and differentiate *in vitro* not only into mesodermal derivatives but also into cells of visceral mesoderm, neuroectoderm, and endoderm.[32]

A very useful alternative approach to isolating enriched HSCs and progenitor populations is by removal of cells expressing antigens that are present on more mature differentiated cells (negative selection). Examples of lineage-specific antigens are CD19 and CD20 for B-cells, CD3 for T-cells, CD14 for monocytes and CD66b for granulocytes. Depletion of these and other lineage-positive cells can be achieved by immunomagnetic

**Table 2.** Progenitor Cells According to Tissue Type

| Tissue | Cellularity | Approximate No. of Progenitor Cells |
|---|---|---|
| Bone Marrow | $40 \times 10^6$ cells/ml | 2000 cells/ml (1 in 20 000 cells) |
| Adipose and Skeletal Muscle | $6 \times 10^6$ cells/cm$^3$ | 1 in 4000 cells |

as well as nonmagnetic cell separation approaches. After harvest of bone marrow, culture adherent cells are expanded and subcultured to increase the number of pluripotent cells. MSCs derived from human bone marrow have been shown to retain their undifferentiated phenotype through an average of 38 doublings, resulting in over a billion-fold expansion.[33] The use of minimal medium, which is not supplemented by cytokines, facilitates the growth of colony forming units-fibroblasts (CFUF) that proliferate to form macroscopic fibroblastoid colonies *in vitro*. These can be further propagated by serial passage to become cell strains while still retaining the capacity to differentiate into several mesodermal directions.

There exists substantial variation in the number of progenitor cells and cellularity among tissues (Table 2).[34,35,36] The mean prevalence of alkaline phosphatase positive CFUs after placement of the bone marrow derived cells into the tissue culture medium was 36 per 1 million nucleated cells, with the number of CFUs showing an increase as the aspiration volume increased.[34] Due to contamination with peripheral blood, not more than 2 ml of aspiration volume is recommended from any one particular site. Significant patient to patient difference in the cellularity of the bone marrow and the prevalence of osteoblast progenitor cells are also observed.

## Genetic modification of MSCs

Delivery of osteoinductive factors, such as bone morphogenetic proteins (BMPs), has been successfully applied to stimulate local bone repair and several formulations are available for selected clinical applications.[37,38] However, the widespread clinical efficacy of these treatment modalities continues to be hampered by inadequate carriers, release kinetics, dosage, and potency.[38,39] Several research groups have hence focused on genetic engineering to repair or regenerate bone. The trend has been to concentrate on a particular molecular pathway and produce targeted intervention to create enhancements in functionality and/or architecture of the repaired or regenerated bone. However, because almost every known cell-cell and cell-matrix interaction involves multiple genes acting at specific tissue locations at specific times, understanding the entire complex system requires a detailed knowledge of gene functionalities and cellular behaviors as well as the spatial and temporal integration at the tissue level. It can be summarized that genetic engineering strategies focusing on osteoinductive factors have emerged as efficient approaches to enhance bone formation and generally consist of two modalities: (a) direct *in vivo* delivery of gene constructs; and (b) *ex vivo* transduction and subsequent transplantation of cells expressing the osteoinductive factor. The choice of the gene delivery method depends on several factors, including the particular gene of interest, indication targeted, desired duration of gene expression, and delivery vector. Gene delivery to musculoskeletal systems has been reviewed in detail elsewhere.[37,40,41,42]

## Cell-host interactions

The question of the host response to implanted MSCs is critical and has been receiving attention as these cells are being considered in a variety of clinical applications. There are several aspects to the implant cell–host interactions that need to be addressed as we attempt to understand the mechanisms underlying stem cell therapies. These are: (a) the host immune response to implanted cells; (b) the homing mechanisms that guide delivered cells to a site of injury; and (c) differentiation of implanted cells under the influence of local signals. By virtue of their distinct immunophenotype, associated with the absence of HLA Class II expression, as well as low expression of co-stimulatory molecules, MSCs may be nonimmunogenic or hypoimmunogenic.[43] HLA Class II expression is also absent from the surface of undifferentiated and differentiated MSCs and these cells do not elicit an alloreactive lymphocyte proliferative response, suggesting that MSCs can be transplantable between HLA-incompatible individuals.[44]

## Bone Tissue Engineering

Bone formation within the body, as part of a development, healing and/or repair process, is a complex event in which cell populations in combination with extracellular matrix self-assemble into functional units. There is intense academic and commercial interest in finding methods to stimulate and control these events and eventually to replicate these events outside the body as close as possible. This interest has accelerated, resulting in bone tissue engineering becoming a well recognized research area in the arena of regenerative medicine.[45]

## *In vitro* differentiation of MSCs

Cells are generally cultured in basal medium such as Dulbecco's modified Eagle's medium (high glucose) in the presence of 10% fetal bovine serum (FBS). The optimal expansion of MSCs from bone marrow requires the pre-selection of FBS.[16,46] MSCs in culture have a fibroblastic morphology and adhere to the tissue culture substrate. Primary cultures are usually maintained for 12–16 days, during which time the nonadherent hematopoietic cell fraction is depleted. The addition of growth factor supplements such as fibroblast growth factor-2 (FGF-2) to primary cultures of human MSCs has been reported.[47] As MSCs are expanded in large-scale culture for human applications, it will be important to identify defined growth media, without or with reduced FBS, to ensure more reproducible culture techniques and enhanced safety.

Stem cells show the tendency to spontaneously differentiate into multiple lineages when transplanted *in vivo*.[48] Hence, only a small subfraction of the transplanted stem cells may differentiate into the tissue type of interest, which could in turn reduce the clinical efficacy of transplantation therapy. Established *in vitro* differentiation protocols aimed at inducing the MSCs are therefore used to direct the osteogenic differentiation of the MSCs before *in vivo* implantation. The MSCs could either be osteogenically induced while in culture or after seeding onto the 3D porous scaffolds. Differentiation protocols aimed at elucidating the molecular mechanisms involved in plasticity of MSCs have been established in both instances.

The most used culture environment for stem cell differentiation *in vitro* should be chemically defined, and either be serum-free or utilizing synthetic serum replacements with the possible supplementation of specific recombinant cytokines and growth factors.[49,50] Currently, for osteogenic differentiation, the maintenance media (DMEM) is supplemented with 10% or 20% FBS, 0.1 mM dexamethasone; 50 mM ascorbate-2-phosphate; and 10 mM beta-glycerophosphate.[51] The cultures are maintained for four weeks. Initial collagen matrix accumulation precedes, and is essential for, sequential expression of differentiation-related proteins, e.g. alkaline phosphatase (ALP); parathyroid hormone receptor (PTHRP); bone sialoprotein (BSP); and osteocalcin (OC). BMSCs are highly proliferative and express low levels of bone-specific proteins, while secretory osteoblasts stop dividing and produce large amounts of bone-specific ECM. Osteogenic maturation is then evaluated by measuring bone specific ALP and OC concentrations in the culture supernatants with quantitative immunoassays. RT-PCR and western blot techniques are used to confirm the presence of osteogenic markers at the genomic and protein level.

The developmental sequence of bone has been described as consisting of three phases: proliferation with matrix secretion, maturation and mineralization.[52] Osteopontin (OP) and ALP are early osteogenic markers and characterize the matrix maturation phase. ALP is a membrane-bound enzyme abundant in early bone formation. Shedding releases the membrane-bound ALP into the serum or supernatant, where it can be measured by ELISA. Histomorphometrical methods has shown that increasing ALP levels correlate with increasing bone formation; even so, it needs to be noted that ALP is expressed in different amounts by several cell types. The expression of OP, an important protein for matrix-cell interaction, was observed two weeks after induction. For the bone-derived cells, the ALP levels are increased around day 12 after induction. Although these results strongly suggest osteogenesis, these proteins are not considered as specific markers for osteogenic differentiation. Osteocalcin is a late bone marker which is specific and determines terminal osteoblast differentiation. Trabecular- and periosteal-derived cell populations show increasing OC concentrations three weeks after induction.[53]

The major components of the osteogenic induction media include dexamethasone, beta glycerophosphate and ascorbic acid. Dexamethasone, a synthetic steroid, has been shown to promote osteogenic differentiation of both embryonic and adult stem cells.[54,55] Ascorbic acid treatment is accompanied by a 5-fold increase in binding of Osf2/Cbfa1 to OSE2 in MC3T3-E1 cells and the osteocalcin and bone sialoprotein promoters are up-regulated.[56] Ascorbic acid also promotes relative collagen synthesis by increasing the procollagen stability and secretion. It is required for matrix maturation and subsequent mineralization. Other biologic modulators which play a role include insulin, TGF-b, EGF, LIF, FGF-4, PDGF, calcitropic hormone and 1,25-dihydroxyvitamin D3. Type I collagen, one of the most commonly occurring extracellular matrix molecules, has been reported to slow down cellular senescence within *in vitro* culture,[57] and has therefore been proposed for use in reducing the rate of ageing during the *in vitro* expansion of stem cells for tissue engineering applications. Integrin-activated signaling pathways have been implicated in the differentiation of bone marrow stromal precursors into functional osteoblasts.[58] Laminin, an integrin ligand, has been shown to direct osteogenic differentiation *in vitro*.[59] Another major class of extracellular matrix molecules are the glycosaminoglycans (GAGs)-like heparin sulphate, which have been reported to maintain CD34$^+$ expression *in vitro* during the proliferation and expansion of CD34$^+$ enriched cord blood cells.[60] Heparan sulfate has also been proposed to have an important role in osteogenic induction of MSCs. A variety of methods have been developed to incorporate these induction molecules in either the culture media or the biomimetic scaffold.

Osteogenic differentiation induced by mechanical stimulation has been widely documented. The differentiation of human osteoblastic periodontal ligament cells has been reported to be enhanced in response to mechanical stress,[61] while a similar result was observed with bone marrow osteogenic progenitors.[62] In the absence of exogenous growth and differentiation factors, mechanical stimuli alone could induce the differentiation of mesenchymal stem cells into the osteogenic lineage.[63] Further investigations showed that there existed multiple competing signaling pathways for osteogenic differentiation in response to mechanical stimuli.[64]

It must, however, be kept in mind that one of the ultimate clinical objectives of stem cell transplantation therapy is to develop well-defined and efficient *in vitro* protocols for stem cell expansion and differentiation, utilizing chemically defined culture media completely devoid of animal and human products, and possibly supplemented with recombinant cytokines. This will then provide the stringent levels of safety and quality control that would make the routine clinical applications of stem cell transplantation therapy realizable.

## Tissue engineered constructs

A wide variety of natural and synthetic devices and matrices have been used as carriers to deliver MSCs, including poly L-lactic acid (PLLA), poly ε-caprolactone (PCL), etc. A comprehensive report on the work on the biomaterial properties and design and fabrication of scaffolds for bone engineering is beyond the scope of this chapter. Comprehensive reviews on the topic can be found elsewhere.[65,66] There is ample evidence that the nature and properties of the scaffold and cell carrier play an important role in bone engineering. It is important to emphasize, at the outset, that the field is still young and many different approaches are under experimental investigation. Thus, it is by no means clear what defines an ideal scaffold-cell or scaffold-neo-tissue construct, even for a specific tissue type. Indeed, since some tissues perform multiple functional roles, it is unlikely that a single scaffold would serve as a universal foundation for the regeneration of even a single tissue.[65]

## Cell Sheet Engineering

An attractive alternative to current strategies, including TEC, is the use of cell sheets. Cell sheets have been used successfully for many years in the regeneration of two-dimensional tissues such as the epidermis[67] and ocular surfaces.[68] In recent years, tissue engineers have extended the use of cell sheets to produce three-dimensional (3D) neo-tissues. Using cell sheets has the advantage that an entirely natural neo-tissue assembled by the cells, with mature extracellular matrix (ECM), can be produced, thereby enhancing the biological potency of the constructs. L'Heureux *et al.*[69] have successfully built a tissue engineered blood vessel by wrapping layers of fibroblasts and smooth muscle cells around a mandrel. This vessel exhibited a well defined 3-layered organization and had a burst-strength comparable to that of native human vessels. Advancement in the culture of cell sheets came with the development of culture dishes coated with temperature-responsive polymers, introduced by Okano's group. These culture surfaces become hydrophilic at temperatures below a critical point, and hydrophobic at temperatures above the critical point.[70] As a result, cells attach and proliferate on this surface above the critical temperature, and easily detach when the temperature is reduced.[71] It was shown that four layers of cardiomyocytes cultured using this technique formed electrically communicative and pulsatile myocardial tissue both *in vitro* and *in vivo*.[72]

However, although cell sheets are strong enough to allow careful manipulation in a laboratory to produce stacked or wrapped constructs, they contract extensively upon detachment from culture surfaces. Thus engineering large-size tissues, of specific shape and size, with cell sheets alone is a challenge, and external supports such as stainless steel rings and nondegradable polymeric membranes are therefore used.[69,72] However, these supports will be removed at the point of application, upon which mechanical support for the constructs will be lost, resulting in the loss of size, shape and even biological functions which are dependent on mechanical stimuli. The aim of a group at NUS is to devise a novel technique to withstand the contraction of cell sheets *in vitro*, maintain the size of the desired graft, and apply it for tissue engineering of bone. The group hypothesized that those

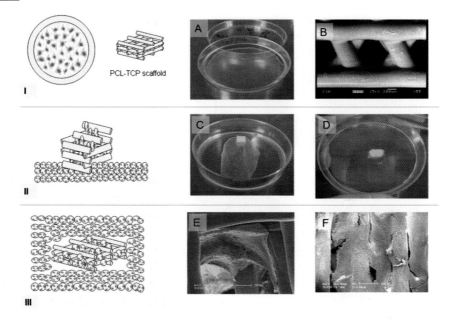

**Figure 2.** Diagrammatic representation of cell-sheet and scaffold components I-III. **A**. shows cell sheet at the bottom of the culture plate; **B**. SEM picture showing PCL-CaP scaffold architecture; **C**. & **D**. cell sheet encapsulation of scaffold; **E**. SEM picture showing cell proliferation and ECM formation throughout the scaffold architecture (inside view); **F**. lateral view showing cell sheet attached around the outer part of the scaffold.

high density cultures of osteogenically differentiated BMSCs, together with hybrid matrices, will produce mature TECs with the mineralized matrices integrated within. A diagrammatic representation of cell sheet and scaffold components is shown in Figs. 2 and 3.

## Clinical Bone Engineering

The first generation of clinically applied tissue engineering concepts in the area of skin, cartilage and bone regeneration was based on the isolation, expansion and implantation of cells from the patient's own tissue. Although successful in selective treatments, bone tissue engineering needs to overcome major challenges to allow widespread clinical application with predictable outcomes. This includes the isolation and expansion of cells with the highest potential to form bone-like tissue following the harvesting of cells without extensive donor site morbidity and to direct and maintain the phenotypic differentiation of the cells while being cultured in a 2D or 3D environment. Another major challenge is to present the cells in a

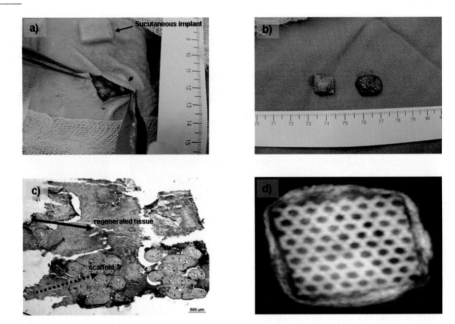

**Figure 3.** (a) Subcutaneous cell-sheet constructs (after 6 weeks) implantation in nude rats show-ing good integration and absence of foreign body reaction (*arrow*); 1st construct already explanted (*arrow head*) (see b); (**b**) nonseeded constructs (*left*) and scaffold-cell con-struct+cell sheet (*right*); (**c**) cryosection of explanted construct shown in Fig. b (*right*) showing scaffold and tissue regeneration (H&E); (**d**) microCT image of the construct shown in Fig. b (*right*), outer part showing the density of cortical bone, whereas tissue inside scaffold shows density of cancellous bone as well as fibrous tissue.

matrix to the implantation site to allow the cells to survive the contraction forces of wound healing and specifically, the remodeling and biomechanical loading in tissues such as bone. Although many generalities regarding the behavior of BMSCs are conserved across the phylogenetic tree, great care must be exercised in extrapolating the details gleaned from animal BMSCs to the case of their application in human clinical-based tissue engineering concepts.

## Orthopedic surgery

Outstanding progress has been made in the last two decades in the area of skeletal reconstruction and treatment of osseous defects. However, nonunion fractures remains critical to numerous orthopedic and trauma treatment concepts. The lack of techniques and approaches in reconstruc-tive surgery emphasizes the number of clinical applications that would ben-efit from tissue-engineered bone. Although commonplace in orthopedic surgery, the current approaches which include autografting and allografting of cancellous bone, bone transport methods like Ilizarov's technique, and applying vascularized grafts of the fibula and iliac crest, have a number of limitations as described earlier.

Vacanti *et al.*[73] reported the replacement of an avulsed phalanx with tissue-engineered bone which resulted in a partially functional and biomechanically stable thumb of normal length. Periosteal cells, which have been shown to shed osteoblastic cells, were obtained from the sections of the distal radius and were seeded onto a coral scaffold. A stable calcium alginate hydrogel which encapsulated the cells saturated the coral implant. MRI examination showed evidences of vascular perfusion and biopsy revealed new bone formation with a lamellar architecture. Bone formation is not observed when the porous hydroxyl apatite is not in direct contact with bone, whereas new bone formation is observed when seeded with marrow cells[74] or with cells derived from the periosteum[75] and placed in subcutaneous tissue that is not adjacent to native bone.

Cancedda and coworkers have reported the use of cell-based tissue engineering approaches to treat large bone defects.[76] The osteoprogenitor cells were isolated from the bone marrow and expanded *in vivo* and placed on macroporous scaffolds. The radiographs on follow-up reported abundant callus formation along the implants and good integration at the interfaces with the host bones, and all patients reported recovery earlier than expected with a traditional bone graft approach. The group had earlier reported the use of BMSC — ceramic composites — in the treatment of full-thickness gaps of tibial diaphysis in sheep.[77] Autologous cells isolated from the bone marrow and expanded *in vitro* were loaded onto highly porous ceramic cylinders and implanted in critical sized stable segmental defects. Similar results in animal models were also reported where cultured marrow cells were implanted with a coral scaffold in large segmental defects in sheep tibia.[78] Bruder and colleagues have also successfully treated an experimentally induced nonunion defect in a adult dog femur with autologous cells loaded on a hydroxyapatite-beta tricalcium phosphate scaffold.[79]

## Oral and maxillofacial surgery

The paucity of techniques in cranial reconstructive surgery emphasizes the need for alternative bone formation strategies. The repair of large cranial defects is often unsuccessful due to the limited amount of autogenous bone available. Calvarial bone is hard and brittle, which makes the contouring of the graft extremely difficult. In instances as in where the corticocancellous bone from the ilium is used, they have appeared to be more prone to resorption than the bone of membraneous origin. Calvarial reconstruction strategies have, hence, recently employed the use of adult MSCs for bone regeneration. Adipose-derived stem cells were recently used in the repair of a calvarial defect of a size of 120 sq cm in a 7-year-old girl who sustained multi-fragment fractures following head injury.[80] Autologous fibrin glue was used to attach the cells on to the milled cancellous bone which was used as an osteoconductive scaffold.

Studies done by Schantz JT *et al.*[81] to investigate the differentiation potential of bone marrow-derived progenitor cells and to compare their potential in bone regeneration against trabecular osteoblasts in a rabbit calvarial defect model, showed promising results of site specific differentiation of unconditioned MSCs. The transplanted mesenchymal progenitor cells mostly committed to the osteogenic phenotype formed island of bone tissue and the amount of mineralized tissue increased and filled out the defect sites almost entirely (Fig. 4). In a clinical pilot study by the same

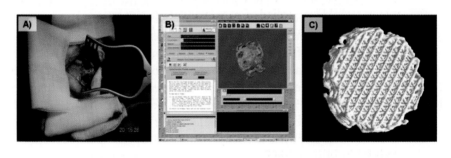

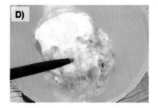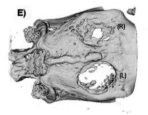

**Figure 4.   A**. Intraoperative picture of critical size calvarial defects (15 mm) created in mature New Zealand white rabbits (age > 6 months); **B**. digital 3D imaging of defects after microCT scanning; **C**. microCT of the PCL-TCP scaffold (burr plug design) (courtesy of Dr Robert Guldberg, Georgia Tech Emory Center, Atlanta); **D**. explanted cell-scaffold construct showing good host graft integration; **E**. microCT image of the rabbit skull 8 weeks post-implantation showing significant bony ingrowth in experimental side (R, scaffold was seeded with 1 million BMSCs) as opposed to control (L, empty scaffold) (courtesy of Dr Mark Knackstedt and Dr Anthony Jones, Nat'l University of Australia).

group, the burr hole of 14 mm diameter resulting from trephination was closed using a novel biodegradable scaffold which is able to entrap the marrow blood clot from the calvarial bone in 10 patients who underwent evacuation of the subdural hematoma.[82] The follow up CT scans showed the implants well integrated with the surrounding calvarial bone, with new bone filling the porous space in the scaffold.

In a case study reported by Terheyden and colleagues,[83] computer-aided design techniques were used. The patient's CT data was used to create a titanium mesh cage that was filled with bone mineral blocks and infiltrated with recombinant human bone morphogenetic protein (BMP). In addition, the patient's bone marrow cells aspirated from the iliac crest were added to provide undifferentiated precursor cells as a target for the recombinant human BMP-7, which is an osteoinductive factor that initiates the conversion of undifferentiated precursor stem cells into osteoprogenitor cells.[84,85] The transplant was implanted into the latissimus dorsi muscle and transplanted 7 weeks later as a free bone-muscle flap to repair the defect. The group reports that the clinical follow-up leads to the conclusion that this method could be applied to more patients in the future.

In yet another pilot study by Sittinger and colleagues,[86] augmentation of the posterior maxilla was carried out in two patients using a TEC derived from mandibular periosteal cells on a polymer fleece. The cell polymer transplants were cultured for a week before implantation with the addition of autologous serum, ascorbic acid, dexamethasone and beta glycerophosphate. Bone biopsies from both patients revealed mineralized trabecular bone with remnants of the biomaterial and the osteocytes were appaent within the bone lacunae.

## Conclusions

Bone formation within the body, as part of a development, healing and/or repair process, is a complex event in which cell populations in combination with extracellular matrix self-assemble into functional units and ultimately into tissues and organs. There is intense academic and commercial interest in finding methods to stimulate and control these events and eventually to replicate these events outside the body as close as possible. This interest has accelerated and resulted in bone tissue engineering increasingly becoming a well recognized research area in the arena of regenerative medicine. The increasing number of papers reporting about formerly unknown function, structure and plasticity of marrow stromal cells has spawned a major switch in the perception of their nature, and the ramifications of their potential surgical application in tissue engineering concepts have been not only envisioned but have already implemented. The ability to isolate a subset of bone marrow stromal cells with the most extensive replication and differentiation potential could naturally be of utmost importance for the regeneration of mesenchymal tissues via bone engineering strategies.

Although early pre-clinical and clinical data demonstrate the safety and effectiveness of TEC-based bone engineering, there are still many questions to be answered as to when, why and how the concept works. Additional information is required concerning the therapeutic efficacy of transplanted TECs and the mechanisms of engraftment, homing and *in vivo* differentiation. There is also a need to carry out appropriately designed toxicology studies to demonstrate the long-term safety of these therapies. The widespread use of bone engineering will also depend upon the availability of validated methods for large-scale culture, storage and distribution. As these areas are addressed, new applications will be developed leading to novel therapeutic opportunities. Much has been learned about bone engineering therapy in the past few years, and much remains to be learned.

# REFERENCES

1. Hutmacher DW, Risbud M, Sittinger M. (2004) Evolution of computer aided design and advanced manufacturing in scaffold research. *Trends Biotechnol* **22**(7):354–362.
2. Davies JE. (1999) Bone Engineering. Toronto: em squared Inc.
3. Keaveny TM, Guo XE, Wachtel EF, *et al.* (1994) Trabecular bone exhibits fully linear elastic behavior and yields at low strains. *J Biomech* **27**(9):1127–1136.
4. Park JB, Lakes, RS. (1992) Biomaterials: An Introduction, 2nd ed. New York: Plenum Press.
5. Martin RB, Burr DB. (1989) *The Structure, Function, and Adaptation of Compact Bone*, New York: Raven Press.
6. Hench LL, Wilson J. (1993) *Introduction to Bioceramics*, Singapore: World Scientific.
7. Owen M. (1985) Lineage of osteogenic cells and their relationship to the stromal system. In *Bone and Mineral Research*, ed. WA Peck, Vol. 3, pp. 1–24, Amsterdam: Elsevier.
8. Owen M, Friedenstein AJ. (1988) Stromal stem cells: Marrow derived osteogenic precursors. *CIBA Found Symp* **136**:42–60.
9. Lakshmipathy U, Verfaillie C. (2005) Stem cell plasticity. *Blood Rev* **19**(1):29–38.

10. Zuk PA, Zhu M, Ashjian P, *et al.* (2002) Human adipose tissue is a source of multipotent stem cells. *Mol Biol Cell* **13**:4279–4295.
11. Strem BM, Hedrick MH. (2005) The growing importance of fat in regenerative medicine. *Trends Biotechnol* **23**(2):64–66.
12. Cao B, Huard J. (2004) Muscle-derived stem cells. *Cell Cycle* **3**(2):104–107.
13. Friedenstein AJ, Chailakhyan RK, Latsinik NV, *et al.* (1974) Stromal cells responsible for transferring the microenvironment of the hemopoietic tissues. Cloning *in vitro* and retransplantation *in vivo*. *Transplantation* **17**:331–340.
14. Caplan AI. (1991) Mesenchymal stem cells. *J Orthop Res* **9**(5):641–650.
15. Prockop DJ. (1997) Marrow stromal cells as stem cells for nonhematopoietic tissues. *Science* **276**(5309):71–74.
16. Pittenger MF, Mackay AM, Beck SC, *et al.* (1999) Multilineage potential of adult human mesenchymal stem cells. *Science* **284**(5411):143–147.
17. Krause DS, Theise ND, Collector MI, *et al.* (2001) Multi-organ, multi-lineage engraftment by a single bone marrow-derived stem cell. *Cell* **105**(3): 369–377.
18. Friedenstein AJ, Petrakova KV, Kurolesova AI, Frolova GP. (1968) Heterotopic of bone marrow. Analysis of precursor cells for osteogenic and hematopoietic tissues. *Transplantation* **6**(2):230–247.
19. Friedenstein AJ, Chailakhjan RK, Lalykina KS. (1970) The development of fibroblast colonies in monolayer cultures of guinea-pig bone marrow and spleen cells. *Cell Tissue Kinet* **3**(4):393–403.
20. Simmons PJ, Torok-Storb B. (1991) Identification of stromal cell precursors in human bone marrow by a novel monoclonal antibody, STRO-1 *Blood* **78**:55–62.
21. Waller EK, Olweus J, Lund-Johansen F, *et al.* (1995) The "common stem cell" hypothesis reevaluated: Human fetal bone marrow contains separate populations of hematopoietic and stromal progenitors *Blood* **85**:2422–2435.
22. Kuznetsov SA, Friedenstein AJ, Robey PG. (1997) Factors required for bone marrow stromal fibroblast colony formation *in vitro*. *Br J Haematol* **97**:561–570.
23. Baddoo M, Hill K, Wilkinson R, *et al.* (2003) Characterization of mesenchymal stem cells isolated from murine bone marrow by negative selection. *J Cell Biochem* **89**(6):1235–1249.
24. Gronthos S, Graves SE, Ohta S, Simmons PJ. (1994) The STRO-1+ fraction of adult human bone marrow contains the osteogenic precursors. *Blood* **84**:4164–4173.
25. Gronthos S, Zannettino AC, Hay SJ, *et al.* (2003) Molecular and cellular characterisation of highly purified stromal stem cells derived from human bone marrow. *J Cell Sci* **116**:1827–1835.
26. Dennis JE, Carbillet JP, Caplan AI, Charbord P. (2002) The STRO-1+ marrow cell population is multipotential. *Cells Tissues Organs* **170**(2–3):73–82.

27. Simmons PJ, Gronthos S, Zannettino A, *et al.* (1994) Isolation, characterization and functional activity of human marrow stromal progenitors in haemopoiesis. *Prog Clin Biol Res* **389**:271–280.
28. Filshie RJ, Zannettino AC, Makrynikola V, *et al.* (1998) MUC18, a member of the immunoglobulin superfamily, is expressed on bone marrow fibroblasts and a subset of hematological malignancies. *Leukemia* **12**:414–421.
29. Haynesworth SE, Baber MA, Caplan AI. (1992) Cell surface antigens on human marrow-derived mesenchymal cells are detected by monoclonal antibodies. *Bone* **13**:69–80.
30. Deschaseaux F, Gindraux F, Saadi R, *et al.* (2003) Direct selection of human bone marrow mesenchymal stem cells using an anti-CD49a antibody reveals their CD45med, low phenotype. *Br J Haematol* **122**:506–517.
31. Jiang Y, Vaessen B, Lenvik T, *et al.* (2002) Multipotent progenitor cells can be isolated from postnatal murine bone marrow, muscle, and brain. *Exp Hematol* **30**:896–904.
32. Jiang Y, Jahagirdar BN, Reinhardt RL, *et al.* (2002) Pluripotency of mesenchymal stem cells derived from adult marrow. *Nature* **418**:41–49.
33. Bruder SP, Jaiswal N, Haynesworth SE. (1997) Growth kinetics, self-renewal, and the osteogenic potential of purified human mesenchymal stem cells during extensive subcultivation and following cryopreservation. *J Cell Biochem* **64**:278–294.
34. Muschler GF, Boehm C, Easley K. (1997) Aspiration to obtain osteoblast progenitor cells from human bone marrow: The influence of aspiration volume. *J Bone Joint Surg Am* **79**(11):1699–1709.
35. Majors AK, Boehm CA, Nitto H, *et al.* (1997) Characterization of human bone marrow stromal cells with respect to osteoblastic differentiation. *J Orthop Res* **5**(4):546–557.
36. Muschler GF, Nitto H, Boehm CA, Easley KA. (2001) Age- and gender-related changes in the cellularity of human bone marrow and the prevalence of osteoblastic progenitors. *J Orthop Res* **19**(1):117–125.
37. Turgeman G, Aslan H, Gazit Z, Gazit D. (2002) Cell-mediated gene therapy for bone formation and regeneration. *Curr Opin Mol Ther* **4**(4):390–394.
38. Yoon ST, Boden SD. (2002) Osteoinductive molecules in orthopaedics: Basic science and preclinical studies. *Clin Orthop* **395**:33–43.
39. Winn SR, Uludag H, Hollinger JO. (1999) Carrier systems for bone morphogenetic proteins. *Clin Orthop* **367** (Suppl):S95–S106.
40. Oligino TJ, Yao Q, Ghivizzani SC, Robbins P. (2000) Vector systems for gene transfer to joints. *Clin Orthop* **379** (Suppl):S17–S30.
41. Hutmacher DW, Garcia A. Scaffold based bone engineering by using genetically modified cells (invited review). *Gene* (in Press).
42. Kofron MD, Laurencin CT. (2005) Orthopaedic applications of gene therapy. *Curr Gene Ther* **5**(1):37–61.

43. Majumdar MK, Keane-Moore M, Buyaner D, *et al.* (2003) Characterization and functionality of cell surface molecules on human mesenchymal stem cells. *J Biomed Sci* **10**(2):228–241.
44. Le Blanc K, Tammik C, Rosendahl K, *et al.* (2003) HLA expression and immunologic properties of differentiated and undifferentiated mesenchymal stem cells. *Exp Hematol* **31**(10):890–896.
45. Hutmacher DW, Lauer G. (2002) Basic science and clinical trends of tissue engineering in oral and maxillofacial surgery. *Implantology* **10/2**:143–156.
46. Digirolamo CM, Stokes D, Colter D, *et al.* (1999) Propagation and senescence of human marrow stromal cells in culture: A simple colony-forming assay identifies samples with the greatest potential to propagate and differentiate. *Br J Haematol* **107**(2):275–281.
47. Martin I, Muraglia A, Campanile G, *et al.* (1997) Fibroblast growth factor-2 supports *ex vivo* expansion and maintenance of osteogenic precursors from human bone marrow. *Endocrinology* **138**(10):4456–4462.
48. Mackenzie TC, Flake AW. (2001) Multilineage differentiation of human MSC after in utero transplantation. *Cytotherapy* **3**:403–405.
49. Wong M, Tuan RS. (1993) Nuserum, a synthetic serum replacement, supports chondrogenesis of embryonic chick limb bud mesenchymal cells in micromass culture. *In Vitro Cell Dev Biol Anim* **29A**:917–922.
50. Goldsborough MD, Tilkins ML, Price PJ, *et al.* (1998) Serum-free culture of murine embryonic stem (ES) cells. *Focus* **20**:8–12.
51. Zuk, PA, Zhu M, Mizuno H, *et al.* (2001) Multilineage cells from human adipose tissue: Implications for cell-based therapies. *Tissue Eng* **7**:211–228.
52. Owen TA, Aronow M, Shalhoub V, *et al.* (1990) Progressive development of the rat osteoblast phenotype *in vitro*: Reciprocal relationships in the expression of genes associated with osteoblast proliferation and differentiation during formation of the bone extracellular matrix. *J Cell Physiol* **143**(3):420–430.
53. Igarashi M, Kamiya N, Hasegawa M, *et al.* (2002) Inductive effects of dexamethasone on the gene expression of Cbfa1, Osterix and bone matrix proteins during differentiation of cultured primary rat osteoblasts. *J Mol Histol* **35**(1):3–10.
54. Buttery LD, Bourne S, Xynos JD, *et al.* (2001) Differentiation of osteoblasts and *in vitro* bone formation from murine embryonic stem cells. *Tissue Eng* **7**:89–99.
55. Rogers JJ, Young HE, Adkison LR, *et al.* (1995) Differentiation factors induce expression of muscle, fat, cartilage, and bone in a clone of mouse pluripotent mesenchymal stem cells. *Am Surg* **61**:231–236.
56. Xiao G, Cui Y, Ducy P, *et al.* (1997) Ascorbic acid-dependent activation of the osteocalcin promoter in MC3T3-E1 preosteoblasts: Requirement for collagen matrix synthesis and the presence of an intact OSE2 sequence. *Mol Endocrinol* **11**:1103–1113.

57. Volloch V, Kaplan D. (2002) Matrix-mediated cellular rejuvenation. *Matrix Biol* **21**:533–543.

58. Gronthos S, Simmons PJ, Graves SE, Robey PG. (2001) Integrinmediated interactions between human bone marrow stromal precursor cells and the extracellular matrix. *Bone* **28**:174–181.

59. Roche P, Goldberg HA, Delmas PD, Malaval L. (1999) Selective attachment of osteoprogenitors to laminin. *Bone* **24**:329–336.

60. Madihally SV, Flake AW, Matthew HW. (1999) Maintenance of CD34 expression during proliferation of CD34+ cord blood cells on glycosaminoglycan surfaces. *Stem Cells* **17**:295–305.

61. Matsuda N, Morita N, Matsuda K, Watanabe M. (1998) Proliferation and differentiation of human osteoblastic cells associated with differential activation of MAP kinases in response to epidermal growth factor, hypoxia, and mechanical stress *in vitro*. *Biochem Biophys Res Commun* **249**:350–354.

62. Yoshikawa Y, Hirayama F, Kanai M, *et al.* (2000) Stromal cell-independent differentiation of human cord blood CD34+CD38− lymphohematopoietic progenitors toward B cell lineage. *Leukemia* **14**:727–734.

63. Altman GH, Horan RL, Martin I, *et al.* (2002) Cell differentiation by mechanical stress. *FASEB J* **16**:270–272.

64. Kapur S, Baylink DJ, Lau KHW. (2003) Fluid flow shear stress stimulates human osteoblast proliferation and differentiation through multiple interacting and competing signal transduction pathways. *Bone* **32**:241–251.

65. Hutmacher DW. (2000) Scaffolds in tissue engineering bone and cartilage. *Biomaterials* **21**(24):2529–2543.

66. Hutmacher DW, Sittinger M. (2003) Periosteal cells in bone tissue engineering. *Tissue Eng* **9**(1):S45–S64.

67. Chester DL, Balderson DS, Papini RP. (2004) A review of keratinocyte delivery to the wound bed. *J Burn Care Rehab* **25**(3):266–275.

68. Dunaief JL, Ng EW, Goldberg MF. (2001) Corneal dystrophies of epithelial genesis: The possible therapeutic use of limbal stem cell transplantation. *Arch Ophthalmol* **119**(1):120–122.

69. L'Heureux N, Paquet S, Labbe R, *et al.* (1998) A completely biological tissue-engineered human blood vessel. *FASEB J* **12**(1):47–56.

70. Yamada N, Okano T, Sakai H, *et al.* (1990) Thermo-responsive polymeric surfaces; control of attachment and detachment of cultured cells. *Makromol Chem Rapid Commun* **11**:571–576.

71. Kwon OH, Kikuchi A, Yamato M, *et al.* (2000) Rapid cell sheet detachment from poly(N-isopropylacrylamide)-grafted porous cell culture membranes. *J Biomed Mater Res* **50**(1):82–89.

72. Shimizu T, Yamato M, Kikuchi A, Okano T. (2003) Cell sheet engineering for myocardial tissue reconstruction. *Biomaterials* **24**:2309–2316.

73. Vacanti CA, Bonassar LJ, Vacanti MP, Shufflebarger J. (2001) Replacement of an avulsed phalanx with tissue-engineered bone. *N Engl J Med* **344**(20):1511–1514.

74. Saitoh A, Tsuda Y, Bhutto IA, *et al.* (1996) Histologic study of living response to artificially synthesized hydroxyapatite implant: 1-year follow-up. *Plast Reconstr Surg* **98**:706–710.

75. Nakahara H, Goldberg VM, Caplan AI. (1992) Culture-expanded periostealderived cells exhibit osteochondrogenic potential in porous calcium phosphate ceramics *in vivo*. *Clin Orthop* **276**:291–298.

76. Quarto R, Mastrogiacomo M, Cancedda R, *et al.* (2001). Repair of large bone defects with the use of autologous bone marrow stromal cells. *N Engl J Med* **344**:385–386.

77. Cancedda R, Mastrogiacomo M, Bianchi G, *et al.* (2003) Bone marrow stromal cells and their use in regenerating bone. *Novartis Found Symp* **249**:133–143.

78. Petite H, Viateau V, Bensaid W, *et al.* (2000) Tissue-engineered bone regeneration. *Nat Biotechnol* **18**(9):959–963.

79. Bruder SP, Kraus KH, Goldberg VM, Kadiyala S. (1998) The effect of implants loaded with autologous mesenchymal stem cells on the healing of canine segmental bone defects. *J Bone Joint Surg Am* **80**(7):985–996.

80. Lendeckel S, Jodicke A, Christophis P, *et al.* (2004) Autologous stem cells (adipose) and fibrin glue used to treat widespread traumatic calvarial defects: Case report. *J Craniomaxillofac Surg* **32**(6):370–373.

81. Schantz JT, Hutmacher DW, Lam CX, *et al.* (2003) Repair of calvarial defects with customised tissue-engineered bone grafts II. Evaluation of cellular efficiency and efficacy *in vivo*. *Tissue Eng* **9**(11):S127–S139.

82. Schantz JT, Lim TC, Chou N, *et al.* (2005) Burr hole cranioplasty using a novel biodegradable polymer implant. *Regenerate*, Atlanta, USA.

83. Warnke PH, Springer IN, Wiltfang J, *et al.* (2004) Growth and transplantation of a custom vascularised bone graft in a man. *Lancet* **364**(9436):766–770.

84. Terheyden H, Wang H, Warnke PH, *et al.* (2003) Acceleration of callus maturation using rhOP-1 in mandibular distraction osteogenesis in a rat model. *Int J Oral Maxillofac Surg* **32**:528–533.

85. Friedlaender GE, Perry CR, Cole JD, *et al.* (2001) Osteogenic protein-1 (bone morphogenetic protein-7) in the treatment of tibial nonunions. *J Bone Joint Surg Am* **83**-A:S151–S158.

86. Schmelzeisen R, Schimming R, Sittinger M. (2003) Making bone: Implant insertion into tissue-engineered bone for maxillary sinus floor augmentation — A preliminary report. *J Craniomaxillofac Surg* **31**(1):34–39.

# 22

# Stem Cells and Cartilage

JB Richardson, JTK Lim, JHP Hui and EH Lee*

## Introduction

As far back as 1743, William Hunter observed that cartilage, "once destroyed, is not repaired."[1] This lack of healing leaves bone exposed in the joint and is a cause of pain. Can stem cell engineering bring a solution?

Osteoarthitis contributes the largest burden to healthcare in the developed world. The extent and involvement of a single cartilage lesion in the development of arthritis is not known. However, in experimental studies, incongruity and articular step-off lead to degenerative changes in the cartilage and subchondral bone that mimic the appearance of arthritis.[2] In the clinical setting, the progression of chondral injuries to arthritis is not validated by large prospective studies but rather by anecdotal evidence and smaller studies. One such study is the observation of radiographic evidence of arthritis within 14 years in greater than 50% of patients who sustained a unipolar, unicompartmental injury as adolescents.[3]

The goal of treating arthritis is to abolish symptoms and restore function. One of the most successful surgical treatments known is artificial joint replacement. Unfortunately, all joint replacements eventually fail if the patient lives long enough. Tissue engineering offers the potential of recreating a "biological joint replacement" which may last a lifetime.

At present, in cartilage repair or regeneration techniques, there are no long term clinical studies as yet. An alternative is to assess how closely the histology or mechanics of a repair tissue resembles normal tissue. The assumption is made that if the normal structure is replicated, then symptom-free joint function can be expected. In the case of cartilage repair it is not merely hyaline cartilage that is the goal but articular cartilage.

---

*Correspondence: Division of Graduate Medical Studies, Faculty of Medicine, Block MD 5, Level 3, 12 Medical Drive, Singapore 117598. E-mail: dosleeeh@nus.edu.sg, Tel.: 65 68746576, Fax: 65 6773 1462.

Defining this end-point has involved a lot of hard work by the scientists of the International Cartilage Repair Society.[4]

Articular cartilage is principally composed of type II collagen, proteoglycans, and chondrocytes forming a complex, partially hydrated, compressed matrix structure. It is anisotropic and typically composed of four layers, each with different orientations of collagen fibers, varying cell morphology, and varying matrix composition and function (Fig. 1). Articular cartilage is a highly adapted structure that provides resistance against compression and shear, and allows frictionless joint movement, shock absorption and lubrication. It has no pain fibers or blood vessels. Metabolism is anerobic and glucose reaches the cells by diffusion both from the joint surface and

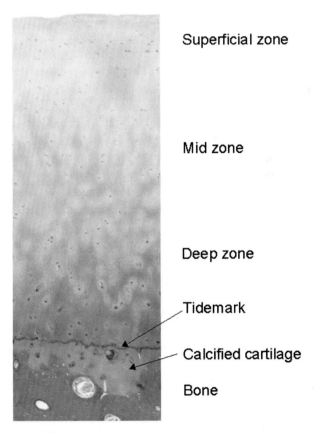

**Figure 1.** H & E stained section of normal adult articular cartilage showing various zones. (Photomicrograph courtesy of S. Roberts, Oswestry, UK.)

the underlying bone. Even fibrocartilage will stain well for collagen type II, but on polarized light, has clearly no collagen orientation. It appears to give good symptomatic relief in patients if all bone is covered. Hyaline is a term that denotes a glass-like appearance and so is not well defined. Hyaline-like cartilage is a term that has been introduced in cartilage repair[5] to denote cartilage that is better than fibrocartilage but not fully normal articular cartilage. An important aspect of histology is that in the normal adult joint, there is variation in the quality of cartilage from one area to another in the same joint.

Fibrous tissue is principally type I collagen and contains vessels and nerves. It does not function well as an articular surface. Success in a repair technique may be usefully measured based on the absence of repair by fibrous tissue.

The best hope for quantification of cartilage repair is in combining absolute measurements of collagen types I and II as well as of proteoglycan with assessments of collagen orientation, integration to bone and to adjacent cartilage and overall thickness. The ability to measure collagen quantity has now been reported on biopsies by Hollander.[6]

Cartilage does not heal if incised or if a partial thickness of the cartilage is removed. It will, however, heal once it is worn down to bone[7] when stem cells are released from the underlying bone. A good repair response will then produce healing by fibrocartilage. This is not true regeneration of hyaline cartilage but a repair process similar to the formation of scar without blood vessels. The collagens formed are the type II, but under polarized light, the bright effect seen in hyaline cartilage due to the orientation of collagen fibres is not evident.

A useful source of clinical data on this question is to be found in trials of osteotomy at the knee.[8] Generally, a third of patients will need total knee replacement within 5 years following osteotomy. Efforts to improve the outcome by adding microfracture or abrasion have found an improvement in histology with a higher proportion of fibrocartilage forming where there was bone-on-bone before.[9-12] This tissue has formed where there will still be significant amounts of load applied while the surface heals. The ability to form new tissue under load is important as this opens the way for tissue engineering techniques that depend on tissue growth in the active patient.

The following section provides a review of experimental studies as well as currently available methods of tissue repair, including the use of stem cells.

# Methods of Repair

Fibrocartilage repair follows exposure of bone. It is not yet clear whether vessels need to be exposed as, theoretically, stem cells from the underlying bone could emerge through small cracks in the subchondral bone plate. Partial thickness injuries that do not reach the underlying bone have not shown any healing over time. The full thickness of the cartilage needs to be removed to allow repair by marrow undifferentiated mesenchymal stem cells as described by Shapiro and co-workers.[14] The authors also noted that the collagen fibrils of the repair tissue were not well integrated with those of the surrounding cartilage. Additionally, the chondrocyte lacunae adjacent to the injury remained empty, and degenerative change appeared as early as 10 weeks and became more advanced by 24 weeks. Although type II collagen was present initially, there were still significant (20%–35%) type I collagen,[13] and between 6 and 12 months the matrix and cells become more typical of fibrocartilage.[14]

## Marrow stimulation

Exposure of bone has been the common theme in a range of surgical techniques that have found success in achieving repair. In a randomized controlled trial, Hubbard[15] demonstrated a significant clinical benefit of debridement over lavage alone in the treatment of isolated symptomatic chondral defects. He undertook meticulous debridement of unstable cartilage in one group, and observation alone in the control group. At 5 years, 65% of the debridement groups were pain-free compared with 20% of the controls.

Drilling of the subchondral bone was used by Pridie and reported in 1959.[16] More recently, the use of microfracture with a sharp curved awl allows easier access to the defect and is advocated by Steadman.[17–20] Steadman takes particular care to arrange continuous passive movement postoperatively, which may be part of the reason for obtaining good results and was also used in a randomized controlled trial from Norway.[21] The repair tissue initially has some hyaline-like cartilage but becomes more fibrous with time.[21–23] Mirroring this, clinical response can be good but is not always predictable and often deteriorates with time.[24–25] Demonstration of the advantages of cortical penetration over simple debridement alone or wash-out has been hampered by a lack of controlled studies.

## Osteochondral grafts

Cartilaginous defects can be treated by transplants of chondral or osteo-chondral grafts. Unlike simple bone grafts, survival of a chondral graft depends on survival of its chondrocytes as it is not possible for new cells to migrate into the cartilage matrix. One problem with many of the techniques is that impaction of the cartilage surface occurs. This impaction can overload the cells and cause apoptosis with resulting loss of proteoglycan turnover and failure of the graft.

The grafts may be autologous or allogenous, each with advantages and disadvantages. Autologous grafts are free of disease transmission, infection, immunorejection risks and have better cell viability, but donor sites are limited and the process of harvesting from the donor site may itself cause symptoms and disability. Allografts can be larger, may be site-specific (e.g. patella graft for patella defect) and are generally from young donors, but have the disadvantages of infection transmission and provocation of an immune response. Freezing reduces immunogenicity but also compromises cell viability.

As a cartilage graft will not inherently bond to a recipient site, current techniques involve the use of osteochondral grafts to allow bone-on-bone healing. Although tissue adhesives such as fibrin glue may be available specifically for cartilage on cartilage grafts in the future, it is currently not practical.[26]

Mohammed *et al.*[27] reported the Toronto experience of fresh allograft osteochondral graft survival rates of 76% at 5 years in 91 patients. Autogenous osteochondral grafts, of course, cannot be orthotopic and must be made to fit the site of the defect. They can be large uniblock grafts as described by Outerbridge *et al.*, who used the lateral facet of the patella press-fit into femoral condyle defects, or multiple small osteochondral cylinders taken from the lateral edge of the lateral femoral condyle and packed into the defect.[29-30] Mosaicplasty appears unsuitable for defects larger than 2 × 2 cm.[31]

## Periosteal graft

Periosteum potentially is an active biomaterial for tissue engineering of cartilage: i) It contains undifferentiated cells with the potential to form cartilage or bone; ii) it can serve as a scaffold or matrix by which other cells or growth factors can adhere; and iii) it produces bioactive factors that are known to be chondrogenic.[32]

Regenerated chondrocytes in the knees of rabbits which had been treated with periosteal transplantation were shown to originate from the periosteal graft. Zarnett *et al.*[33] labeled the grafts with radioactive thymidine before transplantation, and afterwards, were able to see thymidine-labeled cells throughout the regenerated tissue with autoradiography.[33] The same group went on to transplant periosteal allografts from males into females, and karyotyping revealed that in one-third of the female rabbits, all the cells of the neocartilage contained a Y chromosome. The regenerated cells, therefore, originated from the original male periosteal graft. The cells of the other rabbits contained both male and female chromosomes, whose origins were likely from the mesenchymal stem cells (MSCs) of the female hosts and the allografts of the male donor rabbits.[34]

Periosteum contains a cambium layer, that is thought to contain MSCs,[32] and an outer fibrous layer. During chondrogenesis *in vitro*, cartilage formation first appears in the cambium layer; and expression of cartilage-specific markers occurs almost exclusively in the cambium layer. *In vitro* experiments have further revealed that the chondrogenic potential of the periosteum varies at different sites, and is positively correlated with the total number of cells and the thickness of the cambium layer.[35]

## Perichondrial graft

In 1990, Homminga *et al.*[36] reported the use of costal perichondral grafts to treat 30 chondral lesions in 25 patients, with the chondral side facing the joint, fixed with fibrin glue. Arthroscopy at 3 to 12 months postoperatively revealed that 27 out of 30 had filled in completely with tissue resembling cartilage. Subsequent studies observed bone and calcification developing in these grafts and the technique is no longer recommended.

## Autologous chondrocyte implantation (ACI)

In 1994, the publication of early results by the Gotenberg team of Petersen, Lindahl and Brittberg had a catalyzing effect on cell engineering.[37] Orthopedics was launched into a new age where it was claimed that "hyaline-like" cartilage could be regenerated. Controversy raged principally because this claim challenged the old mantra that cartilage could not heal. Since then several other reports have duplicated those initial results.[38-40]

The process involves *in vitro* clonal expansion of harvested autologous chondrocytes after enzymatic digestion, followed by implantation of these cells in suspension into a cartilage defect which has been sutured over using

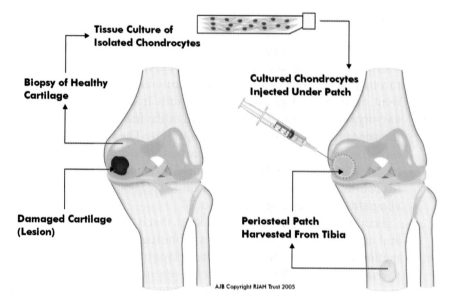

**Figure 2.** Autologous chondrocyte implantation technique.

periosteum. These chondrocytes are later implanted into the defect which has been converted into a contained pouch by a sutured periosteal membrane. The periosteum is harvested with care and speedily transferred to the defect. It is sutured with fine absorbable sutures. This sheet of periosteum is sutured with the bone side downward (Fig. 2). This is contrary to the method of its use by O'Driscoll as discussed above. He is of the opinion that periosteum alone will not be the source of healing of the cartilage when used in this way, and animal studies confirmed that chondrocytes are required for the Gothenburg technique to be successful. It may be the case that without cellular expansion, the numbers of cells from the periosteum alone are insufficient to form hyaline tissue, producing a fibrous repair instead.

Three prospective randomized trials have been reported in 2004. In Norway, a *multicentre study comparison of microfracture to ACI was carried out in 80 patients.*[21] After a 2-year analysis, Knutsen *et al.* favored microfracture, but both groups improved. Histology of the repair tissue type found a range from fibocartilage to hyaline in both groups, with a tendency to better histology in the ACI group. The ACI technique has good reports at 8 years and so it is possible that the results will favor ACI in the long term.

A prospective trial by Horas *et al.*[41] examined 40 patients allocated to either mosaicplasty or ACI and concluded that the results as measured by a surgeon-completed Lysholm score were better with mosaicplasty. That study was, however, not strictly randomized, and the patients treated by ACI had more severe symptoms at the start of the study, and lower initial Lysholm scores. Re-analysis of the results at Oswestry, UK, found little real improvement and also that a previous report by Horas *et al.*[42] in the German literature on the same group of patients was in favor of ACI. It is not a concern that there were these two reports, but on reconsidering the results, Horas came to the conclusion that both techniques were probably of equal benefit.[43]

A third study in 2003 was again a comparison of mosaicplasty with ACI. Professor Bentley and his groups[31] in London studied 100 patients in a mean period of 19 months and found the ACI to be slightly better ($p = 0.032$), although the Cincinatti knee rating showed no significant difference ($p = 0.028$).

Limitations of ACI include a failure to improve patients in 10–20% of all cases (37–39); site specific variability in success rates[43] and limitation to chondral defects rather than the widespread cartilage loss of ostoarthritis. It also cannot be used in inflammatory arthritis. Harvest sites may heal with fibrocartilage if taken down to bone but can themselves be a source of symptoms. Of a group of 10 patients who underwent autologous chondrocyte implantation to the ankle using ipsilateral knee harvests, 3 had Lysholm scores that returned to preoperative levels at one year but 7 had a decrease in Lysholm score of the knee by 15% at 1 year.[44]

Furthermore, the repair tissue in biopsy specimens from ACI does not have the architecture of mature articular cartilage. The histology with this treatment does improve with time. The layers of orientation of collagen seen in the adult knee develop over several years in the adolescent and it remains to be seen whether completely normal patterns will develop with time in the ACI treated areas.

## Second generation ACI

### Alternatives to periosteum

All surgeons will agree that the water-tightness of their repair is less than ideal. Cells are lost either immediately or on return of loading of the ACI patch. Over 10% of patches become detached at one site as early as 3 weeks following suture. Fibrin glue is of help but it is really only a sealant as the tensile strength it can provide does not reach that of glue. There have, therefore, been several developments to attempt improvement of this situation. Chondro-gide, a collagen membrane made by Geistlich Biomaterials, Wolhusen, Switzerland, was used as a substitute for the periosteum. It is a combination of collagen I and III and contains a significant amount of elastin. It handles well from a surgical point of view and remains generally smooth one year later. Cells integrate into all layers but only produce collagen type II in the deeper layers.[45]

Another use for the chondro-gide type of membrane is as a cell carrier and is inserted into the base of a defect with a technique called MACI.[46-47] Clinical trials comparing ACI with MACI are in progress but the initial reports fail to show a benefit of MACI over ACI. It does have the potential for becoming an arthroscopic technique. Another cell carrier that has good clinical results is HYAFF11. This has good laboratory[48] and animal support as chondrocytes retain or regain differentiation. Another option is the use of a collagen gel where good physical results are obtained and the use of minimally expanded chondrocytes may give better results.[49] Transplantation of chondrocytes seeded on a hyaluronan derivative (HYAFF11) into cartilage defects in rabbits have also been reported.[50] This is part of clinical trial currently carried out in Europe.

## Are Chondrocytes the Best Cells to Use?

Chondrocytes kept in monolayer culture at low density seem to lose their phenotype within four passages or approximately one month in culture. They change from a polygonal or round to a more flattened ameboid shape and synthesize type I collagen instead of cartilage collagen II.[51] The relationship between the shape of the cells and the type of collagen produced is not clear. In ACI, cartilage is harvested from the non-weightbearing part of the patient's own knee and chondrocytes are cultured and expanded and returned to the same patient for repair of chondral defects. Martin and Buckwalter[52] have shown that the telomeres shorten with age in chondrocytes as they do in other cells. This implies that in an older patient the autologous chondrocytes that are re-implanted may suffer the fate of earlier degeneration. Animal experiments in our laboratory comparing chondrocyte to mesenchymal stem cell transfer for repair of cartilage defects have clearly demonstrated that the joints repaired with chondrocytes showed degenerative changes in the repaired cartilage within 36 weeks, whereas the joints repaired by stem cells were still intact at that time.[90] As the longevity of the repair of cartilage defects is under question, it may be worthwhile examining the role of stem cells in cartilage repair, as stem cells or progenitor cells may in fact differentiate into chondrocytes which may produce a long lasting repair.

## Stem Cells

By definition, stem cells must have the capacity to divide and self-replicate, as well as the ability to differentiate into many tissue types. There are embryonic stem cells (ESCs) which are believed to be pluripotent, having the ability to give rise to ectodermal, mesodermal and endodermal tissues; and adult stem cells, such as mesenchymal stem cells (MSCs), which are multipotent, having the ability to differentiate into a smaller number of tissue types mainly of mesodermal origin. Bone marrow contains stem cells that can give rise to the hemopoeitic cell lines, as well as the mesenchymal cell lines. Stem cells can also be derived from fetal tissue and cord blood.

## Embryonic Stem Cells (ESCs)

Embryonic stem cells are self-renewing pluripotent cells capable of differentiating into any cell type. Chondrocytes derived from ESCs could have the advantage of forming a stable phenotype compared with those derived from primary cultures, which progressively dedifferentiate.[53] They therefore have the most potential as a tissue engineering material. There are, however, ethical issues involved with the use of ESCs and in some countries, research using ESCs are either prohibited or severely restricted.

Kramer *et al.*[54] showed that ESCs could be differentiated via embryoid bodies into chondrocytes *in vitro*. Their conclusions were evidenced by the expression of cartilage associated genes and proteins, demonstrated by staining with Alcian blue (which stains acidic proteoglycans specifically expressed in cartilage). By immunostaining, they showed that cells in the deeply stained areas of the plates, had cells which produced collagen II and cartilage oligomeric matrix protein (COMP). Reverse transcriptase-polymerase chain reaction analysis also demonstrated the presence of genes encoding transcription factors involved in mesenchymal differentiation. The authors also investigated the effects of addition of TGF-beta-1, BMP-2 and BMP-4 on the number of collagen II-positive areas before embryoid body formation. Treatment with BMP 2 and BMP 4 resulted in a significantly increase in numbers of collagen II areas but these declined slightly after treatment with TGF-beta1. The effect of BMP 2 was found to be dependent on the time of application.

Wakitani et al.[55] conducted a study in mice to determine if the knee intraarticular space would provide a better environment for ESC chondrogenesis. Twenty-five 5-week-old mice with severe combined immunodeficiency (SCID) had ESCs (obtained from 129/Sv/Ev mice), embedded in type 1 collagen, implanted into the right knee as well as the subcutaneous space on the back of each mouse. Eight knees contained a tumor and three of these contained cartilage. In the subcutaneous space of the back, 22 contained tumors of which cartilage was found in six. The tumors in the knee were significantly smaller than those in the back, possibly due to the confined intraarticular space. The ratio of cartilage to tumor was significantly greater than those in the back, but it was observed that when the tumor became larger, the size of the cartilage mass within it remained constant. The higher cartilage to tumor ratio in the knee may therefore relate to tumor size rather than cartilage size. The authors concluded that the tumors were teratomas because the tissues identified originated from ectoderm, mesoderm and endoderm, and that some cells in the histological sections had indeed been derived from the ES cells, due to the presence of a Neo-resistance gene identified by DNA analysis in the sections that was not present in the mice. In two of the knee joints, the tumors extruded from the knee and destroyed the knee structures. Overall, the authors felt that it was currently not possible to use ES cells to treat articular cartilage defects unless a way could be found to optimize ES cells differentiation into cartilage and inhibit tumor growth.

## Mesenchymal Stem Cells (MSCs)

An MSC is a multilineage potential progenitor cell that keeps its capacity to divide and whose progeny differentiates to mesodermal tissue cells, such as cartilage, bone, muscle, fat, tendon, ligament, etc.[56] These multipotent and self-renewing cells have the ability to mobilize to areas of injury or disease, colonize the area, regenerate new tissue and thus potentially represent the most natural, effective and immortal implant. In principle, the goal would be to induce and expand a group of multipotent cells down a signaled pathway into an end-stage phenotype or one that would be capable of further development after implantation, deliver the cells to the repair site using a scaffold and bind it to the edges of the defect.

Two alternative approaches are proposed:

a) Implant cells directly as in the current cell-based techniques, or an appropriate development of this approach such as implantation of a suitable matrix or scaffold, seeded with chondroprogenitor cells and appropriate signaling substances.[57–59] Then, allow the differentiation process to occur *in vivo* (Figs. 3 & 4).
b) Differentiate stem cells *in vitro* and implant a matured construct. The cell-scaffold-bioreactor model is an example of this and more mature tissues can be created in this way (Fig. 5).

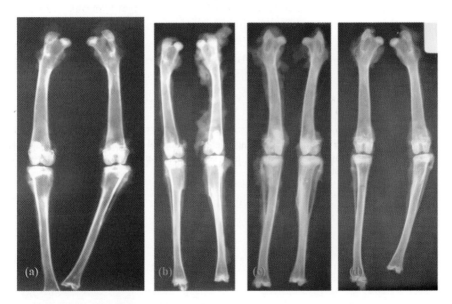

**Figure 3.** Radiographs of rabbits sacrificed at 8 weeks after physeal transplant of the left proximal tibia. (**a**) Varus deformity following iatrogenic growth arrest in the control group; (**b**) after bone marrow derived stem cell transplant into physis; (**c**) after periosteum-derived stem cell transplantation; and (**d**) partial correction with adipose-derived stem cell transplantation.

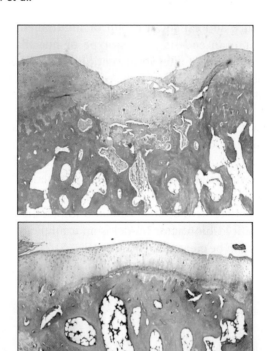

**Figure 4.**  H&E sections of articular cartilage defect in femoral condyle of rabbits transplanted with MSCs showing: (**a**) partial filling at 6 weeks post-transplant; and (**b**) total resurfacing at 36 weeks post-surgery.

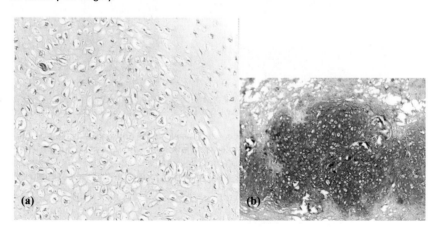

**Figure 5.**  Histological sections of (**a**) *in vitro* chondrogenesis of bone marrow-derived stem cells in pellet form; and (**b**) *in vivo* chondrogenic lineage of MSCs transplanted in nude mice.

## Scaffolds for Cartilage Engineering

Most studies assume that a scaffold is required for regeneration of cartilage. Loads in the joint and fluid movements would simply prevent cells thriving where they are needed. The ideal scaffold for cartilage tissue engineering should be: 1) reproducible; 2) three-dimensional; 3) have a highly porous structure permitting spatially uniform cellular distribution and minimizing diffusional constraints; and 4) biocompatible and biodegradeable.[76] The structure determines transport of nutrients, metabolites and passage of regulatory molecules. Construct composition and mechanical properties are generally better for fibrous meshes[77] than for agarose gels,[78] but mechanical stimulation improved construct composition only for agarose gel cultured chondrocytes.[79] The chondrogenic activity of stem cells, their ability to differentiate and to adhere to scaffolds such as matrices of hyaluronan derivatives[80–81] and gelatin-based resorbable sponge matrices[82] have been investigated (Fig. 6).

Chitin is a natural polymer of shellfish origin and has been investigated as a scaffold in the repair of large growth plate defects in 6-week-old NZW rabbits.[83] Three groups were compared, with the created physeal defects treated with excision of the bony bar and 1) no interposition, 2) chitin alone, and 3) chitin with MSCs. The differences in correction of angular

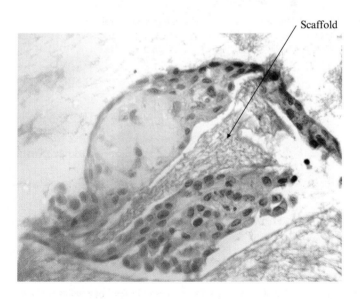

**Figure 6.** Bone marrow derived cells and polyvinylalcohol scaffold (H&E).

deformity and length discrepancy of the tibia were greater between groups 3 and 1, than between groups 2 and 1. Patient studies have been slow as there is no prior history of Chitin being used internally in patients. One development in Canada is its use as a dressing in joints over microfractured chondral defects.

## Bioreactors

Bioreactors affect the development of tissues by controlling density/concentration, temperature, pH, osmolarity, oxygen levels and allow mass transfer of cells and the culture environment. Physical signals in the form of compression, shear, and interstitial blood flow can also be controlled. Figure 7 illustrates an early bioreactor. The composition, morphology and mechanical properties of engineered cartilage grown in mechanically active environments were generally better than in static environments.[84] Hydrodynamic stresses acting on dynamic laminar flow of rotating bioreactors markedly enhanced *in vitro* chondrogenesis.[77,85]

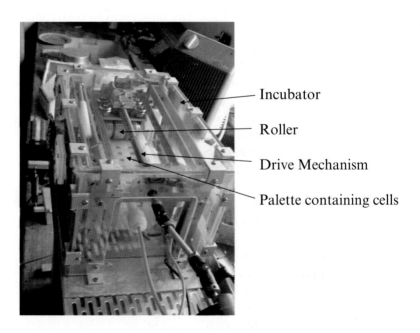

— Incubator

— Roller

— Drive Mechanism

— Palette containing cells

**Figure 7.**  The "Rock-and-Stroller son of the Rock-and-Roller" bioreactor at Oswestry, UK.

## Mature or immature construct?

The duration of cultivation appears to affect both the structure and integratability of the neo-cartilage. With increased cultivation time, the tissue more closely approximated articular cartilage both in biomechanical property and composition (collagen and glycosaminoglycan content).[77,85] However, immature constructs appeared to integrate better than more mature constructs or explants. A high concentration of undifferentiated cells may be important for integration.[76] Integration with progressive cell and cartilaginous tissue proliferation was seen in immature constructs compared to only secretion of extracellular matrix components in mature constructs/native tissue.[86] Higher compressive stiffness was seen with more mature constructs. The conditions and duration of culture will therefore be important considerations and will vary depending on individualized goals.

Neither of these approaches mimics fetal development where a particular biological and mechanical environment is often combined with a sequence of tissues development. It is logical to study the fetal process and apply the controls that are found there.

## Factors affecting repair tissue development

In the embryo, a central cartilaginous core rod can be observed in the developing limb. This is digested and replaced by vascular and marrow tissues, but the cartilage at each end later differentiates to form joint surfaces and hyaline cartilage. Bone formation at the collar "periosteal" zone of the cartilage rods is intimated with the presence of vasculature such that in the absence of vasculature, the same progenitor cells usually differentiate into cartilage.[60] High molecular weight hyaluronan, has been shown to be chondroinductive[61] as well as anti-angiogenic.[62] Mimicking the embryological milieu, Kujawa *et al.*[61] used tissue engineering scaffolds made of porous hyaluronan with marrow mesenchymal progenitor cells in full-thickness femoral condyle defects to produce a chondrocyte population. The dissolution of hyaluronan induced the basal layer of cells to form hypertrophic chondrocytes, which apoptosed. The basal layer was then invaded by vessels brought in by angiogenic hyaluronan oligomers[63] and formed bone, but the surface remained and functioned in weight-bearing for at least 24 weeks. The site where MSCs are simply placed into does not trigger the initiating sequences itself, but induction agents are needed which in this case also formed part of the delivery system. Hyaluronan appears to be a good scaffold for chondrocytes and is used clinically as Hyaff-11. It is already in clinical use with good results in a significant proportion of patients.

The cell-scaffold-bioreactor model of chondrogenesis has been studied both *in vitro*[64] and *in vivo*.[65] Chondrogenic cells are seeded onto scaffolds and cultured in a milieu that promotes formation of the target tissue construct. *In vitro*, chondrogenesis is affected by the activity and age of the cells (embryonic, neonatal, immature or adult); differentiation state (precursor or phenotypically mature); method of preparation (selection, expansion, gene transfer); and cell density. A high concentration of biosynthetically active cells present in immature constructs was associated with rapid integration with native cartilage explants.[66] Gene-transfer of human insulin-like growth factor into bovine articular chondrocytes improved the synthesis rates and wet weight fractions of constituents and the biomechanical properties of cartilage.[67] A high and spatially uniform cell density was associated with improved chondrogenesis on differentiated chondrocytes and bone marrow-derived progenitor cells. Another novel approach is the induction of chondrogenic differentiation through genetic modulation. Epidermal growth factor (EGF), basic fibroblast growth factor (bFGF) and insulin-like growth factor 1 (IGF-1), have stimulating effects on articular chondrocyte metabolism *in vitro*, but have not been shown to have any effect on cartilage defect healing *in vivo*.[68−69]

Bone morphogenic proteins (BMPs) are characterized as members of the TGF-beta superfamily as they share seven highly conserved carboxyl-terminal cysteines. Recombinant human bone morphogenic protein-2 (rhBMP-2) appears to be involved with growth and differentiation of mesenchymal cells to chondroblasts and osteoblasts in developing limb buds, and to enhance the production of articular cartilage matrix *in vivo*.[70]

As early as 1986, TGF-beta has been identified as a factor involved in the induction of chondrogenesis in primitive rat mesenchymal cells.[71] Its use in human studies has been observed.[72] Widespread application of TGF-beta proteins has been limited by its short half-life, and high cost. The short half-life implies short term effect but this has not been fully investigated.

rhBMP-2 has been used with collagen type-1 sponges to treat cartilage defects in rabbits. Although the mean residence time of the rhBMP-2 was eight days with an elimination half-life of 5.6 days, there appeared to be better repair tissue in rhBMP-2 treated defects than in empty controls even at one year. The treated defects were 70% of the thickness of normal cartilage; had better histological features, including chondrocyte-cell numbers; and significantly less collagen type-1. The authors found that in no instance was the repair tissue identical to articular cartilage but demonstrated the longevity of rhBMP-2-induced improvement in cartilage repair.[73–74]

Lee *et al.*[68] tested a cell-mediated TGF-beta1 gene therapeutic approach for regenerating hyaline cartilage in rabbit knee chondral defects. NIH 3T3 fibroblasts containing a transgene, TGF-beta1 were injected into these rabbit knees with surgical partial cartilage defect (3 mm × 6 mm × 1–2 mm deep). At 3 weeks post-injection, most of the defects were filled with a repair tissue and at 6 weeks, the defect was completely repaired. The control defects in contralateral knees injected with NIH 3T3-neo fibroblasts (that had not been transfected with recombinant TGF-beta1 retroviruses) had no new hyaline cartilage. Histological examination of the tissue at 6 weeks showed fully differentiated chondrocytes with morphology and staining patterns similar to adjacent normal hyaline cartilage. The repair cartilage was hyaline and well integrated with the surrounding cartilage. The results indicated that TGF-beta1 produced from the injected cells stimulated the chondrocytes and/or chondrocyte precursers in the surrounding normal cartilage to proliferate and/or differentiate into normal chondrocytes. The authors admit that the mechanical properties of the repair tissue had not been tested but felt that this technique of cell-mediated gene therapy achieved long term production of cytokines, thereby resolving the problem of its short half-life. This approach also enabled treatment by injection rather than by operative placement.

## Do Cells Survive in the Knee?

In the goat knee with simulated arthritis, intraarticularly injected stem cells could be recovered from synovial fluid in viable form and were noted to colonize soft tissue surfaces. Where total meniscectomy had been performed, a neotissue had formed with hyaline-like appearance and focal areas of type 2 collagen similar to developing rabbit meniscus.[87] It would appear that regeneration of a meniscus might be one of the easier tissues to reproduce.

## Do Stem Cells Form Cartilage or Bone?

One of Caplan's classic studies was to demonstrate the ability of stem cells to form cartilage and bone. Autologous marrow or periosteal MSCs with a 0.15% type 1 collagen gel delivery vehicle were implanted into prepared osteochondral medial femoral condyle defects of adult New Zealand White rabbits. The animals were sacrificed at 1, 2, 3 weeks and 1, 3 and 6 months. Regeneration of subchondral bone up to the natural bone-cartilage junction and resurfacing of the cartilage defect were observed. Instead of MSCs at the base forming bone and MSCs at the surface forming cartilage, it seemed that the embryonic sequence was recreated — the MSCs formed embryonic cartilage throughout initially, then the chondrocytes at the base progress to form hypertrophic chondrocytes that were replaced by vasculature, marrow and fresh osteochondral progenitors that formed bone on the remnants of the calcified hypertrophic cartilage. It was postulated that the top cartilage remained articular cartilage, probably from exposure to synovial factors. However, a lack of integration of the repair tissue of around 20–100% of its circumference was noted. 0.005 mg/ml trypsin soaked into a piece of filter paper that was laid into the defect for 5 min before MSCs delivered in type 1 collagen gel were implanted, was found to be optimal for maximal integration of host and repair tissue.[88–89]

## Comparative Studies

Hui *et al.*[90] reported a comparative study where in 20-week-old NZW rabbits, full thickness articular cartilage defects were created and the cases divided into 4 groups treated with: 1) cultured chondrocytes; 2) cultured MSCs; 3) periosteal graft; and 4) mosaicplasty. Gross, histologic and biomechanical analysis at 36 weeks postoperatively showed that cultured chondrocytes and MSCs had comparable repair, whereas mosaicplasty did well initially and periosteal graft did less favorably. Another study comparing MSCs with committed chondrocytes in rabbits was done by Goldberg and Caplan.[91] Both cell types produced repair of the defects with tissue resembling hyaline cartilage. However, MSC repair tissue more closely resembled hyaline cartilage.

## Can Human MSCs Produce Chondrocytes/Cartilage?

Although stem cells have the ability to regenerate new tissue, the theory that depletion of stem cells may be present in degenerative disease has had some evidence in osteoarthritis.[92] Samples taken from patients undergoing total joint replacement showed reduced proliferation and lower differentiation rates *in vitro*, compared to samples from healthy donors, but it is unknown whether this observation is primary or secondary to the disease process.

Mackay *et al.*[93] showed that chondrogenesis could be induced by culturing human MSCs (hMSCs) in micromass pellets in the presence of a medium that included 100 nM dexamethasone and 10 ng/ml TGF Beta3. Bone marrow hMSCs from 60 healthy adult volunteers aged 14–55 years were trypsinized, washed in serum containing medium and suspended in serum-free chondrogenic medium, centrifuged, incubated and left to sediment and form pellet masses at the bottom of the tube. Cell survival was determined by florescence microscopy; DNA content by dyeing and fluorometric assay; and antibodies specific for ECM proteins, collagen I and II used for determining chondrogenic differentiation. The hMSCs yielded positive staining for anionic proteoglycans, as well as type II collagen. No correlation between donor age or gender and the extent of chondrogenic differentiation was seen. Increased chondrogenic activity of hMSCs in high-glucose medium was observed and seemed to be due to increased viability of hMSCs. The hMSCs could be further differentiated to the hypertrophic state by the addition of 50 nM thyroxine, the withdrawal of TGF beta3 and the reduction of dexamethasone concentration to 1 nM. A part of the study involved comparison with human articular chondrocytes from a 15-year-old boy's tibial condyle. The chondrocytes were prepared under conditions that promoted hMSC chondrogenesis and accumulated mass and produced type II collagen considerably faster than the hMSC pellets did. Cultured MSCs do not elaborate a chondrogenic ECM in the absence of dexamethasone and TGF beta and therefore require more time or additional stimuli than dedifferentiated chondrocytes. The effects of the supplements to the medium used were discussed in the paper, with high levels of glucose maybe altering the metabolism of cells in pellet culture so as to lessen their susceptibility to apoptosis. The potential use of hypertrophied chondrocyte differentiation in the formation of bone in the treatment of full thickness defects was also raised.

Similarly human iliac crest bone marrow MSCs have been found to differentiate into chondrocytes and produce collagen II and X and proteoglycans in a chondrogenic medium of Dulbecco's modified Eagles medium, with high glucose, 10 ng/ml TGF beta-I, 1% ITS –Prefix medium, 80 $\mu$M ascorbic acid and 100 nM dexamethasone.[94] Cells from human bone marrow have been proven to differentiate *in vitro* into chondrocytes, osteocytes and fat cells.[95]

## Effect of Donor Age and Limitations of Number of Cell Expansions

To determine the effects of donor age and number of cell expansions, human knee cartilage from donors aged 10 months, 6-, 9- and 25-years were studied by Huckle *et al.*[96] Constructs produced from cells propagated to passage 3 contained 20–40% S-GAG but a lower content (15–30%) from cells harvested between passages 4 and 6. The S-GAG content was highest from younger donors. Histology of articular cartilage cells showed predominately a hyaline-like cartilage matrix from donors aged 2 months to 25 years when expanded to passage 3, but with younger donors only in cells expanded to passage 5.

## Clinical Studies

Thus far, only one clinical study using MSCs in the treatment of osteoarthritis has been reported from Japan by Wakitani *et al.*[97] In an elegant study, he and his colleagues implanted autologous bone marrow mesenchymal cells in 12 of the 24 patients in their study group who underwent high tibial osteotomy for medial unicompartmental osteoarthritis. The two groups were randomized and there were no differences in age or disease severity between the cell-treated and control (cell-free) groups. The defects of the cell-treated group were abraded and perforated and received the cells in a gel-cell composite, sutured over with autologous periosteum. The control group received the collagen gel-sheet cell-free, periosteal cover. Six from the control group received multiple perforations and the other six received abrasion with multiple perforation. At mean follow-up of 16 months, the symptoms of both groups had significantly improved using the Hospital for Special Surgery knee rating scale, from 65 to 81.3 in the cell-group and 66.3 to 79.2 in the cell free-group. First and second look arthroscopies were undertaken at a mean of 6.7 weeks and 42 weeks post-op, coinciding with removal of the osteotomies' Steinmann pins and staples, respectively. In the cell-treated group, at first-look in three patients where the periosteum was partially detached, the observed defects were filled with opaque white to pink tissue, at the same level as normal cartilage but softer. At second-look arthroscopy, in the nine cell-treated patients who underwent this, cartilage defects were covered with a white tissue with metachromatasia in almost all areas and in some parts, hyaline-like cartilage histologically. With the control group, the repair tissue was more irregular, red-to-yellow in color at first look, and white at second look, but metachromatasia was weak when examined histologically. An arthroscopic and histological score out of 20 was used, with the cell-treated group scoring 15.4 and the control group 10.0 (multiple perforation) and 11.3 (abrasion and multiple perforation) at second look. Although the clinical scores were almost the same between the two groups, arthroscopic and histological evaluations favored the cell-treated group. Further follow-up is needed but the study demonstrated that MSCs were capable of regenerating repair tissue for large chondral defects. The advantages of this new procedure were cited as:

1) ease of cell collection,
2) stem cells can be proliferated without losing their capacity to differentiate into cartilage, therefore allowing application to large articular defects.

Other potential advantages not demonstrated by the study include the ability of MSCs to regenerate not only the cartilage surface but also the underlying subchondral bone.[89] Osteochondral defects may be particularly suitable for such treatment rather than bone grafting plus autologous chondrocyte implantation.

## Can Allogeneic Stem Cells Be Used?

Horwitz *et al.*[98] used allogeneic unmanipulated HLA identical or single-antigen mismatched sibling bone marrow transplants in three patients with osteogenesis imperfecta to markedly reduce their fracture rate and skeletal growth. Proof that new osteoblasts were derived from grafted bone marrow was made by identification of Y chromosome in female recipient patient 1 and DNA polymerism in patient 2. This is surprising as the stem cells do have HLA type 2 antigens, but these are the least antigenic type.

Cultured allogeneic embryonal chondrocytes were implanted into defects in an avian model and the results were good.[99–100] As human embryonal tissue is difficult to obtain and ethical problems are raised, MSCs from bone marrow represent an alternative source of pluripotential stem cells. These cells have been cultured *in vitro* and induced to form cartilage prior to implantation into the chondral defects in rabbits.[89,101] Goat articular cartilage has also been resurfaced by chondrocytes derived from bone marrow.[102] In this study, bone marrow was aspirated from the iliac crest of 12 adult white goats, and chondrogenesis of the MSCs was induced. Four goats were transplanted with allogeneic implants, four with autogeneic implants; as all the implants had been impregnated in a hyaluronic acid preparation, the four goats which served as controls had their defects treated with hyaluronic acid only. Macroscopic evaluation indicated that the defects treated with autogeneic implants all filled with a hyaline-like tissue, while the defects from the allogeneic group were only partially filled, and the control groups were not filled at all. This would suggest that allogeneic MSCs can induce tissue regeneration but not as successfully as autologous MSCs.

## Summary and Conclusions

A simple summary of the situation is that there is now a wide range of treatments for a large proportion of symptomatic chondral defects and osteoarthritis. The chondrocyte cell treatments are now well established with several randomized clinical trials reporting satisfactory outcomes at two years. However, longer term reports will be important. Histology of ACI in patients shows predominantly fibrocartilage, but areas of cartilage typical of articular cartilage are frequently observed, especially at the two year point. Development of the technique to produce higher quality histology and at an earlier time-point may follow adjustments in the technique, or may require a new approach.

Adjustments are being made by the use of chondrocytes that have not been in extended subculture *in vivo*. Using a different matrix is another approach and hyaluronan gels or collagen type II carriers may be improvements. More comparative studies are needed.

Currently much of the effort is directed at the healing of chondral defects. In osteoarthritis, large portions of the joint surface is involved. ACI and similar approaches will not be able to repair or regenerate whole joints. It is natural to try and predict how best tissue engineering will develop to meet the needs of tissue regeneration. In practice, it is probable that a variety of methods will be used. The most progress may be made through evolution and development. One approach will be from further developments of existing cellular engineering, and the other through joints grown in bioreactors. It is not inconceivable to fabricate and seed scaffolds that will serve as osteochondral grafts for filling larger defects and eventually resurface whole joints.

The results of ACI have shown that we are still a long way from regenerating articular cartilage. Martin and Buckwalter[52] have raised concerns about the use of aging chondrocytes as a graft. Perhaps using stem cells may result in a more long-lasting repair. Currently, MSCs have been shown to be as good if not better in the repair of chondral lesions.[89,90] ESCs may have the potential of giving the best repair in terms of the phenotype as well as longevity of the regenerated cartilage. However, there are ethical and scientific issues involved with the use of ESCs in humans at the moment. If and when these issues are resolved, the potential for ESCs in cartilage repair and regeneration may be tremendous.

Much needs to be done in this very exciting area of research. The answers will eventually come from collaborations amongst scientists, engineers and clinicians.

# REFERENCES

1. Hunter W. (1743) On the structure and diseases of articulating cartilages. *Philos Trans R Soc Lond* **42B**: 514–521.
2. Lefkoe TP, Trafton PG, Dennehy DT, *et al.* (1992) A new model of articular step-off for the study of post-traumatic arthritis. *Proceedings from the 38th Annual Meeting of the Orthopaedic Research Society*, Washington DC, p. 207.
3. Messner K, Maletius W. (1996) The long-term prognosis for severe damage to weight-bearing cartilage in the knee: A 14-year clinical and radiographic follow-up in 28 young athletes. *Acta Orthop Scand* **67**(2): 165.
4. Mainil-Varlet P, Aigner T, Brittberg M. (2003) International Cartilage Repair Society. Histological assessment of cartilage repair: A report by the Histology Endpoint Committee of the International Cartilage Repair Society (ICRS). *J Bone Joint Surg Am* **85A**(2): 45.
5. Lindahl A, Brittberg M, Peterson L. (2003) Cartilage repair with chondrocytes: Clinical and cellular aspects. *Novartis Found Symp* **249**: 175; discussion 186, 234, 239.
6. Hollander AP, Dickinson SC, Sims TJ, *et al.* (2003) Quantitative analysis of repair tissue biopsies following chondrocyte implantation. *Novartis Found Symp* **249**: 218; discussion 229, 234, 239.
7. Sir James Paget, Bart MD. (1969) The classics. II. Healing of cartilage. *Clin Orthop* **64**: 7–8.
8. Naudie D, Bourne RB, Rorabeck CH. (1999) Survivorship of the high tibial valgus osteotomy. A 10- to -22-year follow-up study. *Clin Orthop* **18**.
9. Akizuki S, Yasukawa Y, Takizawa T. (1997) Does arthroscopic abrasion arthroplasty promote cartilage regeneration in osteoarthritic knees with eburnation? A prospective study of high tibial osteotomy with abrasion arthroplasty versus high tibial osteotomy alone. *Arthroscopy* **13**: 9.

10. Bergenudd H, Johnell O, Redlund-Johnell I, Lohmander LS. (1992) The articular cartilage after osteotomy for medial gonarthrosis. Biopsies after 2 years in 19 cases. *Acta Orthop Scand* **63**: 413.

11. Schultz W, Gobel D. (1999) Articular cartilage regeneration of the knee joint after proximal tibial valgus osteotomy: A prospective study of different intra- and extra-articular operative techniques. *Knee Surg Sports Traumatol Arthrosc* **7**: 29.

12. Wakabayashi S, Akizuki S, Takizawa T, Yasukawa Y. (2002) A comparison of the healing potential of fibrillated cartilage versus eburnated bone in osteoarthritic knees after high tibial osteotomy: An arthroscopic study with 1-year follow-up. *Arthroscopy* **18**: 272.

13. Furukawa T, Eyre DR, Koide S, Glimcher MJ. (1980) Biochemical studies on repair cartilage resurfacing experimental defects in the rabbit knee. *J Bone Joint Surg* **62**(1): 79.

14. Shapiro F, Koide S, Glimcher MJ. (1993) Cell origin and differentiation in the repair of full thickness defects of articular cartilage. *J Bone Joint Surg* **75A**: 532.

15. Hubbard MJ. (1996) Articular debridement versus washout for degeneration of the medial femoral condyle. A five-year study. *J Bone Joint Surg Br* **78**(2): 217.

16. Muller B, Kohn D. (1999) Indication for and performance of articular cartilage drilling using the Pridie method. *Orthopade* **28**(1): 4.

17. Ficat RP, Ficat C, Gedeon P, Toussaint JB. (1979) Spongialization: A new treatment for diseased patellae. *Clin Orthop* **144**: 74.

18. Steadman JR, Rodkey WG, Briggs KK. (2002) Microfracture to treat full-thickness chondral defects: Surgical technique, rehabilitation, and outcomes. *J Knee Surg* **15**(3): 170–176.

19. Sledge SL. (2001) Microfracture techniques in the treatment of osteochondral injuries. *Clinic Sports Med* **20**: 365–377.

20. Steadman JR, Rodkey WG, Briggs KK, Rodrigo JJ. (1999) The microfracture technic in the management of complete cartilage defects in the knee joint. *Orthopade* **28**(1): 26.

21. Knutsen G, Engebretsen L, Ludvigsen TC, *et al.* (2004) Autologous chondrocyte implantation compared with microfracture in the knee. A randomized trial. *J Bone Joint Surg Am* **86A**(3): 45.

22. Johnson LL. (1986) Arthroscopic abrasion arthroplasty. Historical and pathological perspective: present status. *Arthroscopy* **2**: 54–69.

23. Minas T, Nehrer S. (1997) Current concepts in the treatment of articular cartilage defects. *Orthopaedics* **29**: 525–538.

24. Buckwalter JA, Lohmander S. (1994) Operative treatment of osteoarthrosis. Current practice and future development. *J Bone Joint Surg Am* **76**(9): 1405.

25. Buckwalter JA, Mankin HJ. (1997) Articular cartilage. Part II: Degeneration and osteoarthritis, repair, regeneration and transplantation. *J Bone Joint Surg* **79A**: 612–632.
26. Brittberg M, Sjogren-Jansson E, Lindahl A, Peterson L. (1997) Influence of fibrin sealant (Tisseel) on osteochondral defect repair in the rabbit knee. *Biomaterials* **18**: 235–242.
27. Mahomed MN, Beaver RJ, Gross AE. (1992) The long term success of fresh small fragment osteochondral allografts used for intra-articular post-traumatic defects in the knee joint. *Orthopaedics* **15**: 1191–1199.
28. Outerbridge KK, Outerbridge AR, Outerbridge AR. (1995) The use of a lateral patellar autologous graft for the repair of a large osteochondral defect in the knee. *J Bone Joint Surg* **77A**: 65–72.
29. Bobic V. (1996) Arthroscopic osteochondral autograft transplantation in anterior cruciate ligament reconstruction. A preliminary clinical study. *Knee Surg Sports Traumatol Arthrosc* **3**: 262–264.
30. Hangody L, Feczko P, Bartha L, *et al.* (2001) Mosaicplasty for the treatment of articular defects of the knee and ankle. *Clin Orthop* **391**: S328–S336.
31. Bentley G, Biant LC, Carrington RW, *et al.* (2003) A prospective, randomised comparison of autologous chondrocyte implantation versus mosaicplasty for osteochondral defects in the knee. *J Bone Joint Surg Br* **85**(2): 223.
32. O'Driscoll SW, Fitzsimmons JS. (2001) The role of periosteum in cartilage repair. *Clin Orthop* **391**: S190.
33. Zarnett R, Delaney JP, Driscoll SW. (1987) Cellular origin and evolution of neochondrogenesis in major full-thickness defects of a joint surface treated by free autogenous periosteal grafts and subjected to continuous passive motion in rabbits. *Clin Orthop* **222**: 267.
34. Zarnett R, Salter RB. (1989) Periosteal neochondrogenesis for biologically resurfacing joints: Its cellular origin. *Can J Surg* **32**(3): 171.
35. Ito Y, Fitzsimmons JS, Sanyal A, *et al.* (2001) Localization of chondrocyte precursors in periosteum. *Osteoarthritis Cartilage* **9**(3): 215.
36. Homminga GN, Bulstra SK, Bouwmeester PS, van der Linden AJ. (1990) Perichondral grafting for cartilage lesions of the knee. *J Bone Joint Surg* **72B**: 1003.
37. Brittberg M, Lindahl A, *et al.* (1994). Treatment of deep cartilage defects in the knee with autologous chondrocyte transplantation. *N Engl J Med* **331**: 889.
38. Minas T. (1999) The role of cartilage repair techniques, including chondrocyte transplantation, in focal chondral knee damage. *Instr Course Lect* **48**: 629.
39. Richardson JB, Caterson B, Evans EH, *et al.* (1989) Repair of human articular cartilage after implantation of autologous chondrocytes. *J Bone Joint Surg Br* **81**(6): 1064.

40. Henderson IJ, Tuy B, Connell D, *et al.* (2003) Prospective clinical study of autologous chondrocyte implantation and correlation with MRI at three and 12 months. *J Bone Joint Surg Br* **85**(7): 1060.

41. Horas U, Pelinkovic D, Herr G, *et al.* (2003) Autologous chondrocyte implantation and osteochondral cylinder transplantation in cartilage repair of the knee joint. A prospective, comparative trial. *J Bone Joint Surg Am* **85A**(2): 185.

42. Horas U, Schnettler R, Pelinkovic D, *et al.* (2000). Knorpelknochentransplantation versus autogene Chondrocytentransplantation, Eine prospective vergleichende klinische Studie. *Chirug* **71**: 1090.

43. Smith GD, Richardson JB, Brittberg M, *et al.* (2003) Autologous chondrocyte implantation and osteochondral cylinder transplantation in cartilage repair of the knee joint. *J Bone Joint Surg Am* **85A**(12): 2487.

44. Whitaker JP, Smith G, Makwana N, *et al.* (2005) Early results of autologous chondrocyte implantation in the talus. *J Bone Joint Surg* **87B**(2): 179–183.

45. Haddo O, Mahroof S, Higgs D, *et al.* (2004) The use of chondrogide membrane in autologous chondrocyte implantation. *Knee* **11**(1): 51–55.

46. Behrens P, Ehlers EM, Kochermann KU, *et al.* (1999) New therapy procedure for localised cartilage defects. Encouraging results with autologous chondrocyte implantation. *MMW Fortschr Med* **141**: 49.

47. Russlies M, Behrens P, Wunsch l, *et al.* (2002) A cell seeded biocomposite for cartilage repair. *Ann Anat* **184**: 317.

48. Grigolo B, Roseti L, Fiorini M, *et al.* (2002) Evidence of redifferentiation of human chondrocytes grown on a hyaluronan based biomaterial (HYAFF11): Molecular, immunohistochemical and ultrastructural analysis. *Biomaterials* **23**: 1187.

49. Katsube K, Ochi M, Uchio Y, *et al.* (2000) Repair of articular cartilage defects with cultured chondrocytes in Atelocollagen gel. Comparison with cultured chondrocytes in suspension. *Arch Orthop Trauma Surg* **120**(3–4): 121–127.

50. Grigolo B, Roseti L, Fiorini M, *et al.* (2001) Transplantation of chondrocytes seeded on a hyaluronan derivative (hyaff-11) into cartilage defects in rabbits. *Biomaterials* **22**(17): 2417.

51. Muller P, Lemmen C, Gay S, *et al.* (1976) In *Extracellular Matrix Influences on Gene Expression*, eds. Slavkin HC and Gruelich RG, San Francisco: Academic, pp. 293–302; and Mayne R, Vail MS, Mayne PM, Muller EJ. (1976) *PNAC USA* **73**: 1674.

52. Martin JA, Buckwalter JA. (2002) Aging, articular cartilage chondrocyte senescence and osteoarthritis. *Biogerontology* **3**(5): 257.

53. Von der Mark K, Gauss V, Von der Mark H, Muller P. (1977) Relationship between cell shape and type of collagen synthesized as chondrocytes lose their cartilage phenotype in culture. *Nature* **267**: 531.

54. Kramer J, Hegert C, Guan K, *et al.* (2000) Embryonic stem-cell derived chondrogenic differentiation *in vitro*: Activation of BMP-2 and BMP-4. *Mech Dev* **92**: 193.

55. Wakitani S, Takaoka K, Hattori T, *et al.* (2003) Embryonic stem cells injected into the mouse knee joint form teratomas and subsequently destroy the joint. *Rheumatology* **42**: 162.

56. Caplan AI. (1991) Mesenchymal stem cells. *J Orthop Res* **9**: 641.

57. Jackson DW, Simon TM. (1999) Tissue engineering principles in orthopaedic surgery. *Clin Orthop* **367**: S31.

58. Glowacki J. (2000) *In vitro* engineering of cartilage. *J Rehabil Res Dev* **37**(2): 171.

59. Huang Q, Goh JC, Hutmacher DW, Lee EH. (2002) *In vivo* mesenchymal cell recruitment by a scaffold loaded with transforming growth factor beta1 and the potential for *in situ* chondrogenesis. *Tissue Eng* **8**(3): 469.

60. Goshima J, Goldberg, Caplan AI. (1991) The origin of bone formed in composite grafts of porous calcium phosphate ceramic loaded with marrow cells. *Clin Orthop* **269**: 274.

61. Kujawa MJ, Carrino DA, Caplan AI. (1986) Substrate-bonded hyaluronic acid exhibits a size-dependent stimulation of chondrogenic differentiation of stage 24 limb mesenchymal cells in culture. *Dev Biol* **114**: 519.

62. Feinberg RN, Beebe DC. (1983) Hyaluronate in vasculogenesis. *Science* **220**: 1177.

63. Deed R, Rooney P, Kumar P, *et al.* (1997) Early response gene-signaling is induced by angiogenic oligosaccharides of hyaluronan in endothelial cells. *Int J Cancer* **71**: 251.

64. Freed LE, Vunjak-Novakovic G. (2000) Tissue engineering of cartilage. In *The Biomedical Engineering Handbook*, ed. Bronzino JD, Boca Raton: CRC Press, Chap. 124, pp. 1–26.

65. Schaefer D, Martin I, Jundt G, *et al.* (2002) Tissue engineered composite for the repair of large osteochondral defects. *Arthritis Rheum* **46**: 2524.

66. Obradovic B, Martin I, Padera RF, *et al.* (2001) Integration of engineered cartilage. *J Orthop Res* **19**: 1089.

67. Madry H, Padera R, Seidel J, *et al.* (2002) Gene transfer of a human insulin-like growth factor I cDNA enhances tissue engineering of cartilage. *Hum Gen Ther* **13**: 1621.

68. Lee KH, Song SU, Hwang TS, *et al.* (2001) Regeneration of hyaline cartilage by cell mediated gene therapy using transforming growth factor Beta1-producing fibroblasts. *Hum Gene Ther* **12**: 1805.

69. Neidel JJ. (1992) Keine Verbesserung der Gelenkknorpelheilung nach trauma durch Gabe von insulinartigem Wachstumsfaktor 1, epidermalem Wachstumfaktor und Fibroblasten-Wachstumfaktor beim Kanichen. *Z Orthop Ihre Grenzgeb* **130**: 73.

70. Sailor LZ, Hewick RM, Morris EA. (1996) Recombinanat human bone morphogenic protein –2 maintains the articular chondrocyte phenotype in long term culture. *J Orthop Res* **14**: 937.

71. Seyedin SM, Thompson AY, Bentz H, *et al.* (1986) Cartilage-inducing factor-A. Apparent identity to transforming growth factor-beta. *J Biol Chem* **261**: 5693.

72. Johnstone B, Yoo J, Barry FP. (1996) *In vitro* chondrogenesis of bone-marrow derived mesenchymal cells. *Trans Orthop Res Soc* **21**: 65.

73. Sellers RS, Peluso D, Morris EA. (1997) The effect of recombinant human bone morphogenic protein –2 on the healing of full thickness defects of articular cartilage. *J Bone Joint Surg* **79A**: 1452.

74. Sellers RS, Zhang R, Glasson SS, *et al.* (2000) Repair of articular cartilage defects one year after treatment with recombinant human bone morphogenic protein –2 (rhBMP2). *J Bone Joint Surg* **82A**(2): 151.

75. Cook SD, Patron PP, Salkeed SL, Rueger DC. (2003) Repair of articular cartilage defects with osteogenic protein-1 (BMP-7) in dogs. *J Bone Joint Surg* **85A**(3): 116.

76. Vunjak-Novakovic G. (2002) The fundamental of tissue engineering: Scaffolds and bioreactors. In *Tissue Engineering of Cartilage and Bone.*, eds. Bock G and Goode J, Novartis Foundation Symposium 249, John Wiley & Sons Publication, p. 34–46.

77. Vunjak-Novakovic G, Martin I, Obradovic B, *et al.* (1999) Bioreactor cultivation conditions modulate the composition and mechanical properties of tissue-engineered cartilage. *J Orthop Res* **17**: 130.

78. Buschmann MD, Gluzband YA, Grodzinsky AJ, *et al.* (1992) Chondrocytes in agarose culture synthesize a mechanically functional extracellular matrix. *J Orthop Res* **10**: 745.

79. Mauck RL, Soltz MA, Wang CC, *et al.* (2000) Functional tissue engineering of articular cartilage through dynamic loading of chondrocyte-seeded agarose gels. *J Biomech Eng* **122**: 252.

80. Murphy JM, Barry FP. (2000) Chondrogenic differentiation of mesenchymal stem cells on matrices of hyaluronan derivatives. In *New Frontiers in Medical Sciences: Redefining Hyaluronan*, eds. Abatangelo G, Weigel PH, Amsterdam: Elsevier, pp. 247–254.

81. Solchaga LA, Yoo JU, Lundberg M, *et al.* (2000) Hyaluronan-based polymers in the treatment of osteochondral defects. *J Orthop Res* **18**: 773.

82. Ponticello MS, Schinagl RM, Kadiyala S, Barry FP. (2000) Gelatin-based resorbable sponge as a carrier matrix for human mesenchymal stem cells in cartilage regeneration therapy. *J Biomed Mater Res* **52**: 246.

83. Li L, Hui JH, Goh JC, *et al.* (2004) Chitin as a scaffold for mesenchymal stem cells transfers in the treatment of partial growth arrest. *J Paediatr Orthop* **24**(2): 205.

84. Freed LE, Vunjak-Novakovic G. (2000) Tissue engineering bioreactors. In *Principles of Tissue Engineering*, 2nd edn, eds. Lanza R, Langer R, Vacanti J, San Diego: Academic Press, Chap. 13, pp. 143–156.

85. Freed LE, Hollander AP, Martin I, *et al.* (1998) Chondrogenesis in a cell-polymer-reactor system. *Exp Cell Res* **240**: 58.

86. Obradovic B, Martin I, Padera RF, *et al.* (2001) Integration of engineered cartilage. *J Orthop Res* **19**: 1089.

87. Barry FP. (2002) Mesenchymal stem cell therapy in joint disease. In *Tissue Engineering of Cartilage and Bone*, eds. Bock G, Goode J, Novartis Foundation Symposium 249, John Wiley & Sons Publication, pp. 86–966.

88. Caplan AI, Elyaderani M, Mochizuki Y, *et al.* (1997) Principles of cartilage repair and regenration. *Clin Orthop* **342**: 254.

89. Wakitani S, Goto T, Young RG, *et al.* (1994) Mesenchymal cell-based repair of large full thickness defects of articular cartilage. *J Bone Joint Surg* **76A**: 579.

90. Hui JH, Chen F, Thambyah A, Lee EH. (2004) Treatment of chondral lesions in advance osteochondritis dissecans: A comparative study of the efficacy of chondrocyte mesenchymal stem cells, periosteal graft, and mosaicplasty (osteochondrall autograft) in animal models. *J Paediatr Orthop* **24**(4): 427.

91. Goldberg VM, Caplan AI. (1995) Cellular repair of articular cartilage. In eds. Patrick CW Jr, Mikos AG, Goldberg VM. Osteoarthritic Disorders Workshop, Monterey, California, April 1994. Rosemont, IL: American Academy of Orthopaedic Surgeons, pp. 357–364.

92. Murphy JM, Dixon K, Beck S, *et al.* (2002) Reduced chondrogenic and adipogenic activity of mesenchymal stem cells from patients with advanced osteoarthritis. *Arthritis Rheum* **46**(3): 704.

93. Mackay AM, Beck SC, Murphy JM, *et al.* (1998) Chondrogenic differentiation of cultured human mesenchymal stem cells. *Tiss Eng* **4**: 415.

94. Naumann A, Dennis J, Staudenmaier R, *et al.* (2002) Mesenchymal stem cells — A new pathway for tissue engineering in reconstructive surgery. *Laryngo-Rhino-Otologie* **81**: 521.

95. Pittenger MF, Mackay AM, Beck SC, *et al.* (1999) Multilineage potential of adult human mesenchymal stem cells. *Science* **284**: 143.

96. Huckle J, Dootson G, Medcalf N, *et al.* Differentiated chondrocytes for cartilage tissue engineering. In *Tissue Engineering of Cartilage and Bone*, eds. Bock G, Goode J, Novartis Foundation Symposium 249, John Wiley & Sons Publication, pp. 103–117.

97. Wakitani S, Imoto K, Yamamoto T, *et al.* (2002) Human autologous culture expanded bone marrow mesenchymal cell transplantation for repair of cartilage defects in osteoarthritic knees. *Osteoarthr Cartil* **10**: 199.

98. Horwitz EM, Prockop DJ, Fitzpatrick LA, *et al.* (1999) Transplantability and therapeutic effects on bone marrow-derived mesenchymal cells in children with osteogenesis imperfecta. *Nature Med* **5**: 309.

99. Itay S, Abramovici A, Nevo Z. (1987) Use of cultured embryonal chick epiphyseal chondrocytes as grafts for defects in chick articular cartilage. *Clin Orthop* **220**: 284.

100. Robinson D, Halperin N, Nevo Z. (1989) Fate of allogeneic embryonal chick chondrocytes implanted orthotopically, as determined by the host's age. *Mech Ageing Dev* **50**(1): 71.

101. Im GI, Kim DY, Shin JH, *et al.* (2001) Repair of cartilage defect in the rabbit with cultured mesenchymal stem cells from bone marrow. *J Bone Joint Surg Br* **83**(2): 289.

102. Butnariu-Ephrat M, Robinson D, Mendes DG, *et al.* (1996) Resurfacing of goat articular cartilage by chondrocytes derived from bone marrow. *Clin Orthop* **330**: 234.

# 23

# Ligament and Tendon Repair with Adult Stem Cells

James Cho Hong Goh* and Hong Wei Ouyang

## Introduction

The composition and ultra-structure of tendon and ligament have been well documented. Similarly, extensive studies on the biomechanics of tendon and ligament have been reported. These studies have shown that tendons and ligaments are subjected to high physiological loads. As such, injuries due to high impact activities are common. Due to the avascular nature of these connective tissues, the healing rate is rather slow. Currently the therapeutic options to treat tendon and ligament injuries are autografts, allografts and prosthetic devices. However, there are many disadvantages in using biological grafts, as well as concerns over the long-term performance of synthetic prostheses.[1,2] Therefore, further research for alternative repair methods are being pursued relentlessly. One of these methods is the use of tissue engineering strategy in tendon and ligament repair and regeneration. A key factor in the tissue engineering approach to tissue repair and regeneration is the availability of appropriate cells. The presence of cells is crucial for tissue formation, especially their proliferation potential, cell-to-cell signaling, bio-molecule production and formation of extracellular matrix.

## Cells Responsible for Natural Tendon Repair

There is still controversy over the identity and location of the cells responsible for collagen synthesis during tendon repair. Some believe that tendon has the necessary cells,[3] while others believe that the cells are recruited from outside of the tendon.[4] It is currently thought that both intrinsic (tenocyte) and extrinsic (neighboring tissue and the systemic circulation)

---

*Correspondence: Department of Orthopaedic Surgery, National University of Singapore, 5 Kent Ridge Road, Singapore 119074.

cells contribute to tendon healing.[5] One recent study which examined the specific ancestry of cells that participate in tendon healing is noteworthy. Glaser *et al.*[6] used transgenic mice to follow cells of a specific origin in the tendon healing by identifying the cell type specific promoters. The study suggests that smooth muscle cells from vessels or pericytes are a major source of responding cells in the fibroproliferative stage of tendon healing, while satellite cells and bone marrow origin cells have no contribution to the fibroproliferative response. On the other hand, questions remain as to whether there is a sufficient pool of connective tissue progenitor cells to be recruited to the wound site. If the cell number is insufficient, the extent to which the reparative process can occur maybe limited. This uncertainty warrants the application of sufficient exogenous cells in tendon and ligament repair.

## The Effect of Initial Cell Number on Tendon Regeneration

Many cell-mediated processes related to the generation of skeletal tissue depend on the number of cells involved, both in the rate and magnitude of the effect. For example, in the *in vitro* experimental model linked to the production of connective tissue, the rate of collagen gel contraction by fibroblasts embedded in the gel is dependent on the number of cells.[7] The extent of fibroblast orientation in cultures grown on collagen gel has been demonstrated to relate directly to the initial cell density.[8] This cell orientation effect has been correlated with the observation of an "organization center" in the culture, the number of which has been suggested to be a direct indicator of morphogenetic capacity at the molecular and cellular level.[9] In addition, cell density-dependent differentiation was demonstrated in the culture of limb bud cells.[10] The cells exhibited osteogenesis and chondrogenesis at higher density cultures. Therefore, the amount of cells initially present at site, will strongly influence the nature of cell-mediated processes involve in tissue formation and the rate at which these developmental and physiological processes occur. If these observations are applied to the reparative processes of skeletal tissue, it seems that some minimum threshold of cell number may be required at the repair site before formation of normal neo-tissue can occur. Muschleret *et al.*[11] estimated that for one cubic centimeter of new bone formation, one would require 70 million osteoprogenitors. So, it seems to suggest that higher cells density at seeding will influence skeletal tissue formation. This concept can be extended to tendon and ligament regeneration. There must be a minimum threshold quantity of cells that is required at the repair site for normal neo-tissue formation.[12] As such, exogenous cells are essential in facilitating tissue regeneration.

## Fibroblast for Tendon and Ligament Regeneration

There are a number of cell sources that are potential venues for cell-mediated tissue regeneration. These can be classified into either parenchymal cells (e.g. fibroblasts from tendon); or progenitor cells (e.g. bone marrow stromal cells) according to the cell potential. These cell types can be either autologous cells (e.g. cell from one's own body) or allogeneic cells (e.g. cells from a donor). For parenchymal cell source, several groups[13,14,15,16] have used fibroblast-seeded collagen scaffold for tendon/ligament regeneration. The viability and proliferation of the fibroblasts were investigated on the scaffolds. Lin *et al.*[17] cultured fibroblasts on PGA fiber scaffolds *in vitro* and achieved tissue formation after five weeks of culture. The *in vitro* studies exhibited the possibility of engineering tendon or ligament analogs. However, the functional properties of these tendon/ligament analogs were not evaluated. The *in vivo* studies by Bellincampi *et al.*[16] showed that fibroblast-seeded collagen scaffolds were viable for at least eight weeks after re-implantation in the knee joint or subcutaneous tissues of the donor rabbits. However, the investigators did not observe any tendon or ligament-like tissue formation. Cao *et al.*[18] isolated fibroblasts from newborn calves and seeded the cells onto non-woven meshes of PGA fibers. After one week of culture, the cell/PGA composite were implanted subcutaneously in nude mice for up to 10 weeks. Histology at 10 weeks showed linear and parallel organization of collagen bundles throughout the samples and the tensile strength was two-thirds of the natural tendon. These promising findings serve as a basis for further efforts to improve tendon and ligament tissue engineering. Further studies will be needed to determine whether adult fibroblasts can achieve comparable results. Generally, large numbers of parenchymal cells are unavailable. This is because the donor source is limited. Furthermore, parenchymal cells are terminally differentiated and have limited lifespan. These disadvantages will inhibit the clinical application of parenchymal cells for tendon and ligament tissue engineering.

## Autologous Mesenchymal Stem Cells (MSCs) for Tendon and Ligament Regeneration

Bone marrow is a source for mesenchymal progenitor cells. These cells are in abundance and have been shown to possess the potential in cell therapy for replacing, repairing, or enhancing the biologic function of damaged tissue. Bone marrow stromal cells are often referred to as mesenchymal stem cells

(MSCs) or mesenchymal progenitor cells (MPCs), the reason being that they are derived from separate pluripotent cells in the bone marrow,[19] with the potential to differentiate into various mesenchymal derived cell lines *in vitro* and *in vivo*.[20] Most recently, Altman *et al.*[21] reported that *in vitro* mechanical stimulation induces the differentiation of mesenchymal progenitor cells from the bone marrow down the tendon/ligament cell lineage. For *in vivo* studies, Young *et al.*[22] implanted the bone marrow-derived MSCs with collagen gel into 1 cm long defects in the lateral gastrocnemius tendons of rabbits for three months. The MSCs were mixed with collagen solution and incubated for 36–40 hours in the presence of biodegradable sutures, such that the gel contracted around the suture to form an "integrate implant." During the culture period, tensile loading was applied to the sutures in order to align the cells seeded in the incorporated gel. The bio-implant was then sutured into the gap of rabbit gastrocnemius tendons. After three months, significantly greater load-related structural and material properties were seen in the MSCs-treated tendons than in the sutures-alone treated control repairs. The MSCs-treated tissues had a significantly larger cross-sectional area and had better aligned collagen matrix. Awad *et al.*[23] also studied the use of autologous MSCs for tendon repair. In their study, MSCs were suspended in type I collagen gel and implanted into a surgically induced 3.8 mm by 3.8 mm window defect in the rabbit patellar tendon. Four weeks after surgery, MSCs-treated tendon had better biomechanical properties. In another of their studies,[24] they used the MSC-gel-suture composite at three different MSC concentrations (1, 4, and 8 million cells per ml) for the repair of central patellar tendon defects in rabbit. The maximum force and stiffness of cell-repaired tendon was 50% greater than the naturally healed tissues at 12 weeks. However, there is no significant difference among the three cell density conditions. In any case, these results demonstrated the potential application of MSCs for the improvement of tendon repair *in vivo*.

## Allogeneic MSCs for Tendon and Ligament Regeneration

In comparison to the use of autogenous cells, using allogeneic cells has several commercial and regulatory advantages, including the ability of the latter to (1) provide a ready-to-use implantable biological device; (2) produce large cell bank inventories necessary for quality control and safety testing; and (3) manufacture a reproducible and reliable product. Of course, the use of allogeneic cells may raise the problem of immune response. However, the paradigm regarding allogeneic cell initiated immune rejection is changing. In the living skin construct, such as the commercial product "Apligraf" (Streit 2000, Bello 2001), studies have shown that the seeded allogeneic cells did not elicit an immune response in patients. This finding suggests that certain allogeneic cell types can be used to manufacture engineered tissue without the intervention of immunosuppression or immunomodulation, which has positive implications in the use of multipotent cells from allogenic donors. Devine *et al.*[25] recently reported the infusion of allogeneic MSCs in primates. They observed that it did not cause an immunogenic response. Even human MSCs were able to engraft, and demonstrate site specific differentiation in fetal sheep with immunologic competence.[26] It has also been shown that human MSCs express class I human leukocyte antigen but not class II, which may limit immune recognition.[27] All of these results raise the possibility of using allogeneic MSCs for tissue engineering, including tendon and ligament tissue engineering.

Ouyang *et al.*[28] recently investigated the *in vivo* fate of allogeneic bone marrow stromal cells (bMSCs) at different time points after implantation into patella tendon defects (i.e. at 2, 3, 5, and 8 weeks). The protocol involved the labeling of bMSCs with green fluorescent protein (GFP) or carboxyfluorescein diacetate (CFDA) before implantation. A window defect (5 mm by 5 mm) was created at the central portion of rabbit patella tendon and which was subsequently treated with GFP or CFDA marked bMSCs. The marked bMSCs were loaded into the window defect with fibrin glue. Upon sacrifice of the rabbits at the different time-points, the implant site of the patellar tendon was immediately retrieved and, the viability of the labeled cells was assessed under confocal microscopy. The results showed that the seeded bMSCs remained viable within the tendon wound site for at least 8 weeks after implantation. The cell morphology was changed from a circular shape at two weeks to a spindle shape at five weeks after implantation (see Fig. 1). This study demonstrated that the allogenic bMSCs remained viable for prolonged periods after implantation and therefore has the potential to influence the formation and remodeling of neo-tendon tissue, thus enhancing tendon repair.

Further studies[29,30] evaluated the morphology and biomechanical function of Achilles tendons regenerated using knitted poly-lactide-co-glycolide (PLGA) loaded with allogeneic bone marrow stromal cells (bMSCs). The animal model used was that of an adult female New Zealand White rabbit

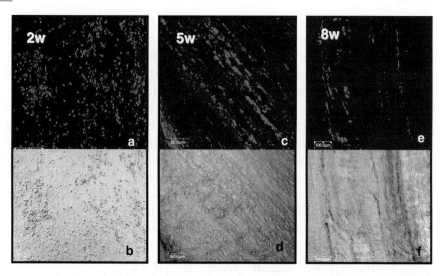

**Figure 1.** Fate of allogeneic bMSCs at tendon site after local delivery (CFDA labelling). (**a, c, e**) confocal micrograph, and (**b, d, f**) transmission micrograph, of rabbit patellar tendon repairs that have been treated with CFDA labeled bone marrow stromal cells, for (**a, b**) 2 weeks, (**c, d**) 5 weeks, and (**e, f**) 8 weeks. (Ouyang *et al., Cell Transplantation* 2004.)

with a 10-mm gap defect of the Achilles tendon. In group I, 19 hind legs with the created defects were bridged and treated with allogeneic bMSCs seeded on knitted PLGA scaffold; in group II, the Achilles tendon defects in 19 hind legs were repaired using the knitted PLGA scaffold alone; while in group III, 6 hind legs were used as normal control. Postoperatively, at two and four weeks, the histology of group I specimens exhibited a higher rate of tissue formation and remodeling as compared with group II, whereas at eight and 12 weeks post-operation, the histology of both group I and group II was similar to that of the native tendon tissue. The wound sites of group I healed well and there was no apparent lymphocyte infiltration (see Fig. 2). Immunohistochemical analysis showed that the regenerated tendons were composed of collagen types I and type III fibers. The tensile stiffness and modulus of group I were 87 and 62.6% of normal tendon, respectively, whereas those of group II were about 56.4 and 52.9% of normal tendon, respectively. These results suggest that the knitted PLGA biodegradable scaffold loaded with allogeneic bone marrow stromal cells has the potential to regenerate and repair gap defect of Achilles tendon and to effectively restore structure and function.

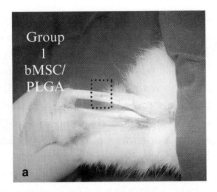
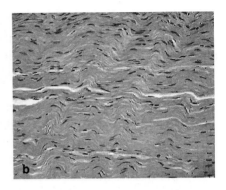

**Figure 2.** The effects of bMSCs/knitted Scaffold on tendon repair. (**a**) Macromorphology, and (**b**) histology of the bMSCs treated group at 12 weeks. (Ouyang *et al.*, *Ann NY Acad Sci* 2002.)

## MSCs for Tendon-to-Bone Interface Healing

Tendon to bone insertion is one of the important components in the muscle-tendon-bone unit. The mechanical strength of the tendon insertion is crucial for the muscle-tendon-bone unit to possess normal function. Previous studies have reported that a normal structure of tendon insertion could not be formed during the natural healing process. Therefore, researchers have turned to the use of growth factors and cells for tendon-to-bone healing. BMP-2[31] and TGF-β1[32] have been shown to have the ability to accelerate the healing process. The treated site possessed higher pull-out strength at an early time point. It was observed that BMP-12[33] was capable of promoting the fibrocartilage zone formation at the tendon-bone interface in the mice model. Ohtera *et al.*[34] studied the effect of fresh periosteum wrapped around the tendon in a bone tunnel. The pull-out strength of transplanted tendons was significantly greater in fresh periosteal grafts than in frozen grafts at four weeks postoperatively, and the 4-week specimens with fresh periosteal grafts demonstrated fibrocartilage formation in the tendon-bone interface. It seems that the mechanical and histological difference between fresh and frozen periosteum treated groups is due to the presence of viable cells in the fresh periosteum.

Bone marrow stromal cells have been shown to have the potential to differentiate into multiple mesenchymal cell lineages.[35] So it is logical to hypothesize that MSCs may help to restore the normal structure of tendon insertions. The efficacy of using bone marrow stromal cells (bMSCs) to enhance tendon to bone healing was evaluated in skeletally mature female

New Zealand White rabbits.[36] The hallucis longus tendons of 18 legs were translated into 2.5 mm diameter calcaneal bone tunnels. Bone marrow stromal cells immobilized in fibrin glue were injected into the bone tunnel of the right leg and fibrin glue without cells was injected into the left leg. At four weeks after surgery, the specimens with bone marrow stromal cells exhibited more perpendicular collagen fibers formation and increased proliferation of cartilage-like cells, which was indicated by positive collagen type II immuno-staining of the tendon-bone interface (see Fig. 3). In contrast, the specimens without the bone marrow stromal cells demonstrated progressive maturation and re-organization of fibrous tissue aligned along the load axis. Furthermore, Lim *et al.*[37] showed that MSCs enhanced the strength of tendon graft-to-bone healing in an anterior cruciate ligament reconstruction model. These studies indicated that large number of bone marrow stromal cells improved the interface of tendon to bone healing in rabbit model by producing the format of fibrocartilagenous attachment at early time points.

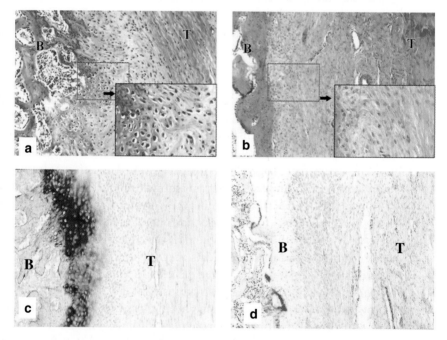

**Figure 3.** The effects of bMSCs in the biological restoration of tendon (T)-to-bone (B) insertion. (**a**) histology, and (**c**) collagen II staining of experimental group (with bMSCs), (**b**) histology and collagen II staining of control group (without cells) at 4 weeks after surgery (100×). (Ouyang *et al. Am J Sport Med* 2004.)

# The Tendon and Ligament Specific Progenitors and Differentiation Factors

Many of the recent studies have demonstrated that the bMSCs accelerate and improve tendon repair. However, these studies also verified that it is difficult to fully regenerate a tendon. The neo-tendons were similar to the native tendon but were not identical to the native tendon. This outcome may be due to two reasons. One is that the regeneration ability of tendon tissue is limited, unlike bone which can heal by regenerating normal bone in most instances, as an injured tendon often heal by scar tissue; the other reason is the lack of knowledge concerning the tendon tissue specific differentiation factors, say for example in bone regeneration, one of the specific differentiation factors is BMP-2. In order to achieve success in tendon and ligament regeneration, further studies need to be performed to gain more knowledge on the tendon and ligament biology and healing mechanism, as well as their progenitors and specific differentiation factors.

Although many aspects of the biology and biomechanics of tendons and ligaments have been known for some time, there are still basic aspects of tendon and ligaments, such as the developmental process of the tissue, that has not been fully understood.[38] It is known that development of the skeleton is a stepwise set of processes, which include the migration of cells to the site of future skeletogenesis, tissue interaction, cell condensation and overt differentiation. So it is suggested that tendons and ligaments develop in similar ways, i.e. from the same initial population of mesodermal cells as the skeleton. In fact, nothing was known about the embryonic origin of tendons, until a recent paper by Brent and colleagues in which they track the origin of tendon progenitors of the body axis and reveal the molecular events as well as the tissue interactions leading to their commitment. They demonstrated that the tendons associated with the axial skeleton derive from an unappreciated, fourth compartment of the somites. *Scleraxis (Scx)*, a bHLH transcription factor, marks this somitic tendon progenitor population. Two earlier-formed somitic compartments, i.e. the sclerotome and myotome, interact to establish this fourth *Scx*-positive compartment.[39,40] Numerous studies have tried to define the role of growth factors in ligament and tendon healing and determine appropriate strategies for the use of growth factors in tissue engineering for tendon and ligament. Various growth factors have been shown to have the ability to improve tendon and ligament cells proliferation or matrix formation. However, little is known about the regulatory signals involved in tendon and ligament differentiation. Wolfman *et al.*[41] reported that GDF-5, 6 (BMP-12 and 13) induced ectopic tendon and ligament-like tissue formation in mice, but this phenomenon will have to be investigated in larger animal models.[42,43]

## Future Research

Treatments of the injuries, especially of ruptured tendons and ligaments, have not had good long-term results.[44] Cell-based tissue engineering approaches promise to offer advantageous solutions. However, there are still specific questions that need further investigations, such as, the limited knowledge of tendon and ligament tissue specific progenitors and differentiation factors when using stem cells. As more and more studies are designed and executed, finding an ideal cell population and specific differentiation factors for tendon/ligament repair might revolutionize the therapeutic approach to tendon and ligament repair and regeneration.

## REFERENCES

1. Jorgensen U, Bak K, Ekstrand J, Scavenius M. (2001) Reconstruction of the anterior cruciate ligament with the iliotibial band autograft in patients with chronic knee instability. *Knee Surg Sports Traumatol Arthrosc* **9**(3):137.
2. Kato YP, Dunn MG, Zawadsky JP, *et al.* (1991) Regeneration of Achilles tendon with a collagen tendon prosthesis. Results of a one-year implantation study. *J Bone Joint Surg Am* **73**(4):561–574.
3. Mass DP, Tuel RJ, (1990) Participation of human superficialis flexor tendon segments in repair *in vitro. J Orthop Res* **8**(1):21–34.
4. Potenza AD. (1962) Tendon healing within the flexor digital sheath of the dog. *J Bone J Surg* **44A**:49–64.
5. Russell JE, Manske PR. (1990) Collagen synthesis during primate flexor tendon repair *in vitro. J Orthop Res* **8**(1):13–20.
6. Glaser DL, Ramachandran R, Shore EM, *et al.* (2003) The origin of cells within a healing tendon. 49th Annual Meeting of the Orthopaedic Research Society Poster 0340.
7. Bell E, Ivarsson B, Merrill C. (1979) Production of a tissue-like structure by contraction of collagen lattices by human fibroblasts of different proliferative potential *in vitro. Proc Natl Acad Sci USA* **76**(3):1274–1278.
8. Klebe RJ, Caldwell H, Milam S. (1989) Cells transmit spatial information by orienting collagen fibers. *Matrix* **9**(6):451–458.

9. Bab I, Howlett CR, Ashton BA, Owen ME. (1984) Ultrastructure of bone and cartilage formed *in vivo* in diffusion chambers. *Clin Orthop* **187**:243–254.

10. Caplan AI. (1970) Effects of the nicotinamide-sensitive teratogen3-acetylpyridine on chick limb cells in culture. *Exp Cell Res* **62**(2):341–355.

11. Muschler GF, Midura RJ. (2002) Connective tissue progenitors: Practical concepts for clinical applications. *Clin Orthop* **395**:66–80.

12. Caplan AI, Fffink DJ, Goto T, *et al.* (1993) Mesenchymal stem cells and tissue repair. In *The Anterior Cruciate Ligament: Current and Future Concepts*. ed. Jackson DW, *et al* (eds.), pp. 405–417. New York: Raven Press Ltd.

13. Huang D, Chang TR, Aggrawal A, *et al.* (1993) Mechanism and dynamics of mechanical strengthening in ligament-equivalent fibroblast-populated collagen matrices. *Ann Biomed Eng* **21**:289.

14. Dunn MG, Liesch JB, Tiku ML, Zawadsky JP. (1995) Development of fibroblast-seeded ligament analogs for ACL reconstruction. *J Biomed Mater Res* **29**(11):1363.

15. Goulet F, Germain L, Rancourt D, *et al.* (1997) Tendons and ligaments. In *Principles of Tissue Engineering*, Lanza RP, Langer R, Chick WL (eds.), pp. 639–645. Austin, TX: RG Landes Academic.

16. Bellincampi LD, Closkey RF, Prasad R, *et al.* (1998) Viability of fibroblast-seeded ligament analogs after autogenous implantation. *J Orthop Res* **16**(4):414.

17. Lin VS, Lee MC, O'Neal S, *et al.* (1999) Ligament tissue engineering using synthetic biodegradable fiber scaffolds. *Tissue Eng* **5**(5):443.

18. Cao Y, Vacanti JP, Ma PX. (1995) Tissue engineering of tendon. *Mat Res Soc Symp Proc* **394**:83.

19. Lazarus HM, Haynesworth SE, Gerson SL, *et al.* (1995) *Ex vivo* expansion and subsequent infusion of human bone marrow-derived stromal progenitor cells (mesenchymal progenitor cells): Implications for therapeutic use. *Bone Marrow Transplant* **16**(4):557–564.

20. Prockop DJ. (1997) Marrow stromal cells as stem cells for nonhematopoietic tissues. *Science* **276**(5309):71–74.

21. Altman GH, Horan RL, Martin I, *et al.* (2002) Cell differentiation by mechanical stress. *FASEB J* **16**(2):270–272.

22. Young RG, Butler DL, Weber W, *et al.* (1998) Use of mesenchymal stem cells in a collagen matrix for Achilles tendon repair. *J Orthop Res* **16**(4):406.

23. Awad HA, Butler DL, Boivin GP, *et al.* (1999) Autologous mesenchymal stem cell-mediated repair of tendon. *Tissue Eng* **5**(3):267.

24. Awad HA, Boivin GP, Dressler MR, *et al.* (2003) Repair of patellar tendon injuries using a cell-collagen composite. *J Orthop Res* **21**(3):420–431.

25. Devine SM, Bartholomew AM, Mahmud N, *et al.* (2001) Mesenchymal stem cells are capable of homing to the bone marrow of non-human primates following systemic infusion. *Exp Hematol* **29**(2):244.

26. Liechty KW, MacKenzie TC, Shaaban AF, *et al.* (2000) Human mesenchymal stem cells engraft and demonstrate site-specific differentiation after in utero transplantation in sheep. *Nature Med* **6**(11):1282.

27. Young HE, Steele TA, Bray RA, *et al.* (2001) Human reserve pluripotent mesenchmal stem cells are present in the connective tissue of skeletal muscle and dermis derived from fetal, adult, and geriatric donors. *The Anat Rec* **264**:51.

28. Ouyang HW, Goh JCH, Lee EH. (2004) Viability of allogeneic bone marrow stromal cells following local delivery into patella tendon in rabbit model. *Cell Transplantation* **13**(6):649–657.

29. Ouyang HW, Goh JC, Mo XM, *et al.* (2002) The efficacy of bone marrow stromal cell-seeded knitted PLGA fiber scaffold for Achilles tendon repair. *Ann NY Acad Sci* **961**:126–129.

30. Ouyang HW, Goh JC, Thambyah A, *et al.* (2003) Knitted poly-lactide-co-glycolide scaffold loaded with bone marrow stromal cells in repair and regeneration of rabbit Achilles tendon. *Tissue Eng* **9**:431–439.

31. Rodeo SA, Suzuki K, Deng XH, *et al.* (1999) Use of recombinant human bone morphogenetic protein-2 to enhance tendon healing in a bone tunnel. *Am J Sports Med* **27**(4):476–488.

32. Yamazaki S, Yasuda K, Tohyama H, *et al.* (2001) The effect of transforming growth factor — beta1 on intraosseous healing of the flexor tendon autograft in anterior cruciate ligament reconstruction. 48th Annual Meeting of the Orthopaedic Research Society, Poster 0627.

33. Hattersley G, Cox K, Soslowsky LJ, *et al.* (1998) Bone morphogenic proteins 2 and 12 alter the attachment of tendon to bone in rat model: A histological and biomechanical investigation. *Trans Orthop Res Soc* **23**:96.

34. Ohtera K, Yamada Y, Aoki M, *et al.* (2000) Effects of periosteum wrapped around tendon in a bone tunnel: A biomechanical and histological study in rabbits. *Crit Rev Biomed Eng* **28**(1 & 2):115.

35. Pittenger MF, Mackay AM, Beck SC, *et al.* (1999) Multilineage potential of adult human mesenchymal stem cells. *Science* **284**(5411):143.

36. Ouyang HW, Goh JC, Lee EH. (2004) Use of bone marrow stromal cells for tendon graft-to-bone healing: Histological and immunohistochemical studies in a rabbit model. *Am J Sport Med* **32**:321–327.

37. Lim JK, Hui JHP, Li L, *et al.* (2004) Enhancement of tendon graft osteointegration using mesenchymal stem cells in a rabbit model of anterior cruciate ligament reconstruction. *Arthroscopy* (in Press).

38. Goh JC, Ouyang HW, Teoh SH, *et al.* (2003) Tissue-engineering approach to the repair and regeneration of tendons and ligaments. *Tissue Eng.* **9**(Suppl 1):31–44.

39. Brent AE, Schweitzer R, Tabin CJ. (2003) A somitic compartment of tendon progenitors. *Cell* **113**(2):235–248.

40. Dubrulle J, Pourquie O. (2003) Welcome to syndetome: A new somitic compartment. *Dev Cell* **4**(5):611–612.
41. Wolfman NM, Hattersley G, Cox K, *et al.* (1997) Ectopic induction of tendon and ligament in rats by growth and differentiation factors 5, 6, and 7, members of the TGF-beta gene family. *J Clin Invest* **100**(2):321.
42. Lou J, Tu Y, Burns M, *et al.* (2001) BMP-12 gene transfer augmentation of lacerated tendon repair. *J Orthop Res* **19**(6):1199–1202.
43. Fu SC, Wong YP, Chan BP, *et al.* (2003) The roles of bone morphogenetic protein (BMP) 12 in stimulating the proliferation and matrix production of human patellar tendon fibroblasts. *Life Sci* **72**(26):2965–2974.
44. Butler DL, Juncosa N, Dressler MR. (2004) Functional efficacy of tendon repair processes. *Annu Rev Biomed Eng* **6**:303–329.

# 24

# Germ Cell Differentiation from Embryonic Stem Cells

Mark Richards and Ariff Bongso*

## Introduction

Mammalian embryonic stem (ES) cell lines are derived from the isolation and serial sub-culture of inner cell masses (ICMs) from mature blastocysts. In principle, ES cells are capable of differentiating into the three primordial germ layers and subsequently all cell types in the adult animal. Therefore, human ES cells (hESCs)[1,2] have enormous potential to provide a source of tissues for replacement in diseases in which native cell types are inactivated or destroyed. The ability of hESCs to undergo multilineage differentiation *in vitro* and *in vivo* is also very well documented.[3-6] However, most cultured embryonic stem cell lines have been generally considered pluripotent rather than totipotent because of the inability to detect germ line progenitor cells and markers of trophoblast differentiation under *in vitro* culture conditions.

## Germ cell differentiation and specification *in vivo*

Germ cells and gametes are responsible for perpetuating the transmission of genetic information and for reproducing totipotency from generation to generation. Two different developmental programs control the specification of germ cell lineages in organisms. In non-mammalian species and invertebrates such as fruitflies, nematodes and frogs, germ cells of both males and females are specified via the inheritance of germ plasm (microscopically distinct oocyte cytoplasm enriched in RNAs and RNA-binding proteins) which segregates with cells destined to be germ cells.[7] Mammals do not utilize a germ plasm allocation, but instead segregate primordial germ cells

---

*Correspondence: Department of Obstetrics and Gynaecology, National University Hospital, NUS, Kent Ridge, Singapore 119074.

(PGCs) apart from somatic lineages early in development. Subsequently, the oocytes and sperm cells (gametes) are derived from this founder population of primordial germ cells. *In vitro* derived pluripotent embryonic germ (EG) stem cell lines are established from PGCs as well.[8] Therefore, germ cell specification must be linked to cell fate commitment leading concurrently to both gametogenesis and the maintenance of pluripotency.

Primordial germ cells arise from the proximal epiblast, a region of the early mammalian embryo in response to signals from the neighboring extraembryonic ectoderm. Several fate-mapping studies have revealed that mammalian germ cell specification *in vivo* involves the proximal epiblast[9] and BMP-4 and BMP-8b, TGF-β family signaling from the extraembryonic ectoderm.[10] Other studies have also implicated interferon-induced membrane proteins as important mediators of germ cell development.[11,12] Some RNAs and RNA-binding proteins such as Pumilio, Nanos and Dazl involved in germ cell development are highly conserved between invertebrates that specify germ cells via germ plasm inheritance and mammals that form germ cells independently of germ plasm.[13,14,15]

In the mouse model, the first recognizable population of approximately 45 founder PGCs are identified following gastrulation, at 7.2 days post coitum (dpc), as an extraembryonic cluster of cells at the base of the allantois that express tissue non-specific alkaline phosphatase (TNAP), *Oct4* and *STELLA*.[16,17] The proximal epiblast, however, is not predestined to a germ cell fate because transplantation of the distal epiblast to contact the extraembryonic ectoderm also results in germ cell formation.[9] Instead, the fate of the proximal epiblast cells is to form both germ cells and extraembryonic mesoderm. Extraembryonic ectoderm very likely provides one of the earliest signals for germ cell specification in the epiblast and later produces an as yet uncharacterized cue to distinguish extraembryonic mesoderm from germ cells as well. The embryological period equivalent to mouse E5.5–E7.2 in human embryo development occurs shortly after implantation. The analysis of human germ cell specification *in vivo* is therefore unfeasible due to ethical considerations regarding research during this period.

Deletions of the Y chromosomal gene, the male determining gene, *Sry* (Sex determining region on the Y chromosome), causes XY male-to-female sex reversal, whereas *SRY* translocations to the X chromosome lead to XX female-to-male sex reversal.[18] The role of *Sry* as a candidate testis-determining gene was confirmed by demonstrating testis development after transgenic expression of *Sry* in XX mice.[19] Gonadal development factors described to date are known to act at the transcriptional level, but their functions are incompletely understood. Based on conserved domain homology to other transcription factors some of these transcription factors are proposed to affect DNA binding,[20] modulate chromatin remodeling,[21] form interactive complexes that activate transcription[22] or have a role in progenitor cell type specification.[23] Inhibition of gene expression is another likely and important mechanism to direct cell fate, but less is known about potential transcriptional repressors and gene silencing mechanisms.

The testis comprises three main cell types, viz. sertoli cells, leydig cells, and germ cells. The sertoli cells and germ cells reside in seminiferous tubules where spermatogenesis occurs; leydig cells populate the interstitial compartment and produce testosterone. Spermatogonia are the male germ stem cells that continuously produce sperm for the next generation. Spermatogenesis is a complicated process that proceeds through a mitotic phase of stem cell renewal and differentiation, meiotic phase, and postmeiotic phase of spermiogenesis. A recent study in the medaka fish has shown it possible to fully recapitulate spermatogenesis *in vitro*.[24] After 140 passages during 2 years of culture, the medaka fish spermatogonial cell line derived by Hong and co-workers retained a diploid karyotype, appropriate phenotype and gene expression pattern of spermatogonial stem cells *in vivo*. Furthermore, the cell line was shown to undergo meiosis and spermiogenesis *in vitro* to generate motile sperm.[24]

The ovary also comprises three main cell types: granulosa cells, theca cells, and oocytes. Oocytes that grow and differentiate during reproductive cycles are surrounded by granulosa and theca cells in the follicles. Oocytes and sperm (gametes) in most mammals are derived from a founder population of primordial germ cells that are committed early in embryogenesis. A basic doctrine of reproductive biology is that most mammalian females lose the capacity for germ-cell renewal during fetal life, such that a fixed reserve of oocytes enclosed within the follicles is endowed at birth, while spermatogenesis occurs throughout the life of mammalian males. However, a recent study has shown that juvenile and adult mouse ovaries possess mitotically active germ cells that, based on rates of oocyte degeneration (atresia) and clearance, are needed to continuously replenish the follicle pool.[25] Consistent with this, treatment of prepubertal female mice with the mitotic germ-cell toxicant busulphan eliminated the primordial follicle reserve by early adulthood without inducing atresia. Furthermore, the identified cells expressed the meiotic entry marker synaptonemal complex protein 3 in juvenile and adult mouse ovaries. Wild-type ovaries grafted into transgenic female mice with ubiquitous expression of green fluorescent protein (GFP) became infiltrated with GFP-positive germ cells that form follicles. Collectively, these data established the existence of proliferative germ cells that sustain oocyte and follicle production in the postnatal mammalian ovary.

# Induction of Germ Cell Differentiation from mESCs *In Vitro*

## Female germ cell and gamete differentiation *in vitro*

Four recent studies have demonstrated that germ cells can be derived from mouse embryonic stem cells (mESCs) and hESCs (Fig. 1). The first study demonstrating the potential of mESCs to form germ cells *in vitro* came from the laboratory of Hans Scholer.[26] mESCs were found to generate both early germ-like cells and more mature oocyte-like cells from both male and female mESC lines.

The proper isolation, identification and enrichment of desired cells from impure populations of differentiated embryonic stem cell derivatives are a major challenge in differentiation studies: Therefore, to visualize germ cell formation, Hubner *et al.*[26] used green floursecent protein (GFP) expression from a modified truncated Oct-4 promoter to delineate a germ-line specific differentiation signal. mESCs differentiated in two-dimensional monolayers on tissue culture plastic exhibited high GFP expression, with almost 40% of cells expressing the GFP signal at day 8. This large percentage of GFP expressing cells either strongly suggest a significant predisposition to germ cell fate in mESC cultures that were differentiating *in vitro* under the conditions outlined by Hubner *et al.*,[26] or perhaps the lack of specificity of this reporter construct in the context of *in vitro* mESC differentiation into germ cell lineages. Nevertheless, these early germ-like cells displayed several markers representative of different stages in germ cell development such as Vasa and c-kit.

## Characterization of follicular structures, oocyte-like and embryo-like elements derived from mESCs *in vitro*

Hubner *et al.*[26] identified well-organized structures that were morphologically similar to early ovarian follicles in their extended culture of mESCs in two-dimensional monolayers for a period of 12 days or more.

Aggregates of cells formed spontaneously and lifted themselves off tissue-culture plates and collected aggregates maintained at high cell densities proceeded to form follicle-like structures that extruded oocytes. Estradiol was

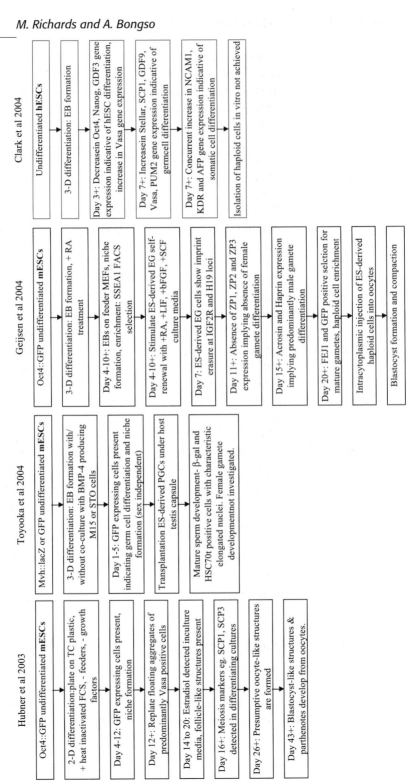

**Figure 1.** Summary of various differentiation strategies used to derive germ cells and gametes from ES cells.

also detected in these cultures, suggesting some degree of functional activity of the somatic cells in these follicle-like structures. Prolonged maintenance of the differentiated mESC cultures led to the formation of oocyte-like cells enclosed in a coat resembling a primitive zona pellucida that was very fragile. Since meiosis is a major hallmark of germ cell development, molecular markers of meiosis, for example SCP3 and SYN1, were assayed and detected by antibody staining in differentiating cultures.

The oocyte-like cells described by Scholer's group were about 50 to 70 microns in diameter and these figures are typical of the size range of natural murine oocytes. Cytoplasmic staining and RT-PCR indicated the presence of GDF-9, zona pellucida proteins 2 and 3. Although confirmation of meiosis by chromosomal imaging was not performed, presumptive oocytes appeared to undergo activation, leading to the formation of blastocyst-like structures after 43 days in culture. Most of these blastocyst-like structures that resembled pre-implantation embryos were probably parthenotes. Additionally, a blastocyst-like structure that resembled a 16-cell morula with an intact zona was identified in differentiating two-dimensional cultures. Nuclear staining with Hoescht, cytoplasm to nucleus ratio, compacted morphology and Oct-4 protein expression and distribution in this blastocyst-like structure were typical of bona fide mouse morulae.

The derivation of oocytes and blastocyst-like structures could be accomplished with both female and male mESC lines,[26] perhaps reflecting the absence of appropriate SRY expression in differentiating cultures that would function as an obligatory signal for PGCs to enter male germ cell lineage commitment. Hubner *et al.* (2003) also suggest that their blastocyst-like structures derived from differentiating mESC cultures looked better than most embryos obtained via nuclear transfer of somatic nuclei. Furthermore, in contrast to cloned embryos, most of these mESC derived blastocyst-like structures displayed correct Oct-4 expression patterns. It remains to be seen if sperm can fertilize these mESC-derived oocytes. A major challenge will also be to arrest these cells in meiosis, thereby preventing their spontaneous activation and parthenogenetic development.

## Male germ cell and gamete differentiation

Shortly after the demonstration of oocyte differentiation from mESCs, two studies documented the derivation of sperm-like cells from mESCs differentiated *in vitro*.[27,28] The first study by Toyooka *et al.* [27] from the Mitsubishi Kagaku Institute of Life Sciences showed that primitive germ cells and sperm could be derived from three-dimensional mESC aggregates called embryoid bodies (EBs).

The EB environment is postulated to mimic the environment of the early differentiating embryo. To visualize germ cell formation, the mouse Vasa homologue (Mvh) gene was tagged with GFP. Mvh is a germ-cell-specific ATP-dependent RNA helicase that is not expressed in ESCs.[29] This reporter enabled the tracking of germ cell differentiation in culture. Mvh-positive germ cells were greatly enhanced within EBs when in a co-culture system with BMP-4 producing transgenic cell lines, an observation that is consistent with previous studies demonstrating the important role of BMP4 in germ cell development.[30,31,32] On the other hand, BMP8b had little effect on the emergence of PGCs.[27]

Toyooka et al.[27] then FACS-sorted primitive germ cells on the basis of GFP expression, co-cultured these cells with gonadal cells and transplanted these cells under a host testis capsule where they showed engraftment in the lumens of tubule structures. Although mature sperm were reported to form, demonstration of an mESC origin of these cells by donor-specific markers was absent and no functional data concerning the fertilization capacity of these mESC-derived sperm were provided.

Independently, Geijsen et al.[28] demonstrated EG cell and male gamete formation from mESCs. EB differentiation was also employed to obtain cells of the germ lineage. Remarkably, both primitive germ cells and gamete-like haploid cells were found present in differentiating EBs, suggesting that a supportive niche for spermatogenesis develops within EBs.

After mESCs had undergone differentiation for 4 to 10 days, cells were sorted on the basis of the surface marker SSEA1, which is expressed by both germ cells and mESCs. Cultures supplemented with retinoic acid (RA) and germ cell growth factors such as basic fibroblast growth factor and stem cell factor were found to further stimulate ES cell differentiation and promote germ cell self-renewal, enabling the isolation of continuously growing clonal lines of EG cells.[28]

## Characterization of male gametes, germ cell and sperm-like cells derived from mESCs *in vitro*

mESC-derived germ cells exhibited imprint erasure and TNAP staining, both important characteristics of germ cell development. If left in the environment of the EB for longer time periods, more mature, round, sperm-like cells developed and these were isolated using an antibody against the sperm acrosome. Cell populations purified with the antibody were enriched for haploid cells.[28]

Geijsen *et al.*[28] failed to detect expression of zona pellucida proteins 1 and 2 in their RA treated differentiating EB cultures, suggesting that differentiation was occurring predominantly along the male germ cell lineage. Restriction digestion and southern blotting conclusively showed that EG clones derived from RA treated EBs displayed imprint erasure for both IGFR2 and H19 loci.

The biological function of the EB-derived haploid male gametes was also further investigated for their capacity to fertilize oocytes. Isolated FE-J1/GFP haploid cells from day 20 EBs were intracytoplasmically injected into recipient oocytes. Five separate microinjection experiments performed in two independent laboratories produced comparable results. Significantly, approximately 50% of the injected oocytes supported cleavage to the 2-cell with up to 20% displaying progression to blastocysts.[28]

Therefore, the injection of purified haploid mESC-derived sperm into oocytes led to successful fertilization and in some cases blastocyst formation ensued. It will be interesting to know if these blastocysts can develop into live pups when transferred to a surrogate mother. Further characterization is required to demonstrate that these mESC-derived germ cells have undergone normal meiotic progression and if they have successfully reestablished paternal imprints.

## Germ Cell Differentiation from hESCs

More recently, Clark *et al.*[33] from the University of San Francisco provided the first evidence that hESCs can differentiate into germ cells. Aggregations of hESCs into three-dimensional EBs were again used to provide the proper conditions for germ cell formation *in vitro*. Microarray gene expression profiling of EBs suggested the concomitant onset of both immature primordial germ cells and more mature gonocyte differentiation.

Using real-time quantitative RT-PCR and western blots, they found that undifferentiated hESCs expressed several pre-meiotic germ cell markers such as STELLAR, DAZL, cKIT (marker of pre-meiotic migrating germ cells), NANOS1, PUM1 and PUM2 (markers of pre-meiotic germ cells of the fetal gonads). In contrast to these genes whose expression is diagnostic of pre-meiotic germ cells, the authors did not observe expression of genes known to be restricted in expression to later stages of germ cell differentiation. Thus, the expression of genes like VASA, SCP1, GDF9 and SCP3 was not detected, and these findings clearly indicated that undifferentiated hESCs express early markers of, but not later markers of, germ cell development. In contrast, human ICMs were found to only express NANOS1 and STELLAR, while human testis samples expressed VASA, SCP1 and other later markers of germ cell development. The authors also suggest that the onset of VASA gene expression is a conclusive and early marker of germ cell differentiation in hESC cultures.

In addition, markers for both female and male germ cells were found in differentiating EBs regardless of the sex of the hESC line used. Although genes involved in meiosis were also detected, no formal isolation of haploid cells was accomplished.[33] Nevertheless, this pioneering study clearly implicated germ cell development from differentiating hESC EB cultures. Further work is needed to accurately identify, characterize and isolate these germ cell derivatives.

## Summary

Two independent reports have employed a three-dimensional EB differentiation strategy for mESCs, concomitant with RA treatment and/or BMP4 stimulation to effectively support a program for male germ cell differentiation.[27,28] The subsequent isolation of differentiated mESC EB sub-fractions with a reporter construct under Oct-4 or Vasa promoter control or enrichment of the EB derived SSEA1 positive cell fractions has led to the isolation of haploid male gametes with the potential to fertilize oocytes. Alternatively, an extended two-dimensional differentiation strategy for both male and female mESCs on tissue culture plastic employed by Scholer and colleagues[26] resulted in predominantly female germ cell and oocyte differentiation. Recently, hESC EBs were also shown to express several gene transcripts and proteins that were indicative of germ cell differentiation.[33]

Proper sperm development requires supportive sertoli cells, while oocyte maturation requires the nourishment from both the theca and granulosa cells. The finding that oocytes and sperm-like cells can be derived from ES cells suggests either that a supportive niche develops from ES cells during differentiation or that *in vitro*-created germ cells do not need the same support as they do *in vivo*. Also, to date no adult stem cell has been shown to be able to differentiate along a germ cell lineage.

In all the studies of germ cell differentiation from ES cells, it has been shown that *in vitro* germ cell differentiation takes place at an accelerated pace. This is in contrast to tightly regulated *in vivo* differentiation events.

A possible reason for this difference may lie in the fact that ES cells in culture have already acquired the capacity to form PGCs as indicated by the expression of several PGC founder genes such as *STELLA, DAZL* and *FRAGILIS* that have been detected in undifferentiated mESCs and hESCs but not in human or mouse ICMs. Another possibility is that some of the factors that tightly regulate PGC development *in vivo* are absent in these *in vitro* culture systems.

Studies to identify the presence of the germ cell niche or a means of enhancing its formation from ES cells could contribute towards more efficient and directed strategies for germ cell development. Furthermore, proper media conditions, including correct nutritional supplements, could also help in the derivation of germ cells from ES cells.[34,35]

The emergence of techniques that allow the *in vitro* derivation of both germ cells and mature gametes will allow investigations into processes guiding germ cell formation and into academic and therapeutic uses for these fascinating cells. ES cell derived gametes could find several applications in animal breeding and assisted reproductive technology. Additionally, ES cell-derived gametes and germ cells may provide an important culture system to examine the roles of key genes and complex signaling processes involved in genomic imprinting and reductive division events during meiosis.

Perhaps the most provocative applications are applicable to the more mature gametes that have been reported to develop from ES cells. Synthetic sperm derived from hESCs could also perhaps be used to treat male infertility. Additionally, if male germ cells derived from ES cells can engraft in host testes and produce viable sperm, these techniques could provide insights into spermatogenesis and possibly provide new treatments for male infertility. Indeed, past studies have shown that testis cell transplants can functionally engraft in the testes of infertile recipients[36,37,38] and preliminary work with germ cells derived from mESCs suggests that these cells exhibit the same potential.[29]

A limitless supply of human eggs derived from hESC lines could have a radical impact on medicine. Can hESC-derived oocytes be fertilized and can they be fertilized with hESC-derived sperm? More studies are needed to determine whether *in vitro*-derived mouse gametes can successfully produce healthy pups. The random parthenogenetic activation of ES cell-derived oocytes needs to be controlled *in vitro* so that these oocytes can be used for fertilization and nuclear transfer technology. Nevertheless, an ability to efficiently develop ES cell-derived oocytes for nuclear transfer studies would be a significant advance and may provide an unlimited source of oocytes. If these oocytes provide all the cues necessary to allow reprogramming of donor nuclei and to successfully develop to the blastocyst stage, patient-specific hESC lines resulting from nuclear transfer could be created, helping circumvent the major obstacle of donor-oocyte availability in the construction of patient-specific hESC lines via nuclear transfer.

# REFERENCES

1. Bongso A, Fong CY, Ng SC, Ratnam S. (1994) Isolation and culture of inner cell mass cells from human blastocysts. *Hum Reprod* **9**:2110–2117.
2. Thomson JA, Itskovitz-Eldor J, Shapiro SS, *et al.* (1998) Embryonic stem cell lines derived from human blastocysts. *Science* **282**:1145–1147.
3. Reubinoff BE, Pera MF, Fong CY, *et al.* (2000) Embryonic stem cell lines from human blastocysts: Somatic differentiation *in vitro*. *Nat Biotechnol* **18**:399–404.
4. Assady S, Maor G, Amit M, *et al.* (2001) Insulin production by human embryonic stem cells. *Diabetes* **50**:1691–1697.
5. Kaufman DS, Hanson ET, Lewis RL, *et al.* (2001) Hematopoietic colony-forming cells derived from human embryonic stem cells. *Proc Natl Acad Sci USA* **98**:10716–10721.
6. Mummery C, Ward D, van den Brink CE, *et al.* (2003) Differentiation of human embryonic stem cells to cardiomyocytes: Role of coculture with visceral endoderm-like cells. *Circulation* **107**:2733–2734.
7. Houston DW, King ML. (2000) Germ plasm and molecular determinants of germ cell fate. *Curr Top Dev Biol* **50**:155–181.
8. Shamblott MJ, Axelman J, Wang S, *et al.* (1998) Derivation of pluripotent stem cells from cultured human primordial germ cells. *Proc Natl Acad Sci USA* **95**:13726–13731.
9. Tam P, Zhou S. (1996) The allocation of epiblast cells to ectodermal and germ-line lineages is influenced by position of the cells in the gastrulating mouse embryo. *Dev Biol* **178**:124–132.
10. Fujiwara T, Dunn NR, Hogan BL. (2001) Bone morphogenetic protein 4 in the extraembryonic mesoderm is required for allantois development and the localization and survival of primordial germ cells in the mouse. *Proc Natl Acad Sci USA* **98**:13739–13744.
11. Saitou M, Barton SC, Surani MA. (2002) A molecular programme for the specification of germ cell fate in mice. *Nature* **418**:293–300.
12. Lange UC, Saitou M, Western PS, *et al.* (2003) The fragilis interferon-inducible gene family of transmembrane proteins is associated with germ cell specification in mice. *BMC Dev Biol* **3**:1.

13. Moore FL, Jaruzelska J, Fox MS, *et al.* (2003) Human Pumilio-2 is expressed in embryonic stem cells and germ cells and interacts with DAZ (Deleted in AZoospermia) and DAZ-Like proteins. *Proc Natl Acad Sci USA* **100**: 538–543.

14. Jaruzelska J, Kotecki M, Kusz K, *et al.* (2003) Conservation of a Pumilio–Nanos complex from Drosophila germ plasm to human germ cells. *Dev Genes Evol* **213**:120–126.

15. Tsuda M, Sasaoka Y, Kiso M, *et al.* (2003) Conserved role of nanos proteins in germ cell development. *Science* **301**:1239–1241.

16. Wylie C. (1996) Germ cells. *Cell* **96**:165–174.

17. Lawson KA, Hage WJ. (1994) Clonal analysis of the origin of primordial germ cells in the mouse. *Ciba Found Symp* **182**:68–91.

18. Harley VR, Clarkson MJ, Argentaro A. (2003) The molecular action and regulation of the testis-determining factors, SRY (sex-determining region on the Y chromosome) and SOX9 [SRY-related high-mobility group (HMG) box 9]. *Endocr Rev* **24**:466–487.

19. Koopman P, Gubbay J, Vivian N, *et al.* (1991) Male development of chromosomally female mice transgenic for Sry. *Nature* **351**:117–121.

20. Phillips NB, Nikolskaya T, Jancso-Radek A, *et al.* (2004) Sry-directed sex reversal in transgenic mice is robust with respect to enhanced DNA bending: Comparison of human and murine HMG boxes. *Biochemistry* **43**:7066–7081.

21. Katoh-Fukui Y, Tsuchiya R, Shiroishi T, *et al.* (1998) Male-to-female sex reversal in M33 mutant mice. *Nature* **393**:688–692.

22. Nachtigal MW, Hirokawa Y, Enyeart-VanHouten DL, *et al.* (1998) Wilms' tumor 1 and Dax-1 modulate the orphan nuclear receptor SF-1 in sex-specific gene expression. *Cell* **93**:445–454.

23. Wilhelm D, Englert C. (2002) The Wilms tumor suppressor WT1 regulates early gonad development by activation of Sf1. *Genes Dev* **16**:1839–1851.

24. Hong Y, Liu T, Zhao H, *et al.* (2004) Establishment of a normal medaka fish spermatogonial cell line capable of sperm production *in vitro*. *Proc Natl Acad Sci USA* **101**:8011–8016.

25. Johnson J, Canning J, Kaneko T, *et al.* (2004) Germline stem cells and follicular renewal in the postnatal mammalian ovary. *Nature* **429**:145–150.

26. Hubner K, Fuhrmann G, Christenson LK, *et al.* (2003) Derivation of oocytes from mouse embryonic stem cells. *Science* **300**:1251–1256.

27. Toyooka Y, Tsunekawa N, Akasu R, Noce T. (2003) Embryonic stem cells can form germ cells *in vitro*. *Proc Natl Acad Sci USA* **100**:11457–11462.

28. Geijsen N, Horoschak M, Kim K, *et al.* (2004) Derivation of embryonic germ cells and male gametes from embryonic stem cells. *Nature* **427**:148–154.

29. Toyooka Y, Tsunekawa N, Takahashi Y, *et al.* (2000) Expression and intracellular localization of mouse Vasa-homologue protein during germ cell development. *Mech Dev* **93**:139–149.

30. Di Carlo A, Travia G, De Felici M. (2000) The meiotic specific synaptonemal complex protein SCP3 is expressed by female and male primordial germ cells of the mouse embryo. *Int J Dev Biol* **44**:241–244.

31. Chuma S, Nakatsuji N. (2001) Autonomous transition into meiosis of mouse fetal germ cells *in vitro* and its inhibition by gp130-mediated signaling. *Dev Biol* **229**:468–479.

32. Larsson SH, Charlieu JP, Miyagawa K, *et al.* (1995) Subnuclear localization of WT1 in splicing or transcription factor domains is regulated by alternative splicing. *Cell* **81**:391–401.

33. Clark AT, Bodnar MS, Fox M, *et al.* (2004) Spontaneous differentiation of germ cells from human embryonic stem cells *in vitro*. *Hum Mol Genet* **13**:727–739.

34. Wong WY, Thomas CM, Merkus JM, *et al.* (2000) Male factor subfertility: Possible causes and the impact of nutritional factors. *Fertil Steril* **73**:435–442.

35. Oishi K, Barchi M, Au AC, *et al.* (2004) Male infertility due to germ cell apoptosis in mice lacking the thiamin carrier, Tht1. A new insight into the critical role of thiamin in spermatogenesis. *Dev Biol* **266**:299–309.

36. Brinster RL, Zimmermann JW. (1994) Spermatogenesis following male germ-cell transplantation. *Proc Natl Acad Sci USA* **91**:11298–11302.

37. Brinster RL, Avarbock MR. (1994) Germline transmission of donor haplotype following spermatogonial transplantation. *Proc Natl Acad Sci USA* **91**:11303–11307.

38. Brinster RL. (2002) Germline stem cell transplantation and transgenesis. *Science* **296**:2174–2176.

# 25

# Stem Cell Therapies in Animal Models: Their Outcome and Possible Benefits in Humans

F. Fändrich* and M. Ruhnke

## Introduction

When reading this chapter, please note that the authors objective was to outline possible perspectives as to how findings and results gathered from *in vitro* and *in vivo* animal data can possibly be transferred to reliable and solid clinical application strategies, an ultimate goal in the field of translational research and biomedical life sciences. Our view is that to approach such a challenge to generate a source of programmable cells endowed with potent differentiation capacity and inherent tissue-specific biological functions, we need to take into consideration current knowledge about stem cell plasticity and related developmental biology. In addition, we attempt to open the reader's mind by outlining an alternative strategy to answer the question how to best satisfy the patient's need to provide cellular and tissue-engineered products useful in the field of regenerative medicine. It is our firm belief that autologous cells, whether single cell expanded cell clones or multiclonal multilineage cells with programmable capacity can be used as long as the biological product is safe and bears reliable *in vivo* function.

## General Considerations

### Requirements of potential cellular products intended for clinical use

Successful restoration of damaged tissue, healing of degenerative joints or organ compartments, and compensating for metabolic or genetic deficiencies

*Correspondence: Department of General Surgery and Thoracic Surgery, University of Kiel, Germany. Tel: +49 431 597 4306, fax: 49 439 597 5023, e-mail: ffändrich@surgery.uni.kiel.de

can be considered an ultimate goal in the field of stem cell biology and related tissue engineering. Heated debates have evolved in the scientific community concerning the most appropriate stem cell source or "working material" to be used in order to solve the obvious dilemma between optimal cellular function, its related biological integrity and minimal patient risk. In this context, the potential advantages and disadvantages between stem cells of embryologic and adult provenance are major issues of academic controversy. Further profound and intensive discussions dealing with issues such as stem cell plasticity, clonality, carcinogenicity and related functional characteristics must be taken into consideration. From the clinician's perspective, various stem cell entities might meet none, several, or all the biological criteria outlined hereafter. In order to fulfil clinical and therapeutic requirements, future customized cellular products envisaged to treat human diseases should: (i) not harm the patient; (ii) reliably satisfy the biological needs they are given for; (iii) avoid host-mediated rejection in order to circumvent immunosuppressive regimens; (iv) be at hand in excess amounts; and (v) be given within reasonable timing in case of acute organ failures.

## Biological Considerations

### Stem cell plasticity and programmability

The architectural framework underlying normal development of a human being consists of more than 200 histological distinct cell types, each of them possessing an identical set of chromosomal material comprising over 40 000 genes. Current knowledge regarding the complex mechanisms of genetic networks which control early developmental processes arises from an orderly selection of genes within the maternal and paternal genome while the rest are switched off, the latter also referred to as epigenetic inheritance. The process of epigenetic modification involves faithful reprogramming during embryonic development after formation of the zygote. It involves heritable but potentially reversible modifications of maternal and paternal gene segments such as DNA-methylation, histone modifications and binding of protein complexes to chromatin structures to ensure normal embryonic development, respectively.[1,2] Unimpaired epigenetic imprinting of the zygote generates embryonic stem cells which, after isolation from the inner cell mass of a blastocyst, satisfy the characteristic features of pluripotency, namely the ability to differentiate into almost all cells that arise from the three germ layers (ectoderm, mesoderm, and endoderm).[3–5] This inherent plasticity of certain embryonic tissue components has, at least in part, also been ascribed to adult stem cells able to transdifferentiate, a terminus describing the capability of adult stem cells to cross lineage boundaries as defined by the embryologic trilayer.

## Transdifferentiation: An important biological feature of adult stem cells

Evidently, the feature of pluripotency and related plasticity of embryonic and non-embryonic stem cells have generated an enormous interest in academic sciences. The option to produce cell types comprising all three germ line layers, has fostered intensive investigations on various sources of stem cells with a view to their origin, harvest, plasticity profile, their ability to be expandable *in vitro*, and related issues of biological programmability both *in vitro* and *in vivo*. Originally, our understanding of tissue development suggested the specified segregation of cells into defined tissue entities. This view was based on the assumption that maturation of undifferentiated totipotent embryonic stem cells (giving rise to all embryonic and extra-embryonic cell types, including their capacity to constitute for germ line competence) into somatic terminally differentiated target cells can be considered a "one-way road", as, after differentiation, cell-type specification is maintained irreversibly throughout the life span of a given cell. Along this line of thinking, normal development suggests that the progeny of totipotent stem cells undergoes some degree of differentiation resulting in cells with pluripotent properties, whereas pluripotent descendants produce multipotent stem cells (able to generate only a subset of cell lineages). Oligopotent stem cells exhibit even more restricted differentiation plasticity. These tissue-resident stem cells procure only those mature cell types which correspond to their tissue of origin without crossing germ line boundaries.

This inherent lineage commitment of adult (non-embryonic) stem cells, e.g. hematopoietic or bone-marrow-derived stem cells, was put into question by a number of recent animal experiments which suggested that depending on certain circumstances, a process called "transdifferentiation" will allow programming of adult stem cells into a much wider array of differentiated progeny than first assumed. By definition, transdifferentiation describes conversion of a cell with defined lineage commitment into cell types of entirely different lineages. During this procedure, concomitant tissue-specific markers and biological properties of the original cell type are lost and replaced by appropriate phenotypical and functional appearance of the newly formed cell type. A large number of *in vitro* and experimental *in vivo* experiments have been conducted to demonstrate the differentiation plasticity of bone marrow cells, including their transdifferentiation into endothelium,[6,7] neurons,[8] epithelial cells of the lung,[9] gut,[9,10] skin,[9,11] liver,[12–14] and muscle cells.[6,15,16]

## Fusion versus transdifferentiation

Identification of the exact mechanisms underlying the process of transdifferentiation is key to the field of stem cell biology and should provide important clues for the use of stem cells in organ repopulation and regeneration. With respect to the various reports quoted above, the concept of "transdifferentiation" has lately been challenged for several reasons. The failure of reproducibility of the initial reports,[17–19] the low frequency of transdifferentiated cells detected *in vivo* in various animal models, and the phenomenon of cellular fusion as an underlying mechanism of transdifferentiation[20–22] have raised serious doubts about the physiological significance of adult stem cell plasticity.

By using mice bearing genetic deficiency in the enzyme fumarylacetoacetate hydrolase (FAH$^{-/-}$), both Willenbring *et al.*[23] and Camargo *et al.*[24] were able to demonstrate that reconstitution of these mice with wild-type (FAH$^{+/+}$) bone marrow-derived hepatocytes (BMH) or hematopoietic stem cells (HSC), respectively, can rescue FAH($^{-/-}$) recipients by introducing normal copies of the FAH gene. Detailed analysis of the HSC progeny cell types in these experiments revealed fusion of donor BMH or HSC with recipient hepatocytes, respectively. Elimination of lymphocytes from whole bone marrow nonetheless supported liver regeneration, thus pointing to myeloid cells within the BM to account for the fusiogenic potential of the inoculum. Both studies, using genetically traced reporter genes and the Cre/lox recombinase, an enzyme that can induce DNA excision, were able to document that highly differentiated myeloid cells, namely macrophages, can fuse with hepatocytes and genetically complement the FAH enzyme defect. In contrast, Jang *et al.* postulated that a highly distinctive source of HSCs, defined as CD45$^+$ cells from CD45.2 male Rosa26 mice, upon transplantation into liver-injured female recipients transdifferentiated into hepatocytes directly.[25] Their *in vivo* findings are substantiated by *in vitro* experiments showing that exposure of HSCs to damaged liver tissue results in transdifferentiation of "neo-hepatocytes" expressing both the blood marker CD45, albumin, and other hepatocyte markers detected immunohistochemically and via RT-PCR analysis. They speculate that damaged or dying hepatocytes liberate factors that induce blood stem cells to transdifferentiate into hepatocytes *in vitro*.

## Dedifferentiation: A complementary mechanism to supplement for stem cell plasticity

Inherent plasticity of a differentiated state of a cell is a fact and is clearly evident in cases of therapeutic cloning, denoting the introduction of a somatic nucleus into an enucleated oocyte.[26] As aforementioned, the process of differentiation in mammals is generally thought to be unidirectional, however, reversal of the underlying epigenetic modifications has been reported for oligodentrocyte precursor cells.[27] Along this line, it seems plausible that certain culture conditions or a given microenvironmental cellular milieu can reprogram the nucleus of a terminally differentiated cell and it can be surmised that stem cells are particularly prone to such a resetting procedure. Whereas *trans*differentiation describes a direct activation of an otherwise dormant differentiation program to alter lineage specificity of a given stem cell, lineage conversion via *de*differentiation of a somatic, terminally differentiated cell type to a more primitive state of programmability and subsequent *re*differentiation along a new lineage pathway inferably offers a theoretical alternative.[28] The phenomenon of dedifferentiation is a frequently occurring event in amphibian life, inasmuch as lower invertebrates such as earthworms or newts readily regenerate whole body parts when cut apart.[29] Remodeling of a new limb, for instance, requires factors within the newt cytoplasm, which, in response to extracellular signals, are able to reprogram the somatic nucleus to allow DNA replication and cell division. The fact that somatic nuclei of mouse cells underwent dedifferentiation when exposed to factors derived from newt cytoplasm indicates existing similarities as to the ability of resolving an arrested nucleic state between newts and mammals. Recently, reprogramming of permeabilized fibroblasts after exposure to human T cell protein extracts was performed and yielded gene expression pattern to some extent indicative of hematopoietic function.[30] Underlying mechanisms of dedifferentiation in mammals have not yet been clarified in detail. Still a matter of debate is the question whether dedifferentiation results in true uncommitted pluripotent stem cell quality or, alternatively, whether a state of programmability is reached which retains memory of the cell's original identity. All operational criteria defined for true pluripotent stem cells, such as: (i) clonogenic unlimited self-renewal by symmetric replication; (ii) asymmetrical division, giving rise to one identical daughter and multiple differentiated cell types; and (iii) originating from a defined source of embryonic or adult stem cells itself, might not necessarily be met by dedifferentiated somatic cells if used for further programming to substitute as cellular products with restorative and regenerative capacities in various tissues of the human body.

## Non-fusiogenic programmable cells of monocyte origin: A supplementary source in tissue engineering

Embryonic stem (ES) cells are known for their ability to differentiate into various cell types *in vivo* and *in vitro*.[9,31-33] Although ES cells have been isolated from humans, their use in research as well as in clinical applications is encumbered by the relatively small number of existing ES cell lines[34] and by general ethical considerations.[28] Likewise, the use of pluripotent adult stem cells, including tissue-specific stem cells of mesenchymal origin,[35] is hampered by their scarcity whereas their propagation *in vitro*, if possible at all, may require long time periods until sufficient quantities of cells are obtained. Upon application *in vivo* their proliferative capacity may be difficult to control thus raising the question of their safety in regard to potential tumorigenicity.[36]

In order to circumvent the ethical concerns and to solve the lack of "starting material" inherently linked to embryonic and adult stem cells, we sought after a cell that is easily accessible, that can be isolated, expanded and reprogrammed *in vitro* and, subsequently, can be differentiated into the desired cell type. Therefore we addressed an alternative approach which investigated the potential of peripheral blood monocytes for dedifferentiation, e.g. the ability to be reprogrammed for further differentiation into cells of all three germ layers. By now, we have strong evidence that CD14+ cells can undergo a dedifferentiation process which renders them susceptible to overcome embryonically defined lineage boundaries when stimulated with specific growth media (macrophage-colony stimulating factor, M-CSF) for a 6-day period, becoming programmable cells of monocyte origin (PCMO). Much is known about the actions of M-CSF in promoting the proliferation and differentiation of monocytes,[37,38] and also about the effects of IL-3 on the differentiation of monocytes and the self-renewal of pluripotent stem cells.[39,40]

## Programming of Neohepatocytes and Neoislets

In view of the urgent need for autologous cell replacement strategies in the treatment of diabetes or liver diseases and for drug testing/screening purposes, we focussed on the generation and functional evaluation of pancreatic islet and hepatocytes-like cells. We go on to show that these monocyte-derived "neo-islets" and "neo-hepatocytes" acquire highly specialized functions *in vitro*, otherwise only displayed by primary hepatocytes and islet cells, respectively. Moreover, we demonstrate that "neoheps" are able to repopulate critically resected liver organs and restore liver function in animals otherwise succumbing to acute liver failure.

## Neohepatocytes derived from PCMO

Human PCMO-derived "neohepatocytes," briefly "neoheps," bear the property to exhibit a typical hexagonal shape and form confluent layers with cell-cell contacts closely resembling those in primary human hepatocytes after conditioning with a hepatocyte conditioning medium (HCM) for 10 to 14 days (Fig. 1A). RT-PCR analysis reveals that these "neohepatocytes," in contrast to PCMOs, express hepatocyte markers, e.g. $\alpha$-fetoprotein, carbamyl phosphate synthetase I and coagulation factor II (Fig. 1B). "Neoheps" possess unique hepatocyte-like functional activities *in vitro* (Figs. 1C and 1D). Besides their morphological appearance and the expression of highly specific hepatocyte markers, "neoheps" also acquire specialized functional characteristics. An in-depth functional analysis of "neoheps" reveals sustained albumin synthesis and a detoxifying capacity similar in quantity and duration to that in genuine human hepatocytes. Furthermore, "neoheps" show measurable 7-ethoxycoumarin O-de-ethylase activity that is catalyzed by a broad range of P450 isoenzymes such as CYP1A2, CYP2A6, CYP2B6, CYP2C8-9, CYP2E1, and CYP3A3-5. This can be demonstrated by metabolization of ethoxy coumarin into 7OH-coumarin after stimulation with 3-methylcholanthren. In addition, phase II metabolization through UDP-glucuronidation of 4-methyl-umbelliferone (4-MU) appears to be identical in primary human hepatocytes and "neohepatocytes" (data not shown). Phase II enzymes are located in the cytoplasm and microsomes and are associated with the excretion of resultant metabolites or the inactivation of toxic substances. In order to assess the capability of immature PCMO-derived "neoheps" to regenerate critically resected liver organs (>80% liver resection), LEW rats were

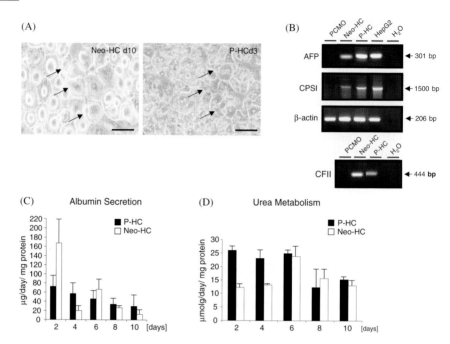

**Figure 1.** (A) Morphology of "neohepatocytes" (Neo-HC, derived from a 10-day incubation of PCMO in HCM) in comparison to primary human hepatocytes (P-HC) cultured for 3 days. The tight junctions between the hexagonally shaped cells are indicated by arrows. (B) RT-PCR analysis of hepatocyte-specific markers α-fetoprotein (AFP), carbamyl phosphate synthetase I (CPS I) and coagulation factor II (CF II). The human hepatoma cell line HepG2 was used as control. (C, D) Functional analysis of "neohepatocytes" *in vitro*. Day-10 "neohepatocytes" cultured for an additional time (as indicated) in hepatocyte-conditioning medium (HCM) were analyzed for (a) albumin secretion and (b) urea metabolism. Data are the mean ± S.D. from 5 to 10 "neohepatocyte" preparations.

treated with the alkaloid retrorsine, a chemical component which blocks hepatocyte proliferation within the liver. After 2 courses or 30 mg/kg × b.w. retrorsine, i.p., over a 4-week period, 80% of the rat liver organs were resected. Experimental groups received $10^6$ immature "neoheps" (PCMO stimulated for 2 days with HCM) intraportally into the remaining liver, or alternatively, unprocessed bone marrow cells or naïve monocytes. Untreated animals received no intraportal injection (see Fig. 2A). As evident from Fig. 2B, "neoheps" and crude bone marrow cells rescued 80% and 60% of experimental animals, respectively, whereas "naïve" non-manipulated monocytes or untreated animals succumbed to acute liver failure.

In brief, PCMO-derived "neoheps" are possible candidate cells for an autologous *in vivo* application to restore liver function in patients suffering

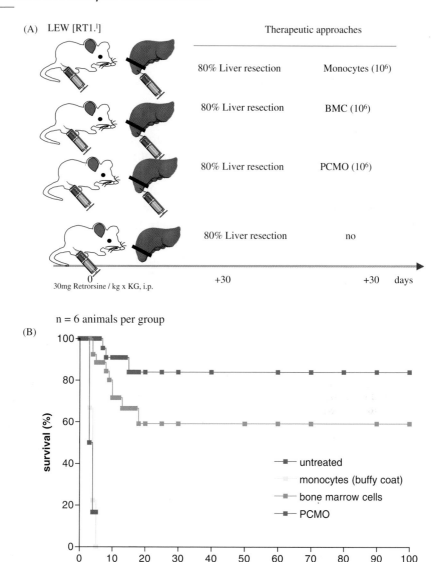

**Figure 2.** (A) Experimental set up to test for *in vivo* capacity of PCMO-derived "neoheps" to restore liver function in critically resected retrorsine pre-conditioned LEW rats. (B) Neohep-treatment rescued >80% of critically resected animals (8/10), crude bone marrow cell grafts sustained liver function in 6 of 10 animals, whereas naïve monocytes and untreated animals succumb to acute liver failure within 8 days.

from extended liver resection or for patients with known liver cirrhosis undergoing liver surgery.

## Islet transplantation models

In animal models, the most widely established protocol is injection of stem cells, pancreatic islet cells, or other programmable cells under the capsule of one of the recipient's kidneys. Injection of islet transplants beneath the kidney capsule represents a common way to test their biological function. After correction of the hyperglycemic state, explantation of the treated kidney organ allows one to distinguish between the functioning transplant and restoration of the recipients' own islets. In addition, islet release from non-viable islets can be ruled out by immunohistochemistry and morphologic examination of the explanted islet/kidney graft. In addition, islets of autologous, allogeneic, and xenogeneic origin can be used in this model as hyperacute rejection is significantly delayed (8 to 10 days after graft injection).

In contrast, the Edmonton protocol for human islet cell transplantation describes intraportal injection of the cellular islet grafts. This application site, however, is minor to other transplantation sites because[41] glucagon synthesis is impaired intrahepatically. Mattson et al.[42] showed that previously transplanted islets, given intraportally into the liver, exhibited minor biological in vitro function when compared to primarily isolated islets. These results support the notion that inside the liver, β-cells do not encounter an in vivo environment supportive for β-cell function and islet survival. Theoretically, intraperitoneal or intraomental administration should constitute optimal environmental conditions for cellular transplants due to good vascularization and nutritional supply of the peritoneal cavity. The disadvantage of intraperitoneal cell injection is linked with its lack of graft localization and the implicated difficulty of graft removal to demonstrate graft function. Alternatively, cells can be encapsulated in alginate capsules[43] as performed in xenogenic islet transplantation models. For clinical application, however, models without graft encapsulation should be preferred as fibrotic and sclerotic alterations of the capsule may hamper long-term graft function.

Hori et al.[44] reported the normalization of blood glucose levels in NOD mice after injection of embryonic stem cell derived β-cells. They treated mouse embryonic stem cells with inhibitors of phosphoinositide

3-kinase *in vitro* and generated cellular aggregates that exhibited glucose-dependent insulin release *in vitro*. Transplantation of these cellular aggregates increased circulating insulin levels, reduced weight loss, improved glycemic control, and completely rescued survival in mice with diabetes mellitus. Graft removal resulted in rapid relapse and death. Graft analysis revealed that transplanted insulin-producing cells remained differentiated, enlarged, and did not form detectable tumors. Just recently, Zorina *et al.*[45] transplanted allogeneic bone marrow (BM) into NOD mice which either suffered from diabetes at a preclinical stage or which exhibited overt clinical disease. This experimental design used lethal and non-lethal doses of radiation for recipient conditioning and was able to show that as low as 1% of initial allogeneic chimerism successfully reversed diabetogenic mechanisms in pre-diabetic NOD mice. In addition, this strategy restored endogenous β-cell function and normalized blood glucose levels if applied after clinical onset of diabetes.

## "Neoislets" derived from PCMO

"Neoislets" are monocyte derived insulin, glucagons, somatostatin and pancreatic-peptide containing endocrine units which, morphologically, share distinct three-dimensional features of cellular cluster aggregates. Programming of "neoislets" requires PCMO as a starting material and addition of an islet-specific conditioning medium containing nictotinamide, epidermal growth factor and glucose. Their principal use is envisioned as intraperitoneal application into the omentum of the abdominal cavity in order to circumvent exogenous insulin treatment for patients suffering from insulin-dependent diabetes mellitus.

*In vitro* and *in vivo* experiments have been conducted to assess the reliability, physiological features and safety of PCMO-derived "neoislets." "Neoislets" express Glut-2 receptor molecules allowing insulin binding. Moreover, "neoislets" are able to store insulin within intracellular vesicles, a zinc-dependent mechanism, demonstrable by dithizon staining. Transcription and protein expression of carboxypeptidase E and associated prohormone convertase in PCMO-derived "neoislets" safeguards insulin release which is glucose concentration dependent, as tested *in vitro*. Moreover, "neoislets" bear the capacity to normalize blood glucose levels within a physiological range between 80 and 140 mg/dl *in vivo* as demonstrated in a diabetic mouse model. Concomitantly, "neoislets" release human C-peptide which allows monitoring islet-function upon transplantation into C-peptide-negative animals or patients. These experiments led to the assumption that "neoislets" will warrant physiological blood glucose levels when used in clinical trials.

## Human *in vitro* studies

Treatment of PCMOs with islet-conditioning medium (ICM) containing EGF, HGF and nicotinamide (IM) for 4–10 days resulted in the appearance of characteristic cell aggregates resembling pancreatic islets (Fig. 3A). Immuno-histochemical analysis of these three-dimensional cellular cluster formations reveals exhibition of all four endocrine cell entities defining normal human islets. As depicted in Fig. 3B, glucagon, insulin, somatostatin and pancreatide-peptide staining cell types can be distinguished within these "neoislets." In addition Pdx-1, a known homeobox gene which directs stem cells to differentiate into islet forming cell clusters is positive in PCMO as soon as two days after addition of ICM (data not shown). We next tested the *in vitro* function of "neoislets." After a 4-day treatment with ICM, the

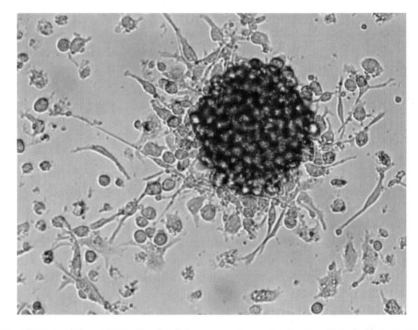

**Figure 3A.** The typical three-dimensional cellular aggregates containing β-cells is generated after a 4- to 6-day period of adding insulin-specific conditioning medium (ICM) to PCMO. These clusters comprise various endocrine cell types and are morphologically similar to human islets of Langerhans.

Insulin-positive  Glucagon-positive

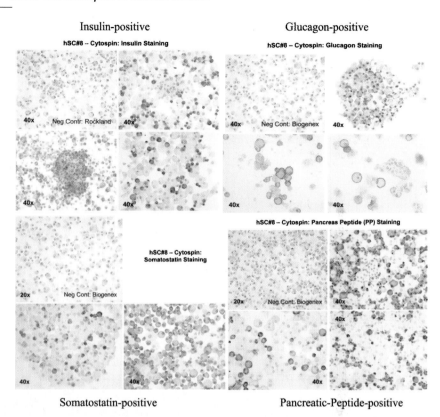

**Figure 3B.** Immunohistochemical analysis reveals four phenotypically different staining patterns within "neoislet" specimens derived from human PCMO. Insulin (*upper left*), glucagon (*upper right*), somatostatin (*lower left*) and pancreatic-peptide (*lower right*) panels. In addition, RT-PCR analysis revealed that these cells contain insulin and glucagon-specific mRNA (data not shown).

insulin and C-peptide content of the supernatant was $0.87 \pm 0.10$ ($n = 5$) and $0.89 \pm 0.07$ ($n = 4$) pg/μg protein, respectively. The subsequent incubation in Kreb's buffer with 3 mM and 22 mM glucose each for 1 hour stimulated the secretion of insulin and C-peptide from undetectable levels to $179.4 \pm 23$ pg/μg protein/60 min ($n = 5$) and to $83.7 \pm 27.5$ pg/μg protein/60 min ($n = 4$), respectively (Fig. 3C), indicating that under these culture conditions the cells showed a glucose-dependent insulin and C-peptide secretion.

The ability to release insulin upon glucose stimulation can be monitored *in vitro* over a minimum period of 14 days as shown (Fig. 3D).

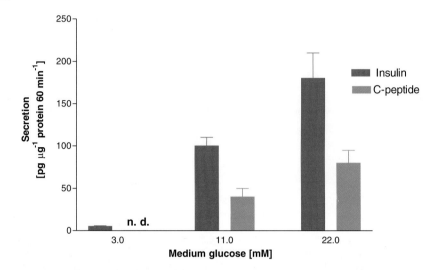

**Figure 3C.** Human insulin and C-peptide levels were measured after stimulation of "neoislets" with 3.0 mM, 11.0 mM and 22 mM glucose-containing medium for 60 min. As evident, "neoislets" respond to glucose stimulation in a linear dose-dependent fashion.

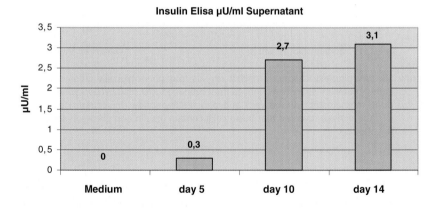

**Figure 3D.** *In vitro* insulin release over time.

## Rodent *in vivo* studies

The potential of PCMO-derived insulin-secreting cells to restore normal glucose values in streptozocin (STZ)-diabetic mice was demonstrated by transplanting $5 \times 10^6$ human "neoislets," cultured in ICM for 4 days, under the kidney capsule of fully immunecompetent Balb/c mice. The cell

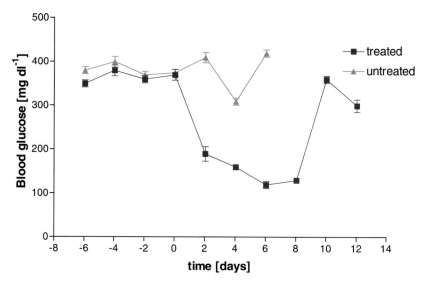

**Figure 3E.** *In vivo* administration of 5 million "neoislets" beneath the kidney capsule of streptozocin-induced diabetic mice (*n* = 5) normalizes blood glucose levels within two days and warrants physiologic glucose concentrations for up to 9 days. After that period, human "neoislets" underwent acute rejection due to xenogeneic immune responses elicited by host mice.

implantation led to correction of hyperglycemia within 2 days (*n* = 7) and the transplanted animals retained normal blood glucose levels up to day 9 post-transplantation (*n* = 5) (Fig. 3E).

## Discussion

An essential introductory remark alluded to the unidirectional pathway thought to determine human cell fate once nuclear programming of a given cell has undergone a still unknown number of molecular events determining cell differentiation. To the best of our knowledge, we here give first hand evidence, that cell fate of blood or spleen derived monocytes can be altered and coaxed to gain plasticity of pluripotent quality. As outlined before, it takes a two-step procedure to turn around a monocyte's destiny. Known physical stressors such as adherence conditions and properties of the plastic surface oxygen and carbon supply, and physical cell-to-cell interactions within the specific cellular microenvironment containing monocytes, lymphocytes, and granulocytes had a strong impact on PCMO's susceptibility to respond to M-CSF and IL-3, *in vitro*. It is noteworthy that non-adherent monocytes cultured in plastic bags in the presence of M-CSF and

IL-3 were unable to develop programmability, in the same way as mere adherence on petri dishes without addition of M-CSF and IL-3 produced negative results (data not shown). The adhesion-enriched monocyte preparations used in this study were $66 \pm 1.5\%$ pure and thus contained a small fraction of non-monocytic cells. It was therefore conceivable that differentiated PCMOs originated from "contaminating" hematopoietic cells or lymphocytes rather than CD14$^+$ cells despite positivity of neohepatocytes for CD14. To exclude lymphocytes and granulocytes as the source of albumin producing neohepatocytes, we determined the methylation status of the colony-stimulating factor-receptor (CSF-R) gene. The second intron of CSF-R has been shown to be hypomethylated in monocytes whereas in lymphocytes and granulocytes from peripheral blood this region is hypermethylated.[46] The epigenetic methylation pattern can thus be used to trace the monocytic origin of a given cell population.[47] We employed a newly developed method based on a previously published protocol[48] that enabled us to quantitatively determine CSF-R methylation. To this end, we found that CSF-R was hypomethylated in monocytes but not in non-monocytic cells, e.g. lymphocytes and granulocytes (not shown). Notably, CD14$^+$ preparations purified before and after various times of HCM treatment showed a stable CSF-R hypomethylation up to the time when "neohepatocytes" expressed hepatocyte-specific markers. The hypomethylation of CSF-R correlated with a strong M-CSF-dependent expression of its protein product, CD115, on PCMOs. Proper CSF-R function, on the other hand, was required for the ability of the cells to resume DNA synthesis as cells infected with a retrovirus encoding a CSF-R point mutant (CSF-R$^{Y809F}$) failed to proliferate in response to M-CSF.[38] Together, these data strongly support the assumption that CD14 and albumin positive "neohepatocytes" originated from peripheral blood monocytes. Finally, if the low number of hematopoietic stem cells in the peripheral blood would have expanded to give rise to "neohepatocytes," this should become evident by the presence of telomerase activity in these cells. However, we could not detect telomerase activity at any time during *in vitro* culture time and the subsequent differentiation in hepatocyte conditioning medium (HCM) (data not shown). Substantial stimuli associated with the property of monocytes to gain stem cell plasticity included both physical attachment and biochemical signaling. Successful resolution of nuclear cell arrest was linked with gene transcription and expression of stem cell markers. Since the cellular integrity of culture-dish attached monocytes was not impaired *in vitro*, extracellular signaling via cell-adhesion mechanisms plus fms receptor activation and IL-3 ligation synergized to trigger downstream transcription factors to initiate monocyte dedifferentiation. To further confirm that PCMOs do not represent a totally dedifferentiated entity, the expression of transcription factors that govern commitment to the monocytic lineage (PU.1, PRDI-BF1/BLIMP-1, ICSBP/IRF8)[48–52] was assayed by semi-quantitative RT-PCR. During the derivation of PCMO from monocytes, the transcript levels of both the β-isoform of PRDI-BF1, and its upstream activator ICSBP[53] declined over the 6-d culture period. Strikingly, the most pronounced decrease was observed between d 3 and 4, coinciding with the onset of cell division. Monocytes cultured for 6d without M-CSF and IL-3 did not lose expression of PRDI-BF1 and ICSBP (data not shown).

In line with the dedifferentiation mechanisms in newt, nuclear reprogramming of terminally differentiated monocytes was also linked with their capacity to divide again. However, programmability was transient and limited to approximately 10 days.

Apparently, reprogrammed monocytes expressing stem cell markers like CD90 or CD117 still coexpressed the inherent lineage marker CD14. By definition, PCMOs thus do not meet the biologic criteria of true stem cells which require a lineage negative phenotype. In this context, it is important to note that after programming of PCMO to yield albumin- or insulin-positive cells with hepatocyte or β-cell morphology, respectively, PCMO progeny still coexpressed CD14 within the first 21 days after the initial reprogramming step. The fact that 12- to 16-day old PCMO lose programmability and differentiate back to give rise to normal macrophages and immature dendritic cell types, if not conditioned otherwise, underlines the transient nature of reprogramming, which by itself protects PCMO from unrestrained proliferation and tumor growth, an important clinical aspect when considered for patient use.

These findings clearly distinguish PCMO from naturally occurring stem cells, from bone-marrow derived multipotent adult progenitor cells[31,54] and from recently described blood-borne macrophages resembling fibroblasts, termed f-macrophages,[55] which all exhibit totally different self-renewal and replication patterns.

It could be argued that the results presented here do not exclude the possibility that different types of monocyte subpopulations existed which were able to respond to the various conditioned media used to generate various cell types, thus lacking true clonality and subsequent pluripotency.

The argument here, again, is to emphasize that it does not necessarily need a single cell with the self-renewal capacity of stem cells to yield large amounts of specialized cells with defined biological function. Indeed, given a leukapherese product consisting of $10^{10}$ CD14$^+$ cells, we were able to procure an average of $10^9$ to $5 \times 10^9$ albumin-producing cells with hepatocyte morphology within 10 to 12 days.

In conclusion, whatever cell source is chosen to gain a specified cellular product envisaged for tissue repair and treatment of human diseases, it should comply with important clinical requirements, including patient safety, source, availability, and lack of immunogenicity being autologous in nature.

## REFERENCES

1. Surami AM. (2001) Reprogramming of genome function through epigenetic inheritance. *Nature* **414**:122–128.
2. Rideout WM, Eggan K, Jaenisch R. (2001) Nuclear cloning and epigenetic reprogramming of the genome. *Science* **293**:1093–1098.
3. Nagy A, Gocza E, Diaz EM, *et al.* (1990) Embryonic stem cells alone are able to support fetal development in the mouse. *Development* **110**:815–821.
4. Thomson JA, Itskovitz-Eldor J, Shapiro SS, *et al.* (1998) Embryonic stem cell lines derived from human blastocysts. *Science* **282**:1145–1147.
5. Smith AG. (2001) Embryo-derived stem cells: of mice and men. *Annu Rev Cell Dev Biol* **17**:435–462.
6. Jackson KA, Majka SM, Wang H, *et al.* (2001) Regeneration of ischemic cardiac muscle and vascular endothelium by adult stem cells. *J Clin Invest* **107**:1395–1402.
7. Asahara T, Masuda H, Takahashi T, *et al.* (1999) Bone marrow origin of endothelial progenitor cells responsible for postnatal vasculogenesis in physiological and pathological neovascularization. *Circ Res* **85**:221–228.
8. Mezey E, Chandross KJ, Harta G, *et al.* (2000) Turning blood into brain: cells bearing neuronal antigens generated in vivo from bone marrow. *Science* **290**:1779–1782.
9. Krause DS, Theise ND, Collector MI, *et al.* (2001) Multi-organ, multi-lineage engraftment by a single bone marrow-derived stem cell. *Cell* **105**:369–377.
10. Okamoto R, Yajima T, Yamazaki M, *et al.* (2002) Damaged epithelia regenerated by bone marrow-derived cells in the human gastrointestinal tract. *Nat Med* **8**:1011–1017.
11. Korbling M, Katz RL, Khanna A, *et al.* (2002) Hepatocytes and epithelial cells of donor origin in recipients of peripheral-blood stem cells. *N Engl J Med* **346**:738–746.

12. Theise ND, Nimmakayalu M, Gardner R, *et al.* (2000) Liver from bone marrow in humans. *Hepatology* **32**:11–16.

13. Lagasse E, Connors H, Al-Dhalimy M, *et al.* (2000) Purified hematopoietic stem cells can differentiate into hepatocytes in vivo. *Nat Med* **6**:1229–1234.

14. Petersen BE, Bowen WC, Patrene KD, *et al.* (1999) Bone marrow as a potential source of hepatic oval cells. *Science* **284**:1168–1170.

15. Ferrari G, Cusella-De Angelis G, Coletta M, *et al.* (1998) Muscle regeneration by bone marrow-derived myogenic progenitors. *Science* **279**:1528–1530.

16. Gussoni E, Soneoka Y, Strickland CD, *et al.* (1999) Dystrophin expression in the mdx mouse restored by stem cell transplantation. *Nature* **401**:390–394.

17. Camargo FD, Green R, Capetenaki Y, *et al.* (2003) Single hematopoietic stem cells generate skeletal muscle through myeloid intermediates. *Nat Med* **9**:1520–1527.

18. Castro RF, Jackson KA, Goodell MA, *et al.* (2002) Failure of bone marrow cells to transdifferentiate into neural cells in vivo. *Science* **297**:1299.

19. Wagers AJ, Sherwood RI, Christensen JL, Weissman IL. (2002) Little evidence for developmental plasticity of adult hematopoietic stem cells. *Science* **297**:2256–2259.

20. Wang X, Willenbring H, Akkari Y, *et al.* (2003) Cell fusion is the principal source of bone-marrow-derived hepatocytes. *Nature* **422**:897–901.

21. Vassilopoulos G, Wang PR, Russell DW. (2003) Transplanted bone marrow regenerates liver by cell fusion. *Nature* **422**:901–904.

22. Alvarez-Dolado M, Pardal R, Garcia-Verdugo JM, *et al.* (2003) Fusion of bone-marrow-derived cells with Purkinje neurons cardiomyocytes and hepatocytes. *Nature* **425**:968–973.

23. Willenbring H, Bailey AS, Foster M, *et al.* (2004) Myelomonocytic cells are sufficient for therapeutic cell fusion in liver. *Nat Med* **10**:744–748.

24. Camargo FD, Finegold M, Goodell MA. (2004) Hematopoietic myelomonocytic cells are the major source of hepatocyte fusion partners. *J Clin Invest* **113**:1266–1270.

25. Jang YY, Collector MI, Baylin SB, *et al.* (2004) Hematopoietic stem cells convert into liver cells within days without fusion. *Nat Cell Biol* **6**:532–539.

26. Hochedlinger K, Jaenisch R. (2003) Nuclear transplantation, embryonic stem cells, and the potential for cell therapy. *N Engl J Med* **349**:275–286.

27. Kondo T, Raff M. (2000) Oligodendrocyte precursor cells reprogrammed to become multipotential CNS stem cells. *Science* **289**:1754–1757.

28. Wagers AJ, Weissman IL. (2004) Plasticity of adult stem cells. *Cell* **116**:639–648.

29. Brockes JP, Kumar A. (2002) Plasticity and reprogramming of differentiated cells in amphibian regeneration. *Nat Rev Mol Cell Biol* **3**:566–574.

30. Hakelien AM, Landsverk HB, Robl JM, *et al.* (2002) Reprogramming fibroblasts to express T-cell functions using cell extracts. *Nat Biotech* **20**:460–466.

31. Thomson JA, Itskovitz-Eldor J, Shapiro SS, *et al.* (1998) Embryonic stem cell lines derived from human blastocysts. *Science* **282**:1145–1147.
32. Ruhnke M, Ungefroren H, Zehle G, *et al.* (2003) Long-term culture and differentiation of rat embryonic stem cell-like cells into neuronal, glial, endothelial and hepatic lineages. *Stem Cells.* **21**:428–436.
33. Bjornson C, Retze R, Reynolds B, *et al.* (1999) Turning brain into blood: a hematopoietic fate adopted by neural stem cells in vivo. *Science* **283**:534–537.
34. Preston SL, Alison MR, Forbes SJ, *et al.* (2003) The new stem cell biology: something for everyone. *J Clin Pathol Mol Pathol* **56**:86–96.
35. Jiang Y, Jahagirdar BN, Reinhardt RL, *et al.* (2002) Pluripotency of mesenchymal stem cells derived from adult marrow. *Nature* **418**:41–49.
36. Rosenthal N. (2003) Prometheus's vulture and the stem-cell promise. *N Engl J Med* **349**:267–274.
37. Hass R, Gunji H, Datta R, *et al.* (1992) Differentiation and retrodifferentiation of human myeloid leukemia cells is associated with reversible induction of cell cycle-regulatory genes. *Cancer Res* **52**:1445–1450.
38. Roussel MF. (1997) Regulation of cell cycle entry and G1 progression by CSF-1. *Mol Reprod Dev* **46**:11–18.
39. Mangi MH, Newland AC. (1999) Interleukin-3 in haematology and oncology: current state of knowledge and future directions. *Cytokines Cell Mol Ther* **5**:87–95.
40. Khapli SM, Mangashetti LS, Yogesha SD, Wani MR. (2003) IL-3 acts directly on osteoclast precursors and irreversibly inhibits receptor activator of NF-kappa B ligand-induced osteoclast differentiation by diverting the cells to macrophage lineage. *J Immunol* **171**:142–151.
41. Gupta V, Wahoff DC, Rooney DP, *et al.* (1997) The defective glucagon response from transplanted intrahepatic pancreatic islets during hypoglycemia is transplantation site-determined. *Diabetes* **46**:28–33.
42. Mattsson G, Jansson L, Nordin A, *et al.* (2004) Evidence of functional impairment of syngeneically transplanted mouse pancreatic islets retrieved from the liver. *Diabetes* **53**:948–954.
43. Ulrichs K, Hamelmann W, Buhler C, *et al.* (1999) Transplantation of porcine Langerhans islets for therapy of type I diabetes. The way to clinical application. *A Zentralbl Chir* **124**:628–635.
44. Hori Y, Rulifson IC, Tsai BC, *et al.* (2002) Growth inhibitors promote differentiation of insulin-producing tissue from embryonic stem cells. *PNAS USA* **99**:16105–16110.
45. Zorina TD, Subbotin VM, Bertera S, *et al.* (2003) Recovery of the endogenous beta cell function in the NOD model of autoimmune diabetes. *Stem Cells* **21**:377–388.
46. Felgner J, Kreipe H, Heidorn K, *et al.* (1992) Lineage-specific methylation of the c-fms gene in blood cells and macrophages. *Leukemia* **6**:420–425.

47. Parwaresch MR, Kreipe H, Felgner J, *et al.* (1990) M-CSF and M-CSF-receptor gene expression in acute myelomonocytic leukemias. *Leuk Res* **14**:27–37.
48. Lehmann U, Hasemeier B, Lilischkis R, Kreipe H. (2001) Quantitative analysis of promoter hypermethylation in laser-microdissected archival specimens. *Lab Invest* **81**:635–638.
49. Odorico JS, Kaufman DS, Thomson JA. (2001) Multilineage differentiation from human embryonic stem cell lines. *Stem Cells* **19**:193–204.
50. Friedman AD. (2002) Transcriptional regulation of granulocyte and monocyte development. *Oncogene* **21**:3377–3390.
51. Chang DH, Angelin-Duclos C, Calame K. (2000) BLIMP-1: trigger for differentiation of the myeloid lineage. *Nat Immunol* **1**:169–176.
52. Györy I, Fejér G, Ghosh N, *et al.* (2003) Identification of a functionally impaired positive regulatory domain I binding factor 1 transcription repressor in myeloma cell lines. *J Immunol* **170**:3125–3133.
53. Tamura T, Kong HJ, Tunyaplin C, *et al.* (2003) ICSBP/IRF-8 inhibits mitogenic activity of p210 Bcr/Abl in differentiating myeloid progenitor cells. *Blood* **102**:4547–4554.
54. Schwartz RE, Reyes M, Koodie L, *et al.* (2002) Multipotent adult progenitor cells from bone marrow differentiate into functional hepatocyte-like cells. *J Clin Invest* **109**:1291–1302.
55. Zhao Y, Glensne D, Huberman E. (2003) A human peripheral blood monocyte-derived subset acts as pluripotent stem cells. *PNAS USA* **100**:2426–2431.

# 26

# The Challenges of Cell-based Therapy

Sir Roy Calne*

## Introduction

To treat disease with cells is not a new concept. In the 17th century, before the nature of cell structure and function was known, blood transfusion experiments were performed between animals using feather quill needles by Christopher Wren and his friends. For blood transfusions to be of value, rather than a "Russian roulette" for sudden death, a means of preventing clot formation and an understanding of red blood cell groups were necessary. Then blood transfusions became life-saving and opened the door to modern major surgery.

In the 1950s, advances in immunology spearheaded by Peter Medawar and his colleagues revealed an immune system vital to life that could be manipulated by cell injection of animals *in utero* and allow acceptance of skin grafts from the cell donors — called "acquired immunological tolerance."[1]

An important advance in the treatment of hematological diseases followed from the demonstration that animals given "lethal" doses of total body x-irradiation could be rescued by intravenous bone marrow infusions. The grafted bone marrow cells homed to the empty bone marrow spaces where the native marrow had been destroyed by the X-rays.[2] The donated marrow cells conferred on the recipients the immune characteristics of the donor. The closer the matching of the major histomatibility complex (MHC) between donor and recipient, the greater the likelihood of success.

More recently it has been possible to condition leukemia patients to accept bone marrow grafts from well-matched donors without the need

*Correspondence: Department of Surgery, Douglas House Annexe, 18 Trumpington Road, Cambridge CB3 2AH. Tel: 44 1223 361467, fax: 44 1223 301601, e-mail: ryc1000@cam.ac.uk

for complete destruction of the recipient bone marrow. This non-ablative treatment can result in mixed macro-chimaerism, with blood cells of both donor and recipient co-existing in the bone marrow and blood, so that it is possible to have the advantage of graft-versus-leukemia immune reactivity, without excessively harsh treatment of the patient.[3] Moreover, this mixed chimaerism even if only temporary, can result in kidney graft acceptance from the bone marrow donor.[4]

There is therefore a large literature and a long follow-up of clinical experience with therapeutic cell transplantation.

In the past 50 years, since the description of the double helical structure of DNA and an understanding of the mechanisms of protein synthesis, an accelerating advance in our knowledge of the molecular nature of many diseases has occurred. Many of the genes responsible have been identified and suggestions made as to how they might be used as engineering tools for therapeutic purposes.

The proliferation in culture of embryonic stem cells and the cloning of intact animals from adult somatic cells is now a challenge to provide cell therapy for many conditions that currently have inadequate treatment.[5]

In this short chapter, I will consider the difficulties and potential goals for cell treatment, using two diseases as examples, namely hemophilia and diabetes. One is a rare but often severely disabling condition, the other, one of the commonest chronic diseases, a cause of immense human suffering and representing an ever-increasing drain on the healthcare budget of all nations.

# The Clinical Problems

## Hemophilia A

Hemophilia A is an X-linked hemorrhagic disease due to a mutation of the gene encoding for factor VIII. The incidence is 1 in 5000 males and 60% of patients with the mutant gene are severely affected with less than 1% of normal factor VIII activity. Those with 1–5% have moderate disease and over 5% the disorder is mild. Intravenous exogenous factor VIII is the accepted treatment that does not always prevent bleeding episodes and is expensive. Bleeding into joints leads to arthropathy which can be severely disabling and bleeding can occur at any site causing a multiplicity of symptoms. Plasma and blood transfusions have caused severe viral infections, notably HIV and hepatitis B and C.

Hemophilia A is a suitable disease for consideration of gene therapy since an activity of over 5% of normal factor VIII is beneficial. The broad therapeutic index of factor VIII means that overdose is unlikely, and even low levels can be of therapeutic value. Production of factor VIII is not regulated by bleeding episodes.

Thus, in hemophilia A the condition should be managed by gene therapy producing 5–50% of normal blood levels of factor VIII. The mutant gene lacks normal expression; insertion of an engineered gene is a possible approach.[6] Roth *et al.*[7] described an ingenious solution of this objective using a non-viral gene delivery to *ex vivo* cultured fibroblasts of the patient. The gene engineered fibroblasts were selected and re-injected as autografts via a laparoscope into the omentum.

The clinical trial followed successful experiments in mice. Six patients were treated and there were no observed side-effects. In four patients, the plasma levels of factor VIII rose, coinciding with reduced bleeding episodes and the need for exogenous factor VIII. In one patient, activity persisted for 10 months.

This trial showed the feasibility of this approach using autografts and therefore avoiding the need for immunosuppression. The main disadvantages were the short duration of effect in most patients and the need for laparoscopic cell delivery; also, such a bespoke manipulation would be expensive. Nevertheless, this study clearly defines an obvious goal, namely the safe administration of a functioning gene in a patient with disease due to the inability to express that gene.

## Diabetes

There are two forms of diabetes, type I and type II that differ in their pathogenesis. Type I is an autoimmune disease associated with certain genetic HLA configurations most commonly present between infancy and teenage years, but can also present in adults. Often, but not necessarily, the onset follows a viral infection and can be insidious. The β cells in pancreatic islets of Langerhans are singled out for immune destruction by primed T-cells, whose molecular target has not yet been defined. Recovery of the β cell mass cannot occur due to continuing autoimmune activity and insufficient progenitor cells. Before the introduction of insulin in the 1920s, patients died, usually in a distressing, emaciated state, around puberty, before they could have children. Refinements in insulin therapy and a strict diet can restore patients to a relatively normal life, but even with excellent compliance to the regimen of frequent blood sugar estimations and insulin injections, the secondary complications of diabetes can develop in a relentless progressive manner causing blindness, renal failure, gangrene, coronary arterial disease and neuropathy. Inappropriate management of the therapeutic regimen can lead to dangerous and sometimes fatal hypoglycemia, often with no warning for the patient. Insufficient insulin results in hyperglycemic ketosis and diabetic coma.

The diagnosis of type I diabetes in a child is a sentence to a lifelong strict regimen of diet and medication and is a major and continuing trauma to the whole family.

Type II diabetes is a common condition with many patients only mildly affected. The disease usually presents in adults but can present in children. It is especially common in obese people and has reached almost an epidemic scale in India and Southeast Asia. Change from a frugal traditional diet to a liberal western-style of food has been blamed for the sudden increase in incidence of type II diabetes in the Eastern countries. Initially, many patients can be managed by diet and oral hypoglycemic agents. Insulin resistance in the tissues is a feature of type II diabetes and the β cell mass may increase, producing excessive insulin apparently in an attempt to overcome the resistance. Eventually, there is β cell failure and in approximately half of the cases exogenous insulin injections are necessary and the same secondary complications occur as in type I diabetes.

Taken together, type I and type II diabetes result in serious morbidity and mortality in all communities. Diabetes is a major cause of blindness and renal failure. In addition to the cost in human suffering, the financial burden of diabetes on healthcare resources is enormous and accelerating yearly as the incidence of both type I and type II diabetes increases.

## The Role of Insulin

The history of insulin is fascinating and has been told especially well by Michael Bliss in *The Discovery of Insulin*.[8]

In 1889, Minkawski and Von Mering, in Strasbourg found that dogs subjected to pancreatectomy became diabetic. One account of the finding was that the technician raised the suspicion of sugar in the urine to Minkawski, by observing flies settling in large numbers on the puddles of urine passed by the diabetic dogs, in contrast to their relative lack of interest in the urine of normal dogs.

In 1869 Paul Langerhans, a medical student writing his thesis, observed microscopic islands of different structure to the main mass of digestive enzyme secreting pancreas. This seminal observation, perhaps the most perspicacious of any medical student, led to intense study of the islets. They are miniature organs embedded within the pancreas in most creatures, but constituting separate independent organs in some fish. Each islet consists of approximately 1000 cells of four distinct types each with its own secretion task:

α   cells producing glucogon
β   cells producing insulin. They constitute 60–80% of the cells in the islets, i.e. 6–800 cells/islet
δ   cells producing somatostatin
pp  cells producing pancreatic polypeptide.

There is a delicate and profuse capillary network and nerve connections in the islet, somewhat resembling the renal glomerulus. The capillaries of the islets anastomose with the main pancreatic vasculature which may facilitate signaling between endocrine and exocrine pancreatic cells. The interaction of cytokines between the individual cell types may be important attributes that would be lost to separated islets or surrogate β cells. The pancreas contains one million islets and therefore $6–8 \times 10^8$ β cells. The endocrine secretions of the islets enter the portal blood and the first organ they reach is the liver. Insulin is partially metabolized by the liver, which converts glucose to glycogen.

In the 1920s, the connection between removal of the pancreas and diabetes was established, but various oral preparations of pancreas did not ameliorate diabetes. The young orthopedic surgeon, Frederick Banting, working in Toronto, was convinced that an extract of pancreas injected would provide the vital substance missing in diabetes. With the technical assistance and a major intellectual contribution from a medical student, Charles Best, the two rather low profile researchers produced an extract of pancreas that lowered the blood sugar of diabetic dogs and eventually in 1922, they persuaded clinical colleagues to try a similar extract in diabetic patients. Some but not all of the early clinical cases responded, but first the help of a protein chemist, James Collip was needed. There was much opposition from conservative clinicians, but eventually the concept was accepted that a substance from the pancreatic islets called "insulin" could be used as a treatment for diabetic patients. It soon became apparent that a large commercial pharma company, with deep pockets and prepared to accept a risky project, would be required to produce enough of the substance in relative purity to provide lifelong treatment. The Eli Lilly Company stepped in, rose to this challenge, and the lives of diabetics were transformed, albeit with the reservations of the diabetic way of life and the risk of complications to which I have referred.

The molecular structure of the complicated protein insulin was determined in Cambridge in the 1950s at the Laboratory of Molecular Biology by Frederick Sanger in the course of his first Nobel Prize work. The physiology of insulin and the control of glucose metabolism is complex. Before active insulin is available, a non-active molecule called C-peptide must be cleaved from the parent molecular proinsulin. There is an important basal secretion of insulin, but on the intake of food, insulin granules, stored in the β cells, are released in a pulsatile manner simultaneously from a number of β cells, in amounts relating to the ambient blood glucose concentration in the islets. The timing is critical. If released too early or too late, high insulin blood levels will cause inappropriate, possibly dangerous, hypoglycemia. If not enough insulin is available at the appropriate time, normal glucose metabolism cannot take place and the blood sugar level will rise. There is a considerable reserve of β cell function, so after even a large meal not all the β cells exhaust their supply of secreted insulin from within their cell membranes. There is a slow turnover of β cells, perhaps around 5% per annum in man, from progenitor cells present in the islets and/or in the ducts of the exocrine pancreas. In rodents the turnover is much greater.[9]

The chemistry of insulin secretion varies in different species. In man an inactive pro-insulin is the first main synthetic step and this becomes cleaved into the inactive C-peptide, a marker of insulin synthesis and insulin. In mice there are two active insulins, I and II. In diabetic patients, the level of glycosolated hemoglobin rises. The interactions between insulin, glucogon and other endocrine secretions are complicated and in some patients, microangiopathy develops in the retinae, glomeruli, and small blood vessels throughout the body associated with serious complications.

First passage of insulin through the liver is physiological, but release of insulin directly into the caval venous system appears to be well tolerated following vascularized pancreatic transplants.

## The Problem

It will be clear from the preceding short summary that in contrast to a consideration of cell treatment for hemophilia A, cell treatment for diabetes is likely to be orders of magnitude more difficult, but in view of the morbidity, costs and inadequacy of current therapies and advances made in cell biology, it is surely worth attempting.

## Vascularized Pancreas Transplantations

Surgical transplantation of a vascularized whole pancreas or even half a pancreas can give excellent long-term results,[8] with cure of diabetes in many cases. Most patients have suffered from diabetic renal failure and often it has been possible to transplant a kidney and a pancreas from the same donor. Powerful lifelong immunosuppression is necessary, but this would be standard treatment for the kidney graft. The operation is a major surgical procedure with the special danger of leakage of pancreatic digestive enzymes, but results are improving steadily. Unfortunately, the incidence of diabetes is far in excess of the availability of donor pancreata.

## Islet Cell Transplantation

Since islets, when separated, are small enough to survive temporarily in a suitable environment, by simple diffusion of nutrients and oxygen into them and $CO_2$ and waste products out, whilst a new blood supply is established, the idea of transplanting islets based on the same concepts as split skin grafts is an old one. Islets, however, do not part company with their surroundings in the pancreas easily. In rodents, they can be hand-picked under a dissecting microscope, but in large animals, including man, enzymatic digestion and mechanical chopping of the pancreas are necessary. The islets are vulnerable to damage from ischemia and the effects of collagenase and the more refined enzyme "liberase." Dicing the pancreas into small pieces also damages the islets. An elaborate, highly skilful and prolonged process is necessary. Five people working for five hours, with a cooled pancreas removed immediately from a brain-dead cadaver may, in the best circumstances, produce about 3–400,000 or 1/3 of the total number of islets in a tolerably well-preserved state suitable for transplantation. Yet, twice that number are required to release a patient from the need for insulin injections. The islet isolation procedure has some fanciful resemblance to digging for potatoes on a dark night with a sharp spade.

The next questions are:

- Should the islets be cultured before transplantation?
- Can they be safely frozen and thawed?
- Most importantly, where to transplant them.

In mice, under the kidney capsule is a good site to inject islets despite the caval drainage of insulin. In man, the portal blood stream has been most favored, the islets hopefully lodging as microemboli in the liver sinusoids, where they take up residence and after a few days acquire a new blood supply, mainly from recipient capillaries growing into the transplanted islets. Islets floating in the blood are in an abnormal environment and may activate complement, causing local platelet aggregation and clot formation precluding rapid neovascularization and endangering liver parenchyma to ischemia.[11,12] An optimal site for islet transplantation has yet to be found; in the meantime, the report of clinical islet transplantation by Shapiro *et al.* in Edmonton has marked a halt to the extensive scepticism that prevailed in the transplant community for clinical islet grating.[13]

Using usually two cadaveric pancreas donors per recipient and immunosuppression designed to try and avoid diabetogenic toxicity, the Edmonton workers obtained 80% one-year independence from exogenous insulin and 70% at two years in type I diabetic patients with brittle disease, usually involving hypoglycemic unawareness but without other serious diabetic complications. Repeating their results has only been possible in a few of the specialised centres that have made the attempt.

The shortage of suitable human cadaveric pancreata and the huge numbers of diabetics would make it reasonable to view the Edmonton experience as an extremely important "proof of principle" that the procedure is possible but at great cost of healthcare resource and skilled technical ability, with the lucky patients no longer requiring insulin but nevertheless having to take full doses of immunosuppressive drugs indefinitely. No doubt better yields of islet extraction will be achieved and safer immunosuppressions developed, but the disadvantages outlined above remain.

## Xeno-islet Grafting

Pig insulin differs from human insulin only in one amino acid. Porcine insulin has been used successfully therapeutically in patients for many years. Porcine glucose homeostasis is similar to man's and pig islets are potentially available in large numbers and can be extracted in a similar manner to that used for human islets. The pig, however, is a different species, separated from man in evolution for many millions of years and of the hundreds or even thousands of proteins produced by pig cells, each is different to the human equivalent and most are capable of eliciting immune destructive reactions following transplantation.

To date results of xeno-islet transplantation to primate species have been disappointing, but Bernard Hering has recently obtained encouraging results of pig to monkey islet grafts using powerful immunosuppression with agents that could be used in patients.[14] The question again arises: Does the immunosuppression justify the procedure? There are hopes that genetic engineering of pigs by "knock out" and "knock-in" genes to make pigs more like humans or at least make their tissues more acceptable as grafts to man may one day be successful but how soon that will be cannot be predicted. Many transplant researchers have sympathy with Norman Shumway's comment that "xenografting is the future of organ transplantation and always will be!"

## Other Approaches

### 1. Large-scale proliferative culture of β cell progenitor cells in pancreatic ducts or from islet β cells

This is attractive in that these are the cells that normally produce β cells, but to date there has been a severe shortfall in numbers of β cells that can be produced in culture and, so far, the numbers are far below the threshold of therapeutic use.[15] Also the site of origins of the precursor cells is disputed.[16]

### 2. Transdifferentation of liver cells to islet cells

Both liver and pancreas develop from the same embryological rudiment so, by the use of certain growth factors and cultural procedures, workers have succeeded in taking this step in experimental settings.

### 3. *In vivo* "cultural" growth of embryonic pancreas rudiments

Hammerman in St. Louis[17] and Reissner in Israel[18] have achieved considerable progress in this endeavor, but any clinical application would seem to require an excessively costly "bespoke" individual approach for each patient. The use of fetal tissue would raise worrying ethical dilemmas.[19] The justification of using a fetus to treat a patient with diabetes might be difficult to sustain.

### 4. Guide or Engineer

Guide or engineer undifferentiated or differentiated cells to act as surrogate β cells.

### i. Embryonic stem cells (ESCs)

Since ESCs can and do turn into every cell type in the body, their use for producing β cells has received much publicity and Soria has been successful in introducing the human insulin gene into mouse ESCs (MESCs) and selecting the cells producing insulin to treat diabetic mice successfully.[20]

This was an important achievement, but may be difficult to translate in the context of HESCs, which grow more slowly and are more vulnerable to culture than MESCs. Monkey ESCs have been differentiated into pancreatic cell phenotypes.[21]

If an *in vitro* process using HESCs was successful, it would be of vital importance to eliminate every undifferentiated cell from the innoculum to be given to patients because of the risk that such cells might differentiate into teratomata.[22]

### ii. Adult "Stem Cells"

Multipotent cells have been identified in a number of adult tissues and in umbilical cord blood. They are the source of successful bone marrow grafts and may have the potential to differentiate into other cell lineages, though such claims are disputed.

Blood monocytes have been shown to de-differentiate under certain cultural conditions, into cells which can be persuaded with growth factors and certain cultural conditions to proliferate some 5- to 6-fold and then differentiate into liver-like cells producing albumen; islet-like cells producing insulin and glucogen and fat cells; or return back to monocytes.[23–25]

### iii. Transfecting adult cells with the human insulin gene with or without a glucose sensing promoter

This approach can use non-viral electroporition to introduce the insulin gene plasmid into cells *in vitro* or *in vivo*.[26] Alternatively, viral vectors can be used which are more efficient, but some viruses have the danger of unmasking oncogenes.[27]

## Discussion

An attractive theoretical concept is to introduce the glucose sensing human insulin gene into stable autologous cells of the patient, so avoiding the need for immunosuppression, provided the immune target in the case of type I diabetes does not reside in these cells. The goal is similar to that outlined for the treatment of hemophilia A. Unanswered is the question, whether enough insulin could be produced at the right time for long enough to be of therapeutic use. These matters are likely to be more problematic in the case of diabetes, because of the greater demands and increased complexity. Nevertheless, the stage is sufficiently set for an attempt to be made to recruit the actors, learn the lines and put on an acceptable play, that might bring a successful new treatment to many thousands of patients.

How can a cell be persuaded to adopt a new specialized role and act the part energetically for a long time? Every nucleated cell with the exception of gametes has a "full house" of genes in their nuclear DNA, yet until the end of the 20th century, most biologists believed that cells evolved with special functions had "crossed the Rubicon" of plasticity and were settled in their designated role. Deviation would lead to metaplasia or tumors, both pathological states.

With the cloning of "Dolly" subsequently followed by many other examples of mammalian cloning in a variety of species from adult cells, it was shown that in extreme conditions of the laboratory it was possible to reawaken the nucleus of an adult cell to behave with the totipotency of a zygote.

It is not yet clear which adult cell nuclei will behave in this way; the plasticity could be limited to certain cell types; nor has cloning yet been possible in all species that have been investigated. Moreover, a successful cloning outcome is a chancy business, current success of 1 in 200 attempts would be an average score.

No matter how one considers the data, it is difficult to escape the conclusion that adult mammals harbor cells that can be persuaded by extreme means to fully de-differentiate into embryonic stem cell behavior. From this hypothesis, the task of finding cells in adult animals that could be programmed to change their behavior to concentrate their efforts to synthese a single protein, e.g. factor VIII or insulin, would not seem to be an impossible dream. To achieve this goal would be the first step towards cell-based therapy. To produce enough of the specific protein in response to appropriate stimuli for a long period of time would be the next hurdle to overcome.

For an assault on the first goal, there are a variety of possibilities each with its own attractive features and disadvantages.

## 1. Cultural Persuasion

Grow cells in culture in the presence of differentiation factors that are known to be present in the embryonic development of the tissue in question. The sources of such cells could vary, for example:

### i. Embryonic Stem Cells (ESCs)

A popular choice with the media attracted by the idea that ES cells can change into any cell type. To take an ESC all the way to a β cell may involve numerous changing differentiation environments in normal development, probably influenced by what is going on in adjacent tissues. Currently, we have only fragmentary knowledge of these processes so a deliberate, dedicated approach to reproduce them *in vitro* is not possible. Some advocates would hopefully believe in lucky shortcuts, but there is little to support such optimism. Recent publications using MESCs have so far produced only modest progress. The selected programmed cells may produce only 1/50 of insulin that a normal β cell will synthesize. Human embryonic stem cells grow slowly and are delicate. They have a tendency to differentiate spontaneously in an uncontrolled manner. If a culture could be produced that resembled β cells, it would be essential but difficult to ensure that every undifferentiated cell was eliminated before using the cell culture therapeutically, since in theory, one undifferentiated ESC could produce a malignant teratoma *in vivo*.

No doubt there would be voracious opposition by a minority to the therapeutic use of HESCs, but if the result was a major step in helping sick patients it is likely that there would eventually be acceptance of the procedure, just as occurred historically with blood transfusions.

HESCs would be expected to elicit an immune response in HLA mismatched recipients, but the immunogenicity of HESCs as allografts is not known and it was encouraging that rat ES-like inner cell mass cells, spontaneously produced allograft tolerance (as well as teratomata in some recipient animals).[28]

We do not know if immunosuppressive therapy would be needed in recipients of HESCs. In theory, nuclear transfer from the recipient into the ES cells might overcome the danger of allograft rejection, but such a "bespoke" process would be excessively expensive using current techniques.

## ii. Cultured cell lines

Cultured cell lines would overcome the quantity difficulties to be expected with most other approaches, but most cell lines are derived from neoplastic sources and would be unsuitable for therapeutic use.

## iii. Adult and neonatal "stem" cells

Clinical experience with bone marrow transplantation relies on the multipotency of "stem" cells in the marrow and umbilical cord blood. Stem cells can be mobilized into the peripheral blood of normal adults treated with growth factors. It has been suggested that the normal differentiation of these marrow stem cells into cells of red and white cell lineage might be deviated by cultural techniques similar to those discussed for ESCs. The strict hierarchy of the three primary stem cell lines, mesoderm, endoderm and ectoderm seems to be more malleable than was believed. Ruhnke *et al.*[25] have reported apparent de-differentiation of normal peripheral blood buffy coat monocytes, which can proliferate up to six divisions and then be programmed to differentiate into liver-like cells producing albumin, and fat cells and islet-like cells producing insulin and glucogen that temporarily reversed the diabetic state when grafted into streptocytocin treated mice. "Stem"-like cells have been found in many tissues besides blood, including brain, fat and skeletal muscle. Attempts have been made to use adult stem cells to repair damaged cardiac muscle and to treat streptocytosin-treated diabetic animals. In some experiments, the claims of the authors have been disputed because cell fusion rather than true stem cell replacement has been demonstrated, but this objection has not always been sustained and the attraction of approaches using adult stem cells is that in theory, they could be autologous, the cells persuaded to adopt a new role and then returned to the patient, avoiding the danger of allograft rejection and the need for immunosuppressive drug treatment.

## iv. Transdifferentiation of adult cells

There are reports of transdifferentiation of adult pancreas cells to liver cells and vice versa. In these experiments already differentiated tissues, derived from the same embryonic primordia, have been made to switch roles by culture and/or genetic engineering.[29-32]

In all of the above examples, the changing role of cells to adopt an unnatural synthetic task depends on cultural techniques and/or genetic engineering. In both approaches, significant proliferation to provide large numbers of redirected cells will be necessary to be of use in treatment of hemophilia and diabetes. Moreover for the treatment of diabetes the cells should produce insulin in sufficient quantities in response to appropriate stimuli and the effect would need to be long-lasting.

It is attractive to view cultural persuasions to express the wanted gene as the most physiological approach, but as discussed above, we do not yet have the knowledge or the reagents to do this even in a semi-physiological manner.

## 2. Genetic Engineering

In the diseases under consideration, hemophilia A and diabetes, the relevant human genes have been identified, characterized and cloned, but introducing the plasmid to be active in the chosen target cell is still problematic.

i. *Electroporition* — the use of an electric current to make the cell membrane porous will allow insertion of plasmids into some cells, but the process is not very efficient in cells *in vitro* or in tissues.

ii. *Viral Vectors.* One of the main attributes of virus behavior is to gain entry into target cells and either reside there or kill the cells, having made use of their nuclear material. To act as a vector, the virus must be big enough for the construct in question. Most studies have been with two classes of virus — the adeno and adeno-like viruses and the lente-modified HIV viruses. Early clinical trials of both classes have sometimes led to modest clinical improvement, but three disasters have been reported. In one case in Philadelphia, the adeno virus proliferated with fatal consequences.[33,34] In the other two cases in Paris, it would appear that the lente virus used had unmasked nuclear oncogenes leading to leukemia. These tragedies have alerted researchers to the dangers and have also led to sharp and often aggressive criticism of the workers. Despite this background, I believe that in the foreseeable future cultural techniques alone will not be sufficient and vector help will be needed.

Currently, we are working with a modified herpes I virus as a vector for the human insulin gene. A variant of this virus has been used as local treatment for glioblastoma and injected into the brain. There has been no evidence of systemic disease in the six patients treated with a follow-up of five years.[35]

The theoretical advantages of the herpes virus are:

a) the large capacity to accommodate a construct.
b) the availability of the virus to infect primary and second cell lines *in vitro*.
c) although the virus enters the nucleus it does not integrate with the host DNA and is therefore not likely to unmask oncogenes.
d) most patients have already had contacts with the herpes I virus, which normally resides in a quiescent state in neurological tissue.
e) immune reaction against the virus is relatively mild.
f) established anti-viral treatment against the herpes virus is available.

We are engaged in experiments to determine which cell line or tissue might be appropriate for engineered viral infection and whether it is preferable to work *in vitro* with autologous cells to be returned to the recipient or should the virus be injected directly into recipient tissue. We need to study the longevity of gene activity in the virus and what factors may limit its continued protein synthesis.

A method of determination of insulin secretion of the viral vector compared with normal β cells is needed. Measurements are required for protein synthesis and secretion of the kinetics of insulin release and activation in response to ambient glucose levels must be defined.

Our modest approach is one of many that are being studied worldwide in an accelerating interest in responding to the challenge of providing safe and useful treatment of disease by cell-based therapy.

# REFERENCES

1. Billingham RE, Brent L, Medawar PB. (1953) Actively acquired tolerance of foreign cells. *Nature* 172:603–606.
2. Main JM, Prehn RT. (1955) Successful skin homografts after the administration of high dosage X radiation and homologous bone marrow. *J Natl Cancer Inst* 15:1023.
3. Storb R, Yu C, McSweeney P. (1999) Mixed chimerism after transplantation of allogeneic hematopoietic cells. In *Hematopoietic Cell Transplantation*, 2nd edn. ED Thomas, KG Blume, SJ Forman (eds), p. 287, Boston: Blackwell.
4. Juanita MS, Tatsuo K, Yasuhiro F, *et al.* (2004) Mechanisms of donor-specific unresponsiveness in tolerant recipient of combined non-myeloablative HLA-mismatched bone marrow and kidney transplantation. *AMJT* 8(4):303.
5. Colman A. (2004) Making new beta cells from stem cells. *Semin. Cell Dev. Biol.* 15:337–345.

6. Kay MA, Manno CS, Ragni MV, *et al.* (2000) Evidence for gene transfer and expression of factor IX in haemophilia B patients treated with an AAV vector. *Nat Gene* **24**(3):257–261.

7. Roth DA, Tawa NE, O'Brien JM, *et al.* (2001) Nonviral transfer of the gene encoding coagulation factor VIII in patients with severe haemophilia A. *NEJM* **344**:1735.

8. Bliss M. (1982) *The Discovery of Insulin.* University of Chicago Press.

9. Bonner-Weir S. (2000) Life and death of the pancreatic beta cells. *TEM* **11**(9):375–378.

10. Sutherland DER. (1997) *Newsletter Int Pancreas Transplant Reg* **9**:1.

11. Bennet W, Sundberg B, Lundgren AT, *et al.* (2000) Damage to porcine islets of Langerhans after exposure to human blood *in vitro,* or after intraportal transplantation to cynomologus monkeys. *Transplantation* **69**:711–719.

12. Goto M, Johansson H, Maeda A, *et al.* (2004) Low molecular weight dextran sulphate prevents the instant blood-mediated inflammatory reaction induced by adult porcine islets. *Transplantation* **77**:741–747.

13. Shapiro AM, Lakey JR, Ryan EA, Korbutt GS. (2000) Islet transplantation in seven patients with type 1 diabetes mellitus using a glucocorticoid-free immunosuppressive regimen. *N Engl J Med* **343**:230–238.

14. Wijkstrom M, Kirchhof N, Clemmings S. (2003) Prolonged pig islet xenograft survival and function in cynomolgus monkeys immunosuppressed with basiliximab, Rad and FTY720. Abstract no. 75. *Transplantation* **76**(4).

15. Bonner-Weir S, Taneja M, Weir GC, *et al.* (2000) *In vitro* cultivation of human islets from expanded ductal tissue. *Proc Natl Acad Sci USA* **97**(14):7999–8004.

16. Dor Y, Bornbw J, Martinex OI, Melton DA. (2004) Adult pancreatic β-cells are formed by self-duplication rather than stem cell differentiation. *Nature* **429**:41–46.

17. Hammerman MR. (2001) Growing kidneys. *Curr Opin Nephrol Hyperten* **10**:13–17.

18. Dekel B, Burakova T, Arditti FD, *et al.* (2003) Human and porcine early kidney precursors as a new source for transplantation. *Nat Med* **9**(1):55–60.

19. Castaing M, Peault B, Basmaciogullari A, *et al.* (2001) Blood glucose normalization upon transplantation of human embryonic pancreas into beta-cell-deficient SCID mice. *Diabetologia* **44**:2066–2076.

20. Soria B, Roche E, Berna G, *et al.* (2000) Insulin-secreting cells derived from embryonic stem cells normalize glycemia in streptozotocin-induced diabetic mice. *Diabetes* **49**(2):157–62.

21. Lester BL, Kuo HC, Andrews L. (2004) Directed differentiation of rhesus monkey ES cells into pancreatic cell phenotypes. *Reprod Biol Endocrinol* **2**:42.

22. Soria B. (2001) *In vitro* differentiation of pancreatic beta cells. *Differentiation* **68**:205–219.

23. Zhao Y, Glesne D, Huberman EA. (2003) Human peripheral blood monocyte-derived subset acts as pluripotent stem cells. *Proc Natl Acad Sci* **100**(5):2426–2431.
24. Abuljadayel IS. (2003) Induction of stem cell-like plasticity in mononuclear cells derived from unmobilised adult human peripheral blood. *Curr Med Res Opin* **19**(5):355–375.
25. Ruhnke M, Ungefroren H, Nussler A, *et al*. Reprogramming of human peripheral blood monocytes into functional hepatocyte and pancreatic islet-like cells. (in press)
26. Chen NKF, Sivalingam J, Tan SY, Kon OL. Plasmid-electroporated primary hepatocytes acquire quasi-physiological secretion of human insulin and restore euglycemia in diabetic mice. (in press)
27. McCormack MP, Rabbitts TH. (2004) Activation of the T-cell oncogene LM02 after gene therapy for X-linked severe combined immunodeficiency. *NEJM* **50**:913–992.
28. Fandrich F, Lin X, Gui, Chai GX, *et al*. (2002) Preimplantation-stage stem cells induce long-term allogenic graft acceptance without supplementary host conditioning. *Nat Med* **8**:171.
29. Horb ME, Shen CN, Tosh D, Slack JM. (2003) Experimental conversion of liver to pancreas. *Curr Biol* **13**(2):105–115.
30. Alam T, Sollinger HW. (2002) Glucose-regulated insulin production in hepatocytes. *Transplantation* **74**(12):1781–1787.
31. Nakajima-Nagata N, Sakurai T, Mitaka T, *et al*. (2004) *In vitro* induction of adult hepatic progenitor cells into insulin-producing cells. *Biochem Biophy Res Commun* **318**:625–630.
32. Le Ber I, Shternhall K, Perl S, *et al*. (2003) Functional, persistent and extended liver to pancreas transdifferentiation. *J Biol Chem* **34**:31950–7.
33. Raper SE, Chirmule N, Lee FS, *et al*. (2003) Fatal systemic inflammatory response syndrome in an ornithine transcarbamylase deficient patient following adenoviral gene transfer. *Mol Genet Metab* **80**:148–158.
34. Marshall E. (1999) Gene therapy death prompts review of adenovirus vector. *Science* **286**:2244–2245.
35. Rampling R, Cruickshank G, Papanatassiou V, *et al*. (2000) Toxicity evaluation of replication-competent herpes simplex virus (ICP 34.5 null mutant 171) in patients with recurrent malignant glioma. *Gene Ther* **7**:859–866.

# Index